T0350308

Logic, Mathematics, and Computer Science

Yves Nievergelt

Logic, Mathematics, and Computer Science

Modern Foundations with Practical Applications

Second Edition

 Springer

Yves Nievergelt
Department of Mathematics
Eastern Washington University
Cheney, WA, USA

Cover art excerpted from triadic truth tables from Charles Sanders Peirce's *Logic Notebook*.

ISBN 978-1-4939-3222-1   ISBN 978-1-4939-3223-8   (eBook)
DOI 10.1007/978-1-4939-3223-8

Library of Congress Control Number: 2015949237

Mathematics Subject Classification (2010): Primary 03-01; Secondary: 68-01, 91-01

Springer New York Heidelberg Dordrecht London
© Springer Science+Business Media New York 2002, 2015
This work is subject to copyright. All rights are reserved by the Publisher, whether the whole or part of
the material is concerned, specifically the rights of translation, reprinting, reuse of illustrations, recitation,
broadcasting, reproduction on microfilms or in any other physical way, and transmission or information
storage and retrieval, electronic adaptation, computer software, or by similar or dissimilar methodology
now known or hereafter developed.
The use of general descriptive names, registered names, trademarks, service marks, etc. in this publication
does not imply, even in the absence of a specific statement, that such names are exempt from the relevant
protective laws and regulations and therefore free for general use.
The publisher, the authors and the editors are safe to assume that the advice and information in this book
are believed to be true and accurate at the date of publication. Neither the publisher nor the authors or
the editors give a warranty, express or implied, with respect to the material contained herein or for any
errors or omissions that may have been made.

Printed on acid-free paper

Springer Science+Business Media LLC New York is part of Springer Science+Business Media (www.
springer.com)

Preface

This second edition, entitled *Logic, Mathematics, and Computer Science: Modern Foundations with Practical Applications*, has been adapted from *Foundations of Logic and Mathematics: Applications to Computer Science and Cryptography*, © 2002 by Birkhäuser, from which Chapters 1–5 have been retained but extensively revised. Chapters 6 and 7 have been added.

This text discusses the foundations where logic, mathematics, and computer science begin. The intended readership consists of undergraduate students majoring in mathematics or computer science who must learn such foundations either for their own interest or for further studies. For a motivated reader, there are no technical prerequisites: you need not know any technical subject to start reading this text.

Although the text does not focus on the history and philosophy of the foundations, the material cites copious references to the literature, where the reader may find additional historical context. Consulting such references is neither suggested nor necessary to study the theory or to work on the exercises, but individual citations document the material by original sources, and all the citations together provide a guide to the variations and chronological developments of logic, mathematics, and computer science. For example, Chapter 1 traces the origin of Truth tables to Charles Sanders Peirce's unpublished 1909 *Logic Notebook* on philosophy and points out their applications over one half of a century later to the design of computers for use on Earth and on board the Apollo lunar spacecraft.

Along informal arguments, this text also shows the corresponding purely symbolic manipulations of formulae, because they clarify the reasoning [11] and can reveal hitherto hidden logical properties, such as the mutual independence of different patterns of reasoning, or the impossibility of some proofs within some logics:

> As for algebra [of logic], the very idea of the art is that it presents formulae which can be manipulated, and that by observing the effects of such manipulation we find properties not to be otherwise discerned (Charles Sanders Peirce [104, p. 182]).

If professionals are unable to learn some topics by any means other than the pure manipulation of symbols, then it would seem unfair to claim that all learning must be intuitive and hide from students such purely manipulative but successful methods.

The selection of topics also reflects major accomplishments from the twentieth century: the foundation of all of mathematics, and later computer science, as well as computer-assisted proofs of mathematical theorems, on a formal logic based on only a few axioms, transformation rules, and postulates for set theory [47, 50, 54, 105, 139]. Also, while not written in formal logic, Nobel-Prize winning applications to the social sciences rely on the same foundations, as shown in Chapter 7.

To introduce the foundations of logic, the provability theorem in Chapter 1 provides an algorithm to design proofs in propositional logic. Chapter 1 also explains the concept of undecidability with multi-valued ("fuzzy") logic and presents a proof of unprovability. Chapter 2 introduces logical quantifiers. A working knowledge of logical quantifiers is crucial for the study of basic concepts in modern mathematical analysis and topology, such as the uniform convergence of a sequence of equicontinuous functions. Continuing with the foundations of mathematics, Chapter 3 presents a version of the Zermelo–Fraenkel set theory. At the juncture of mathematics and computer science, Chapter 4 develops the concepts of definition and proof by induction. Chapter 4 then uses induction with set theory to define the integers and rational numbers and derive the associative, commutative, and distributive laws, as well as algorithms, for their arithmetics. To give readers some idea of topics at an intermediate level, Chapter 5 shows that in a well-formed theory some paradoxes do not occur, while Chapter 6 completes the foundations of set theory with the axiom of choice.

No extragalactic asteroid has yet been found with the universal laws of logic engraved in it. Consequently, not just one logic but many different logics have been invented. Different logics lead to different mathematics and different computer sciences. However, the acid test for adopting a particular logic is its ability to make predictions that are born by subsequent experiments. Formal logic is thus a mathematical model of rational thought processes. In this aspect, logic, mathematics, and computer science are experimental sciences. Only one logic has passed all such tests, which is the one used throughout this text. Other logics are outlined in Chapter 1 as a pedagogical device and to show some of their shortcomings.

Acknowledgments

I thank Dr. Stephen P. Keeler at the Boeing Company for having corrected many errors in the first edition. New and remaining errors are mine. I also thank Dr. Mary Keeler for guiding me to Charles Sanders Peirce's unpublished work on logic and philosophy. I commend the editors at Birkhäuser and Springer, in particular, Ann Kostant and Elizabeth Loew, for their hard work, patience, encouragements, and positive attitude through the development and production of both texts.

Cheney, WA, USA   Yves Nievergelt
30 June 2015

Contents

Chapter 1
Propositional Logic: Proofs from Axioms and Inference Rules

1.1 Introduction

This chapter introduces propositional logics, which consist of starting formulae called *axioms* and *rules of inference* to derive from the axioms other formulae called *theorems*. Axioms and rules of inference form a mathematical model of rational thinking processes; theorems are their consequences. Different such logics, which are also called *calculi*, rely on different axioms or different rules of inference. For example, the Pure Positive Implicational Propositional Calculus focuses only on the logical implication. The first few sections derive some of its theorems, for instance, the transitivity of the logical implication and the law of commutation, using the rule of *Detachment* with the laws of affirmation of the consequent and self-distributivity of the logical implication taken as axiom schemata from Frege and Łukasiewicz. A preliminary version of the Deduction Theorem for the propositional calculus provides a method for designing proofs. Another section shows the mutual equivalence of these axioms with Kleene's axioms and Tarski's axioms. Adding the converse law of contraposition, subsequent sections focus on the Classical Propositional Calculus, deriving the laws of double negation, reductio ad absurdum, proofs by contradiction, and proofs by cases. Yet another section outlines the equivalence of Frege and Łukasiewicz's axioms with Church's, Kleene's, Tarski's, and Rosser's axioms, respectively. A final section demonstrates the contrast between logics that admit of a recipe for constructing proofs of all "valid" formulae, and logics where some formulae are "valid" but unprovable. The prerequisite for this chapter is the ability to read, compare, and substitute sequences and tables of symbols. The goal of this chapter is merely to develop a working knowledge of propositional logic:

> Young man, in mathematics you don't understand things, you just get used to them. — John von Neumann, cited by Felix Smith, cited by Gary Zukav [148, p. 226].

© Springer Science+Business Media New York 2015
Y. Nievergelt, *Logic, Mathematics, and Computer Science*,
DOI 10.1007/978-1-4939-3223-8_1

Logic, mathematics, and computations can be traced through several millennia to ancient civilizations in Babylonia, China, and India. Documents attributed to them show methods to calculate such items as taxes, the dimensions of altars, and the dates of future solstices or eclipses. More complicated problems arose, for example, the determination of the shapes and sizes of the Earth and the Moon, or the distances from Earth to the Moon and the Sun. (For a survey of these ancient records, consult, for instance, the texts by Dreyer [27], Evans [32], Neugebauer [98–100], Van Brummelen [133], and van der Waerden [134].) The solutions to such problems require methods more sophisticated than mere calculation, and hence arises the need for a study of logic itself, which can be traced to the Greece of a couple of millennia ago. This study of logic continues: the ambiguity of the classical verbal exposition of logic and the need for unambiguity in complex situations led to algebraic and symbolic treatments of logic, for instance, the Truth tables presented here. The point of logic is not only "Truth" but also "relevance" to its users [24, p. 6]. Yet relevance is subjective. Therefore, the following subsection presents an example that is not only claimed but *documented* to be relevant in the real lives of real people.

1.1.1 An Example Demonstrating the Use of Logic in Real Life

For the purpose of an introduction, the following example demonstrates how logic can help in resolving practical issues in real life, how questions arise about the validity of logical methods to reach conclusions, and eventually what thought processes are acceptable or successful in explanations and predictions. Yet an understanding of this example will not be necessary for any of the subsequent material.

1.1 Example. The planetary status of Pluto has been debated in newspapers:

> Is Pluto really a planet? Like all civil wars, this has even split families apart [67].

The public's interest in Pluto's planet-hood is sufficient to devote an entire book explaining the question to children [73]. Textbooks have classified Pluto as a planet since its discovery in 1930 [69, p. 213], but the question remains whether Pluto might rather be an asteroid [75]. Various answers rely on various definitions and on logic [101]. For instance, one definition states that planets are bigger than moons. This definition can be stated in terms of a **hypothesis** (abbreviated by H),

> hypothesis H: a celestial object P is a planet,

a **conclusion** (abbreviated by C),

> conclusion C: the celestial object P must be bigger than every moon,

and a logical **implication** of the form "if H, then C":

> If a celestial object P is a planet (if H),
> then the celestial object P must be bigger than every moon (then C).

The logical implication "if H, then C" (also worded as "H implies C") is true by this definition of planets. If the hypothesis H is also true, then the conclusion C is true, too: C can be "detached from the implication" [80, p. 124].

With a logical implication ("if H, then C") there are two other useful statements: its **converse** ("if C, then H") and its **contraposition** ("if *not* C, then *not* H").

The hypothesis H can be tested by the contraposition "if *not* C, then *not* H":

If a celestial object *P* is *not* bigger than every moon (if *not* C),
then *P* is *not* a planet (then *not* H).

In 1978 measurements revealed that Pluto was smaller than the Moon [69, p. 213]; consequently, Pluto would no longer be a planet, by the foregoing definition.

The definition "if H, then C" can also be tested in practice. For instance, textbooks classify Mercury as a planet, but they also classify Ganymede as a moon (of Jupiter), even though Mercury is smaller than Ganymede [69, p. 182 & 203]. Thus the statement "if Mercury is a planet, then Mercury is bigger than every moon" is false. Therefore the foregoing definition "if H, then C" is false, and Pluto can remain a planet. Logic has thus resolved the issue by revealing that the question pertains not to the status of Pluto but to the definition of planets.

Other definitions and logical arguments have also been debated [73, 101]. Very shortly thereafter, all existing definitions of planets were again put into question:

Scientifically, we are unable to define a planet in a sufficiently meaningful way such as to include Pluto without including many other objects [...] we are also unable to develop a definition based on principles of astronomy and physics that excludes Pluto in any nonarbitrary way [137, p. 219, summarizing chapter 14, p. 185–221].

The definition of planet will not be settled here, but some patterns of logical reasoning will. For instance, the preceding arguments show that detachment and contraposition are valid popular and scientific modes of reasoning; converse is not.

The preceding discussion contains one logical principle that has been so successful that it remains widely accepted in theory and in practice: the **law of contraposition**, which states that if an implication "if H, then C" holds, then its contraposition "if *not* C, then *not* H" also holds.

The **converse law of contraposition** — which states that if the contraposition "if *not* C, then *not* H" holds, then the implication "if H, then C" also holds — was not used in the preceding discussion, and it is not a part of some logical systems.

In this example, the **converse** statement "if C, then H" (also worded as "C implies H") is false, because the Sun is bigger than every moon (*C* is true), but the Sun is not a planet (*H* is false). Nevertheless, implications of the form "if H, then C" and their converse "if C, then H" have been confused by professional scientists, so that the difference between an implication and its converse bears being emphasized:

Leontovich explained to me why the paper could not be published in ZETP ["the main Russian physics journal"]:

the paper claims that "A implies B" while every physicist knows examples showing that B does not imply A; [...]

An author, claiming that A implies B, *must* say whether the converse holds, otherwise the reader who is not spoiled by the mathematical slang would interpret the claim as "A is equivalent to B." If mathematicians do not follow this rule, they are wrong [3, p. 619–620].

Such confusion among world-class scientists shows the necessity of specifying patterns of rational thought processes exactly, for instance, as done in this chapter.

1.2 Remark (difficulties with real examples). Some examples of uses of logic may enhance the effectiveness of the exposition. Such examples might consist of English sentences, for example, "Pluto is a planet." However, difficulties arise in determining whether and why such a practical sentence is true or false. Indeed, a statement as simple as "Pluto is a planet" can immediately be challenged to no end, as demonstrated in example 1.1. Moreover, the word "planet" comes from the Greek word for "wanderer" and therefore in antiquity the Sun was also considered a planet [133, p. 3]. Thus the argument about such an elementary practical logical issue as Pluto's planetary status really has no ends in any direction, without any mention of other more complex practical questions. Therefore, to focus on logic, mathematics, and computing, instead of endlessly debatable issues, the following discussion will also use "toy" examples with truth or falsity decided in advance.

1.2   The Pure Propositional Calculus

Propositional logic focuses on the derivation of conclusions from hypotheses, by means of *rules of inference* and initial hypotheses called *axioms* that state patterns of rational thinking precisely. Different axioms or different rules of inference may lead to different logics or to mutually equivalent logics, for instance, the Pure Positive Implicational Propositional Logic, which is a part of the full Classical Propositional Logic.

Different readers may prefer prose or formulae. Jan Łukasiewicz stated concisely the disadvantage of prose in formulating and solving logical problems:

Alles zerfließt in vagen philosophischen Spekulationen [80, p. 125]
("everything melts into vague philosophical speculations").

In contrast, formulae provide greater precision in complex situations [11]. To this end, the *Pure Propositional Calculus* (also called *Sentential Calculus*) presented here can be traced to Gottlob Frege [38, 39]. The adjective "pure" means that the simplest ("atomic") formulae are letters or symbols that do not denote anything:

mathematical logic is a *meaningless game with symbols* [108, p. xi].

Yet such symbols may later denote various types of atomic formulae that apply to such various fields as algebra, arithmetic, geometry, or set theory, and therein lies the power of the pure propositional calculus. Furthermore, purely symbolic manipulations of formulae can reveal hitherto hidden logical properties:

> As for algebra [of logic], the very idea of the art is that it presents formulae which can be manipulated, and that by observing the effects of such manipulation we find properties not to be otherwise discerned (Charles Sanders Peirce, [104, p. 182]).

If famous logicians find manipulations of formulae indispensable in clarifying logic, then such manipulations might also help nonspecialists in studying logic.

1.2.1   Formulae, Axioms, Inference Rules, and Proofs

Common instances of logical reasoning consist of sentences. For example, the first axiom of Euclidean geometry is a sentence (paraphrased from [61, p. 3]):

> For each pair of distinct points there exists exactly one line passing through both points.

Similarly, propositional logic starts with certain logical formulae called **axioms**. The word "axiom" can mean "self-evident truth" but in the present context, which focuses on patterns of reasoning, the word "axiom" means "initial" or "starting" logical pattern [110, p. 55]. Different selections of axioms can lead to different kinds of logic, but the present chapter focuses mainly on **classical logic**, which has been successful for several millennia. Several choices for the initial axioms and formulae lead to the same classical logic. Because the principal concepts of logic consist of "negation" and "implication" several common choices of initial axioms and formulae involve only the connectives for negation ($\neg$) and implication ($\Rightarrow$). Also, to allow for applications in various areas, the *pure* propositional calculus replaces the "atomic formulae" from algebra, arithmetic, geometry, or set theory by general symbols called propositional variables or sentence symbols.

1.3 Definition (Well-formed formulae).   Select *two* disjoint lists of symbols.

Every symbol from the *first* list of symbols, which may consist of one or more letter(s) from a specified alphabet, P, Q, ..., optionally with subscript(s) $P_\flat$, $P_{\flat\flat}$, ..., superscript(s) $P^\sharp$, $P^{\sharp\sharp}$, ..., or "middlescript(s)" $P|$, $P||$, ..., is called a **formulaic letter**. Such formulaic letters are not parts of the propositional calculus, but they help in describing the following rules to define well-formed formulae.

Also, every symbol from the *second* list of symbols, which may consist of one or more letter(s) from a specified alphabet, A, B, ..., optionally with subscript(s) $A_\flat$, $A_{\flat\flat}$, ..., superscript(s) $A^\sharp$, $A^{\sharp\sharp}$, ..., or "middlescript(s)" $A|$, $A||$, ..., is called a **propositional variable** or a **sentence symbol** [31, p. 17]. (Propositional variables may later be replaced by atomic formulae specific to applications.)

Every propositional variable is a **well-formed formula**. For all well-formed formulae P and Q, the following strings of symbols are also well-formed formulae:

$$(W1) \quad \neg(P) \quad \text{(read "not } P\text{"),}$$
$$(W2) \ (P) \Rightarrow (Q) \ \text{(read "} P \text{ implies } Q\text{" or "if } P\text{, then } Q\text{").}$$

Furthermore, only strings of symbols built from letters or variables through applications of the rules W1 and W2 can be well-formed formulae. Equivalent definitions apply to other connectives and to prefix and postfix notations.

The use of parentheses in definition 1.3 conforms to [22, p. 7]. and [114, p. 185]. Parentheses without rules of precedence reflect the motto

> more parentheses but less memorizing [114, p. 216].

By definition 1.3, a string of symbols such as $(P) \Rightarrow (Q)$ is not yet a well-formed formula of the propositional calculus. It only becomes so after P and Q have been replaced by propositional variables or well-formed formulae, for instance, $(A) \Rightarrow (B)$. In this section, any capital letter may denote a propositional letter, variable, or formula. In subsequent sections, however, the distinction may matter.

From definition 1.3, propositional letters or variables are "atoms" or "atomic" propositional formulae in the sense that they are the simplest well-formed propositional formulae [72, p. 5], in contrast to more elaborate "composite" or "compound" formulae, also called **propositional forms**, built from rules W1 and W2.

Several choices of well-formed propositional formulae can serve as axioms. A system that remains concise and differentiates the roles of separate connectives consists of the following three axioms. Subsection 1.3.10 explains their popularity [18, §20, p. 119], [84, p. 49], [81, p. 31], [85, p. 165], [122, p. 165].

1.4 Definition (Jan Łukasiewicz's axioms). A logical formula is an **axiom** of the version of the classical propositional calculus considered here if and only if it is one of the following three formulae, attributed to Łukasiewicz [62, p. 29], where P, Q, and R may be any well-formed propositional formulae. The first two axioms are also all the axioms of the Pure Positive Implicational Propositional Calculus:

Axiom P1 $(P) \Rightarrow [(Q) \Rightarrow (P)]$.

Axiom P2 $\{(P) \Rightarrow [(Q) \Rightarrow (R)]\} \Rightarrow \{[(P) \Rightarrow (Q)] \Rightarrow [(P) \Rightarrow (R)]\}$.

Axiom P3 $\{[\neg(Q)] \Rightarrow [\neg(P)]\} \Rightarrow [(P) \Rightarrow (Q)]$.

The first two axioms are also in Frege's work [38], [39, p. 137, eq. (1) & (2)], where they reflect a common mathematical model of rational thinking:

- Axiom P1, $(P) \Rightarrow [(Q) \Rightarrow (P)]$, is called the **law of affirmation of the consequent**. In Frege's (translated) words, axiom P1 states that

  > If a proposition [P] holds, it also holds in case an arbitrary proposition [Q] holds [39, p. 137].

- Axiom P2, $\{(P) \Rightarrow [(Q) \Rightarrow (R)]\} \Rightarrow \{[(P) \Rightarrow (Q)] \Rightarrow [(P) \Rightarrow (R)]\}$, is called the **law of self-distributivity of implication**. In Frege's words, axiom P2 states that

  > If a proposition [R] is the necessary consequence of two propositions ([Q] and [P]), that is, if $[(P) \Rightarrow \{(Q) \Rightarrow (R)\}]$, and if the first term [Q] is again the necessary consequence of the other [P], then the proposition [R] is the necessary consequence of the last proposition [P] alone [39, p. 139].

- Axiom P3, $\{[\neg(Q)] \Rightarrow [\neg(P)]\} \Rightarrow [(P) \Rightarrow (Q)]$, called the **converse law of contraposition** [18, §20, p. 119], states that a contraposition, $[\neg(Q)] \Rightarrow [\neg(P)]$ suffices to establish a classical logical implication $(P) \Rightarrow (Q)$.

The converse law of contraposition distinguishes classical logic from several other systems of logic, for instance, Hilbert's Positive Propositional Calculus, Brouwer & Heyting's Intuitionistic Logic, and Kolmogorov & Johansson's Minimal Logic [18, §26, p. 140–146]. Still, all these logical systems include Łukasiewicz's first two axioms, P1 and P2.

The propositional calculus includes the following concepts of theorem and proof.

1.5 Definition. A well-formed propositional form is a **theorem** of a propositional calculus if and only if it is obtained by the following **rules of inference**:

1.6 Rule (Axioms).

Every axiom (of a logic) is a theorem (of the same logic).

1.7 Rule ("Modus Ponens" (abbreviated by M. P.), or "Detachment").

For all propositional forms H and C,
if H is a theorem and
if $(H) \Rightarrow (C)$ is a theorem,
then C is a theorem.

The name of this rule will be printed here as "*Detachment*" to avoid unintended awkward sentences. With *Detachment*, H is the **minor premiss** while $(H) \Rightarrow (C)$ is the **major premiss** (so spelled to distinguish its plural from "premises" [18, p. 1, footnote 3]). Rule 1.7 states that if a hypothesis H and an implication $(H) \Rightarrow (C)$ hold, then the conclusion C may be "detached from the implication" ("von der Implikation abgetrennt" in Łukasiewicz's language [80, p. 124]).

1.8 Remark. Each use of the rule of *Detachment* requires *two* previously proved well-formed formula: a proved hypothesis H and a proved implication $(H) \Rightarrow (C)$.

Definitions 1.3 and 1.4 allow P, Q, and R to denote any propositional letters or well-formed propositional formulae, so that axioms P1–P3 are templates, or schemas, to generate axioms. Alternatively, allowing only propositional variables in axioms P1–P3 but introducing the substitution rule 1.9 gives equivalent axioms:

1.9 Rule (Substitution).

For each propositional variable K in a theorem R (which is a propositional form),
and for each well-formed propositional form L,
the propositional form obtained by replacing in R every occurrence of K by L is again a theorem.

A **proof**, or **deduction**, of a theorem R is a finite sequence of logical formulae $P, Q, \ldots, R$, in which each formula is either a substitution in an axiom or in a previously proven formula, or results from the rule of *Detachment* (*Modus Ponens*).

1.10 Example (Substitution). The formula $(A) \Rightarrow [(B) \Rightarrow (A)]$ is an instance of axiom P1; hence it is a theorem. Substituting $\neg(C)$ for A in $(A) \Rightarrow [(B) \Rightarrow (A)]$

yields $[\neg(C)] \Rightarrow \{(B) \Rightarrow [\neg(C)]\}$, which is another instance of axiom P1, and hence also a theorem. Because such substitutions in an axiom yield other axioms, each axiom is also called an **axiom schema**. Thus both axioms $(A) \Rightarrow [(B) \Rightarrow (A)]$ and $[\neg(C)] \Rightarrow \{(B) \Rightarrow [\neg(C)]\}$ result from the axiom schema P1.

1.11 Definition. For every logical formula R, the symbol $\vdash$ (called a "turnstyle" and read "yield(s)" [110, p. 57]) and the notation

$$\vdash R$$

means that there exists a proof of R. An alternative notation, $P1, P2, P3 \vdash R$, also specifies the list of axioms, here $P1, P2, P3$, from which R is a theorem.

More generally, for all logical formulae P and R, the notation $P \vdash R$ means that with P added to the list of axioms, there exists a proof of R. The corresponding alternative notation, $P1, P2, P3, P \vdash R$, again specifies the list of axioms. In other words, R is a **theorem** for the logic with axioms $P1, P2, P3, P$. With a different terminology, $P \vdash R$ means that R is **derivable** from P and the axioms.

Yet more generally, for all logical formulae $P, Q, \ldots, R$, either notation $P, Q, \ldots \vdash R$ or $P1, P2, P3, P, Q, \ldots \vdash R$, means that with $P, Q, \ldots$ added to the list of axioms, there exists a proof of R. In other words, R is a **theorem** for the logic with axioms $P1, P2, P3, P, Q, \ldots$ The formula R is then **derivable** from $P, Q, \ldots$, if and only if $P, Q, \ldots \vdash R$. In the notation of Smullyan [117, p. 17] and Stolyar [123, p. 63], $P, Q, \ldots \vdash R$ is also denoted by

$$\frac{P, Q, \ldots}{R}.$$

Verifying a proof reduces to checking that each step conforms to the foregoing definition of proof. In contrast, *constructing* a proof may require some creativity, which may involve trying some rules and some axioms in various combinations, some of which may fail whereas others may succeed [72, p. 55, lines 1–3], [114, p. 31]. For the propositional calculus presented here, there is an algorithm (a recipe) to design proofs, but its justification first requires most of the proofs presented here [123, p. 193–197]. Moreover, the algorithm is cumbersome and would generate proofs longer than the ones explained here. Nevertheless, the collection of all the proofs shown here will demonstrate the steps that the algorithm would involve. With such an understanding of the algorithm, a user might then automate the algorithm with a computer [47, 50, 54, 139]. Furthermore, the following proofs also provide some practice in creating proofs without using an algorithm, a practice that corresponds more closely to the situation in mathematics, and for which there can be no algorithm [46].

1.3   The Pure Positive Implicational Propositional Calculus

With the two rules of inference (*Detachment* and substitution), the first two axioms of classical propositional calculus (P1 and P2) pertain only to logical implications ($\Rightarrow$); they form the **Pure Positive Implicational Propositional Calculus**, which is common to other logics [18, § 29, p. 161]. In contrast, the concept of negation ($\neg$) does not belong to the Positive Implicational Propositional Calculus. Axioms about the negation, for instance, axiom P3, differentiate classical logic from other logics.

1.3.1   Examples of Proofs in the Implicational Calculus

The following three theorems provide examples of proofs about or in the Pure Positive Implicational Propositional Calculus, which is the propositional calculus with implications but no negations. The first theorem is called a "derived rule" (of inference) because it involves a hypothesis, T, which can be any axiom or previous theorem. Such a derived rule of inference is a theorem *about* rather than *in* the implicational calculus, but it provides a recipe to shorten subsequent proofs. Specifically, theorem 1.12 shows a derivation of $(S) \Rightarrow (T)$ from T and axiom P1. The proof of theorem 1.12 is also a building block of the Deduction Theorem 1.22.

1.12 Theorem (derived rule). *For each well-formed formula S and for each theorem T, the implication* $(S) \Rightarrow (T)$ *is a theorem: P1, $T \vdash (S) \Rightarrow (T)$.*

Proof. Apply axiom P1 and *Detachment* as follows:

$\vdash T$ hypothesis (minor premiss),

$\vdash (T) \Rightarrow [(S) \Rightarrow (T)]$ substitution in axiom P1 (major premiss),

$\vdash (S) \Rightarrow (T)$ *Detachment* and preceding two formulae.

Thus $(S) \Rightarrow (T)$ is a theorem derivable from axiom P1, the hypothesis T, and *Detachment*. $\qquad\square$

The second theorem uses axiom P2 and also involves a hypothesis, $(H) \Rightarrow [(K) \Rightarrow (L)]$, which can be any formula of this form that has already been proved.

1.13 Theorem (derived rule). *For all well-formed formulae H, K, L, if*

$(H) \Rightarrow [(K) \Rightarrow (L)]$

is a theorem, then

$[(H) \Rightarrow (K)] \Rightarrow [(H) \Rightarrow (L)]$

is also a theorem: P2, $(H) \Rightarrow [(K) \Rightarrow (L)] \vdash [(H) \Rightarrow (K)] \Rightarrow [(H) \Rightarrow (L)]$.

Proof. Apply axiom P2 and *Detachment*:

$\vdash (H) \Rightarrow [(K) \Rightarrow (L)]$ hypothesis,

$\vdash \{(H) \Rightarrow [(K) \Rightarrow (L)]\} \Rightarrow \{[(H) \Rightarrow (K)] \Rightarrow [(H) \Rightarrow (L)]\}$ axiom P2,

$\vdash [(H) \Rightarrow (K)] \Rightarrow [(H) \Rightarrow (L)]$ *Detachment.*

$\qquad\square$

The third theorem (1.14) involves no hypotheses other than the axioms. Thus theorem 1.14 is a theorem of the Pure Implicational Propositional Calculus. More accurately, theorem 1.14 is a theorem *schema* representing a different theorem for each different formula *P*. The proof of theorem 1.14 is also a building block of the Deduction Theorem 1.22.

1.14 Theorem (reflexive law of implication). *For each well-formed propositional formula P, the formula*

$$(P) \Rightarrow (P)$$

is a theorem derivable from axioms P1, P2, and Detachment.

Proof. Apply axioms P1, theorem 1.13, and *Detachment*, with lines numbered for clarity (the line numbers are not parts of the proof):

$$1 \quad \vdash \underset{H}{(\underset{}{\;P\;})} \Rightarrow \{\overset{Q}{\overbrace{\underset{K}{[(P) \Rightarrow (P)]}}} \Rightarrow \underset{L}{(\underset{}{\;P\;})}\} \qquad\qquad \text{substitution in axiom P1,}$$

$$2 \quad \vdash \{\overset{H}{\overbrace{(\;P\;)}} \Rightarrow \overset{K}{\overbrace{[(P) \Rightarrow (P)]}}\} \Rightarrow [\overset{H}{\overbrace{(\;P\;)}} \Rightarrow \overset{L}{\overbrace{(\;P\;)}}] \quad \text{theorem 1.13,}$$

$$3 \quad \vdash (P) \Rightarrow [(P) \Rightarrow (P)] \qquad\qquad\qquad\qquad \text{substitution in axiom P1,}$$

$$4 \quad \vdash (P) \Rightarrow (P) \qquad\qquad\qquad\qquad\qquad\qquad\quad \text{lines 2, 3, } Detachment.$$

A formal proof would replace line 2 by the details of the proof of theorem 1.13, labeled here as lines 2a and 2b, which gives the following proof:

$$1 \quad \vdash \underset{S}{\underbrace{\overset{P}{\overbrace{(\;P\;)}} \Rightarrow \{[(P) \Rightarrow (P)] \Rightarrow \overset{R}{\overbrace{(\;P\;)}}\}}} \qquad\qquad\qquad\qquad\qquad\qquad \text{P1,}$$

$$2a \quad \vdash [\underset{S}{\underbrace{\overset{P}{\overbrace{(\;P\;)}} \Rightarrow \{\overset{Q}{\overbrace{[(P) \Rightarrow (P)]}} \Rightarrow \overset{R}{\overbrace{(\;P\;)}}\}}}] \Rightarrow (\underset{U}{\underbrace{\{\overset{P}{\overbrace{(\;P\;)}} \Rightarrow \overset{Q}{\overbrace{[(P) \Rightarrow (P)]}}\} \Rightarrow [\overset{P}{\overbrace{(\;P\;)}} \Rightarrow \overset{R}{\overbrace{(\;P\;)}}]}}) \quad \text{P2,}$$

$$2b \quad \vdash \overset{U}{\overbrace{\{(P) \Rightarrow [(P) \Rightarrow (P)]\}}} \Rightarrow \underset{V}{\underbrace{[(P) \Rightarrow (P)]}}... $$

$$2b \quad \vdash \{\underset{V}{\underbrace{(P) \Rightarrow [(P) \Rightarrow (P)]}}\} \Rightarrow \underset{W}{\underbrace{[(P) \Rightarrow (P)]}} \qquad\qquad\qquad 1, 2a, M.P.,$$

$$4 \quad \vdash \overset{V}{\overbrace{(P) \Rightarrow [(P) \Rightarrow (P)]}} \qquad\qquad\qquad\qquad\qquad\qquad\qquad \text{P1,}$$

$$5 \quad \vdash \underset{W}{\underbrace{(P) \Rightarrow (P)}} \qquad\qquad\qquad\qquad\qquad\qquad\qquad\qquad\quad 2b, 4, M.P.$$

Thus, P1, P2 $\vdash (P) \Rightarrow (P)$.   □

1.3.2   Derived Rules: Implications Subject to Hypotheses

The following theorems *about* the Pure Positive Implicational Propositional Calculus are variants of the transitivity of implications: they shorten sequences of implications into one implication, from the initial hypothesis to the last conclusion. The same theorems also allow for the substitution of any well-formed propositional formula by any logically equivalent formula. Moreover, their proofs constitute steps of an algorithm known as the Deduction Theorem 1.22. For instance, the following **derived rules of inference** extend the rule of *Detachment* to situations where K and $(K) \Rightarrow (L)$ might hold only under some hypothesis H: then the conclusion L also holds under the same hypothesis H.

1.15 Theorem (derived rule).   *For all well-formed formulae H, K, L, if*
$(H) \Rightarrow (K)$ and
$(H) \Rightarrow [(K) \Rightarrow (L)]$
are theorems, then
$(H) \Rightarrow (L)$
is also a theorem. Thus,
$(H) \Rightarrow (K), \ (H) \Rightarrow [(K) \Rightarrow (L)] \vdash (H) \Rightarrow (L)$.

Proof. Apply theorem 1.13 and *Detachment*:

1	$\vdash (H) \Rightarrow [(K) \Rightarrow (L)]$	hypothesis;
2	$\vdash [(H) \Rightarrow (K)] \Rightarrow [(H) \Rightarrow (L)]$	theorem 1.13, major premiss;
3	$\vdash (H) \Rightarrow (K)$	hypothesis, minor premiss;
4	$\vdash (H) \Rightarrow (L)$	lines 3, 2, and *Detachment*.

□

The following theorem simplifies the use of theorem 1.15 if one of the logical implications is a theorem. In particular, in an implication $(K) \Rightarrow (L)$, theorem 1.16 allows for the substitution of a stronger hypothesis H implying (or equivalent to) K. Alternatively, in an implication $(H) \Rightarrow (K)$, theorem 1.16 allows for the substitution of a weaker conclusion L implied by (or equivalent to) K.

1.16 Theorem (derived rule).   *For all well-formed formulae H, K, L, if*
$(H) \Rightarrow (K)$ *and*
$(K) \Rightarrow (L)$
are theorems, then
$(H) \Rightarrow (L)$
is also a theorem. Thus, $(H) \Rightarrow (K), (K) \Rightarrow (L) \vdash (H) \Rightarrow (L)$.

Proof. Apply theorems 1.12 and 1.15:

$\vdash (K) \Rightarrow (L)$	hypothesis,
$\vdash (H) \Rightarrow [(K) \Rightarrow (L)]$	theorem 1.12,
$\vdash (H) \Rightarrow (K)$	hypothesis,
$\vdash (H) \Rightarrow (L)$	theorem 1.15.

□

Similarly, the following theorem simplifies the use of theorem 1.15 if one of the components is already a theorem.

1.17 Theorem (derived rule). *For all well-formed formulae H, K, L, if*

(K) *and*
$(H) \Rightarrow [(K) \Rightarrow (L)]$
are theorems, then
$(H) \Rightarrow (L)$
is also a theorem. Thus, $K, (H) \Rightarrow [(K) \Rightarrow (L)] \vdash (H) \Rightarrow (L).$

Proof. Apply theorems 1.12 and 1.15:

$\vdash K$ hypothesis,
$\vdash (H) \Rightarrow (K)$ theorem 1.12,
$\vdash (H) \Rightarrow [(K) \Rightarrow (L)]$ hypothesis,
$\vdash (H) \Rightarrow (L)$ theorem 1.15.

$\square$

The following theorem demonstrates how the rule of *Detachment* extends to a sequence of several logical implications.

1.18 Theorem (derived rule). *For all well-formed formulae P, Q, R, S, if*

$(P) \Rightarrow (Q),$
$(P) \Rightarrow [(Q) \Rightarrow (R)],$ *and*
$(P) \Rightarrow [(R) \Rightarrow (S)]$
are theorems, then
$(P) \Rightarrow (S)$
is also a theorem. Thus,

$$(P) \Rightarrow (Q),\ (P) \Rightarrow [(Q) \Rightarrow (R)],\ (P) \Rightarrow [(R) \Rightarrow (S)] \vdash (P) \Rightarrow (S).$$

Proof. Apply theorem 1.15 twice:

$\vdash (P) \Rightarrow (Q)$ hypothesis,
$\vdash (P) \Rightarrow [(Q) \Rightarrow (R)]$ hypothesis,
$\vdash (P) \Rightarrow (R)$ theorem 1.15,
$\vdash (P) \Rightarrow [(R) \Rightarrow (S)]$ hypothesis,
$\vdash (P) \Rightarrow (S)$ theorem 1.15.

$\square$

Similar theorems hold for a sequence of more than three consecutive implications, but their need will not arise here. The following theorem simplifies the use of theorem 1.18 if two of the logical implications are already theorems.

1.19 Theorem (derived rule). *For all well-formed formulae P, Q, R, S, if*

$(P) \Rightarrow (Q),$
$(Q) \Rightarrow (R),$ *and*
$(R) \Rightarrow (S)$

are theorems, then
   $(P) \Rightarrow (S)$
is also a theorem. Thus,

$$(P) \Rightarrow (Q), \ (Q) \Rightarrow (R), \ (R) \Rightarrow (S) \ \vdash \ (P) \Rightarrow (S).$$

Proof. Apply theorem 1.16 twice:
   $\vdash (P) \Rightarrow (Q)$ hypothesis,
   $\vdash (Q) \Rightarrow (R)$ hypothesis,
   $\vdash (P) \Rightarrow (R)$ theorem 1.16,   □
   $\vdash (R) \Rightarrow (S)$ hypothesis,
   $\vdash (P) \Rightarrow (S)$ theorem 1.16.

   The following theorems demonstrate the transitivity of implications subject to a common hypothesis.

1.20 Theorem (derived rule). *For all well-formed formulae H, K, L, M, if*
   $(H) \Rightarrow [(K) \Rightarrow (L)]$ *and*
   $(H) \Rightarrow \{(K) \Rightarrow [(L) \Rightarrow (M)]\}$
are theorems, then
   $(H) \Rightarrow [(K) \Rightarrow (M)]$
is also a theorem. Thus,

$$(H) \Rightarrow [(K) \Rightarrow (L)], \ (H) \Rightarrow \{(K) \Rightarrow [(L) \Rightarrow (M)]\} \ \vdash \ (H) \Rightarrow [(K) \Rightarrow (M)].$$

Proof. Use axiom P2 with theorems 1.16 and 1.15:
   $\vdash (H) \Rightarrow \{(K) \Rightarrow [(L) \Rightarrow (M)]\}$ hypothesis,
   $\vdash \{(K) \Rightarrow [(L) \Rightarrow (M)]\} \Rightarrow \{[(K) \Rightarrow (L)] \Rightarrow [(K) \Rightarrow (M)]\}$ axiom P2,
   $\vdash (H) \Rightarrow \{[(K) \Rightarrow (L)] \Rightarrow [(K) \Rightarrow (M)]\}$ theorem 1.16;
   $\vdash (H) \Rightarrow [(K) \Rightarrow (L)]$ hypothesis,
   $\vdash (H) \Rightarrow [(K) \Rightarrow (M)]$ theorem 1.15.
                                                                                    □

1.21 Theorem (derived rule). *For all well-formed formulae H, K, L, M, if*
   $(H) \Rightarrow (K)$,
   $(H) \Rightarrow (L)$,   *and*
   $(K) \Rightarrow [(L) \Rightarrow (M)]$
are theorems, then
   $(H) \Rightarrow (M)$
is also a theorem. Thus,

$$(H) \Rightarrow (K), \ (H) \Rightarrow (L) \ (K) \Rightarrow [(L) \Rightarrow (M)] \ \vdash \ (H) \Rightarrow (M).$$

Proof. Apply theorems 1.12, 1.13, and 1.15:

1	$\vdash (K) \Rightarrow [(L) \Rightarrow (M)]$	hypothesis,
2	$\vdash (H) \Rightarrow \{(K) \Rightarrow [(L) \Rightarrow (M)]\}$	theorems 1.12,
3	$\vdash [(H) \Rightarrow (K)] \Rightarrow \{(H) \Rightarrow [(L) \Rightarrow (M)]\}$	theorem 1.13, major premiss,
4	$\vdash (H) \Rightarrow (K)$	hypothesis, minor premiss,
5	$\vdash (H) \Rightarrow [(L) \Rightarrow (M)]$	lines 4, 3, and *Detachment*,
6	$\vdash (H) \Rightarrow (L)$	hypothesis,
7	$\vdash (H) \Rightarrow (M)$	theorem 1.15.

□

1.3.3 A Guide for Proofs: an Implicational Deduction Theorem

In general the question arises, how to find a proof of a theorem. One guide to design a proof of an implication $(H) \Rightarrow (C)$, where H and C denote well-formed propositional formulae, begins by deriving a proof for a derivation $H \vdash C$ of the conclusion C from the hypothesis H and the axioms. For instance, *all the proofs of derived rules in subsection 1.3.2 are examples of such derivations.* The method for designing a proof then proceeds to "discharge" the hypothesis H to produce a proof of $(H) \Rightarrow (C)$, as described in the proof of the Deduction Theorem 1.22, which is not *in* but *about* the implicational calculus. More generally, from any proof that a logical proposition S is derivable from proved hypotheses $H, K, \ldots, M, N$ the Deduction Theorem provides a recipe to turn that proof into a proof of

$$(H) \Rightarrow \{(K) \Rightarrow \ldots (M) \Rightarrow [(N) \Rightarrow (S)] \ldots\}.$$

The Deduction Theorem presented here is also a part of an algorithm to design proofs within the full Classical Propositional Calculus.

1.22 Theorem (Deduction Theorem for the Pure Classical Propositional Calculus, preliminary version). *With any axiom system for which axioms P1, P2, and $(P) \Rightarrow (P)$ are axioms or theorems (or schema thereof), there is an algorithm to transform any proof of*

$$H, K, \ldots, M, N \vdash S$$

within the Classical Propositional Calculus into a proof of

$$(H) \Rightarrow \{(K) \Rightarrow \ldots (M) \Rightarrow [(N) \Rightarrow (S)] \ldots\}.$$

Proof (Outline). The Deduction Theorem removes the hypotheses one at a time, for instance, beginning with the last one listed, here N, from all the steps in the proof.

(D1) If the step $\vdash P$ in the initial proof is the hypothesis N being removed, then in the new proof the Deduction Theorem replaces the old step

$\vdash N$ (current hypothesis)

by a complete proof of $(N) \Rightarrow (N)$, for instance, that of theorem 1.14:

$\vdash (N) \Rightarrow \{[(N) \Rightarrow (N)] \Rightarrow (N)\}$ *axiom P1,*

$\vdash \big[(N) \Rightarrow \{[(N) \Rightarrow (N)] \Rightarrow (N)\}\big]$ *axiom P2,*
   $\Rightarrow \big(\{(N) \Rightarrow [(N) \Rightarrow (N)]\} \Rightarrow [(N) \Rightarrow (N)]\big)$

$\vdash \{(N) \Rightarrow [(N) \Rightarrow (N)]\} \Rightarrow [(N) \Rightarrow (N)]$ *Detachment,*

$\vdash (N) \Rightarrow [(N) \Rightarrow (N)]$ *axiom P1,*

$\vdash (N) \Rightarrow (N)$ *Detachment.*

(D2) If a step $\vdash P$ of the initial proof is a substitution in one of the axioms, or in a previously proved theorem, or one of the hypotheses other than the one N being removed here, then in the new proof the Deduction Theorem replaces the old step

$\vdash P$ (axiom or hypothesis)

by a complete proof of $(N) \Rightarrow (P)$, for instance, as in theorem 1.12:

$\vdash P$ axiom or hypothesis,

$\vdash (P) \Rightarrow [(N) \Rightarrow (P)]$ axiom P1,

$\vdash (N) \Rightarrow (P)$ Detachment.

(D3) If the step $\vdash P$ is derived in the initial proof by *Detachment* from previously proved propositions M and $(M) \Rightarrow (P)$, then (D2) and (D1) allow for their replacement by complete proofs of $(N) \Rightarrow (M)$ and $(N) \Rightarrow [(M) \Rightarrow (P)]$ respectively. Specifically, in the new proof, the Deduction Theorem then replaces the old steps

$\vdash (M) \Rightarrow (P)$ (previously proven True),

$\vdash M$ (previously proven True),

$\vdash P$ (*Detachment*),

by a complete proof that $(N) \Rightarrow (P)$, for instance, as in theorem 1.15:

$\vdash (N) \Rightarrow [(M) \Rightarrow (P)]$ theorem 1.12,

$\vdash \{(N) \Rightarrow [(M) \Rightarrow (P)]\}$ axiom P2 ...
   $\Rightarrow \{[(N) \Rightarrow (M)] \Rightarrow [(N) \Rightarrow (P)]\}$ ... continued,

$\vdash \{[(N) \Rightarrow (M)] \Rightarrow [(N) \Rightarrow (P)]\}$ *Detachment,*

$\vdash (N) \Rightarrow (M)$ theorem 1.12,

$\vdash (N) \Rightarrow (P)$ *Detachment,*

with the proof of each instance of theorem 1.12 completely written out.

(D4) If a step $\vdash P$ results from a previously proved derived rule, then it may be necessary first to replace the step $\vdash P$ by a complete proof of the derived rule, and only then to replace each step of the proof of the derived rule as instructed by directives (D1), (D2), (D3).

Still with the hypothesis N, after the completion of any operation (D1)–(D3) on step P, the Deduction Theorem then performs the same operations (D1)–(D3) on each of the following steps, $Q, \ldots, R$. After the completion of operations (D1)–(D3) on all the steps $P, Q, \ldots, R$, for the hypothesis N, the Deduction Theorem gives a proof of

$$H, K, \ldots, M \vdash [(N) \Rightarrow (S)].$$

Then the Deduction Theorem repeats the whole process with the preceding hypotheses, $H, \ldots, M$. The Deduction Theorem terminates with a proof of

$$(H) \Rightarrow \{(K) \Rightarrow \ldots (M) \Rightarrow [(N) \Rightarrow (S)] \ldots\}.$$

The general case follows by several applications of the previous cases, in a way that may be specified more explicitly after the availability of the Principle of Mathematical Induction in chapter 4.   □

Example 1.23 shows how to use theorem 1.22.

1.23 Example (Tarski's axiom II). To prove Tarski's axiom II, $\{(P) \Rightarrow [(P) \Rightarrow (Q)]\} \Rightarrow [(P) \Rightarrow (Q)]$, define H and C by

$$\underbrace{\{(P) \Rightarrow [(P) \Rightarrow (Q)]\}}_{H} \Rightarrow \underbrace{[(P) \Rightarrow (Q)]}_{C}.$$

Phase 1: a proof of $H \vdash C$.

For $H \vdash C$, in other words, to derive C from H, substitute H in axiom P2:

1 $\vdash \underbrace{(P) \Rightarrow [(P) \Rightarrow (Q)]}_{H}$ hypothesis,

2 $\vdash \underbrace{\{(P) \Rightarrow [(P) \Rightarrow (Q)]\}}_{\underset{L}{H}} \Rightarrow \underbrace{\{[(P) \Rightarrow (P)] \Rightarrow [(P) \Rightarrow (Q)]\}}_{L}$ axiom P2,

3 $\vdash \underbrace{[(P) \Rightarrow (P)]}_{K} \Rightarrow \underbrace{[(P) \Rightarrow (Q)]}_{C}$ 1, 2, *Detachment*,

4 $\vdash \underbrace{(P) \Rightarrow (P)}_{K}$ theorem 1.14,

5 $\vdash \underbrace{(P) \Rightarrow (Q)}_{C}$ 3, 4, *Detachment*,

which completes the derivation $H \vdash C$ of C from H.

Phase 2: a proof of $(H) \Rightarrow (C)$ from $H \vdash C$.

To transform this derivation $(H \vdash C)$ into a proof of $(H) \Rightarrow (C)$, apply to each step of phase 1 the procedure described in the Deduction Theorem (1.22):

$$\vdash \underbrace{\{(P)\Rightarrow[(P)\Rightarrow(Q)]\}}_{\substack{H \\ H}}\Rightarrow\underbrace{\{(P)\Rightarrow[(P)\Rightarrow(Q)]\}}_{H} \qquad \text{(D1)},$$

$$\vdash \overbrace{\{(P)\Rightarrow[(P)\Rightarrow(Q)]\}}^{H} \\ \qquad \Rightarrow\Big(\underbrace{\{(P)\Rightarrow[(P)\Rightarrow(Q)]\}}_{H}\Rightarrow\underbrace{\{[(P)\Rightarrow(P)]\Rightarrow[(P)\Rightarrow(Q)]\}}_{L}\Big) \qquad \text{(D2), P2},$$

$$\vdash \underbrace{\{(P)\Rightarrow[(P)\Rightarrow(Q)]\}}_{H}\Rightarrow\{\underbrace{[(P)\Rightarrow(P)]}_{K}\Rightarrow\underbrace{[(P)\Rightarrow(Q)]}_{C}\} \qquad \text{(D3)},$$

$$\vdash \underbrace{\{(P)\Rightarrow[(P)\Rightarrow(Q)]\}}_{H}\Rightarrow\underbrace{[(P)\Rightarrow(P)]}_{K} \qquad \text{(D2), 1.12},$$

$$\vdash \overbrace{\{(P)\Rightarrow[(P)\Rightarrow(Q)]\}}^{H}\Rightarrow\underbrace{[(P)\Rightarrow(Q)]}_{C} \qquad \text{(D3)},$$

which completes the proof of $\{(P) \Rightarrow [(P) \Rightarrow (Q)]\} \Rightarrow [(P) \Rightarrow (Q)]$. A completely formal proof would also expand each of the steps just listed into its own proof, as specified in the proof of the Deduction Theorem 1.22, for instance, expanding each use of the directive (D3), which uses theorem 1.15, into a complete proof of theorem 1.15.

Although theorem 1.22 can produce lengthy proofs, the resulting long proofs can also suggest shorter proofs, for instance, as in theorem 1.24.

1.24 Theorem.   *The formula $\{(P) \Rightarrow [(P) \Rightarrow (Q)]\} \Rightarrow [(P) \Rightarrow (Q)]$ is a theorem.*

Proof. Apply axiom P2 with theorems 1.14 and 1.17:

$$\vdash \underbrace{(P) \Rightarrow (P)}_{K} \qquad\qquad\qquad\qquad \text{theorem 1.14},$$

$$\vdash \underbrace{\{(P) \Rightarrow [(P) \Rightarrow (Q)]\}}_{H} \Rightarrow \{\underbrace{[(P) \Rightarrow (P)]}_{K} \Rightarrow \underbrace{[(P) \Rightarrow (Q)]}_{L}\} \quad \text{axiom P2},$$

$$\vdash \underbrace{\{(P) \Rightarrow [(P) \Rightarrow (Q)]\}}_{H} \Rightarrow \underbrace{[(P) \Rightarrow (Q)]}_{L} \qquad\qquad \text{theorem 1.17}.$$

$$\square$$

Theorem 1.24 is also used in the form of a derived rule.

1.25 Theorem (derived rule).   *If $(P) \Rightarrow [(P) \Rightarrow (Q)]$ is a theorem, then $(P) \Rightarrow (Q)$ is also a theorem.*

Proof. Apply theorem 1.24 and *Detachment*:

$\vdash \underbrace{\{(P) \Rightarrow [(P) \Rightarrow (Q)]\}}_{H} \Rightarrow \underbrace{[(P) \Rightarrow (Q)]}_{C}$ theorem 1.24,

$\vdash \underbrace{(P) \Rightarrow [(P) \Rightarrow (Q)]}_{H}$ hypothesis,

$\vdash \underbrace{(P) \Rightarrow (Q)}_{C}$ *Detachment.*

$\square$

The design of proofs of formulae of the form $(A) \Rightarrow [(B) \Rightarrow (C)]$ can start with a derivation $A, B \vdash C$ of C from two hypotheses A and B. With A treated as an axiom, and B as the hypothesis H, a first application of the Deduction Theorem (1.22) leads to a derivation $A \vdash (B) \Rightarrow (C)$. Then, with A as the hypothesis H, a second application of the Deduction Theorem (1.22) yields a proof of $(A) \Rightarrow [(B) \Rightarrow (C)]$.

The general case with several hypotheses follows by several applications of the Deduction Theorem (1.22), in a way that may be specified more explicitly after the availability of the Principle of Mathematical Induction in chapter 4.

1.3.4   Example: Law of Assertion from the Deduction Theorem

The following proof shows the use of the Deduction Theorem in designing proofs.

1.26 Theorem.   *The law of assertion* $(A) \Rightarrow \{[(A) \Rightarrow (B)] \Rightarrow (B)\}$ *is a theorem.*

Proof. A finished proof can proceed as follows:

$\vdash [(A) \Rightarrow (B)] \Rightarrow [(A) \Rightarrow (B)]$ theorem 1.14,
$\vdash (A) \Rightarrow \{[(A) \Rightarrow (B)] \Rightarrow [(A) \Rightarrow (B)]\}$ theorem 1.12,
$\vdash (A) \Rightarrow \{[(A) \Rightarrow (B)] \Rightarrow (A)\}$ axiom P1,
$\vdash (A) \Rightarrow \{[(A) \Rightarrow (B)] \Rightarrow (B)\}$ theorem 1.20.

The following considerations explain how to *design* such a proof.

The formula $(A) \Rightarrow \{[(A) \Rightarrow (B)] \Rightarrow (B)\}$ has the pattern $(H) \Rightarrow [(K) \Rightarrow (S)]$ of the Deduction Theorem, with A for H, $(A) \Rightarrow (B)$ for K, and B for S.

Step 1.

As in the Deduction Theorem, assume first that the hypotheses H and K are proved, and from them derive the conclusion S by proving $H, K \vdash S$. Here, assume that the hypotheses A and $(A) \Rightarrow (B)$ are both proved, and prove $A, [(A) \Rightarrow (B)] \vdash B$:

$\vdash A$ first temporary hypothesis,
$\vdash (A) \Rightarrow (B)$ second temporary hypothesis,
$\vdash B$ *Detachment.*

The foregoing derivation shows that if A and $(A) \Rightarrow (B)$ are proved, then B is proved. Still under the first hypothesis A, the Deduction Theorem allows for the removal of the second hypothesis, $(A) \Rightarrow (B)$, as follows.

Step 2.

Step 2.1

The *first line in step 1* consists of the other hypothesis, A, which is assumed proved, whence instructions (D2) in the Deduction Theorem replace A with a complete proof of $[(A) \Rightarrow (B)] \Rightarrow (A)$ as in theorem 1.12. In other words, replace the first line, $\vdash A$, by the following three lines:

	Proof of theorem 1.12:
$\vdash A$	temporary hypothesis,
$\vdash (A) \Rightarrow \{\underbrace{[(A) \Rightarrow (B)]}_{Q} \Rightarrow (A)\}$	axiom P1,
$\vdash [(A) \Rightarrow (B)] \Rightarrow (A)$	*Detachment.*
	End of proof of theorem 1.12.

Step 2.2

Similarly, the *second line in step 1* consists of the hypothesis K being currently removed, here $(A) \Rightarrow (B)$, which instructions (D1) in the Deduction Theorem replace with a complete proof of $(K) \Rightarrow (K)$, here $[(A) \Rightarrow (B)] \Rightarrow [(A) \Rightarrow (B)]$, as in theorem 1.14. Thus, replace the second line, $\vdash (A) \Rightarrow (B)$, by the following lines:

	Proof of theorem 1.14 with $[K]$ for $[(A) \Rightarrow (B)]$:
$\vdash [K] \Rightarrow (\{[K] \Rightarrow [K]\} \Rightarrow [K])$	axiom P1,
$\vdash \{[K] \Rightarrow (\{[K] \Rightarrow [K]\} \Rightarrow [K])\}$ $\Rightarrow [([K] \Rightarrow \{[K] \Rightarrow [K]\}) \Rightarrow \{[K] \Rightarrow [K]\}]$	axiom P2 ... ... continued,
$\vdash ([K] \Rightarrow \{[K] \Rightarrow [K]\}) \Rightarrow \{[K] \Rightarrow [K]\}$	*Detachment,*
$\vdash [K] \Rightarrow \{[K] \Rightarrow [K]\}$	axiom P1,
$\vdash [K] \Rightarrow [K]$	*Detachment,*
$\vdash [(A) \Rightarrow (B)] \Rightarrow [(A) \Rightarrow (B)]$	substitution.
	End of proof of theorem 1.14.

Step 2.3

Finally, the *third line in step 1* invokes *Detachment,* which instructions (D3) replace by an instance of (the proof of) theorem 1.15:

$$\vdash [(A) \Rightarrow (B)] \Rightarrow (A) \qquad\qquad \text{step 2.1,}$$
$$\vdash [(A) \Rightarrow (B)] \Rightarrow [(A) \Rightarrow (B)] \quad \text{step 2.2,}$$
$$\vdash [(A) \Rightarrow (B)] \Rightarrow (B) \qquad\qquad \text{theorem 1.15.}$$

Hence the proof no longer assumes $(A) \Rightarrow (B)$ as a hypothesis, but it still assumes A as a hypothesis, thus proving that

$$A \vdash \{[(A) \Rightarrow (B)] \Rightarrow (B)\}.$$

Step 3.

Finally, the Deduction Theorem allows for the removal of the first hypothesis, A, from step 2. Here step 2 consists of steps 2.1, 2.2, and 2.3.

Step 3.1

In step 2.1 the first line consists of this hypothesis, A, whence instructions (D1) replace A with $(A) \Rightarrow (A)$ by a complete proof of theorem 1.14. In other words, replace the first line in step 2, $\vdash A$, by the following lines:

Proof of theorem 1.14:

$$\vdash [A] \Rightarrow (\{[A] \Rightarrow [A]\} \Rightarrow [A]) \qquad\qquad \text{axiom P1,}$$
$$\vdash \{[A] \Rightarrow (\{[A] \Rightarrow [A]\} \Rightarrow [A])\} \qquad\qquad \text{axiom P1 \ldots}$$
$$\qquad \Rightarrow \big[([A] \Rightarrow \{[A] \Rightarrow [A]\}) \Rightarrow \{[A] \Rightarrow [A]\}\big] \quad \text{\ldots continued,}$$

$$\vdash ([A] \Rightarrow \{[A] \Rightarrow [A]\}) \Rightarrow \{[A] \Rightarrow [A]\} \qquad \textit{Detachment,}$$
$$\vdash [A] \Rightarrow \{[A] \Rightarrow [A]\} \qquad\qquad\qquad \text{axiom P1,}$$
$$\vdash [A] \Rightarrow [A] \qquad\qquad\qquad\qquad\qquad \textit{Detachment.}$$

End of proof of theorem 1.14.

Step 3.2

The second line in step 2.1 is an instance of axiom P1, which instructions (D2) replace by $(A) \Rightarrow \big[(A) \Rightarrow \{[(A) \Rightarrow (B)] \Rightarrow (A)\}\big]$.

Step 3.3

The third line in step 2.1 yields $[(A) \Rightarrow (B)] \Rightarrow (A)$, from *Detachment,* which instructions (D3) replace by a complete proof of $(A) \Rightarrow \{[(A) \Rightarrow (B)] \Rightarrow (A)\}$, as in theorem 1.15. In this case, however, such a proof would be correct but not necessary, because $(A) \Rightarrow \{[(A) \Rightarrow (B)] \Rightarrow (A)\}$ is merely an instance of axiom P1. Because it is an axiom, all the preceding lines also become superfluous.

Step 3.4

The result of step 2.2, $[(A) \Rightarrow (B)] \Rightarrow [(A) \Rightarrow (B)]$, is proved by theorem 1.14. Hence, instructions (D2) replace it by $(A) \Rightarrow [(A) \Rightarrow (B)] \Rightarrow [(A) \Rightarrow (B)]$.

Step 3.5

Fully written out, the remaining lines in step 2.3 would follow the proof of theorem 1.15. Removing the hypothesis A then amounts to theorem 1.20, which forms the last line of the final proof:

$\vdash (A) \Rightarrow \{[(A) \Rightarrow (B)] \Rightarrow (A)\}$ axiom P1 (from 3.3, replacing 2.1),

$\vdash [(A) \Rightarrow (B)] \Rightarrow [(A) \Rightarrow (B)]$ theorem 1.14 (from step 2.2),

$\vdash (A) \Rightarrow \{[(A) \Rightarrow (B)] \Rightarrow [(A) \Rightarrow (B)]\}$ theorem 1.12 (from 3.4, replacing 2.2),

$\vdash (A) \Rightarrow \{[(A) \Rightarrow (B)] \Rightarrow (B)\}$ Theorem 1.20 (from 3.5, replacing 2.3).

   Thus the Deduction Theorem has provided some guidance for the construction of a proof of the theorem $(A) \Rightarrow \{[(A) \Rightarrow (B)] \Rightarrow (B)\}$.   □

1.3.5   More Examples to Design Proofs of Implicational Theorems

In patterns of deductive reasoning involving two premises, the following theorems confirm that the order of the premises does not matter.

1.27 Theorem (transitive law of implication).   *The following formula is a theorem:*

$$[(Q) \Rightarrow (R)] \Rightarrow \{[(P) \Rightarrow (Q)] \Rightarrow [(P) \Rightarrow (R)]\}.$$

Proof. The formula to be proved has the form $(A) \Rightarrow [(B) \Rightarrow (C)]$, with A denoting $(Q) \Rightarrow (R)$, B denoting $(P) \Rightarrow (Q)$, and C denoting $(P) \Rightarrow (R)$:

$$\overbrace{[(Q) \Rightarrow (R)]}^{A} \Rightarrow \{\overbrace{[(P) \Rightarrow (Q)]}^{B} \Rightarrow \overbrace{[(P) \Rightarrow (R)]}^{C}\}.$$

Phase 1: deriving $A, B \vdash C$ and discharging B.

Designing a proof of $(A) \Rightarrow [(B) \Rightarrow (C)]$ can start with a derivation $A, B \vdash C$, in this case $(Q) \Rightarrow (R), (P) \Rightarrow (Q) \vdash (P) \Rightarrow (R)$, which is exactly theorem 1.16.

Expanding the proof of theorem 1.16 which invokes theorems 1.12 and 1.15, helps discharging the hypothesis $(P) \Rightarrow (Q)$:

$\vdash (Q) \Rightarrow (R)$ hypothesis A,

$\vdash (P) \Rightarrow [(Q) \Rightarrow (R)]$ theorem 1.12,

$\vdash (P) \Rightarrow (R)$ conclusion C, by theorem 1.15.

The preceding three steps form a derivation $A, B \vdash C$ of C from A and B, but without invoking B, which plays a hidden rôle in the proof of theorem 1.15. Replacing the citation of theorem 1.15 by its proof leads to a derivation of $(B) \Rightarrow (C)$ from A:

$\vdash (Q) \Rightarrow (R)$ hypothesis A,

$\vdash (P) \Rightarrow [(Q) \Rightarrow (R)]$ theorem 1.12,

$\vdash [(P) \Rightarrow (Q)] \Rightarrow [(Q) \Rightarrow (R)]$ $(B) \Rightarrow (C)$ by theorem 1.13.

Replacing the citation of theorem 1.13 by its proof, which uses only axioms and *Detachment*, gives a derivation of $(B) \Rightarrow (C)$ from A directly from the axioms:

1   $\vdash (Q) \Rightarrow (R)$ hypothesis A,

2   $\vdash (P) \Rightarrow [(Q) \Rightarrow (R)]$ theorem 1.12,

3   $\vdash \{(P) \Rightarrow [(Q) \Rightarrow (R)]\} \Rightarrow \{[(P) \Rightarrow (Q)] \Rightarrow [(P) \Rightarrow (R)]\}$ axiom P2,

4   $\vdash [(P) \Rightarrow (Q)] \Rightarrow [(P) \Rightarrow (R)]$ 2, 3, *Detachment*.

Lines 1–4 complete the derivation of $[(P) \Rightarrow (Q)] \Rightarrow [(Q) \Rightarrow (R)]$ from the first hypothesis $(Q) \Rightarrow (R)$ with axiom P2 and *Detachment*. Since the second hypothesis, $(P) \Rightarrow (Q)$, has not been used, it need not be discharged.

Phase 2: discharging A.

An application of the Deduction Theorem (1.22) discharges the first hypothesis and yields a proof of $(A) \Rightarrow [(B) \Rightarrow (C)]$. The resulting proof can be shortened, or alternative proofs may result from trial and error. To this end, H, K, L refer to theorem 1.15:

$\vdash \underbrace{[(Q) \Rightarrow (R)]}_{H} \Rightarrow \underbrace{\{(P) \Rightarrow [(Q) \Rightarrow (R)]\}}_{K}$ axiom P1,

$\vdash \underbrace{\{(P) \Rightarrow [(Q) \Rightarrow (R)]\}}_{K} \Rightarrow \underbrace{\{[(P) \Rightarrow (Q)] \Rightarrow [(P) \Rightarrow (R)]\}}_{L}$ axiom P2,

$\vdash \underbrace{[(Q) \Rightarrow (R)]}_{H} \Rightarrow \underbrace{\{[(P) \Rightarrow (Q)] \Rightarrow [(P) \Rightarrow (R)]\}}_{L}$ theorem 1.15.

                                                             □

Swapping the premisses $(P) \Rightarrow (Q)$ and $(Q) \Rightarrow (R)$ also yields $(P) \Rightarrow (R)$:

1.28 Theorem (transitive law of implication, law of the hypothetical syllogism). *The following formula is a theorem:*

$$[(P) \Rightarrow (Q)] \Rightarrow \{[(Q) \Rightarrow (R)] \Rightarrow [(P) \Rightarrow (R)]\}.$$

Proof. With notation as in the proof of theorem 1.27, the formula to be proved has the form $(B) \Rightarrow [(A) \Rightarrow (C)]$, with A denoting $(Q) \Rightarrow (R)$, B denoting $(P) \Rightarrow (Q)$, and C denoting $(P) \Rightarrow (R)$:

$$\overbrace{[(P) \Rightarrow (Q)]}^{B} \Rightarrow \{\overbrace{[(Q) \Rightarrow (R)]}^{A} \Rightarrow \overbrace{[(P) \Rightarrow (R)]}^{C}\}.$$

Thus designing a proof of $(B) \Rightarrow [(A) \Rightarrow (C)]$ can start with a derivation $B, A \vdash C$, in this case $(P) \Rightarrow (Q), (Q) \Rightarrow (R) \vdash (P) \Rightarrow (R)$, which is exactly theorem 1.16. Hence steps as in the proof of theorem 1.27 discharge the hypotheses, but in the reverse order. The resulting proof might then be shortened or give clues for a shorter alternative proof. For example, apply axiom P1 with theorems 1.27, 1.13, and 1.16:

$$\vdash \underbrace{\overbrace{[(P) \Rightarrow (Q)]}^{U}}_{K} \Rightarrow \{\overbrace{\underbrace{[(Q) \Rightarrow (R)]}_{H} \Rightarrow \underbrace{[(P) \Rightarrow (Q)]}_{K}}^{V}\} \qquad \text{axiom P1,}$$

$$\vdash \underbrace{\{(Q) \Rightarrow (R)\}}_{H} \Rightarrow \{\underbrace{[(P) \Rightarrow (Q)]}_{K} \Rightarrow \underbrace{[(P) \Rightarrow (R)]}_{L}\} \qquad \text{1.27,}$$

$$\vdash \underbrace{\{\overbrace{[(Q) \Rightarrow (R)]}^{H} \Rightarrow \overbrace{[(P) \Rightarrow (Q)]}^{K}\}}_{V} \Rightarrow \underbrace{\{\overbrace{[(Q) \Rightarrow (R)]}^{H} \Rightarrow \overbrace{[(P) \Rightarrow (R)]}^{L}\}}_{W} \qquad \text{1.13,}$$

$$\vdash \overbrace{[(P) \Rightarrow (Q)]}^{U} \Rightarrow \overbrace{\{[(Q) \Rightarrow (R)] \Rightarrow [(P) \Rightarrow (R)]\}}^{W} \qquad \text{1.16.}$$

$\square$

1.3.6   Another Guide for Proofs: Substitutivity of Equivalences

Besides the Deduction Theorem (theorem 1.22), another guide to design proofs consists of replacing any occurrence of a formula by an equivalent formula, thanks to theorem 1.29 [18, p. 101, 124, 189], [108, p. 48].

1.29 Theorem (Substitutivity of Equivalence in the Pure Positive Implicational Propositional Calculus, preliminary version). *For all well-formed implicational logical formulae U and V, if $\vdash (U) \Rightarrow (V)$ and $\vdash (V) \Rightarrow (U)$, and if a formula Q results from substituting any (zero, one, several, or all) occurrence(s) of U by V in a well-formed formula P, then $\vdash (P) \Rightarrow (Q)$ and $\vdash (Q) \Rightarrow (P)$.*

Proof (Outline). This proof proceeds by cases and subcases.

In all cases, if Q is P, which results by substituting none of the occurrences of U by V, then each of $(P) \Rightarrow (Q)$ and $(Q) \Rightarrow (P)$ is $(P) \Rightarrow (P)$, which is theorem 1.14.

1. If P is U, then Q is either U or V. If Q is V, then $(P) \Rightarrow (Q)$ and $(Q) \Rightarrow (P)$ become $(U) \Rightarrow (V)$ and $(V) \Rightarrow (U)$, which are the hypotheses.

2. If P is $(U) \Rightarrow (W)$, then Q is either $(U) \Rightarrow (W)$, which is P, or Q is $(V) \Rightarrow (W)$. If Q is $(V) \Rightarrow (W)$, then the hypothesis $\vdash (V) \Rightarrow (U)$ yields $\vdash (P) \Rightarrow (Q)$:

$\vdash (V) \Rightarrow (U)$ hypothesis,
$\vdash [(V) \Rightarrow (U)] \Rightarrow \{[(U) \Rightarrow (W)] \Rightarrow [(V) \Rightarrow (W)]\}$ theorem 1.28,
$\vdash \underbrace{[(U) \Rightarrow (W)]}_{P} \Rightarrow \underbrace{[(V) \Rightarrow (W)]}_{Q}$ Detachment.

Similarly, the hypothesis $\vdash (U) \Rightarrow (V)$ yields $\vdash (Q) \Rightarrow (P)$:

$\vdash (U) \Rightarrow (V)$ hypothesis,
$\vdash [(U) \Rightarrow (V)] \Rightarrow \{[(V) \Rightarrow (W)] \Rightarrow [(U) \Rightarrow (W)]\}$ theorem 1.28,
$\vdash \underbrace{[(V) \Rightarrow (W)]}_{Q} \Rightarrow \underbrace{[(U) \Rightarrow (W)]}_{P}$ Detachment.

3. If P is $(W) \Rightarrow (U)$, then Q is either $(W) \Rightarrow (U)$ or $(W) \Rightarrow (V)$. If Q is $(W) \Rightarrow (V)$, then P with $(U) \Rightarrow (V)$ yield Q, and Q with $(V) \Rightarrow (U)$ yield P, by transitivity (theorem 1.16).

If P is $(W) \Rightarrow (U)$, then Q is either $(W) \Rightarrow (U)$, which is P, or Q is $(W) \Rightarrow (V)$. If Q is $(W) \Rightarrow (V)$, then the hypothesis $\vdash (V) \Rightarrow (U)$ yields $\vdash (Q) \Rightarrow (P)$:

$\vdash (V) \Rightarrow (U)$ hypothesis,
$\vdash [(V) \Rightarrow (U)] \Rightarrow \{[(W) \Rightarrow (V)] \Rightarrow [(W) \Rightarrow (U)]\}$ theorem 1.27,
$\vdash \underbrace{[(W) \Rightarrow (V)]}_{Q} \Rightarrow \underbrace{[(W) \Rightarrow (U)]}_{P}$ Detachment.

Similarly, the hypothesis $\vdash (U) \Rightarrow (V)$ yields $\vdash (P) \Rightarrow (Q)$:

$\vdash (U) \Rightarrow (V)$ hypothesis,
$\vdash [(U) \Rightarrow (V)] \Rightarrow \{[(W) \Rightarrow (U)] \Rightarrow [(W) \Rightarrow (V)]\}$ theorem 1.27,
$\vdash \underbrace{[(W) \Rightarrow (U)]}_{P} \Rightarrow \underbrace{[(W) \Rightarrow (V)]}_{Q}$ Detachment.

The general case follows by several applications of the previous cases, in a way that may be specified more explicitly after the availability of the Principle of Mathematical Induction in chapter 4.   □

1.30 Example. If U denotes $(H) \Rightarrow (K)$, and V denotes $(H) \Rightarrow [(H) \Rightarrow (K)]$, then $\vdash (U) \Rightarrow (V)$ and $\vdash (V) \Rightarrow (U)$:

$\vdash \overbrace{[(H) \Rightarrow (K)]}^{U} \Rightarrow \overbrace{\{(H) \Rightarrow [(H) \Rightarrow (K)]\}}^{V}$ axiom P1,
$\vdash \underbrace{\{(H) \Rightarrow [(H) \Rightarrow (K)]\}}_{V} \Rightarrow \underbrace{[(H) \Rightarrow (K)]}_{U}$ theorem 1.24.

Also, if P denotes $(L) \Rightarrow [(H) \Rightarrow (K)]$, which is $(L) \Rightarrow (U)$, then $\vdash (P) \Rightarrow (Q)$ and $\vdash (Q) \Rightarrow (P)$ become

$\vdash \overbrace{\{(L) \Rightarrow [(H) \Rightarrow (K)]\}}^{P} \Rightarrow \overbrace{\left[(L) \Rightarrow \{(H) \Rightarrow [(H) \Rightarrow (K)]\}\right]}^{Q}$ theorem 1.29,
$\vdash \underbrace{\left[(L) \Rightarrow \{(H) \Rightarrow [(H) \Rightarrow (K)]\}\right]}_{Q} \Rightarrow \underbrace{\{(L) \Rightarrow [(H) \Rightarrow (K)]\}}_{P}$ theorem 1.29,

by axiom P1 with theorems 1.24 and 1.29.

1.3.7   More Derived Rules of Inference

The following derived rules allow for substitutions within implications subject to hypotheses, for instance, a substitution within an intermediate hypothesis.

1.31 Theorem (derived rule). *For all well-formed formulae H, K, L, M, if*
$$(H) \Rightarrow [(L) \Rightarrow (M)] \quad and$$
$$(K) \Rightarrow (L)$$
are theorems, then
$$(H) \Rightarrow [(K) \Rightarrow (M)]$$
is also a theorem.

Proof. Apply theorems 1.28 and 1.16:

$\vdash (K) \Rightarrow (L)$	hypothesis,
$\vdash [(K) \Rightarrow (L)] \Rightarrow \{[(L) \Rightarrow (M)] \Rightarrow [(K) \Rightarrow (M)]\}$	theorem 1.28,
$\vdash [(L) \Rightarrow (M)] \Rightarrow [(K) \Rightarrow (M)]$	Detachment.
$\vdash (H) \Rightarrow [(L) \Rightarrow (M)]$	hypothesis,
$\vdash (H) \Rightarrow [(K) \Rightarrow (M)]$	theorem 1.16.

$\square$

The second theorem allows for a substitution in the conclusion.

1.32 Theorem (derived rule). *For all well-formed formulae H, L, M, N, if*
$$(H) \Rightarrow [(L) \Rightarrow (M)] \quad and$$
$$(M) \Rightarrow (N)$$
are theorems, then
$$(H) \Rightarrow [(L) \Rightarrow (N)]$$
is also a theorem.

Proof. Apply theorems 1.28, 1.16, and 1.17:

$\vdash (H) \Rightarrow [\underbrace{(L) \Rightarrow (M)}_{P}]$ hypothesis,

$\vdash [\underbrace{(L) \Rightarrow (M)}_{P}] \Rightarrow \{[\underbrace{(M) \Rightarrow (N)}_{Q}] \Rightarrow [(L) \Rightarrow (N)]\}$ theorem 1.28,

$\vdash (H) \Rightarrow \{\underbrace{[(M) \Rightarrow (N)] \Rightarrow [(L) \Rightarrow (N)]}_{Q}\}$ theorem 1.16,

$\vdash (M) \Rightarrow (N)$ hypothesis,

$\vdash (H) \Rightarrow [(L) \Rightarrow (N)]$ theorem 1.17.

$\square$

The following three derived rules of inference will simplify subsequent proofs.

1.33 Theorem (derived rule). *If* $(K) \Rightarrow (L)$ *is a theorem, then the following formula is also a theorem:* $[(H) \Rightarrow (K)] \Rightarrow [(H) \Rightarrow (L)]$.

Proof. Apply the transitivity of implication in the form of theorem 1.28:

$\vdash (K) \Rightarrow (L)$ hypothesis,

$\vdash [(K) \Rightarrow (L)] \Rightarrow \{[(H) \Rightarrow (K)] \Rightarrow [(H) \Rightarrow (L)]\}$ theorem 1.28,

$\vdash [(H) \Rightarrow (K)] \Rightarrow [(H) \Rightarrow (L)]$ *Detachment.*

□

1.34 Theorem (derived rule). *If* $(I) \Rightarrow (H)$ *is a theorem, then the following formula is also a theorem:* $[(H) \Rightarrow (K)] \Rightarrow [(I) \Rightarrow (K)]$.

Proof. Apply the transitivity of implication in the form of theorem 1.27:

$\vdash (I) \Rightarrow (H)$ hypothesis,

$\vdash [(I) \Rightarrow (H)] \Rightarrow \{[(H) \Rightarrow (K)] \Rightarrow [(I) \Rightarrow (K)]\}$ theorem 1.27,

$\vdash [(H) \Rightarrow (K)] \Rightarrow [(I) \Rightarrow (K)]$ *Detachment.*

□

1.35 Theorem (derived rule). *If* $(A) \Rightarrow (B)$ *and* $(C) \Rightarrow (D)$ *are theorems, then the following formula is also a theorem:* $[(B) \Rightarrow (C)] \Rightarrow [(A) \Rightarrow (D)]$.

Proof. Apply theorems 1.34 and 1.33, with H, K, L as in theorem 1.16:

$\vdash (A) \Rightarrow (B)$ hypothesis,

$\vdash \underbrace{[(B) \Rightarrow (C)]}_{H} \Rightarrow \underbrace{[(A) \Rightarrow (C)]}_{K}$ theorem 1.34,

$\vdash (C) \Rightarrow (D)$ hypothesis,

$\vdash \underbrace{[(A) \Rightarrow (C)]}_{K} \Rightarrow \underbrace{[(A) \Rightarrow (D)]}_{L}$ theorem 1.33,

$\vdash \underbrace{[(B) \Rightarrow (C)]}_{H} \Rightarrow \underbrace{[(A) \Rightarrow (D)]}_{L}$ theorem 1.16.

□

In contrast to the preceding derived rules, theorem 1.36 reveals a different pattern.

1.36 Theorem (derived rule). *If* $[(H) \Rightarrow (L)] \Rightarrow (M)$ *and* $(H) \Rightarrow (K)$ *are theorems, then the following formula is also a theorem:* $[(K) \Rightarrow (L)] \Rightarrow (M)$.

Proof. Apply theorems 1.27, 1.13, 1.12, and 1.16:

$\vdash \underbrace{[(K) \Rightarrow (L)]}_{P} \Rightarrow \{\underbrace{[(H) \Rightarrow (K)]}_{Q} \Rightarrow \underbrace{[(H) \Rightarrow (L)]}_{R}\}$ 1.27,

$\vdash \{\underbrace{[(K) \Rightarrow (L)]}_{P} \Rightarrow \underbrace{[(H) \Rightarrow (K)]}_{Q}\} \Rightarrow \{\underbrace{[(K) \Rightarrow (L)]}_{P} \Rightarrow \underbrace{[(H) \Rightarrow (L)]}_{R}\}$ 1.13,

$\vdash \underbrace{[(H) \Rightarrow (K)]}_{Q}$ hypothesis,

$\vdash \underbrace{[(K) \Rightarrow (L)]}_{P} \Rightarrow \underbrace{[(H) \Rightarrow (K)]}_{Q}$ 1.12,

$$\vdash \underbrace{[(K) \Rightarrow (L)]}_{P} \Rightarrow \underbrace{[(H) \Rightarrow (L)]}_{R} \qquad \textit{Detachment,}$$

$$\vdash \underbrace{[(H) \Rightarrow (L)]}_{R} \Rightarrow (M) \qquad\qquad \text{hypothesis,}$$

$$\vdash \underbrace{[(K) \Rightarrow (L)]}_{P} \Rightarrow (M) \qquad\qquad 1.16.$$

□

1.3.8   The Laws of Commutation and of Assertion

The following "Law of Commutation" allows for yet another change in the order of hypotheses:

$$\{(P) \Rightarrow [(Q) \Rightarrow (R)]\} \Rightarrow \{(Q) \Rightarrow [(P) \Rightarrow (R)]\}.$$

The Law of Commutation is one of Frege's axioms [39, p. 146, eq. (8)]. In Frege's 1879 words, the Law of Commutation states that

> If a proposition is the consequence of two propositions, their order is immaterial [39, p. 147].

In 1935, however, Łukasiewicz showed that the Law of Commutation is a theorem derivable from *Detachment* with axioms P1 and P2 [80, p. 127], as proved by theorem 1.37. (The time lapse between 1879 and 1935 indicates that recognizing whether a formula is a theorem can also be difficult for specialists.)

1.37 Theorem (law of commutation).   *The following formula is a theorem:*

$$\{(P) \Rightarrow [(Q) \Rightarrow (R)]\} \Rightarrow \{(Q) \Rightarrow [(P) \Rightarrow (R)]\}.$$

Proof. Apply theorem 1.31:

$$\vdash \underbrace{\{(P) \Rightarrow [(Q) \Rightarrow (R)]\}}_{H} \Rightarrow \{\underbrace{[(P) \Rightarrow (Q)]}_{L} \Rightarrow \underbrace{[(P) \Rightarrow (R)]}_{M}\} \qquad \text{axiom P2,}$$

$$\vdash (\underbrace{Q}_{K}) \Rightarrow \underbrace{[(P) \Rightarrow (Q)]}_{L} \qquad\qquad\qquad \text{axiom P1,}$$

$$\vdash \underbrace{\{(P) \Rightarrow [(Q) \Rightarrow (R)]\}}_{H} \Rightarrow \{(\underbrace{Q}_{K}) \Rightarrow \underbrace{[(P) \Rightarrow (R)]}_{M}\} \qquad \text{theorem 1.31.}$$

□

With a shorter proof relying on the law of commutation, the following theorem combines the rule of *Detachment* into a single formula.

1.38 Theorem (law of assertion). *The formula* $(H) \Rightarrow \{[(H) \Rightarrow (C)] \Rightarrow (C)\}$ *is a theorem.*

Proof. Apply theorems 1.14 and 1.37 with *Detachment*:

$$\vdash \overbrace{[(H) \Rightarrow (C)]}^{P} \Rightarrow [\overbrace{(\underbrace{H}) \Rightarrow (\underbrace{C})}^{P}] \quad\quad \text{theorem 1.14,}$$

$$\vdash (\underset{Q}{\underbrace{H}}) \Rightarrow \{[\underset{P}{\underbrace{(H) \Rightarrow (C)}}] \Rightarrow (\underset{R}{\underbrace{C}})\} \quad\quad \text{commutation (1.37) and } \textit{Detachment.}$$

$\square$

Also relying on the law of commutation, the following derived rule will shorten the proof of subsequent results in particular, theorem 1.50.

1.39 Theorem (derived rule). *If* $\vdash (U) \Rightarrow [(V) \Rightarrow (W)]$
  and $\vdash (H) \Rightarrow [(W) \Rightarrow (R)]$,
  then $\vdash (H) \Rightarrow \{(U) \Rightarrow [(V) \Rightarrow (R)]\}$.

Proof. Apply the transitivity of implication (theorem 1.27):

$\vdash [(V) \Rightarrow (W)] \Rightarrow \{[(W) \Rightarrow (R)] \Rightarrow [(V) \Rightarrow (R)]\}$	theorem 1.27,
$\vdash (U) \Rightarrow ([(V) \Rightarrow (W)] \Rightarrow \{[(W) \Rightarrow (R)] \Rightarrow [(V) \Rightarrow (R)]\})$	theorem 1.12,
$\vdash (U) \Rightarrow [(V) \Rightarrow (W)]$	hypothesis,
$\vdash (U) \Rightarrow \{[(W) \Rightarrow (R)] \Rightarrow [(V) \Rightarrow (R)]\}$	theorem 1.15,
$\vdash \{(U) \Rightarrow [(W) \Rightarrow (R)]\} \Rightarrow \{(U) \Rightarrow [(V) \Rightarrow (R)]\}$	theorem 1.13,
$\vdash (H) \Rightarrow [(W) \Rightarrow (R)]$	hypothesis,
$\vdash (U) \Rightarrow \{(H) \Rightarrow [(W) \Rightarrow (R)]\}$	theorem 1.12,
$\vdash (H) \Rightarrow \{(U) \Rightarrow [(W) \Rightarrow (R)]\}$	commutation (1.37),
$\vdash (H) \Rightarrow \{(U) \Rightarrow [(V) \Rightarrow (R)]\}$	transitivity.

$\square$

1.3.9 Exercises on the Classical Implicational Calculus

The foregoing theorems involve only implications but no negation, and their proofs do not involve any negation either. Nevertheless, there exist other theorems involving only $\Rightarrow$ but not $\neg$ for which there does not exist any proof involving only implications. Examples of such theorems are hidden in the exercises, to be revealed later. Investigate whether the formulae in the following exercises are theorems, using any of the axioms P1 and P2, any rules of inference, and any of the theorems just proved.

1.1 . $[(H) \Rightarrow (L)] \Rightarrow \{(H) \Rightarrow [(K) \Rightarrow (L)]\}$

1.2 . $[(K) \Rightarrow (L)] \Rightarrow \{(H) \Rightarrow [(K) \Rightarrow (L)]\}$

1.3 . $[(A) \Rightarrow (B)] \Rightarrow \big[\{[(A) \Rightarrow (B)] \Rightarrow (B)\} \Rightarrow (B)\big]$

1.4 . $[(A) \Rightarrow (B)] \Rightarrow \big[\{[(A) \Rightarrow (B)] \Rightarrow (A)\} \Rightarrow (A)\big]$

1.5 . $\{[(P) \Rightarrow (P)] \Rightarrow (P)\} \Rightarrow (P)$

1.6 . $(P) \Rightarrow \{[(P) \Rightarrow (P)] \Rightarrow (P)\}$

1.7 . $\{[(P) \Rightarrow (Q)] \Rightarrow (P)\} \Rightarrow (P)$ (Peirce's law.)

1.8 . $[(P) \Rightarrow (R)] \Rightarrow \big[\{[(P) \Rightarrow (Q)] \Rightarrow (R)\} \Rightarrow (R)\big]$

1.9 . $\{[(R) \Rightarrow (Q)] \Rightarrow (P)\} \Rightarrow \{[(R) \Rightarrow (Q)] \Rightarrow [(S) \Rightarrow (P)]\}$

1.10 . $[(R) \Rightarrow (Q)] \Rightarrow \big(\{[(R) \Rightarrow (Q)] \Rightarrow (P)\} \Rightarrow [(S) \Rightarrow (P)]\big)$

1.3.10   Equivalent Implicational Axiom Systems

The Classical Implicational Calculus just presented rests on the rules of *Detachment* and *Substitution* with Frege's axioms P1 and P2:

Frege's axiom P1 $(P) \Rightarrow [(Q) \Rightarrow (P)]$.
Frege's axiom P2 $\{(P) \Rightarrow [(Q) \Rightarrow (R)]\} \Rightarrow \{[(P) \Rightarrow (Q)] \Rightarrow [(P) \Rightarrow (R)]\}$.

Other selections of axioms exist, for instance, Stephen Cole Kleene's [72, p. 15],

Kleene's axiom 1a $(A) \Rightarrow [(B) \Rightarrow (A)]$,
Kleene's axiom 1b $[(A) \Rightarrow (B)] \Rightarrow (\{(A) \Rightarrow [(B) \Rightarrow (C)]\} \Rightarrow [(A) \Rightarrow (C)])$,

and Tarski's [129, p. 147],

Tarski's axiom I $(P) \Rightarrow [(Q) \Rightarrow (P)]$,
Tarski's axiom II $\{(P) \Rightarrow [(P) \Rightarrow (Q)]\} \Rightarrow [(P) \Rightarrow (Q)]$,
Tarski's axiom III $[(P) \Rightarrow (Q)] \Rightarrow \{[(Q) \Rightarrow (R)] \Rightarrow [(P) \Rightarrow (R)]\}$.

Frege's, Kleene's, and Tarski's implicational axiom systems are mutually equivalent, in the sense that each system leads to the same Pure Positive Implicational Propositional Calculus.

Indeed, their first axioms, P1, 1a, and I are mutually identical.

Second, Kleene's axiom 1b results from applying the law of commutation (theorem 1.37) to Frege's axiom P2. Consequently, both of Kleene's axioms 1a and 1b are theorems derivable from Frege's axioms P1 and P2. Thus, prepending derivations of Kleene's axioms from Frege's axioms to any proof of any theorem from Kleene's axioms yields a proof of the same theorem from Frege's axioms. Therefore, every theorem derivable from Kleene's axioms is also derivable from Frege's axioms.

Similarly, Tarski's axiom II is theorem 1.24, while Tarski's axiom III is theorem 1.28. Consequently, all three of Tarski's axioms I, II, and III are theorems derivable from Frege's axioms P1 and P2. Therefore, every theorem derivable from Tarski's axioms is also derivable from Frege's axioms.

The exercises establish the converse derivations. From Kleene's axioms, exercises 1.11 and 1.13 derive Tarski's axiom II, whereas exercises 1.14–1.22 derive Tarski's axiom III. Thus exercises 1.11–1.22 show that every theorem derivable from Tarski's axioms is also derivable from Kleene's axioms. Hence from Tarski's axioms, exercises 1.23–1.32 derive the law of commutation following Tarski's outline [129, p. 148–149], and thence Frege's axioms P2 from Tarski's axiom III. Thus exercises 1.23–1.36 show that every theorem derivable from Frege's axioms is derivable from Tarski's axioms, and thus also from Kleene's axioms.

However, after the proof of $(P) \Rightarrow (P)$ in theorem 1.14, Frege's axioms lead immediately to the Deduction Theorem (1.22), whereas the exercises reveal that several intermediate theorems stand between Kleene's or Tarski's axioms and the Deduction Theorem, which may explain the popularity of Frege's axioms, already announced before definition 1.4 on page 6.

1.3.11 Exercises on Kleene's Axioms

For the following exercises, use *only* Kleene's axioms 1a and 1b with Substitution and *Detachment*.

1.11. Prove $(P) \Rightarrow (P)$.

1.12. Establish the derived rule of inference that if T is a theorem, then $(S) \Rightarrow (T)$ is also a theorem.

1.13. Prove $\{[(P) \Rightarrow [(P) \Rightarrow (Q)]\} \Rightarrow [(P) \Rightarrow (Q)]$.

1.14. Establish the derived rule of inference that if $(H) \Rightarrow (K)$ and $(K) \Rightarrow (L)$ are theorems, then $(H) \Rightarrow (L)$ is also a theorem.

1.15. Establish the derived rule of inference that if T is a theorem, then $\{(A) \Rightarrow [(T) \Rightarrow (C)]\} \Rightarrow [(A) \Rightarrow (C)]$ is also a theorem.

1.16. Prove $[(B) \Rightarrow (C)] \Rightarrow \{(A) \Rightarrow [(B) \Rightarrow (C)]\}$.

1.17. Prove

$$\left\{[(B) \Rightarrow (C)] \Rightarrow \left(\{(A) \Rightarrow [(B) \Rightarrow (C)]\} \Rightarrow [(A) \Rightarrow (C)]\right)\right\}$$
$$\Rightarrow \{[(B) \Rightarrow (C)] \Rightarrow [(A) \Rightarrow (C)]\}.$$

1.18. Prove

$$\{[(B) \Rightarrow (C)] \Rightarrow [(A) \Rightarrow (B)]\}$$
$$\Rightarrow \left[\left([(B) \Rightarrow (C)] \Rightarrow \{[(A) \Rightarrow (B)] \Rightarrow (\{(A) \Rightarrow [(B) \Rightarrow (C)]\} \Rightarrow [(A) \Rightarrow (C)])\}\right)\right.$$
$$\left.\Rightarrow \{[(B) \Rightarrow (C)] \Rightarrow (\{(A) \Rightarrow [(B) \Rightarrow (C)]\} \Rightarrow [(A) \Rightarrow (C)])\}\right].$$

1.19 . Prove

$$\{[(B) \Rightarrow (C)] \Rightarrow [(A) \Rightarrow (B)]\}$$
$$\Rightarrow \Big\{[(B) \Rightarrow (C)] \Rightarrow (\{(A) \Rightarrow [(B) \Rightarrow (C)]\} \Rightarrow [(A) \Rightarrow (C)])\Big\}$$

1.20 . Prove

$$\{[(B) \Rightarrow (C)] \Rightarrow [(A) \Rightarrow (B)]\} \Rightarrow \{[(B) \Rightarrow (C)] \Rightarrow [(A) \Rightarrow (C)]\}.$$

1.21 . Prove $[(A) \Rightarrow (B)] \Rightarrow \{[(B) \Rightarrow (C)] \Rightarrow [(A) \Rightarrow (B)]\}$.

1.22 . Prove $[(A) \Rightarrow (B)] \Rightarrow \{[(B) \Rightarrow (C)] \Rightarrow [(A) \Rightarrow (C)]\}$.

1.3.12   Exercises on Tarski's Axioms

For the following exercises, use *only* Tarski's axioms I, II, and III, with Substitution and *Detachment*.

1.23 . Establish the derived rule of inference that if $(H) \Rightarrow (K)$ and $(K) \Rightarrow (L)$ are theorems, then $(H) \Rightarrow (L)$ is also a theorem.

1.24 . Establish the derived rule of inference that if T is a theorem, then $(S) \Rightarrow (T)$ is also a theorem.

1.25 . Prove $(P) \Rightarrow \{[(P) \Rightarrow (Q)] \Rightarrow (P)\}$.

1.26 . Prove

$$\{[(P) \Rightarrow (Q)] \Rightarrow (P)\} \Rightarrow ([(P) \Rightarrow (Q)] \Rightarrow \{[(P) \Rightarrow (Q)] \Rightarrow (Q)\}).$$

1.27 . Prove

$$(P) \Rightarrow ([(P) \Rightarrow (Q)] \Rightarrow \{[(P) \Rightarrow (Q)] \Rightarrow (Q)\}).$$

1.28 . Prove

$$([(P) \Rightarrow (Q)] \Rightarrow \{[(P) \Rightarrow (Q)] \Rightarrow (Q)\}) \Rightarrow \{[(P) \Rightarrow (Q)] \Rightarrow (Q)\}.$$

1.29 . Prove the law of assertion:

$$(P) \Rightarrow \{[(P) \Rightarrow (Q)] \Rightarrow (Q)\}.$$

1.30 . Prove

$$\{(P) \Rightarrow [(Q) \Rightarrow (R)]\} \Rightarrow \big(\{[(Q) \Rightarrow (R)] \Rightarrow (R)\} \Rightarrow [(P) \Rightarrow (R)]\big).$$

1.31 . Prove

$$\big(\{[(Q) \Rightarrow (R)] \Rightarrow (R)\} \Rightarrow [(P) \Rightarrow (R)]\big) \Rightarrow \{(Q) \Rightarrow [(P) \Rightarrow (R)]\}.$$

1.32 . Prove the law of commutation:

$$\{(P) \Rightarrow [(Q) \Rightarrow (R)]\} \Rightarrow \{(Q) \Rightarrow [(P) \Rightarrow (R)]\}.$$

1.33 . Prove Frege's axiom 2:

$$[(Q) \Rightarrow (R)] \Rightarrow \{[(P) \Rightarrow (Q)] \Rightarrow [(P) \Rightarrow (R)]\}.$$

1.34 . Prove

$$\{(Q) \Rightarrow [(P) \Rightarrow (R)]\} \Rightarrow \big([(P) \Rightarrow (Q)] \Rightarrow \{(P) \Rightarrow [(P) \Rightarrow (R)]\}\big).$$

1.35 . Prove

$$[(P) \Rightarrow (Q)] \Rightarrow \big(\{(P) \Rightarrow [(Q) \Rightarrow (R)]\} \Rightarrow [(P) \Rightarrow (R)]\big).$$

1.36 . Prove

$$\{(P) \Rightarrow [(Q) \Rightarrow (R)]\} \Rightarrow \{[(P) \Rightarrow (Q)] \Rightarrow [(P) \Rightarrow (R)]\}.$$

1.4   Proofs by the Converse Law of Contraposition

The Pure Positive Implicational Propositional Calculus belongs to several logical systems, which differ from one another by their different axioms about negation. For instance, classical logic defines its concept of negation by the converse law of contraposition:

Axiom P3: $\{[\neg(Q)] \Rightarrow [\neg(P)]\} \Rightarrow [(P) \Rightarrow (Q)].$

1.4.1   Examples of Proofs in the Full Propositional Calculus

The following proofs demonstrate the use of the converse law of contraposition.

1.40 Theorem (law of denial of the antecedent).   *For all well-formed formulae P and Q, the following formula is a theorem:* $[\neg(P)] \Rightarrow [(P) \Rightarrow (Q)]$.

Proof. Apply axioms P1 and P3 with the transitivity of implication (theorem 1.16):

$\vdash [\neg(P)] \Rightarrow \{[\neg(Q) \Rightarrow [\neg(P)]\}$ substitution in axiom P1,

$\vdash \{[\neg(Q) \Rightarrow [\neg(P)]\} \Rightarrow [(P) \Rightarrow (Q)]$ axiom P3,

$\vdash [\neg(P)] \Rightarrow [(P) \Rightarrow (Q)]$ theorem 1.16.   □

1.41 Theorem.   *The formula* $(P) \Rightarrow \{[\neg(P)] \Rightarrow (Q)\}$ *is a theorem (schema).*

Proof. Apply theorem 1.40 and the law of commutation (theorem 1.37):

$\vdash [\neg(P)] \Rightarrow [(P) \Rightarrow (Q)]$ theorem 1.40,

$\vdash \{[\neg(P)] \Rightarrow [(P) \Rightarrow (Q)]\} \Rightarrow \big[(P) \Rightarrow \{[\neg(P)] \Rightarrow (Q)\}\big]$ commutation (1.37),

$\vdash (P) \Rightarrow \{[\neg(P)] \Rightarrow (Q)\}$ *Detachment.*   □

The following two theorems establish the complete law of double negation.

1.42 Theorem (law of double negation).   *The formula* $[\neg\neg(P)] \Rightarrow (P)$ *is a theorem (schema).*

Proof. Apply the transitivity of implication (theorem 1.19) and theorem 1.25:

$\vdash \{\neg[\neg(P)]\} \Rightarrow \{[\neg\neg\neg\neg(P)] \Rightarrow [\neg\neg(P)]\}$ axiom P1,

$\vdash \big(\{\neg[\neg\neg\neg(P)]\} \Rightarrow \{\neg[\neg(P)]\}\big) \Rightarrow \{[\neg(P)] \Rightarrow [\neg\neg\neg(P)]\}$ axiom P3,

$\vdash \big([\neg(P)] \Rightarrow \{\neg[\neg\neg(P)]\}\big) \Rightarrow \{[\neg\neg(P)] \Rightarrow (P)\}$ axiom P3,

$\vdash [\neg\neg(P)] \Rightarrow \{[\neg\neg(P)] \Rightarrow (P)\}$ theorem 1.19,

$\vdash [\neg\neg(P)] \Rightarrow (P)$ theorem 1.25.   □

1.43 Theorem (converse law of double negation).   *The formula* $(P) \Rightarrow [\neg\neg(P)]$ *is a theorem (schema).*

Proof. Apply the law of double negation (theorem 1.42) and contraposition (P3):

$\vdash \big(\neg\{\neg[\neg(P)]\}\big) \Rightarrow [\neg(P)]$ theorem 1.42,

$\vdash \{\big(\neg\{\neg[\neg(P)]\}\big) \Rightarrow [\neg(P)]\} \Rightarrow \big[(P) \Rightarrow \{\neg[\neg(P)]\}\big]$ axiom P3,

$\vdash (P) \Rightarrow \{\neg[\neg(P)]\}$ *Detachment.*   □

With axiom P3, the following theorem gives the complete law of contraposition.

1.44 Theorem (law of contraposition, principle of transposition).   *The following formula is a theorem (schema):* $[(P) \Rightarrow (Q)] \Rightarrow \{[\neg(Q)] \Rightarrow [\neg(P)]\}$.

Proof. Apply transitivity (theorem 1.16) with A, B, C, D as in theorem 1.35:

$\vdash \underbrace{\{\neg[\neg(P)]\}}_{A} \Rightarrow (\underbrace{\quad P \quad}_{B})$ theorem 1.42,

$\vdash (\underbrace{\quad Q \quad}_{C}) \Rightarrow \underbrace{\{\neg[\neg(Q)]\}}_{D}$ theorem 1.43,

$\vdash [(\underbrace{\quad P \quad}_{B}) \Rightarrow (\underbrace{\quad Q \quad}_{C})] \Rightarrow \{[\underbrace{\neg\neg(P)}_{A}] \Rightarrow [\underbrace{\neg\neg(Q)}_{D}]\}$ theorem 1.35,

$\vdash \{[\neg\neg(P)] \Rightarrow [\neg\neg(Q)]\} \Rightarrow \{[\neg(Q)] \Rightarrow [\neg(P)]\}$ axiom P3,

$\vdash [(P) \Rightarrow (Q)] \Rightarrow \{[\neg(Q)] \Rightarrow [\neg(P)]\}$ theorem 1.16.

□

Theorem 1.44 is the theoretical basis for reasoning by contraposition. Theorem 1.45 syncopates several steps for later use.

1.45 Theorem. *For all well-formed formulae P and Q, the following formula is a theorem:* $\{[\neg(P)] \Rightarrow (Q)\} \Rightarrow \{[\neg(Q)] \Rightarrow (P)\}$.

Proof. Apply the principle of transposition and the law of double negation:

$\vdash \{[\neg(P)] \Rightarrow (Q)\} \Rightarrow ([\neg(Q)] \Rightarrow \{\neg[\neg(P)]\})$ transposition (1.44) ,

$\vdash \{\neg[\neg(P)]\} \Rightarrow (P)$ double negation (1.42) ,

$\vdash \{[\neg(P)] \Rightarrow (Q)\} \Rightarrow \{[\neg(Q)] \Rightarrow (P)\}$ derived rule (1.32).

□

1.4.2 Guides for Proofs in the Propositional Calculus

Besides the Deduction Theorem (theorem 1.22), an extension of the substitutivity of equivalence in the implicational calculus (theorem 1.29) also allows for the replacement of any occurrence of a formula by an equivalent formula containing negations, thanks to theorem 1.46 [18, p. 101, 124, 189], [108, p. 48].

1.46 Theorem (Substitutivity of Equivalence in the Pure Propositional Calculus, preliminary version). *For all well-formed propositional formulae U and V, if* $\vdash (U) \Rightarrow (V)$ *and* $\vdash (V) \Rightarrow (U)$, *and if a formula Q results from substituting any (zero, one, several, or all) occurrence(s) of U by V in a well-formed formula P, then* $\vdash (P) \Rightarrow (Q)$ *and* $\vdash (Q) \Rightarrow (P)$.

Proof (Outline). Theorem 1.29 has already established the conclusions for implications.

For negations, if P is ¬(U), then Q is either ¬(U) or ¬(V). If Q is ¬(V), then $(P) \Rightarrow (Q)$ and $(Q) \Rightarrow (P)$ become $[\neg(U)] \Rightarrow [\neg(V)]$ and $[\neg(V)] \Rightarrow [\neg(U)]$, which hold by the law of contraposition (theorem 1.44):

$\vdash (U) \Rightarrow (V)$ hypothesis ,

$\vdash [(U) \Rightarrow (V)] \Rightarrow \{[\neg(V)] \Rightarrow [\neg(U)]\}$ theorem 1.44 ,

$\vdash [\underbrace{\neg(V)}_{Q}] \Rightarrow [\underbrace{\neg(U)}_{P}]$ *Detachment*;

$\vdash (V) \Rightarrow (U)$ hypothesis ,

$\vdash [(V) \Rightarrow (U)] \Rightarrow \{[\neg(U)] \Rightarrow [\neg(V)]\}$ theorem 1.44 ,

$\vdash [\underbrace{\neg(U)}_{P}] \Rightarrow [\underbrace{\neg(V)}_{Q}]$ *Detachment*;

The general case follows by several applications of the previous case and the cases in theorem 1.29, in a way that may be specified more explicitly after the availability of the Principle of Mathematical Induction in chapter 4.   □

1.4.3   Proofs by Reductio ad Absurdum

Within classical logic, a proposition and its negation together form an "absurdity" that cannot hold. In particular, if a hypothesis implies a conclusion and its negation — an absurdity — then the hypothesis may be rejected. The following theorems establish the validity of such a pattern of reasoning, called **reduction to the absurd**.

1.47 Theorem (special law of reductio ad absurdum). *For each well-formed formula P, the following formula is a theorem:* $\{(P) \Rightarrow [\neg(P)]\} \Rightarrow [\neg(P)]$.

Proof. Start with theorem 1.44 and the denial of the antecedent (theorem 1.40):

$\vdash \{(P) \Rightarrow [\neg(P)]\} \Rightarrow (\{\neg[\neg(P)]\} \Rightarrow [\neg(P)])$ (1.44),

$\vdash \{\neg[\neg(P)]\} \Rightarrow ([\neg(P)] \Rightarrow \{\neg[(P) \Rightarrow (P)]\})$ (1.40),

$\vdash [\{\neg[\neg(P)]\} \Rightarrow ([\neg(P)] \Rightarrow \{\neg[(P) \Rightarrow (P)]\})]$

       $\Rightarrow [(\{\neg[\neg(P)]\} \Rightarrow [\neg(P)]) \Rightarrow (\{\neg[\neg(P)]\} \Rightarrow \{\neg[(P) \Rightarrow (P)]\})]$ (P2),

$\vdash (\{\neg[\neg(P)]\} \Rightarrow [\neg(P)]) \Rightarrow (\{\neg[\neg(P)]\} \Rightarrow \{\neg[(P) \Rightarrow (P)]\})$ (M.P.),

$\vdash (\{\neg[\neg(P)]\} \Rightarrow \{\neg[(P) \Rightarrow (P)]\}) \Rightarrow \{[(P) \Rightarrow (P)] \Rightarrow [\neg(P)]\}$ (P3),

$\vdash \{(P) \Rightarrow [\neg(P)]\} \Rightarrow \{[(P) \Rightarrow (P)] \Rightarrow [\neg(P)]\}$ (1.19),

$\vdash (P) \Rightarrow (P)$ (1.14),

$\vdash \{(P) \Rightarrow [\neg(P)]\} \Rightarrow [\neg(P)]$ (1.17).

                                                      □

1.48 Theorem (law of reductio ad absurdum). *For all well-formed formulae P and Q, the following formula is a theorem:*

$$[(P) \Rightarrow (Q)] \Rightarrow (\{(P) \Rightarrow [\neg(Q)]\} \Rightarrow [\neg(P)]).$$

Proof. Apply theorems 1.44, 1.12, 1.37, 1.16, 1.47, 1.32:

$\vdash [(P) \Rightarrow (Q)] \Rightarrow \{[\neg(Q)] \Rightarrow [\neg(P)]\}$ (1.44),

$\vdash (P) \Rightarrow ([(P) \Rightarrow (Q)] \Rightarrow \{[\neg(Q)] \Rightarrow [\neg(P)]\})$ (1.12),

$\vdash [(P) \Rightarrow (Q)] \Rightarrow [(P) \Rightarrow \{[\neg(Q)] \Rightarrow [\neg(P)]\}]$ (1.37),

$\vdash [(P) \Rightarrow \{[\neg(Q)] \Rightarrow [\neg(P)]\}] \Rightarrow (\{(P) \Rightarrow [\neg(Q)]\} \Rightarrow \{(P) \Rightarrow [\neg(P)]\})$ (P2),

$\vdash [(P) \Rightarrow (Q)] \Rightarrow (\{(P) \Rightarrow [\neg(Q)]\} \Rightarrow \{(P) \Rightarrow [\neg(P)]\})$ (1.16),

$\vdash \{(P) \Rightarrow [\neg(P)]\} \Rightarrow [\neg(P)]$ (1.47),

$\vdash [(P) \Rightarrow (Q)] \Rightarrow (\{(P) \Rightarrow [\neg(Q)]\} \Rightarrow [\neg(P)])$ (1.32).   □

Theorem 1.48 is the theoretical basis for the pattern of reasoning by **reduction to the absurd:** if $(P) \Rightarrow (Q)$ and $(P) \Rightarrow [\neg(Q)]$ are theorems, then theorem 1.48 and *Detachment* twice prove that $\neg(P)$ is a theorem. As a special case of theorem 1.48, theorem 1.47 shows that if a statement P implies its negation $\neg(P)$, then $\neg(P)$ is a theorem.

1.4.4 Proofs by Cases

Theorem 1.49 provides a variation on the theoretical foundation of proofs by cases: if, in the first case, a conclusion R follows from a hypothesis P, and if, in the second case, the same conclusion R also follows from the negation $\neg(P)$ of the same hypothesis P, then R is a theorem, derivable from the axioms and inference rule, so that the hypothesis is superfluous.

1.49 Theorem (proof by cases). *For all well-formed formulae P and R,*
$$P1, P2, P3, \vdash \{[\neg(P)] \Rightarrow (R)\} \Rightarrow \{[(P) \Rightarrow (R)] \Rightarrow (R)\}.$$

Proof. Apply the laws of double negation (theorem 1.42), contraposition (theorem 1.44), and reduction to the absurd (theorem 1.48):

$\vdash \{[\neg(R)]\Rightarrow(P)\}\Rightarrow\big(\{[\neg(R)]\Rightarrow[\neg(P)]\}\Rightarrow\{\neg[\neg(R)]\}\big)$	theorem 1.48,
$\vdash \{\neg[\neg(R)]\}\Rightarrow(R)$	double negation (1.42),
$\vdash \{[\neg(R)]\Rightarrow(P)\}\Rightarrow[\{[\neg(R)]\Rightarrow[\neg(P)]\}\Rightarrow(R)]$	derived rule (1.32),
$\vdash [(P)\Rightarrow(R)]\Rightarrow\{[\neg(R)]\Rightarrow[\neg(P)]\}$	transposition (1.44),
$\vdash \{[\neg(R)]\Rightarrow(P)\}\Rightarrow[\{[(P)\Rightarrow(R)]\}\Rightarrow(R)]$	derived rule (1.31),
$\vdash \{[\neg(P)]\Rightarrow(R)\}\Rightarrow\{[\neg(R)]\Rightarrow(P)\}$	theorem 1.45,
$\vdash \{[\neg(P)]\Rightarrow(R)\}\Rightarrow\{[(P)\Rightarrow(R)]\Rightarrow(R)\}$	derived rule (1.16).

$\square$

Theorem 1.50 generalizes theorem 1.49 to the situation with an intermediate hypothesis [18, p. 205, footnote 355].

1.50 Theorem (proof by cases subject to hypotheses). *For all well-formed formulae P, Q, and R,*
$$P1, P2, P3, \vdash \{[\neg(P)] \Rightarrow (R)\} \Rightarrow \big[[(Q) \Rightarrow (R)] \Rightarrow \big(\{[(P) \Rightarrow (Q)] \Rightarrow (R)\}\big)\big].$$

Proof. Apply the law of proof by cases (theorem 1.49), the transitivity of implication (theorem 1.27), and a derived rule (theorem 1.39):

$$\vdash \underbrace{\{[\neg(P)] \Rightarrow (R)\}}_{H} \Rightarrow \underbrace{\{[(P) \Rightarrow (R)]}_{W} \Rightarrow (R)\} \qquad \text{theorem 1.49,}$$

$$\vdash \underbrace{[(Q) \Rightarrow (R)]}_{U} \Rightarrow \{\underbrace{[(P) \Rightarrow (Q)]}_{V} \Rightarrow \underbrace{[(P) \Rightarrow (R)]}_{W}\} \qquad \text{theorem 1.27,}$$

$$\vdash \underbrace{\{[\neg(P)] \Rightarrow (R)\}}_{H} \Rightarrow \big[\underbrace{[(Q) \Rightarrow (R)]}_{U} \Rightarrow \big(\{\underbrace{[(P) \Rightarrow (Q)]}_{V} \Rightarrow (R)\}\big)\big] \qquad \text{theorem 1.39.}$$

$\square$

1.4.5 *Exercises on Frege's and Church's Axioms*

For the following four exercises, use the rules of inference and only the following six axioms, due to Frege [39]:

Axiom F1 *[39, p. 137, eq. (1)]:* $(P) \Rightarrow [(Q) \Rightarrow (P)]$.

Axiom F2 *[39, p. 137, eq. (2)]:* $\{(P) \Rightarrow [(Q) \Rightarrow (R)]\} \Rightarrow \{[(P) \Rightarrow (Q)] \Rightarrow [(P) \Rightarrow (R)]\}$.

Axiom F3 *[39, p. 146, eq. (8)]:* $\{(P) \Rightarrow [(Q) \Rightarrow (R)]\} \Rightarrow \{(Q) \Rightarrow [(P) \Rightarrow (R)]\}$.

Axiom F4 *[39, p. 154, eq. (28)]:* $[(P) \Rightarrow (Q)] \Rightarrow \{[\neg(Q)] \Rightarrow [\neg(P)]\}$.

Axiom F5 *[39, p. 156, eq. (31)]:* $\{\neg[\neg(P)]\} \Rightarrow (P)$.

Axiom F6 *[39, p. 158, eq. (41)]:* $(P) \Rightarrow \{\neg[\neg(P)]\}$.

1.37 . Prove that $F1$–$F6 \vdash (P) \Rightarrow \{[\neg(P)] \Rightarrow [\neg(Q)]\}$.

1.38 . Prove that $F1$–$F6 \vdash [\neg(P)] \Rightarrow \{(P) \Rightarrow [\neg(P)]\}$.

1.39 . Prove that $F1$–$F6 \vdash \{[\neg(Q)] \Rightarrow [\neg(P)]\} \Rightarrow [(P) \Rightarrow (Q)]$.

1.40 . Prove that Frege's six axioms F1–F6 are logically equivalent to Łukasiewicz's three axioms P1, P2, P3.

The following exercises outline a proof that the axioms P1, P2, P3 of classical logic are logically equivalent to the following three axioms C1, C2, C3, used by Church [18, §10, p. 72] and Robbin [108, p. 14]. Because the two logical systems have the same first two axioms, they also have the same implicational calculus. The two logical systems differ from each other only by their third axiom, where F stands for False, so that $\neg(F)$ is a theorem.

Axiom C1 $(P) \Rightarrow [(Q) \Rightarrow (P)]$.

Axiom C2 $\{(P) \Rightarrow [(Q) \Rightarrow (R)]\} \Rightarrow \{[(P) \Rightarrow (Q)] \Rightarrow [(P) \Rightarrow (R)]\}$.

Axiom C3 $\{[(P) \Rightarrow (F)] \Rightarrow (F)\} \Rightarrow (P)$.

1.41 . This exercise establishes the converse law of contraposition in Church's system. Prove the tautology $\{[(B) \Rightarrow (F)] \Rightarrow [(A) \Rightarrow (F)]\} \Rightarrow [(A) \Rightarrow (B)]$ within Church's system, using only results from the *implicational* calculus (axioms C1 and C2) and axiom C3 (*not* axiom P3). Hint: start from axiom C2, with $(B) \Rightarrow (F)$ for P, with A for Q, and with F for R. Then use the transitivity of implications.

To show the equivalence of Church's logical system and classical logic, the following exercise establishes the equivalence of $\neg(P)$ and Church's $(P) \Rightarrow (F)$.

1.42 . Prove that $[(P) \Rightarrow (F)] \Rightarrow [\neg(P)]$ and $[\neg(P)] \Rightarrow [(P) \Rightarrow (F)]$ are theorems, using the theorem $\neg(F)$, and the classical axioms P1, P2, P3.

1.43 . Define a "negation" $\sim (P)$ to be an abbreviation for $(P) \Rightarrow (F)$. Prove that every theorem of classical logic is a theorem of Church's logic.

1.44 . Using the theorem $\neg (F)$ and axioms P1, P2, P3, prove the theorems $\{[(P) \Rightarrow (F)] \Rightarrow (F)\} \Rightarrow \{\neg[\neg(P)]\}$ and $\{\neg[\neg(P)]\} \Rightarrow \{[(P) \Rightarrow (F)] \Rightarrow (F)\}$.

1.45 . Using axioms P1, P2, P3 and any of the classical theorems already proved, prove $[\neg(P)] \Rightarrow \big[(P) \Rightarrow \{\neg[(S) \Rightarrow (S)]\}\big]$ and $\big[(P) \Rightarrow \{\neg[(S) \Rightarrow (S)]\}\big] \Rightarrow [\neg(P)]$.

1.46 . In classical logic define a constant f to be an abbreviation for $\neg[(S) \Rightarrow (S)]$. Prove that every theorem of Church's logic is a theorem of classical logic.

1.5   Other Connectives

There exist logical connectives other than the negation ($\neg$) and implication ($\Rightarrow$), for example, the conjunction ($\wedge$), disjunction ($\vee$), and equivalence ($\Leftrightarrow$). Such other connectives can be specified by other axioms, for instance, by Tarski's axioms IV–VI to specify the equivalence, as in example 1.87 on page 55. Alternatively, other connectives can be defined as abbreviations of longer formulae in terms of negations and implications, as presented in this section.

1.5.1   Definitions of Other Connectives

The logical connectives $\neg$ and $\Rightarrow$ suffice to define all the other logical connectives, for instance, the conjunction $\wedge$, the disjunction $\vee$, and the equivalence $\Leftrightarrow$, as outlined here.

1.51 Definition (conjunction, disjunction, and equivalence).   In the full Classical Propositional Calculus, the connectives $\wedge$, $\vee$, and $\Leftrightarrow$ may be defined as follows:

$(P) \wedge (Q)$ stands for $\neg\{(P) \Rightarrow [\neg(Q)]\}$;

$(P) \vee (Q)$ stands for $[(P) \Rightarrow (Q)] \Rightarrow (Q)$;

$(P) \Leftrightarrow (Q)$ stands for $[(P) \Rightarrow (Q)] \wedge [(Q) \Rightarrow (P)]$.

1.5.2   Examples of Proofs of Theorems with Conjunctions

The logical conjunction ($\wedge$) can also be introduced into a logic by additional axioms, for instance, as in Hilbert's Positive Propositional Calculus, Brouwer & Heyting's Intuitionistic Logic and Kolmogorov & Johansson's Minimal Logic [18, §26, p. 140–146]:

(AND.1) ⊢ [(P) ∧ (Q)] ⇒ (Q)
(AND.2) ⊢ [(P) ∧ (Q)] ⇒ (P)
(AND.3) ⊢ (P) ⇒ {(Q) ⇒ [(P) ∧ (Q)]}

The following theorems reveal that within the Classical Propositional Calculus, such axioms also follow from definition 1.51.

The first theorems show that if $(P) \wedge (Q)$ holds, then P holds and Q holds.

1.52 Theorem (AND.1). *The formula* $[(P) \wedge (Q)] \Rightarrow (Q)$ *is a theorem (schema).*

Proof. Apply transposition, double negation, transitivity, and the definition:

⊢ [¬(Q)] ⇒ {(P) ⇒ [¬(Q)]}	axiom P1,
⊢ (¬{(P) ⇒ [¬(Q)]}) ⇒ {¬[¬(Q)]}	contraposition (1.44) of P1, and *M.P.*,
⊢ [(P) ∧ (Q)] ⇒ {¬[¬(Q)]}	definition 1.51 of ∧,
⊢ {¬[¬(Q)]} ⇒ (Q)	theorem 1.42,
⊢ [(P) ∧ (Q)] ⇒ (Q)	theorem 1.16.

□

1.53 Theorem (AND.2). *The formula* $[(P) \wedge (Q)] \Rightarrow (P)$ *is a theorem.*

Proof. Apply theorems 1.40, 1.44, 1.42, and 1.16:

⊢ [¬(P)] ⇒ [(P) ⇒ (Q)]	denial of the antecedent (1.40),
⊢ (¬{(P) ⇒ [¬(Q)]}) ⇒ {¬[¬(P)]}	contraposition (1.44) of 1.40,
⊢ [¬¬(P)] ⇒ (P)	double negation (1.42),
⊢ [(P) ∧ (Q)] ⇒ (P)	transitivity (theorem 1.16).

□

The next theorem demonstrates a "converse" to the preceding two theorems. so that if P and Q hold, then their conjunction $(P) \wedge (Q)$ also holds.

1.54 Theorem (AND.3). *If P and Q are both theorems, then* $(P) \wedge (Q)$ *is a theorem; equivalently, for all well-formed formulae P and Q,* $\vdash (P) \Rightarrow \{(Q) \Rightarrow [(P) \wedge (Q)]\}$.

Proof. Apply the law of assertion, contraposition, and transitivity:

⊢ (P) ⇒ ({(P) ⇒ [¬(Q)]} ⇒ [¬(Q)])	assertion (theorem 1.38),

⊢ ({(P) ⇒ [¬(Q)]} ⇒ [¬(Q)])	contraposition . . .
⇒ {(Q) ⇒ (¬{(P) ⇒ [¬(Q)]})}	. . . continued,

⊢ (P) ⇒ {(Q) ⇒ (¬{(P) ⇒ [¬(Q)]})}	theorem 1.16,
⊢ (P) ⇒ {(Q) ⇒ [(P) ∧ (Q)]}	definition of ∧.

Hence, if P and Q are both theorems, then $(P) \wedge (Q)$ is a theorem:

⊢ (P) ⇒ {(Q) ⇒ [(P) ∧ (Q)]}	just proved,
⊢ (P)	hypothesis,
⊢ (Q) ⇒ [(P) ∧ (Q)]	*Detachment*,

$\vdash (Q)$ hypothesis,

$\vdash (P) \wedge (Q)$ *Detachment.*

$\square$

Theorem 1.54 allows for the following derived rule of inference.

1.55 Theorem (derived rule). *For all well-formed formulae H, K, and L, if* $(H) \Rightarrow$ *(K) and $(H) \Rightarrow (L)$ are theorems, then $(H) \Rightarrow [(K) \wedge (L)]$ is a theorem:*

$$\vdash [(H) \Rightarrow (K)] \Rightarrow \big([(H) \Rightarrow (L)] \Rightarrow \{(H) \Rightarrow [(K) \wedge (L)]\}\big);$$

conversely, if $(H) \Rightarrow [(K) \wedge (L)]$ is a theorem, then $(H) \Rightarrow (K)$ and $(H) \Rightarrow (L)$ are theorems.

Proof. Apply theorem 1.54 and transitivity (theorem 1.32) with M for $(K) \wedge (L)$:

$\vdash \big[(H) \Rightarrow \{(K) \Rightarrow [(L) \Rightarrow (M)]\}\big]$
$\quad \Rightarrow \big([(H) \Rightarrow (K)] \Rightarrow \{(H) \Rightarrow [(L) \Rightarrow (M)]\}\big)$ axiom P2,

$\vdash \{(H) \Rightarrow [(L) \Rightarrow (M)]\}$
$\quad \Rightarrow \{[(H) \Rightarrow (L)] \Rightarrow [(H) \Rightarrow (M)]\}$ axiom P2,

$\vdash \big[(H) \Rightarrow \{(K) \Rightarrow [(L) \Rightarrow (M)]\}\big]$
$\quad \Rightarrow \big([(H) \Rightarrow (K)] \Rightarrow \{[(H) \Rightarrow (L)] \Rightarrow [(H) \Rightarrow (M)]\}\big)$ theorem 1.32,

$\vdash (K) \Rightarrow [(L) \Rightarrow (M)]$ theorem 1.54,
$\vdash (H) \Rightarrow \{(K) \Rightarrow [(L) \Rightarrow (M)]\}$ theorem 1.12,
$\vdash [(H) \Rightarrow (K)] \Rightarrow \{[(H) \Rightarrow (L)] \Rightarrow [(H) \Rightarrow (M)]\}$ *Detachment,*

with M for $(K) \wedge (L)$. The converse results from theorems 1.52 and 1.12:

$\vdash [(K) \wedge (L)] \Rightarrow (L)$ theorem 1.52,
$\vdash (H) \Rightarrow \{[(K) \wedge (L)] \Rightarrow (L)\}$ theorem 1.12,
$\vdash \{(H) \Rightarrow [(Q) \Rightarrow (R)]\} \Rightarrow \{[(H) \Rightarrow (Q)] \Rightarrow [(H) \Rightarrow (R)]\}$ axiom P2,
$\vdash \{(H) \Rightarrow [(K) \wedge (L)]\} \Rightarrow [(H) \Rightarrow (L)]$ *Detachment.*

Replacing $[(K) \wedge (L)] \Rightarrow (L)$ (theorem 1.52) by $[(K) \wedge (L)] \Rightarrow (K)$ (theorem 1.53) in the foregoing proof yields a proof of $\{(H) \Rightarrow [(K) \wedge (L)]\} \Rightarrow [(H) \Rightarrow (K)]$. $\square$

For instance, theorem 1.55 yields the following theorem.

1.56 Theorem (idempotency of $\wedge$). *For each well-formed formula P, the formulae* $(P) \Rightarrow [(P) \wedge (P)]$ *and* $[(P) \wedge (P)] \Rightarrow (P)$ *are theorems.*

Proof. Substitute P for each of H, K, L in theorem 1.55:

$\vdash [(H) \Rightarrow (K)] \Rightarrow \big([(H) \Rightarrow (L)] \Rightarrow \{(H) \Rightarrow [(K) \wedge (L)]\}\big)$ theorem 1.55,
$\vdash [(P) \Rightarrow (P)] \Rightarrow \big([(P) \Rightarrow (P)] \Rightarrow \{(P) \Rightarrow [(P) \wedge (P)]\}\big)$ substitutions,

$\vdash (P) \Rightarrow (P)$ theorem 1.14,
$\vdash [(P) \Rightarrow (P)] \Rightarrow \{(P) \Rightarrow [(P) \wedge (P)]\}$ *Detachment,*
$\vdash (P) \Rightarrow [(P) \wedge (P)]$ *Detachment.*
The converse implication is a substitution in theorem 1.53.   □

The next theorem shows that the conjunction $\wedge$ commutes.

1.57 Theorem (commutativity of $\wedge$). *For all well-formed formulae P and Q, the formula $[(P) \wedge (Q)] \Rightarrow [(Q) \wedge (P)]$ is a theorem.*

Proof. Apply theorems 1.52, 1.53, and 1.55:
$\vdash [(P) \wedge (Q)] \Rightarrow (Q)$ theorem 1.52,
$\vdash [(P) \wedge (Q)] \Rightarrow (P)$ theorem 1.53,
$\vdash [(P) \wedge (Q)] \Rightarrow [(Q) \wedge (P)]$ theorem 1.55.
Swapping the roles of P and Q yields the converse: $\vdash [(Q) \wedge (P)] \Rightarrow [(P) \wedge (Q)]$.
□

1.58 Theorem (derived rule). *If $\vdash [(H) \wedge (L)] \Rightarrow (N)$, then $\vdash (H) \Rightarrow [(L) \Rightarrow (N)]$.*

Proof. Apply theorems 1.54 and 1.32.
$\vdash (H) \Rightarrow \{(L) \Rightarrow [(H) \wedge (L)]\}$ theorem 1.54,
$\vdash [(H) \wedge (L)] \Rightarrow (N)$ hypothesis,
$\vdash (H) \Rightarrow [(L) \Rightarrow (N)]$ theorem 1.32.
□

1.59 Theorem (derived rule). *If $\vdash (H) \Rightarrow [(L) \Rightarrow (N)]$, then $\vdash [(H) \wedge (L)] \Rightarrow (N)$.*

Proof. Apply theorems 1.54 and 1.32.

1	$\vdash [(H) \wedge (L)] \Rightarrow (H)$	theorem 1.53,
2	$\vdash (H) \Rightarrow [(L) \Rightarrow (N)]$	hypothesis,
3	$\vdash [(H) \wedge (L)] \Rightarrow [(L) \Rightarrow (N)]$	lines 1, 2, theorem 1.16;
4	$\vdash [(H) \wedge (L)] \Rightarrow (L)$	theorem 1.52,
5	$\vdash [(H) \wedge (L)] \Rightarrow \{[(H) \wedge (L)] \Rightarrow (N)\}$	lines 3, 4, theorem 1.31,
6	$\vdash [(H) \wedge (L)] \Rightarrow (N)$	theorem 1.25.

□

1.5.3   *Examples of Proofs of Theorems with Equivalences*

The logical equivalence ($\Leftrightarrow$) can also be introduced into a logic by additional axioms, for instance, Hilbert's Positive Propositional Calculus, Brouwer & Heyting's Intuitionistic Logic and Kolmogorov & Johansson's Minimal Logic [18, §26, p. 140–146], by Tarski's axioms IV, V, and VI [129, p. 147]:

(EQ.1) Tarski's Axiom IV: $\vdash [(P) \Leftrightarrow (Q)] \Rightarrow [(P) \Rightarrow (Q)]$

(EQ.2) Tarski's Axiom V: $\vdash [(P) \Leftrightarrow (Q)] \Rightarrow [(Q) \Rightarrow (P)]$
(EQ.3) Tarski's Axiom VI: $\vdash [(P) \Rightarrow (Q)] \Rightarrow \{[(Q) \Rightarrow (P)] \Rightarrow [(P) \Leftrightarrow (Q)]\}$

The following theorems reveal that within the Classical Propositional Calculus, such axioms also follow from definition 1.51.

Combining theorem 1.54 with the definition of $\Leftrightarrow$ gives Tarksi's Axiom VI.

1.60 Theorem (Tarski's axiom VI). *For all well-formed formulae U and V, Tarski's axioms VI is a theorem:* $\vdash [(U) \Rightarrow (V)] \Rightarrow \{[(V) \Rightarrow (U)] \Rightarrow [(U) \Leftrightarrow (V)]\}$.

Proof. Combine theorem 1.54 with the definition of $\Leftrightarrow$:

$$\underbrace{[(U) \Rightarrow (V)]}_{P} \Rightarrow \Big(\underbrace{[(V) \Rightarrow (U)]}_{Q} \Rightarrow \overbrace{\{\underbrace{[(U) \Rightarrow (V)]}_{P} \wedge \underbrace{[(V) \Rightarrow (U)]}_{Q}\}}^{(U) \Leftrightarrow (V)}\Big).$$

$\square$

A particular instance of the foregoing theorem allows for any proof of any equivalence $(I) \Leftrightarrow (J)$ to be split into two separate proofs of $(I) \Rightarrow (J)$ and $(J) \Rightarrow (I)$.

1.61 Theorem (derived rule). *If $(I) \Rightarrow (J)$ and $(J) \Rightarrow (I)$ are theorems, then so is $(I) \Leftrightarrow (J)$. Conversely, if $(I) \Leftrightarrow (J)$ is a theorem, then so are $(I) \Rightarrow (J)$ and $(J) \Rightarrow (I)$.*

Proof. Apply theorem 1.54 with $(I) \Rightarrow (J)$ for P, and with $(J) \Rightarrow (I)$ for Q, so that if $(I) \Rightarrow (J)$ and $(J) \Rightarrow (I)$ are theorems, then $[(I) \Rightarrow (J)] \wedge [(J) \Rightarrow (I)]$ is also a theorem, which is $(I) \Leftrightarrow (J)$ by definition, and conversely. $\square$

Theorem 1.62 discharges the hypothesis $(I) \Leftrightarrow (J)$ in theorem 1.61, which yields Tarski's axioms IV and V.

1.62 Theorem (Tarski's axioms IV and V). *For all well-formed formulae I and J, Tarski's axioms IV $[(I) \Leftrightarrow (J)] \Rightarrow [(I) \Rightarrow (J)]$ and Tarski's axioms V $[(I) \Leftrightarrow (J)] \Rightarrow [(J) \Rightarrow (I)]$ are theorems.*

Proof. By definition 1.51, the formula $(I) \Leftrightarrow (J)$ is an abbreviation for $[(I) \Rightarrow (J)] \wedge [(J) \Rightarrow (I)]$. The present theorem results from substitutions in theorems 1.52 and 1.53. $\square$

Hence the following theorem establishes the reflexivity of equivalence.

1.63 Theorem (reflexivity of $\Leftrightarrow$). *For each well-formed formula P,* $\vdash (P) \Leftrightarrow (P)$.

Proof. Apply the definition of $\Leftrightarrow$ and the reflexivity of $\Rightarrow$ (theorem 1.14) with theorem 1.61:
     $\vdash (P) \Rightarrow (P)$ theorem 1.14,
     $\vdash (P) \Rightarrow (P)$ theorem 1.14,
     $\vdash (P) \Leftrightarrow (P)$ theorem 1.61.

$\square$

The following theorem establishes the symmetry of equivalence.

1.64 Theorem (symmetry of $\Leftrightarrow$). *For all well-formed formulae H and K, if $(H) \Leftrightarrow (K)$ is a theorem, then so is $(K) \Leftrightarrow (H)$.*

Proof. Apply the definition of $\Leftrightarrow$ and the commutativity of $\wedge$ (theorem 1.57):

$\vdash (H) \Leftrightarrow (K)$ hypothesis,

$\vdash [(H) \Rightarrow (K)] \wedge [(K) \Rightarrow (H)]$ definition of $\Leftrightarrow$,

$\vdash [(K) \Rightarrow (H)] \wedge [(H) \Rightarrow (K)]$ commutativity of $\wedge$ and *Detachment*,

$\vdash (K) \Leftrightarrow (H)$ definition of $\Leftrightarrow$.

$\square$

Similarly, the following theorem establishes the transitivity of equivalence.

1.65 Theorem (transitivity of $\Leftrightarrow$). *For all well-formed formulae H, K, and L, if $(H) \Leftrightarrow (K)$ and $(K) \Leftrightarrow (L)$ are theorems, then $(H) \Leftrightarrow (L)$ is a theorem.*

Proof. Apply the definition of $\Leftrightarrow$ and theorems 1.61 and 1.16:

$\vdash (H) \Leftrightarrow (K)$ hypothesis,

$\vdash (H) \Rightarrow (K)]$ theorem 1.61,

$\vdash (K) \Leftrightarrow (L)$ hypothesis,

$\vdash (K) \Rightarrow (L)]$ theorem 1.61,

$\vdash (H) \Rightarrow (L)$ theorem 1.16,

By symmetry (theorem 1.64) $(K) \Leftrightarrow (H)$ and $(L) \Leftrightarrow (K)$ are also theorems, whence $(L) \Rightarrow (H)$ is a theorem. From $(H) \Rightarrow (L)$ and $(L) \Rightarrow (H)$ it then follows that $(H) \Leftrightarrow (L)$ is a theorem. $\square$

The following theorem demonstrates the use of the transitivity of equivalence in the proof that the conjunction $\wedge$ is associative.

1.66 Theorem (associativity of $\wedge$). *For all well-formed formulae P, Q, and R,*

$$\vdash \{(P) \wedge [(Q) \wedge (R)]\} \Rightarrow \{[(P) \wedge (Q)] \wedge (R)\}.$$

Proof. Apply the law of commutation (theorem 1.37) and theorem 1.55:

$\vdash [(P) \Rightarrow \{(R) \Rightarrow [\neg(Q)]\}]$ twice theorem 1.37 ...

$\Leftrightarrow [(R) \Rightarrow \{(P) \Rightarrow [\neg(Q)]\}]$ ... and theorem 1.55,

$\Updownarrow$ contrapositions,

$\vdash \{\neg[(P) \Rightarrow \{(Q) \Rightarrow [\neg(R)]\}]\}$

$\Leftrightarrow \{\neg[(R) \Rightarrow \{(P) \Rightarrow [\neg(Q)]\}]\}$

$\Updownarrow$ definition of $\wedge$,

$\vdash \{\neg[(P) \Rightarrow \{\neg[(Q) \wedge (R)]\}]\}$

$\Leftrightarrow \{\neg[(R) \Rightarrow \{\neg[(P) \wedge (Q)]\}]\}$

$\Updownarrow$ definition of $\wedge$,

$$\vdash \{(P) \wedge [(Q) \wedge (R)]\}$$
$$\Leftrightarrow \{(R) \wedge [(P) \wedge (Q)]\}$$
$$\Updownarrow \quad \text{commutativity of } \wedge.$$
$$\vdash \{(P) \wedge [(Q) \wedge (R)]\}$$
$$\Leftrightarrow \{[(P) \wedge (Q)] \wedge (R)\}$$

□

1.5.4   Examples of Proofs of Theorems with Disjunctions

The logical disjunction ($\vee$) can also be introduced into a logic by additional axioms, for instance, as in Hilbert's Positive Propositional Calculus, Brouwer & Heyting's Intuitionistic Logic and Kolmogorov & Johansson's Minimal Logic [18, §26, p. 140–146]:

(Or.1) $\vdash (P) \Rightarrow [(P) \vee (Q)]$
(Or.2) $\vdash (Q) \Rightarrow [(P) \vee (Q)]$
(Or.3) $\vdash [(P) \Rightarrow (R)] \Rightarrow \big([(Q) \Rightarrow (R)] \Rightarrow \{[(P) \vee (Q)] \Rightarrow (R)\}\big)$

The following theorems reveal that within the Classical Propositional Calculus, the first two axioms also follow from definition 1.51. The third axiom is derived in theorem 1.78.

1.67 Theorem (Or.1, Or.2). *For all well-formed formulae P and Q, the formulae* $(Q) \Rightarrow [(P) \vee (Q)]$ *and* $(P) \Rightarrow [(P) \vee (Q)]$ *are theorems.*

Proof. For the first formula, apply axiom P1 and definition of $\vee$:
$\vdash (Q) \Rightarrow \{[(P) \Rightarrow (Q)] \Rightarrow (Q)\}$ axiom P1,
$\vdash (Q) \Rightarrow [(P) \vee (Q)]$ definition of $\vee$.
For the second formula, apply theorem 1.14, the Law of Commutation, and the definitions:
$\vdash [(P) \Rightarrow (Q)] \Rightarrow [(P) \Rightarrow (Q)]$ theorem 1.14,
$\vdash (P) \Rightarrow \{[(P) \Rightarrow (Q)] \Rightarrow (Q)\}$ Law of Commutation (theorem 1.37),
$\vdash (P) \Rightarrow [(P) \vee (Q)]$ definition 1.51 of $\vee$.

□

The next theorem shows that the disjunction $\vee$ is idempotent:

1.68 Theorem (idempotency of $\vee$). *For each well-formed formula Q,*

$$\vdash (Q) \Leftrightarrow [(Q) \vee (Q)].$$

Proof. Substituting Q for P in theorem 1.67 yields $\vdash (Q) \Rightarrow [(Q) \vee (Q)]$. For the converse, apply theorem 1.14, the Law of Assertion (theorem 1.38), and *Detachment*:

$$\vdash \underbrace{[(Q) \Rightarrow (Q)]}_{H} \qquad \text{theorem 1.14,}$$

$$\vdash \underbrace{[(Q) \Rightarrow (Q)]}_{H} \Rightarrow (\{\underbrace{[(Q) \Rightarrow (Q)]}_{H} \Rightarrow \underbrace{(Q)}_{C}\} \Rightarrow \underbrace{(Q)}_{C}) \qquad \text{theorem 1.38,}$$

$$\vdash \{\underbrace{[(Q) \Rightarrow (Q)]}_{H} \Rightarrow \underbrace{(Q)}_{C}\} \Rightarrow \underbrace{(Q)}_{C} \qquad \text{Detachment,}$$

$$\vdash [(Q) \vee (Q)] \Rightarrow (Q) \qquad \text{definition of } \vee.$$

Hence the equivalence $\vdash (Q) \Leftrightarrow [(Q) \vee (Q)]$ results from theorem 1.61.   □

Theorem 1.69 provides an alternative definition of the disjunction in the Classical Propositional Calculus.

1.69 Theorem. *For all well-formed formulae P and Q, the following formulae are theorems:*

$$\vdash \{[\neg(P)] \Rightarrow (Q)\} \Rightarrow \{[(P) \Rightarrow (Q)] \Rightarrow (Q)\},$$

$$\vdash \{[(P) \Rightarrow (Q)] \Rightarrow (Q)\} \Rightarrow \{[\neg(P)] \Rightarrow (Q)\}.$$

Proof. The first formula is a substitution in the proof by cases (theorem 1.49).

The second formula results from the law of denial of the antecedent (theorem 1.40) and a derived rule (theorem 1.33):

$$\vdash \underbrace{[\neg(P)]}_{I} \Rightarrow \underbrace{[(P) \Rightarrow (Q)]}_{H} \qquad \text{theorem 1.40,}$$

$$\vdash \{\underbrace{[(P) \Rightarrow (Q)]}_{H} \Rightarrow (\underbrace{Q}_{K})\} \Rightarrow \{\underbrace{[\neg(P)]}_{I} \Rightarrow (\underbrace{Q}_{K})\} \qquad \text{theorem 1.33.}$$

□

The next theorem establishes the law of excluded middle in classical logic.

1.70 Theorem (law of excluded middle). *For each well-formed formula B, the formula $(B) \vee [\neg(B)]$ is a theorem.*

Proof. Apply theorems 1.14 and 1.69, and the definition of the disjunction ($\vee$):

$\vdash [\neg(B)] \Rightarrow [\neg(B)]$ theorem 1.14,

$\vdash (B) \vee [\neg(B)]$ theorem 1.69 and definition of $\vee$ with $\neg$ and $\Rightarrow$.

□

The following theorem shows that the disjunction $\vee$ is associative:

1.71 Theorem (associativity of $\vee$). *For all well-formed formulae P, Q, and R,*

$$\vdash \{(P) \vee [(Q) \vee (R)]\} \Leftrightarrow \{[(P) \vee (Q)] \vee (R)\}.$$

Proof. Apply the law of commutation (theorem 1.37) and theorem 1.55:

$$\vdash \big([\neg(P)] \Rightarrow \{[\neg(R)] \Rightarrow (Q)\}\big) \qquad \text{twice theorem 1.37} \dots$$
$$\Leftrightarrow \big([\neg(R)] \Rightarrow \{[\neg(P)] \Rightarrow (Q)\}\big) \quad \dots \text{and theorem 1.55,}$$
$$\Updownarrow \quad \text{definition of } \vee,$$
$$\vdash \{(P) \vee [(R) \vee (Q)]\} \Leftrightarrow \{(R) \vee [(P) \vee (Q)]\}$$
$$\Updownarrow \quad \text{commutativity of } \vee, \text{ etc.}$$
$$\vdash \{(P) \vee [(Q) \vee (R)]\} \Leftrightarrow \{[(P) \vee (Q)] \vee (R)\}$$

$\square$

1.5.5 *Examples of Proofs with Conjunctions and Disjunctions*

The following theorems establish De Morgan's laws in classical logic.

1.72 Theorem (De Morgan's first law). *For all well-formed formulae P and Q,*

$$\vdash \{\neg[(P) \wedge (Q)]\} \Leftrightarrow \{[\neg(P)] \vee [\neg(Q)\}.$$

Proof. Apply the definitions of $\wedge$ and $\vee$ and double negations:

$$\neg[(P) \wedge (Q)]$$
$$\Updownarrow \quad \text{definition of } \wedge,$$
$$\big[\neg\big(\neg\{(P) \Rightarrow [\neg(Q)]\}\big)\big]$$
$$\Updownarrow \quad \text{double negation (theorems 1.42 and 1.43),}$$
$$(P) \Rightarrow [\neg(Q)]$$
$$\Updownarrow \quad \text{double negation (theorems 1.42 and 1.43),}$$
$$\{\neg[\neg(P)]\} \Rightarrow [\neg(Q)]$$
$$\Updownarrow \quad \text{definition of } \vee \text{ by theorem 1.69.}$$
$$[\neg(P)] \vee [\neg(Q)].$$

$\square$

1.73 Theorem (De Morgan's second law). *For all well-formed formulae P and Q,*

$$\vdash \{\neg[(P) \vee (Q)]\} \Leftrightarrow \{[\neg(P)] \wedge [\neg(Q)\}.$$

Proof. Apply the definitions of $\wedge$ and $\vee$ and double negations:

$$\neg[(P) \vee (Q)]$$
$$\Updownarrow \quad \text{definition of } \vee \text{ by theorem 1.69,}$$
$$\neg\{[\neg(P)] \Rightarrow (Q)\}$$
$$\Updownarrow \quad \text{double negation (theorems 1.42 and 1.43),}$$
$$\neg\big([\neg(P)] \Rightarrow \{\neg[\neg(Q)]\}\big)$$
$$\Updownarrow \quad \text{definition of } \wedge.$$
$$[\neg(P)] \wedge [\neg(Q)].$$

$\square$

The following theorem shows that disjunctions distribute over conjunctions.

1.74 Theorem (distributivity of ∨ over ∧). *For all well-formed formulae P, Q, and R,*

$$\vdash \{(P) \lor [(Q) \land (R)]\} \Leftrightarrow \{[(P) \lor (Q)] \land [(P) \lor (R)]\}.$$

Proof. Apply the definition of ∨ with De Morgan's laws and theorem 1.55:

$$[(P) \lor (Q)] \land [(P) \lor (R)]$$
$$\Updownarrow \quad \text{definition of } \lor \text{ by theorem 1.69,}$$
$$\{[\neg(P)] \Rightarrow (Q)\} \land \{[\neg(P)] \Rightarrow (R)\}$$
$$\Updownarrow \quad \text{theorem 1.55,}$$
$$[\neg(P)] \Rightarrow [(Q) \land (R)]$$
$$\Updownarrow \quad \text{definition of } \lor \text{ by theorem 1.69.}$$
$$(P) \lor [(Q) \land (R)]$$

□

The following theorem shows that conjunctions distribute over disjunctions.

1.75 Theorem (Distributivity of ∧ over ∨). *The following formula is a theorem:*

$$\{(P) \land [(Q) \lor (R)]\} \Leftrightarrow \{[(P) \land (Q)] \lor [(P) \land (R)]\}.$$

Proof. Apply the definition of ∧ and theorem 1.55:

$$[(P) \land (Q)] \lor [(P) \land (R)]$$
$$\Updownarrow \quad \text{definition of } \land,$$
$$\left(\neg\{(P) \Rightarrow [\neg(Q)]\}\right) \lor \left(\neg\{(P) \Rightarrow [\neg(R)]\}\right)$$
$$\Updownarrow \quad \text{De Morgan's first law,}$$
$$\neg\left(\{(P) \Rightarrow [\neg(Q)]\} \land \{(P) \Rightarrow [\neg(R)]\}\right)$$
$$\Updownarrow \quad \text{theorem 1.55,}$$
$$\neg\left[(P) \Rightarrow \{[\neg(Q)] \land [\neg(R)]\}\right]$$
$$\Updownarrow \quad \text{De Morgan's second law,}$$
$$\neg\left[(P) \Rightarrow \{\neg[(Q) \lor (R)]\}\right]$$
$$\Updownarrow \quad \text{definition of } \land.$$
$$(P) \land [(Q) \lor (R)]$$

□

1.5.6   Exercises on Other Connectives

For the following exercises, prove that the stated formulae are theorems (schemas), using the classical propositional calculus and any of the results just proved.

1.47 . $\{[(P) \Rightarrow (Q)] \Rightarrow (P)\} \Rightarrow (P)$

1.48 . $[(P) \Rightarrow (R)] \Rightarrow \left[\{[(P) \Rightarrow (Q)] \Rightarrow (R)\} \Rightarrow (R)\right]$

1.49 . $\{[(P) \Rightarrow (Q)] \Rightarrow (R)\} \Rightarrow \{[(R) \Rightarrow (P)] \Rightarrow (P)\}$

1.50 . $\{[(P) \Rightarrow (Q)] \Rightarrow (R)\} \Rightarrow \{[(P) \Rightarrow (R)] \Rightarrow (R)\}$

1.51 . $(P) \Rightarrow \big([\neg(Q)] \Rightarrow \{\neg[(P) \Rightarrow (Q)]\}\big)$

1.52 . $[\neg(P)] \Rightarrow \{\neg[(P) \wedge (Q)]\}$

1.53 . $(P) \Leftrightarrow \{[(P) \Rightarrow (Q)] \Rightarrow (P)\}$

1.54 . $[\neg(P)] \Rightarrow \big([\neg(Q)] \Rightarrow \{\neg[(P) \vee (Q)]\}\big)$

1.55 . $[(P) \Rightarrow (Q)] \Leftrightarrow \big(\neg\{(P) \wedge [\neg(Q)]\}\big)$

1.56 . $[(P) \Rightarrow (Q)] \Leftrightarrow \{[\neg(P)] \vee (Q)\}$

1.6 Patterns of Deduction with Other Connectives

The preceding sections have demonstrated the theoretical foundations for patterns of deduction in the implicational calculus, for example, the transitivity of implication and the law of commutation, and with contraposition, for example, the law of reductio ad absurdum. Similarly, this section presents patterns of deduction with conjunctions and disjunctions, for instance, proofs by cases or by contradiction.

1.6.1 Conjunctions of Implications

The first theorem shows yet another form of the transitivity of the logical implication.

1.76 Theorem. *For all well-formed formulae P, Q, and R,*

$$\vdash \{[(P) \Rightarrow (Q)] \wedge [(Q) \Rightarrow (R)]\} \Rightarrow [(P) \Rightarrow (R)].$$

Proof. Apply theorems 1.52, 1.53, 1.27, and 1.21:

$\vdash \{\underbrace{[(P) \Rightarrow (Q)] \wedge [(Q) \Rightarrow (R)]}_{H}\} \Rightarrow \underbrace{[(Q) \Rightarrow (R)]}_{K}$ theorem 1.52,

$\vdash \{\underbrace{[(P) \Rightarrow (Q)] \wedge [(Q) \Rightarrow (R)]}_{H}\} \Rightarrow \underbrace{[(P) \Rightarrow (Q)]}_{L}$ theorem 1.53,

$\vdash \underbrace{[(Q) \Rightarrow (R)]}_{K} \Rightarrow \{\underbrace{[(P) \Rightarrow (Q)]}_{L} \Rightarrow \underbrace{[(P) \Rightarrow (R)]}_{M}\}$ theorem 1.27,

$\vdash \{\underbrace{[(P) \Rightarrow (Q)] \wedge [(Q) \Rightarrow (R)]}_{H}\} \Rightarrow \underbrace{[(P) \Rightarrow (R)]}_{M}$ theorem 1.21.

□

The following theorems give derived rules of inference with conjunctions of implications. The first theorem shows that if either of two hypotheses leads to a conclusion, then the *disjunction* of both hypotheses also leads to the same conclusion.

1.77 Theorem. *If* $\vdash (U) \Rightarrow (W)$ *and* $\vdash (V) \Rightarrow (W)$, *then* $\vdash [(U) \vee (V)] \Rightarrow (W)$:

$$\vdash \{[(U) \Rightarrow (W)] \wedge [(V) \Rightarrow (W)]\} \Rightarrow \{[(U) \vee (V)] \Rightarrow (W)\}.$$

Proof. The first part of the proof assumes the two hypotheses.

$\vdash (U) \Rightarrow (W)$	hypothesis,
$\vdash [\neg(W)] \Rightarrow [\neg(U)]$	contraposition and *Detachment*,
$\vdash (V) \Rightarrow (W)$	hypothesis,
$\vdash [\neg(W)] \Rightarrow [\neg(V)]$	contraposition and *Detachment*,
$\vdash [\neg(W)] \Rightarrow \{[\neg(U)] \wedge [\neg(V)]\}$	theorem 1.55,
$\vdash [\neg(W)] \Rightarrow \{\neg[(U) \vee (V)]\}$	De Morgan's second law,
$\vdash [(U) \vee (V)] \Rightarrow (W)$	axiom P3 and *Detachment*.

The second part of the proof dispenses with the two hypotheses.

$\vdash \{[(U) \Rightarrow (W)] \wedge [(V) \Rightarrow (W)]\} \Rightarrow [(U) \Rightarrow (W)]$	theorem 1.53,
$\vdash [(U) \Rightarrow (W)] \Rightarrow \{[\neg(W)] \Rightarrow [\neg(U)]\}$	transposition,
$\vdash \{[(U) \Rightarrow (W)] \wedge [(V) \Rightarrow (W)]\} \Rightarrow \{[\neg(W)] \Rightarrow [\neg(U)]\}$	theorem 1.16;

$\vdash \{[(U) \Rightarrow (W)] \wedge [(V) \Rightarrow (W)]\} \Rightarrow [(V) \Rightarrow (W)]$	theorem 1.53,
$\vdash [(V) \Rightarrow (W)] \Rightarrow \{[\neg(W)] \Rightarrow [\neg(V)]\}$	transposition,
$\vdash \{[(U) \Rightarrow (W)] \wedge [(V) \Rightarrow (W)]\} \Rightarrow \{[\neg(W)] \Rightarrow [\neg(V)]\}$	theorem 1.16;

$\vdash \{[(U) \Rightarrow (W)] \wedge [(V) \Rightarrow (W)]\}$ $\Rightarrow \left(\{[\neg(W)] \Rightarrow [\neg(U)]\} \wedge \{[\neg(W)] \Rightarrow [\neg(V)]\}\right)$	theorem 1.55;

$\vdash \left(\{[\neg(W)] \Rightarrow [\neg(U)]\} \wedge \{[\neg(W)] \Rightarrow [\neg(V)]\}\right)$ $\Rightarrow \left([\neg(W)] \Rightarrow \{[\neg(U)] \wedge [\neg(V)]\}\right)$	theorem 1.55;

$\vdash \left([\neg(W)] \Rightarrow \{[\neg(U)] \wedge [\neg(V)]\}\right) \Rightarrow \left[\{[(U) \vee (V)]\} \Rightarrow (W)\right]$	contraposition;

$\vdash \{[(U) \Rightarrow (W)] \wedge [(V) \Rightarrow (W)]\} \Rightarrow \left[\{[(U) \vee (V)]\} \Rightarrow (W)\right]$	theorem 1.55.

$\square$

Theorem 1.78 shows that an intuitionistic and minimalist axiom for the disjunction is a theorem in the Classical Propositional Calculus.

1.78 Theorem (OR.3). *For all propositional forms P, Q, and R,*

$$\vdash [(P) \Rightarrow (R)] \Rightarrow \left([(Q) \Rightarrow (R)] \Rightarrow \{[(P) \vee (Q)] \Rightarrow (R)\}\right).$$

Proof. Apply theorems 1.58 and 1.77:

$$\vdash \underbrace{\{[(P) \Rightarrow (R)] \wedge [(Q) \Rightarrow (R)]\}}_{H} \Rightarrow \underbrace{\left[\{[(P) \vee (Q)]\} \Rightarrow (R)\right]}_{N} \quad \text{theorem 1.77,}$$

$$\vdash \underbrace{[(P) \Rightarrow (R)]}_{H} \Rightarrow \left(\underbrace{[(Q) \Rightarrow (R)]}_{L} \Rightarrow \underbrace{\{\{[(P) \vee (Q)]\} \Rightarrow (R)\}}_{N}\right) \quad \text{theorem 1.58.}$$

$\square$

The next theorem shows that the disjunction $\vee$ commutes:

1.79 Theorem (commutativity of $\vee$). *For all well-formed formulae P and Q, the formula* $[(P) \vee (Q)] \Rightarrow [(Q) \vee (P)]$ *is a theorem.*

Proof. Substitute $(Q) \vee (P)$ for R in theorem 1.78 and apply theorem 1.67 with *Detachment*:

$$\vdash \{(P) \Rightarrow [(Q) \vee (P)]\} \Rightarrow (\{(Q) \Rightarrow [(Q) \vee (P)]\}$$
$$\Rightarrow \{[(P) \vee (Q)] \Rightarrow [(Q) \vee (P)]\}) \qquad\qquad \text{theorem 1.78 ,}$$
$$\vdash (P) \Rightarrow [(Q) \vee (P)] \qquad\qquad\qquad\qquad\qquad \text{theorem 1.67,}$$
$$\vdash \{(Q) \Rightarrow [(Q) \vee (P)]\} \Rightarrow \{[(P) \vee (Q)] \Rightarrow [(Q) \vee (P)]\} \quad \textit{Detachment ,}$$
$$\vdash (Q) \Rightarrow [(Q) \vee (P)] \qquad\qquad\qquad\qquad\qquad \text{theorem 1.67, ,}$$
$$\vdash [(P) \vee (Q)] \Rightarrow [(Q) \vee (P)] \qquad\qquad\qquad\qquad \textit{Detachment.}$$

The reverse implication results from swapping the rôles of P and Q. $\square$

Similarly, the second theorem shows that if either of two hypotheses leads to a conclusion, then the *conjunction* of both hypotheses also leads to the same conclusion.

1.80 Theorem. *If* $\vdash (U) \Rightarrow (W)$ *and* $\vdash (V) \Rightarrow (W)$, *then* $\vdash [(U) \wedge (V)] \Rightarrow (W)$:

$$\vdash \{[(U) \Rightarrow (W)] \wedge [(V) \Rightarrow (W)]\} \Rightarrow \{[(U) \wedge (V)] \Rightarrow (W)\}.$$

Proof. This proof relies on theorems 1.77, 1.53, 1.52, 1.67, 1.31:

$$\vdash \underbrace{\{[(U) \Rightarrow (W)] \wedge [(V) \Rightarrow (W)]\}}_{H}$$

$$\Rightarrow \{\underbrace{[(U) \vee (V)]}_{L} \Rightarrow (\underbrace{W}_{M})\} \qquad\qquad\qquad\qquad \text{theorem 1.77,}$$

$$\vdash \underbrace{[(U) \wedge (V)]}_{K} \Rightarrow \underbrace{[(U) \vee (V)]}_{L} \qquad\qquad\qquad \text{1.53, 1.52, 1.67,}$$

$$\vdash \underbrace{\{[(U) \Rightarrow (W)] \wedge [(V) \Rightarrow (W)]\}}_{H} \Rightarrow \{\underbrace{[(U) \wedge (V)]}_{K} \Rightarrow (\underbrace{W}_{M})\} \quad \text{theorem 1.31.}$$

$\square$

The third theorem shows that if one hypothesis leads to either of two conclusions, then the same hypothesis also leads to the disjunction of both conclusions.

1.81 Theorem. *For all well-formed formulae P, Q, and S,*

$$\vdash \{[(P) \Rightarrow (Q)] \vee [(P) \Rightarrow (S)]\} \Leftrightarrow \{(P) \Rightarrow [(Q) \vee (S)]\}.$$

Proof. This proof uses contraposition and previously established equivalences.

$$\neg\{(P) \Rightarrow [(Q) \vee (S)]\}$$

$\quad\quad\quad\quad\quad\quad\quad\quad\updownarrow\quad$ definition of $\wedge$ and equivalences,

$$(P) \wedge \{\neg[(Q) \vee (S)]\}$$

$\quad\quad\quad\quad\quad\quad\quad\quad\updownarrow\quad$ De Morgan's second law,

$$(P) \wedge \{[\neg(Q)] \wedge [\neg(S)]\}$$

$\quad\quad\quad\quad\quad\quad\quad\quad\updownarrow\quad$ idempotence of $\wedge$,

$$[(P) \wedge (P)] \wedge \{[\neg(Q)] \wedge [\neg(S)]\}$$

$\quad\quad\quad\quad\quad\quad\quad\quad\updownarrow\quad$ associativity and commutativity of $\wedge$,

$$\{(P) \wedge [\neg(Q)]\} \wedge \{(P) \wedge [\neg(S)]\}$$

$\quad\quad\quad\quad\quad\quad\quad\quad\updownarrow\quad$ definition of $\wedge$,

$$\{\neg[(P) \Rightarrow (Q)]\} \wedge \{\neg[(P) \Rightarrow (S)]\}$$

$\quad\quad\quad\quad\quad\quad\quad\quad\updownarrow\quad$ De Morgan's first law.

$$\neg\{[(P) \Rightarrow (Q)] \vee [(P) \Rightarrow (S)]\}$$

□

The fourth theorem shows that if two implications hold, then the conjunction of their hypotheses leads to the *conjunction* of their conclusions.

1.82 Theorem. *If* $\vdash (P) \Rightarrow (Q)$ *and* $\vdash (R) \Rightarrow (S)$, *then* $\vdash [(P) \wedge (R)] \Rightarrow [(Q) \wedge (S)]$:

$$\vdash \{[(P) \Rightarrow (Q)] \wedge [(R) \Rightarrow (S)]\} \Rightarrow \{[(P) \wedge (R)] \Rightarrow [(Q) \wedge (S)]\}.$$

Proof. The proof relies repeatedly on theorem 1.53:

$\vdash [(P) \wedge (R)] \Rightarrow (P)$ theorem 1.53,

$\vdash \{[(P) \wedge (R)] \Rightarrow (P)\}$
$\quad\quad \Rightarrow \big([(P) \Rightarrow (Q)] \Rightarrow \{[(P) \wedge (R)] \Rightarrow (Q)\}\big)$ theorem 1.28,
$\vdash [(P) \Rightarrow (Q)] \Rightarrow \{[(P) \wedge (R)] \Rightarrow (Q)\}$ *Detachment*;

$\vdash \{[(P) \Rightarrow (Q)] \wedge [(R) \Rightarrow (S)]\} \Rightarrow [(P) \Rightarrow (Q)]$ theorem 1.53,
$\vdash \underbrace{\{[(P) \Rightarrow (Q)] \wedge [(R) \Rightarrow (S)]\}}_{H} \Rightarrow \underbrace{\{[(P) \wedge (R)] \Rightarrow (Q)\}}_{K}$ theorem 1.15;

$\vdash [(P) \wedge (R)] \Rightarrow (R)$ theorem 1.53,
$\vdash \{[(P) \wedge (R)] \Rightarrow (R)\}$
$\quad\quad \Rightarrow \big([(R) \Rightarrow (S)] \Rightarrow \{[(P) \wedge (R)] \Rightarrow (S)\}\big)$ theorem 1.28,
$\vdash [(R) \Rightarrow (S)] \Rightarrow \{[(P) \wedge (R)] \Rightarrow (S)\}$ *Modus Ponens*;

$\vdash \{[(P) \Rightarrow (Q)] \wedge [(R) \Rightarrow (S)]\} \Rightarrow [(R) \Rightarrow (S)]$ theorem 1.53,
$\vdash \underbrace{\{[(P) \Rightarrow (Q)] \wedge [(R) \Rightarrow (S)]\}}_{H} \Rightarrow \underbrace{\{[(P) \wedge (R)] \Rightarrow (S)\}}_{L}$ theorem 1.15;

$$\vdash \{\underbrace{[(P) \Rightarrow (Q)] \wedge [(R) \Rightarrow (S)]}_{H}\}$$
$$\Rightarrow (\{\underbrace{[(P) \wedge (R)] \Rightarrow (Q)}_{K}\} \wedge \{\underbrace{[(P) \wedge (R)] \Rightarrow (S)}_{L}\}) \qquad \text{theorem 1.55;}$$

$$\vdash (\{\underbrace{[(P) \wedge (R)] \Rightarrow (Q)}_{H}\} \wedge \{\underbrace{[(P) \wedge (R)] \Rightarrow (S)}_{H}\})$$
$$\Rightarrow \{\underbrace{[(P) \wedge (R)] \Rightarrow [(Q) \wedge (S)]}_{H}\} \qquad \text{theorem 1.55;}$$

$$\vdash \{[(P) \Rightarrow (Q)] \wedge [(R) \Rightarrow (S)]\}$$
$$\Rightarrow \{[(P) \wedge (R)] \Rightarrow [(Q) \wedge (S)]\} \qquad \text{theorem 1.15.}$$
$$\square$$

Similarly, the fifth theorem shows that if two implications hold, then the conjunction of their hypotheses leads to the *disjunction* of their conclusions.

1.83 Theorem. *If* $\vdash (P) \Rightarrow (Q)$ *and* $\vdash (R) \Rightarrow (S)$, *then* $\vdash [(P) \wedge (R)] \Rightarrow [(Q) \vee (S)]$:

$$\vdash \{[(P) \Rightarrow (Q)] \wedge [(R) \Rightarrow (S)]\} \Rightarrow \{[(P) \wedge (R)] \Rightarrow [(Q) \vee (S)]\}.$$

Proof. This proof relies on theorem 1.82:
$$\vdash \{\underbrace{[(P) \Rightarrow (Q)] \wedge [(R) \Rightarrow (S)]}_{H}\} \Rightarrow \{\underbrace{[(P) \wedge (R)]}_{L} \Rightarrow \underbrace{[(Q) \wedge (S)]}_{M}\} \qquad 1.82,$$
$$\vdash \underbrace{[(Q) \wedge (S)]}_{M} \Rightarrow \underbrace{[(Q) \vee (S)]}_{N} \qquad 1.53, 1.52, 1.67,$$
$$\vdash \{\underbrace{[(P) \Rightarrow (Q)] \wedge [(R) \Rightarrow (S)]}_{H}\} \Rightarrow \{\underbrace{[(P) \wedge (R)]}_{L} \Rightarrow \underbrace{[(Q) \vee (S)]}_{N}\} \qquad 1.32.$$
$$\square$$

1.84 Theorem. *For all well-formed formulae P, Q, and R,*

$$\vdash \{[(P) \Rightarrow (Q)] \wedge (R)\} \Rightarrow \{(P) \Rightarrow [(Q) \wedge (R)]\}.$$

Proof. Apply theorems 1.14, 1.82 and 1.56:
$$\vdash [(P) \Rightarrow (Q)] \Rightarrow [(P) \Rightarrow (Q)] \qquad \text{theorem 1.14,}$$
$$\vdash (R) \Rightarrow [(P) \Rightarrow (R)] \qquad \text{axiom P1,}$$
$$\vdash \{[(P) \Rightarrow (Q)] \wedge (R)\} \Rightarrow \{[(P) \Rightarrow (Q)] \wedge [(P) \Rightarrow (R)]\} \qquad \text{theorem 1.82,}$$
$$\vdash \{[(P) \Rightarrow (Q)] \wedge [(P) \Rightarrow (R)]\} \Rightarrow \{(P) \Rightarrow [(Q) \wedge (R)]\} \qquad \text{theorems 1.82 and 1.56.}$$
$$\square$$

1.6.2   *Proofs by Cases or by Contradiction*

The following theorems establish further derived rules of inference. The first theorem forms a part of the basis for an algorithm — called the Completeness Theorem — to design proofs within the propositional calculus, as explained in section 1.9.

1.85 Theorem (proof by cases). *If* $(H) \Rightarrow (R)$ *and* $[\neg(H)] \Rightarrow (R)$ *are theorems, then R is a theorem. Hence the following formula is also a theorem (schema):*

$$\big([(H) \Rightarrow (R)] \wedge \{[\neg(H)] \Rightarrow (R)\}\big) \Rightarrow (R).$$

Proof. Apply theorem 1.77 and the law of excluded middle (theorem 1.70):

$\vdash (H) \Rightarrow (R)$	hypothesis,
$\vdash [\neg(H)] \Rightarrow (R)$	hypothesis,
$\vdash \{(H) \vee [\neg(H)]\} \Rightarrow (R)$	theorem 1.77,
$\vdash (H) \vee [\neg(H)]$	theorem 1.70,
$\vdash R$	Detachment;

$\vdash \big([(H) \Rightarrow (R)] \wedge \{[\neg(H)] \Rightarrow (R)\}\big) \Rightarrow \big[\{(H) \vee [\neg(H)]\} \Rightarrow (R)\big]$	theorem 1.77,
$\vdash (H) \vee [\neg(H)]$	theorem 1.70,
$\vdash \big([(H) \Rightarrow (R)] \wedge \{[\neg(H)] \Rightarrow (R)\}\big) \Rightarrow (R)$	theorem 1.17.

$\square$

Theorem 1.85 is the theoretical basis for proofs by cases, in the sense that if a conclusion R is a necessary consequence of a case H, and if R is also a necessary consequence of all the other cases, lumped together into $\neg(H)$, then R is a theorem.

The second theorem establishes the classical principle of proof by contradiction.

1.86 Theorem (proof by contradiction). *If* $[\neg(R)] \Rightarrow (S)$ *and* $[\neg(R)] \Rightarrow [\neg(S)]$ *are theorems, then R is a theorem: for all well-formed formulae R and S,*

$$\vdash \big(\{[\neg(R)] \Rightarrow (S)\} \wedge \{[\neg(R)] \Rightarrow [\neg(S)]\}\big) \Rightarrow (R).$$

Proof. Apply theorem 1.82, De Morgan's first law, and contraposition:

$\vdash [\neg(R)] \Rightarrow (S)$	hypothesis,
$\vdash [\neg(R)] \Rightarrow [\neg(S)]$	hypothesis,

$\vdash \big(\{[\neg(R)] \Rightarrow (S)\} \wedge \{[\neg(R)] \Rightarrow [\neg(S)]\}\big)$ $\Rightarrow \big(\{[\neg(R)] \wedge [\neg(R)]\} \Rightarrow \{(S) \wedge [\neg(S)]\}\big)$	theorem 1.82,

$\vdash [\neg(R)] \Rightarrow \{[\neg(R)] \wedge [\neg(R)]\}$	theorem 1.56,
$\vdash \big(\{[\neg(R)] \Rightarrow (S)\} \wedge \{[\neg(R)] \Rightarrow [\neg(S)]\}\big)$ $\Rightarrow \big([\neg(R)] \Rightarrow \{(S) \wedge [\neg(S)]\}\big)$	theorem 1.31,

$\vdash \big([\neg(R)]\Rightarrow\{(S)\wedge[\neg(S)]\}\big)\Rightarrow\big[\big(\neg\{(S)\wedge[\neg(S)]\}\big)\Rightarrow\{\neg[\neg(R)]\}\big]$ contraposition,

$\vdash \big([\neg(R)]\Rightarrow\{(S)\wedge[\neg(S)]\}\big)\Rightarrow\big[\{[\neg(S)]\vee(S)\}\Rightarrow(R)\big]$ De Morgan,

$\vdash [\neg(S)]\vee(S)$ theorem 1.70 ,

$\vdash \big([\neg(R)]\Rightarrow\{(S)\wedge[\neg(S)]\}\big)\Rightarrow(R)$ theorem 1.17 ,

$\vdash \big(\{[\neg(R)]\Rightarrow(S)\}\wedge\{[\neg(R)]\Rightarrow[\neg(S)]\}\big)\Rightarrow(R)$ theorem 1.16.

$\square$

    Theorem 1.86 is the theoretical basis for proofs by contradiction: if a conclusion S and its negation $\neg(S)$ are both necessary consequences of the negation $\neg(R)$ of a statement R, then R is a theorem.

1.6.3   Exercises on Patterns of Deduction

1.57 . Determine whether the following derived rule holds. "If $(V) \Rightarrow (W)$ and $(R) \Rightarrow (S)$ are theorems, then $[(V) \vee (R)] \Rightarrow [(W) \vee (S)]$ is also a theorem":

$$\{[(V) \Rightarrow (W)] \wedge [(R) \Rightarrow (S)]\} \Rightarrow \{[(V) \vee (R)] \Rightarrow [(W) \vee (S)]\}.$$

1.58 . Determine whether the following derived rule holds. "If $(I) \Rightarrow (R)$ and $(I) \Rightarrow [\neg(S)]$ are both theorems, then $\neg[(R) \Rightarrow (S)]$ is also a theorem":

$$\big([(I) \Rightarrow (R)] \wedge \{(I) \Rightarrow [\neg(S)]\}\big) \Rightarrow \{\neg[(R) \Rightarrow (S)]\}.$$

1.59 . Determine whether the following derived rule holds. "If $(U) \Rightarrow (W)$ or $(V) \Rightarrow (W)$ is a theorem, then $[(U) \vee (V)] \Rightarrow (W)$ is also a theorem":

$$\{[(U) \Rightarrow (W)] \vee [(V) \Rightarrow (W)]\} \Rightarrow \{[(U) \vee (V)] \Rightarrow (W)\}.$$

1.60 . Determine whether the following derived rule holds. "If $(P) \Rightarrow (Q)$ or $(R) \Rightarrow (S)$ is a theorem, then $[(P) \vee (R)] \Rightarrow [(Q) \vee (S)]$ is also a theorem":

$$\{[(P) \Rightarrow (Q)] \vee [(R) \Rightarrow (S)]\} \Rightarrow \{[(P) \vee (R)] \Rightarrow [(Q) \vee (S)]\}.$$

1.61 . Determine whether the following derived rule holds. "If $(P) \Rightarrow (S)$ and $[\neg(Q)] \Rightarrow [\neg(S)]$ is a theorem, then $(P) \Rightarrow (Q)$ is also a theorem":

$$\big([(P) \Rightarrow (S)] \wedge \{[\neg(Q)] \Rightarrow [\neg(S)]\}\big) \Rightarrow [(P) \Rightarrow (Q)].$$

1.62 . Prove that if T is a theorem, then $(R) \Rightarrow [(T) \Rightarrow (R)]$ is a theorem.

1.63 . Prove that if T is a theorem, then $[(T) \Rightarrow (R)] \Rightarrow (R)$ is a theorem.

1.64 . Prove that if $\neg(F)$ is a theorem, then so is $\{[\neg(R)] \Rightarrow (F)\} \Rightarrow (R)$.

1.65. Prove that if $\neg(F)$ is a theorem, then so is $(R) \Rightarrow \{[\neg(R)] \Rightarrow (F)\}$.

1.66. Determine whether the following derived rule of inference holds. "If $\neg(F)$ and $\{(P) \wedge [\neg(Q)]\} \Rightarrow (F)$ are theorems, then $(P) \Rightarrow (Q)$ is also a theorem."

1.6.4 Equivalent Classical Axiom Systems

The Pure Positive Implicational Propositional Calculus just presented rests on the rules of *Detachment* and *Substitution* with axioms P1, P2, and P3. Other equivalent selections of axioms exist, for instance, Tarksi's axioms I–VII listed in example 1.87.

1.87 Example (Tarski's Axioms). Alfred Tarski lists seven axioms [129, p. 147]:

Tarski's Axiom I.	$(P) \Rightarrow [(Q) \Rightarrow (P)]$.
Tarski's Axiom II.	$\{(P) \Rightarrow [(P) \Rightarrow (Q)]\} \Rightarrow [(P) \Rightarrow (Q)]$.
Tarski's Axiom III.	$[(P) \Rightarrow (Q)] \Rightarrow \{[(Q) \Rightarrow (R)] \Rightarrow [(P) \Rightarrow (R)]\}$.
Tarski's Axiom IV.	$[(P) \Leftrightarrow (Q)] \Rightarrow [(P) \Rightarrow (Q)]$.
Tarski's Axiom V.	$[(P) \Leftrightarrow (Q)] \Rightarrow [(Q) \Rightarrow (P)]$.
Tarski's Axiom VI.	$[(P) \Rightarrow (Q)] \Rightarrow \{[(Q) \Rightarrow (P)] \Rightarrow [(P) \Leftrightarrow (Q)]\}$.
Tarski's Axiom VII.	$\{[\neg(Q)] \Rightarrow [\neg(P)]\} \Rightarrow [(P) \Rightarrow (Q)]$.

The exercises show that Tarksi's axioms I–VII are equivalent to axioms P1, P2, and P3. Other equivalent selections of axioms include Kleene's, listed in example 1.88 [72, p. 15].

1.88 Example (Kleene's Axioms). Kleene lists four axioms [72, p. 15]:

Kleene's axiom 1a.	$(A) \Rightarrow [(B) \Rightarrow (A)]$.
Kleene's axiom 1b.	$[(A) \Rightarrow (B)] \Rightarrow (\{(A) \Rightarrow [(B) \Rightarrow (C)]\} \Rightarrow [(A) \Rightarrow (C)])$.
Kleene's axiom 7.	$[(A) \Rightarrow (B)] \Rightarrow (\{(A) \Rightarrow [\neg(B)]\} \Rightarrow [\neg(A)])$.
Kleene's axiom 8.	$\{\neg[\neg(A)]\} \Rightarrow (A)$.

The exercises also show that Kleene's axioms are derivable from axioms P1, P2, and P3. Yet other equivalent selections of axioms include John Barkley Rosser's, listed in example 1.89 [110, p. 55]:

1.89 Example (Rosser's Axioms). Rosser lists three axioms [110, p. 55]:

Axiom R1 $(P) \Rightarrow [(P) \wedge (P)]$.

Axiom R2 $[(P) \wedge (Q)] \Rightarrow (P)$.

Axiom R3 $[(P) \Rightarrow (Q)] \Rightarrow (\{\neg[(Q) \wedge (R)]\} \Rightarrow \{\neg[(R) \wedge (P)]\})$.

The exercises show that Rosser's axioms are derivable from axioms P1, P2, and P3. Rosser's reference [110] establishes the converse.

1.6.5   Exercises on Kleene's, Rosser's, and Tarski's Axioms

For the following exercises, define $\vee$, $\wedge$, and $\Leftrightarrow$ as in definition 1.51.

1.67. Prove that Tarski's axiom IV, $[(P) \Leftrightarrow (Q)] \Rightarrow [(P) \Rightarrow (Q)]$, is derivable from axioms P1, P2, and P3.

1.68. Prove that Tarski's axiom V, $[(P) \Leftrightarrow (Q)] \Rightarrow [(Q) \Rightarrow (P)]$, is derivable from axioms P1, P2, and P3.

1.69. Prove that Tarski's axiom VI, $[(P) \Rightarrow (Q)] \Rightarrow \{[(Q) \Rightarrow (P)] \Rightarrow [(P) \Leftrightarrow (Q)]\}$, is derivable from axioms P1, P2, and P3.

1.70. Prove that every theorem derivable from Tarski's axioms I– VII is also derivable from axioms P1, P2, and P3.

1.71. Prove that every theorem derivable from axioms P1, P2, and P3 is also derivable from Tarski's axioms I– VII.

1.72. Prove that Rosser's axiom R1, $(P) \Rightarrow [(P) \wedge (P)]$, is derivable from axioms P1, P2, and P3.

1.73. Prove that Rosser's axiom R3, $[(P) \Rightarrow (Q)] \Rightarrow (\{\neg[(Q) \wedge (R)]\} \Rightarrow \{\neg[(R) \wedge (P)]\})$, is derivable from axioms P1, P2, and P3.

1.74. Prove that Rosser's axiom R2, $[(P) \wedge (Q)] \Rightarrow (P)$, is derivable from axioms P1, P2, and P3.

1.75. Prove that Kleene's axiom 7, $[(A) \Rightarrow (B)] \Rightarrow (\{(A) \Rightarrow [\neg(B)]\} \Rightarrow [\neg(A)])$, is derivable from axioms P1, P2, and P3.

1.76. Prove that Kleene's axiom 8 $\{\neg[\neg(A)]\} \Rightarrow (A)$, is derivable from axioms P1, P2, and P3.

1.7   Completeness, Decidability, Independence, Provability, and Soundness

This section addresses the question how to determine whether a formula admits of a proof from selected axioms, and, if so, how to find such a proof. The main tool to this end consists of multi-valued logics, also called "fuzzy" logics [102].

1.7.1   Multi-Valued Fuzzy Logics

Multi-valued fuzzy logics are "models" of propositional logics that assign a "value" to each propositional variable, and hence also to each formulaic letter and each well-formed propositional form, by means of a table or a formula.

Table 1.94 Church's logic with values u, w, and distinguished value v [18, p. 113].

$\neg(P)$	P	Q	$(P) \Rightarrow (Q)$	(Q)	$\Rightarrow$	$[(P) \Rightarrow (Q)]$
u	v	v	v	v	v	v
u	v	w	w	w	v	w
u	v	u	u	u	v	u
w	w	v	v	v	v	v
w	w	w	v	w	v	v
w	w	u	u	u	v	u
v	u	v	v	v	v	v
v	u	w	v	w	v	v
v	u	u	v	u	v	v

1.90 Definition. A **value** in propositional logic may be any symbol that is not already allowed in well-formed propositional formulae. Thus a value may not be a connective, formulaic or propositional letter, parenthesis, bracket, or brace.

For the present purposes three values suffice, denoted here by u, v, and w. A logic with exactly three values is also called a **triadic** logic. A multi-valued logic also designates any one of its values as the **distinguished** value, for example, v (Fig. 1.1).

1.91 Definition. A propositional form **holds**, or is **valid** or a **tautology**, in a multi-valued logic if and only if it has the distinguished value regardless of the values of its components. The notation $\models P$ indicates that P is valid [72, p. 12 & p. 14].

1.92 Remark. Such software packages as John Harrison's program [54, 55] and Stephen Wolfram's *Mathematica* [142] provide facilities to calculate and print multi-valued Truth tables of propositional forms, as explained in [102].

1.93 Example (Church's triadic logic). Table 1.94 *defines* (by fiat) the values of the negation $\neg(P)$ and of the implication $(P) \Rightarrow (Q)$ from the values of P and Q in Church's triadic logic [18, p. 113, un-numbered table, penultimate column]. The last column shows how to derive the values of the compound formula $(Q) \Rightarrow [(P) \Rightarrow (Q)]$ from the values of its components. The formula $(Q) \Rightarrow [(P) \Rightarrow (Q)]$ is valid because it has the distinguished value v regardless of the values of P and Q.

1.95 Example (Łukasiewicz's triadic logic). Table 1.96 *defines* (by fiat) the values of the negation $\neg(P)$ and of the implication $(P) \Rightarrow (Q)$ from the values of P and Q in Łukasiewicz's triadic logic [18, 79, p. 113, un-numbered table, last column]. The last column of table 1.96 shows how to derive the values of the compound formula $(Q) \Rightarrow [(P) \Rightarrow (Q)]$ from the values of its components. The formula $(Q) \Rightarrow [(P) \Rightarrow (Q)]$ is valid because it has the distinguished value v regardless of the values of P and Q.

1.7.2 Sound Multi-Valued Fuzzy Logics

In some logics, validity provides a necessary criterion for provability: if there is a proof of a theorem L, then L is valid. Such logics are called *sound*.

Table 1.96 Łukasiewicz's logic with values u, w, and distinguished value v [79].

$\neg(P)$	P	Q	$(P) \Rightarrow (Q)$	(Q)	$\Rightarrow$	$[(P) \Rightarrow (Q)]$
u	v	v	v	v	v	v
u	v	w	w	w	v	w
u	v	u	u	u	v	u
w	w	v	v	v	v	v
w	w	w	v	w	v	v
w	w	u	w	u	v	w
v	u	v	v	v	v	v
v	u	w	v	w	v	v
v	u	u	v	u	v	v

Fig. 1.1 Triadic Truth-tables on page 640 of Charles Sanders Peirce's 1909 *Logic Notebook*: MS Am 1632 (339) Houghton Library, Harvard University, used by permission. (http://pds.lib.harvard.edu/pds/view/15255301?n=640&printThumbnails=no).

1.97 Definition. A multi-valued fuzzy logic is **sound** if and only if all its theorems are valid: for every formula L, if $\vdash L$ (if L is a theorem), then $\models L$ (then L is valid).

Every axiom is also a theorem and thus must be valid in a sound logic.

1.98 Example. Table 1.94 in example 1.93 shows that Frege's axiom P1 (the law of affirmation of the consequent) is valid in Church's triadic logic. Similarly, exercise 1.83 confirms that Frege's axiom P2 (the law of self-distributivity of implication) is also valid in Church's triadic logic. Thus all the axioms of the Pure Positive Implicational Propositional Calculus are valid in Church's triadic logic.

If a sound logic allows for inferences by *Detachment*, then *Detachment* must be "sound" or "preserve" valid formulae, in the sense that from valid premises P and $(P) \Rightarrow (Q)$ *Detachment* must produce a valid conclusion Q.

1.99 Example. Table 1.94 in example 1.93 contains only one line where P and $(P) \Rightarrow (Q)$ both have the distinguished value (v): only in the first line. In that line Q also has the distinguished value (v). Thus *Detachment* preserves valid formulae in Church's triadic logic.

1.100 Example. Table 1.96 in example 1.95 contains only one line where P and $(P) \Rightarrow (Q)$ both have the distinguished value (v): only in the first line. In that line Q also has the distinguished value (v). Thus *Detachment* preserves valid formulae in Łukasiewicz's triadic logic.

Conversely, theorem 1.101 confirms that sound axioms and a sound *Detachment* suffice for a propositional logic to be sound.

1.101 Theorem. *Suppose that a multi-valued propositional logic is such that its axioms are valid and* Detachment *from valid premises produces a valid conclusion: for all propositional forms H and C, if H is valid, and if* $(H) \Rightarrow (C)$ *is valid, then C must be valid. Then such a logic is sound.*

Proof (Outline). This proof shows that each step of a proof is valid. Each step C of a proof is either an axiom or the result of *Detachment*. In particular, every formal proof starts with an axiom.

If C is an axiom, then C is valid by hypothesis.

Suppose that all the steps up to but not including C have already been proved valid. If C results from *Detachment* from previous steps H and $(H) \Rightarrow (C)$, then H and $(H) \Rightarrow (C)$ are valid, and hence C is valid by the hypotheses of the theorem.

Thus every step of a proof is a valid formula. In particular, the last step of a proof is also a valid formula.

A rigorous proof uses the Principle of Mathematical Induction in chapter 4. □

1.102 Example. Examples 1.98 and 1.99 show that in Church's triadic logic *Detachment* is sound and all the axioms of the Pure Positive Implicational Propositional Calculus are valid. Consequently, theorem 1.101 shows that the Pure Positive Implicational Propositional Calculus with Church's triadic logic is sound: every theorem of the Pure Positive Implicational Propositional Calculus is a valid formula in Church's triadic logic.

1.7.3 Independence and Unprovability

In a sound logic, every theorem is valid. By contraposition, if a formula is not valid, then it is not a theorem. Consequently, to prove that there are no proofs of a formula L from specified axioms and rules of inference, it suffices to find a system of logical values where the logic is sound but the formula L is not valid.

1.103 Example. Example 1.102 shows that the Pure Positive Implicational Propositional Calculus is sound with Church's triadic logic defined by table 1.94. However, table 1.104 shows that Peirce's Law is not valid in this logic. Consequently, Peirce's

Table 1.104 Peirce's Law in Church's triadic logic [18, p. 113].

P	Q	$\{[(P) \Rightarrow (Q)]$	$\Rightarrow$	$(P)\}$	$\Rightarrow$	(P)
v	v	v	v	v	v	v
v	w	w	v	v	v	v
v	u	u	v	v	v	v
w	v	v	w	w	v	w
w	w	v	w	w	v	w
w	u	u	v	w	$\mathbf{w}$	w
u	v	v	u	u	v	u
u	w	v	u	u	v	u
u	u	v	u	u	v	u

Law is unprovable in — is not a theorem of — the Pure Positive Implicational Propositional Calculus: there are no proofs of Peirce's Law using only *Detachment* and Frege's axiom P1 (the law of affirmation of the consequent) and axiom P2 (the law of self-distributivity of implication).

Different analyses of different formulae may require different multi-valued fuzzy logics. For instance, the question whether Frege's axiom P2 (the law of self-distributivity of implication) can be proved from axiom P1 (the law of affirmation of the consequent) cannot be answered from Church's triadic logic defined by table 1.94, because both axioms are valid there. However, another multi-valued fuzzy logic answers the question.

1.105 Example. Table 1.96 in example 1.95 shows that Frege's axiom P1 (the law of affirmation of the consequent) and *Detachment* are valid in Łukasiewicz's triadic logic. However, exercise 1.83 reveals that axiom P2 (the law of self-distributivity of implication) is not valid in the same logic. Therefore, axiom P2 is not provable in — is not a theorem of — the implicational propositional logic with *Detachment* and only one axiom: the single axiom P1.

1.106 Definition. A logical formula L is **logically independent** from a system of axioms and rules of inference if and only if L is not provable in — is not a theorem of — the logic defined by the system of axioms and rules of inference.

1.107 Example. Example 1.103 shows that Peirce's Law is independent from the Pure Positive Implicational Propositional Calculus.

1.108 Example. Example 1.105 shows that Frege's axiom P2 (the law of self-distributivity of implication) is independent from the implicational propositional logic defined by Frege's axiom P1 (the law of affirmation of the consequent) and *Detachment*.

In general, the concept of *mathematical* as opposed to practical impossibility seems difficult to explain to the general public [28, 29]. Nevertheless, as just demonstrated, multi-valued logic provides elementary examples of mathematical impossibilities, in the form of rigorous proofs of the nonexistence of proofs of specific formulae, and thereby an explanation of the *concept* — though not the proof — of such impossibilities as Arrow's Impossibility Theorem in voting presented in chapter 7, ruler-and-compass angle trisection, duplication of the cube, or a proof of Euclid's fifth postulate in geometry, and a solution to the decision problem in first-order logic, or Gödel's Incompleteness Theorem in logic.

1.7.4 Complete Multi-Valued Fuzzy Logics

In some logics, validity provides a sufficient criterion for provability: if a formula L is valid, then there is a proof of L. Such logics are called *complete*.

1.109 Definition. A multi-valued logic is **complete** if and only if all its valid formulae are theorems: for every formula L, if $\models L$ (if L is valid), then $\vdash L$ (then L is a theorem). Otherwise, if there are valid formulae that are not theorems, then the logic is called **incomplete**.

1.110 Example. In table 1.94 from example 1.93, deleting all the lines where P or Q may take the value w, thus keeping only the lines where P and Q may take only the value u or v, gives a two-valued logic where Frege's axiom P2 (the law of self-distributivity of implication) and axiom P1 (the law of affirmation of the consequent) and *Detachment* are still valid, because they are valid in the larger table 1.94.

Similarly, in table 1.104 from example 1.103, deleting all the lines where P or Q may take the value w, thus keeping only the lines where P and Q may take only the value u or v, gives table 1.111, where Peirce's Law is valid.

Yet example 1.103 shows that Peirce's Law is not a theorem of the Pure Positive Implicational Propositional Calculus. Consequently, the Pure Positive Implicational Propositional Calculus with the dyadic logic defined by table 1.111 is incomplete.

In a two-valued (**dyadic**) Boolean logic, the distinguished value v is also called True or denoted by 1, while the other value u is also called False or denoted by 0.

1.111 Example. The three axioms formed by Peirce's Law with Axioms P1 and P2 in addition to *Detachment* lead to a complete logic, where every Boolean dyadic tautology has a proof [108, p. 25].

Table 1.111 Peirce's Law is valid in Boolean dyadic logic.

P	Q	$\{[(P) \Rightarrow (Q)]$	$\Rightarrow$	$(P)\}$	$\Rightarrow$	(P)
v	v	v	v	v	v	v
v	u	u	v	v	v	v
u	v	v	u	u	v	u
u	u	v	u	u	v	u

1.7.5   Peirce's Law as a Denial of the Antecedent

Theorems 1.67, 1.68, and 1.78 show that the formula $[(P) \Rightarrow (Q)] \Rightarrow (Q)$ uses only the connective $\Rightarrow$ but is equivalent to $(P) \vee (Q)$ in the Classical Propositional Calculus. Similarly, without negations, using only implications, Charles Sanders Peirce's Law

$$\textbf{(Peirce's Law)} \quad \{[(P) \Rightarrow (Q)] \Rightarrow (P)\} \Rightarrow (P) \tag{1.1}$$

[104, p. 189–190] corresponds to the pattern of reasoning by the Law of the Excluded Middle $(B) \vee [\neg(B)]$ as follows. With Boolean logic, in Peirce's Law (1.1) the formula $(P) \Rightarrow (Q)$ is True for every proposition Q if and only if P is False. For this reason, $(P) \Rightarrow (Q)$ is a form of a *denial* of the antecedent P, as in the Law of Denial of the Antecedent $[\neg(P)] \Rightarrow [(P) \Rightarrow (Q)]$ (theorem 1.40):

> Peirce took '$p \supset \alpha$' as the denial of whatever statement 'p' abbreviates on the ground that to say of a statement 'It implies everything,' was tantamount to saying of it 'It is false.' —[9, p. 157]

With P and Q replaced by $(P) \Rightarrow (Q)$ and P respectively, if the denial $(P) \Rightarrow (Q)$ is False, then $[(P) \Rightarrow (Q)] \Rightarrow (P)$ is True, whence, by Peirce's Law (1.1) and *Detachment*, the consequent P is True. In other words, from the Falsity of the denial of P follows the Truth of P. Equivalently, if $\neg(P)$ is False, then P must be True. In this sense, using only the implicational connective $\Rightarrow$ Peirce's Law (1.1) expresses the law of the excluded middle $(P) \vee [\neg(P)]$ [104, p. 189–190].

1.7.6   Exercises on Church's and Łukasiewicz's Triadic Systems

For each of the following formulae, determine whether it is a triadic tautology with Church's table 1.94, or Łukasiewicz's table 1.96, or both, or neither.

1.77 . $[(P) \Rightarrow [(Q)] \Rightarrow \{[(Q) \Rightarrow (R)] \Rightarrow [(P) \Rightarrow (R)]\}$.

1.78 . $(P) \Rightarrow (P)$.

1.79 . $\{[(P) \Rightarrow (Q)] \Rightarrow (P)\} \Rightarrow (P)$.

1.80 . $[(P) \Rightarrow (Q)] \Rightarrow (\{(P) \Rightarrow [\neg(Q)]\} \Rightarrow [\neg(P)])$.

1.81 . $\{[(P) \Rightarrow (Q)] \Rightarrow (Q)\} \Rightarrow [(Q) \Rightarrow (P)] \Rightarrow (P)\}$.

1.82 . $[\neg(P)] \Rightarrow [(P) \Rightarrow (Q)]$.

1.83 . $\{(P) \Rightarrow [(Q) \Rightarrow (R)]\} \Rightarrow \{[(P) \Rightarrow (Q)] \Rightarrow [(P) \Rightarrow (R)]\}$.

1.84 . $[(P) \Rightarrow (Q)] \Rightarrow (\{(P) \Rightarrow [\neg(Q)]\} \Rightarrow [\neg(P)])$.

1.85 . $\{(P) \Rightarrow [(Q) \Rightarrow (R)]\} \Rightarrow \{(Q) \Rightarrow [(P) \Rightarrow (R)]\}$.

1.86 . $\{[\neg(Q)] \Rightarrow [\neg(P)]\} \Rightarrow [(P) \Rightarrow (Q)]$.

1.8 Boolean Logic

Boolean logic is a **dyadic** logic: a multi-valued logic with exactly two values. The distinguished value may be called "True" and denoted by v, whereas the other value may be called "False" and denoted by u. Table 1.113 defines the Boolean Truth values of compound propositional formulae with selected connectives.

Boolean logic will lead to an algorithm to design proofs in the Full Propositional Calculus in section 1.9.

1.8.1 The Truth Table of the Logical Implication

Based on [101], this subsection clarifies the Truth table of the logical implication.

True logical implications with a False hypothesis are convenient in classical mathematics, but other versions of mathematics do not include them [33]. Moreover, True logical implications with a False hypothesis rarely occur in practical reasoning:

> Actually, the rule that any conditional is true if its antecedent is known to be false has almost no parallel in natural logic. Examples of the type "if snow is black, then $2 \times 2 = 5$," which keep cropping up in textbooks, are only capable of confusing the student, since no natural subsystem in our language has expressions with this semantics [81, p. 36].

Correspondingly, some computers and logical circuits do not include any facility to test the Truth value of logical implications. Therefore the following four examples serve solely to demonstrate the difference between Boolean algebraic logic and practical reasoning. There exist other logics, but they are less used and more complicated than Boolean logic [18, p. 142, p. 146, §26.11].

The present subsection shows that the Truth table for the Boolean logical implication is the *only* Truth table that satisfies certain requirements. Specifically, the present considerations confirm that the complete law of contraposition

$$[(P) \Rightarrow (Q)] \Leftrightarrow \{[\neg(Q)] \Rightarrow [\neg(P)]\}$$

and the nonequivalence of an implication $(P) \Rightarrow (Q)$ with its converse $(Q) \Rightarrow (P)$ hold *only* with implications defined as in table 1.113. To this end, denote by $\Rightarrow$ any candidate connective for a logical implication. To reflect common experience, as in

Table 1.113 Boolean dyadic logic with values u and distinguished value v.

$\neg(P)$	P	Q	$(P) \Rightarrow (Q)$	$(P) \wedge (Q)$	$(P) \vee (Q)$	$[(P) \text{ NOR } (Q)]$	$(P) \Leftrightarrow (Q)$
u	v	v	v	v	v	u	v
u	v	u	u	u	v	u	u
v	u	v	v	u	v	u	u
v	u	u	v	u	u	v	v

Table 1.114 A partial Truth table for a connective $(P) \Rrightarrow (Q)$ conforming to (Implication. 1) & (Implication. 2).

REQUIREMENT	P	Q	$(P) \Rrightarrow (Q)$
(Implication. 1)	T	T	T
(Implication. 2)	T	F	F

Table 1.115 The connectives $\looparrowright$, $\rightsquigarrow$, $\curvearrowright$, and $\Rightarrow$.

P	Q	$(P) \looparrowright (Q)$	P	Q	$(P) \rightsquigarrow (Q)$	P	Q	$(P) \curvearrowright (Q)$	P	Q	$(P) \Rightarrow (Q)$
T	T	T	T	T	T	T	T	T	T	T	T
T	F	F	T	F	F	T	F	F	T	F	F
F	T	T	F	T	F	F	T	F	F	T	T
T	F	F	F	F	F	F	F	T	F	F	T

example 1.1, any concept of logical implication may have to satisfy the following two requirements.

(Implication.1) To allow for reasoning by *Detachment*, if a hypothesis P holds, and if the implication $(P) \Rrightarrow (Q)$ holds, then the conclusion Q also holds.

(Implication.2) To avoid faulty reasoning, if a hypothesis P holds, but if the conclusion Q fails, then the implication $(P) \Rrightarrow (Q)$ also fails.

(Implication.3) To allow reasoning by contraposition, an implication and its *contraposition* must have the *same* Truth table.

(Implication.4) Again to prevent faulty reasoning, an implication and its *converse* must have *different* Truth tables.

The first two requirements, (Implication. 1) & (Implication. 2), dictate the first two lines of Truth table 1.114, where the hypothesis P is True. There remain only four possibilities for $\Rrightarrow$ in the last two rows, where the hypothesis is False. For convenience, denote these four connectives by $\Rightarrow$, $\looparrowright$, $\rightsquigarrow$, and $\curvearrowright$ respectively (these last three symbols are used in this manner only in the present discussion). Table 1.115 shows their Truth values. For comparison, the last two requirements, (Implication. 3) & (Implication. 4), dictate the last two lines of the desired Truth table.

Verifications based on table 1.113. confirm that the logical implication $\Rightarrow$ has the same Truth values as its contraposition has, but not as its converse has.

In contrast, $(P) \looparrowright (Q)$ and its contraposition $[\neg(Q)] \looparrowright [\neg(P)]$ have *different* Truth values, as in table 1.116. Also, for the connective $\looparrowright$, neither the law of contraposition nor its converse holds.

Similarly, $(P) \rightsquigarrow (Q)$ and its contraposition $[\neg(Q)] \rightsquigarrow [\neg(P)]$ do *not* have the same Truth values, as in table 1.117. Moreover, for the connective $\rightsquigarrow$, neither the law of contraposition nor its converse hold.

Table 1.116 The connective ⇸ and its contraposition do *not* have the same Truth values.

P	Q	$(P) ⇸ (Q)$	$[¬(Q)] ⇸ [¬(P)]$
T	T	T	**F**
T	F	F	F
F	T	T	T
F	F	F	**T**

Table 1.117 The connective ↝ and its contraposition do *not* have the same Truth values.

P	Q	$(P) ↝ (Q)$	$[¬(Q)] ↝ [¬(P)]$
T	T	T	**F**
T	F	F	F
F	T	F	F
F	F	F	**T**

Table 1.118 The connective ↷ and its converse have the same Truth values.

P	Q	$(P) ↷ (Q)$	$(Q) ↷ (P)$
T	T	T	T
T	F	F	F
F	T	F	F
F	F	T	T

Finally, $(P) ↷ (Q)$ and its converse $(Q) ↷ (P)$ have the same Truth values, as in table 1.118. However, for the connective ↷, both the law of contraposition and its converse hold.

Thus the Truth table 1.113 specified for $(P) ⇒ (Q)$ is the *only* one that reflects experience. There exist other concepts of logical implication, but they do not lend themselves to Truth tables [18, p. 146, #26.12].

1.8.2 Boolean Logic on Earth and in Space

Some logical connectives — for instance, NOR — can combine so as to play the rôle of every logical connective.

1.119 Definition. A logical connective is called **primitive**, or also **universal**, if and only if every propositional form is logically equivalent to a propositional form containing only that connective.

1.120 Example. The logical connective NOR, defined so that (A) NOR (B) stands for $¬(A ∨ B)$, is universal; in particular, the following equivalences are tautologies:

$$[¬(P)] ⇔ [(P) \text{ NOR } (P)],$$

$$[(P) \land (Q)] \Leftrightarrow \{ [(P) \text{ NOR } (P)] \text{ NOR } [(Q) \text{ NOR } (Q)] \},$$

$$[(P) \lor (Q)] \Leftrightarrow \{[(P) \text{ NOR } (Q)] \text{ NOR } [(P) \text{ NOR } (Q)]\},$$

$$[(P) \Rightarrow (Q)] \Leftrightarrow \{[(P)\text{NOR}(P)]\text{NOR}(Q)\}\text{NOR}\{[(P)\text{NOR}(P)]\text{NOR}(Q)\}.$$

1.121 Example. The logical connective NAND, defined so that (A) NAND (B) stands for $\neg(A \land B)$, is universal; in particular, the following equivalences are tautologies:

$$[\neg(P)] \Leftrightarrow [(P) \text{ NAND } (P)],$$

$$[(P) \land (Q)] \Leftrightarrow \{ [(P) \text{ NAND } (Q)] \text{ NAND } [(P) \text{ NAND } (Q)] \},$$

$$[(P) \lor (Q)] \Leftrightarrow \{[(P) \text{ NAND } (P)] \text{ NAND } [(Q) \text{ NAND } (Q)]\},$$

$$[(P) \Rightarrow (Q)] \Leftrightarrow (P) \text{ NAND } [(Q) \text{ NAND } (Q)].$$

1.122 Example. For logic and arithmetic, Westinghouse DPS-2402 Computers used only NAND gates:

> Its function can be considered fundamental in that all sequential and combinational logic functions can be performed entirely by NAND gates [138, Section 4, § 4–1(7)(a), p. 4–4].

1.123 Example. The Apollo spacecraft contained two electrically identical Apollo Guidance Computers (AGC): a Command Module Computer (CMC) in the Command Module (CM), and a Lunar Module Computer (LMC) in the Lunar Module (LM) [51, § 2.1, p. 23], pictured in figure 1.2.

Fig. 1.2 The logic in the present chapter was used in the Apollo Guidance Computer onboard the Command and Lunar Modules. (Neil A. Armstrong's photograph of Edwin E. Aldrin Jr., 20 July 1969, NASA ID AS11-40-5927.)

Each AGC used only one type of logical circuit: a NOR gate with three variables, such that NOR (A, B, C) is equivalent to $\neg(A \lor B \lor C)$ [51, § 3.2.1, p. 60, fig. 3–1]. Setting C to False ($0V$) shows that NOR (A, B, F) is equivalent to $\neg(A \lor B)$ and hence to (A) NOR (B). By universality of NOR, every logical, arithmetic, reading, writing, and copying operation necessary during space flight was implemented by circuits consisting entirely and exclusively of NOR gates [51, § 3.2.1, p. 62]. The universality of NOR provides a reliability greater than several different connectives would:

> The single logic type simplified packaging, manufacturing, and testing, and gave higher confidence to the reliability predictions [51, § 1.1, p. 10].

1.9   Automated Theorem Proving

Relying on the Deduction Theorem 1.22, the Provability Theorem and the Completeness Theorem will provide not only guidance but an algorithm to design proofs within the propositional calculus.

1.9.1   The Provability Theorem

The full Classical Propositional Calculus based on axioms P1, P2, and P3 is absolutely complete: every tautology is a theorem. Moreover, there are algorithms to determine for each propositional form whether it is a theorem, and, if it is, to design a proof of it. The demonstration relies on the following notation.

1.124 Definition.   For each proposition P, define a proposition P' by

$$P' := \begin{cases} P & \text{if } P \text{ is True,} \\ \neg(P) & \text{if } P \text{ is False.} \end{cases}$$

The following theorem constitutes a first step in a proof of completeness.

1.125 Theorem (Provability Theorem).   *For each propositional form S with propositional variables from a finite list $P, \ldots, R$, there exists a proof of S' from $P', \ldots, R'$:*

$$P', \ldots, R' \vdash S'.$$

Proof (Outline). This proof proceeds by cases, removing from S one connective at each step.

Negation

If S is $\neg(V)$, then V has one fewer connective than S has. Suppose that $P', \ldots, R' \vdash V'$ has already been proved, and consider two cases.

S True   If S is True, then S' is S. However, if S is True, then V is False, and V' is $\neg(V)$, which is S and hence also S'. Thus $P', \ldots, R' \vdash V'$ by the hypothesis on V, and substituting S' for V' yields $P', \ldots, R' \vdash S'$.

S False   In contrast, if S is False, then S' is $\neg(S)$. However, if S is False, then V is True, and V' is V. Thus $P', \ldots, R' \vdash V'$ by the hypothesis on V, and substituting V for V' yields $P', \ldots, R' \vdash V$. Hence, appending a proof of the converse law of double negation (theorem 1.43) produces a proof of $P', \ldots, R' \vdash \{\neg[\neg(V)]\}$, and substituting S for $\neg(V)$ gives $P', \ldots, R' \vdash [\neg(S)]$, whence substituting S' for $\neg(S)$ yields $P', \ldots, R' \vdash S'$.

Implication

If S is $(V) \Rightarrow (W)$, then V and W have fewer connectives than S has. Suppose that $P', \ldots, R' \vdash V'$ and $P', \ldots, R' \vdash W'$ have already been proved, and consider two cases. The first two cases occur if S is True, which occurs if W is True or V is False.

S True, W True   If W is True, then W is W' and by the hypothesis on W there exists a proof of $P', \ldots, R' \vdash W$. Again because W is True, it follows from theorem 1.12 that $(V) \Rightarrow (W)$ is also True, and appending a proof of theorem 1.12 after the proof of $P', \ldots, R' \vdash W$ produces a proof of $P', \ldots, R' \vdash [(V) \Rightarrow (W)]$. However, because $(V) \Rightarrow (W)$ is True and $(V) \Rightarrow (W)$ is S, it also follows that S is S', whence $P', \ldots, R' \vdash [(V) \Rightarrow (W)]$ is $P', \ldots, R' \vdash S'$.

S True, W False   If V is False, then V' is $\neg(V)$ and True. Thus there exists a proof of $P', \ldots, R' \vdash V'$ by the hypothesis on V, which is thus a proof of $P', \ldots, R' \vdash [\neg(V)]$. Hence the law of denial of the antecedent (theorem 1.40) gives a proof of $[\neg(V)] \Rightarrow [(V) \Rightarrow (W)]$, which is $[\neg(V)] \Rightarrow (S)$, and thence the transitivity of implications (theorem 1.16) yields a proof of $P', \ldots, R' \vdash S$, which is also a proof of $P', \ldots, R' \vdash S'$.

S False   The third case occurs if S is False, which occurs if and only if V is True and W is False. Then S' is $\neg(S)$ and W' is $\neg(W)$ but V' is V. By the hypotheses on V and W, there exist proofs of $P', \ldots, R' \vdash V'$ and $P', \ldots, R' \vdash W'$, which are thus proofs of $P', \ldots, R' \vdash V$ and $P', \ldots, R' \vdash [\neg(W)]$. Appending a proof of theorem 1.54 then gives a proof of $P', \ldots, R' \vdash \{(V) \wedge [\neg(W)]\}$, whence the definition (1.51) of $\wedge$ produces a proof of $P', \ldots, R' \vdash \{\neg[(V) \Rightarrow \{\neg[\neg(W)]\}]\}$. Thence the converse law of double negation, transitivity applied to

$$[(V) \Rightarrow (W)] \Rightarrow \{[(W) \Rightarrow \{\neg[\neg(W)]\}] \Rightarrow [(V) \Rightarrow \{\neg[\neg(W)]\}]\}$$

and contraposition yield a proof of $P', \ldots, R' \vdash \{\neg[(V) \Rightarrow (W)]\}$, which is a proof of $P', \ldots, R' \vdash [\neg(S)]$ and hence also a proof of $P', \ldots, R' \vdash S'$.

The general case follows by several applications of the previous cases, in a way that may be specified more explicitly after the availability of the Principle of Mathematical Induction in chapter 4. □

1.9.2 The Completeness Theorem

The **Completeness Theorem** shows that within the full classical propositional calculus every tautology is a theorem, provable from the axioms and the rules of inference.

1.126 Theorem (Completeness Theorem). *Within the full classical propositional calculus, every tautology is a theorem.*

Proof. This proof uses the Deduction Theorem 1.22 and the Provability Theorem 1.125, removing at each step one propositional variable that occurs in a tautology.

For every tautology S with propositional variables $P, \ldots, Q, R$, theorem 1.125 produces a proof of $P', \ldots, Q', R' \vdash S$, because S' is S. Two cases arise with the last variable R.

R True   If R is True, then R' is R, whence from the proof of $P', \ldots, Q', R \vdash S$, the Deduction Theorem gives a proof of $P', \ldots, Q' \vdash [(R) \Rightarrow (S)]$.

R False   If R is False, then R' is $\neg(R)$, whence from the proof of $P', \ldots, Q', R' \vdash S$, the Deduction Theorem gives a proof of $P', \ldots, Q' \vdash \{[\neg(R)] \Rightarrow (S)\}$.

A proof of $P', \ldots, Q' \vdash S$ follows by the principle of proofs by cases (theorem 1.85):

$$\vdash \big([(R) \Rightarrow (S)] \wedge \{[\neg(R)] \Rightarrow (S)\} \big) \Rightarrow (S).$$

Thus the Deduction Theorem reduces the number of propositional variables by 1. Therefore, applying the Deduction Theorem as many times as the number of propositional variables in S yields a proof of $\vdash S$. □

1.9.3 Example: Peirce's Law from the Completeness Theorem

The following considerations demonstrate how to plan the design of a proof by the Completeness Theorem (theorem 1.126), here with the example of Peirce's Law:

$$\{[(P) \Rightarrow (Q)] \Rightarrow (P)\} \Rightarrow (P).$$

To apply the Completeness Theorem, let S designate Peirce's Law. Because S involves only two propositional variables, P and Q, for each of the four combinations of Truth values of P and Q, the Completeness Theorem first invokes the Provability Theorem (theorem 1.125) for a separate proof of $P', Q' \vdash S'$, here

$$P', Q' \vdash \big[\{[(P) \Rightarrow (Q)] \Rightarrow (P)\} \Rightarrow (P) \big]'.$$

In all cases, S has the propositional form $(V) \Rightarrow (W)$, where W is P, and where V is $(H) \Rightarrow (K)$, with $(P) \Rightarrow (Q)$ for H, and P for K:

$$\overbrace{\underbrace{\underbrace{\{[(P) \Rightarrow (Q)]}_{H} \Rightarrow \overbrace{(P)}^{K} \}}_{V} \Rightarrow \underbrace{(P)}_{W}}^{S} \ .$$

P True, *Q* True   If *P* is True, then *W* is also True, because *W* is *P*. Hence the Provability Theorem calls for a proof of $P \vdash W$, which is here $P \vdash P$. Thence the Deduction Theorem provides a proof of $(P) \Rightarrow (W)$, in effect here the proof of theorem 1.14. Because *S* has the form $(V) \Rightarrow (W)$, the proof just obtained gives the following main steps (the final complete proof replaces every theorem cited by a complete proof of that theorem).

$$\vdash (P) \Rightarrow (W) \qquad\qquad\qquad \text{theorem 1.14,}$$
$$\vdash (W) \Rightarrow [(V) \Rightarrow (W)] \qquad\qquad \text{axiom P1,}$$

$$\vdash (P) \Rightarrow [(V) \Rightarrow (W)] \qquad\qquad \text{theorem 1.16,}$$
$$\vdash (P) \Rightarrow \big[\underbrace{\{[(P) \Rightarrow (Q)] \Rightarrow (P)\}}_{V} \Rightarrow \underbrace{(P)}_{W} \big] \ \text{substitutions.}$$

Alternatively axiom P1 yields the conclusion directly, but the foregoing derivation serves to illustrate the use of the Completeness Theorem.

P True, *Q* False   Because *P* is again True, the preceding reasoning remains valid because it does not use the Truth value of *Q*.

P False, *Q* True   If *P* is False, then so is *W*. Hence the Provability Theorem calls for a proof of V'. Here *V* is $[(P) \Rightarrow (Q)] \Rightarrow (P)$, which has the form $(H) \Rightarrow (K)$. With *P* False, *H* is True and *K* is False, whence *V* is False. Consequently, V' is $\neg(V)$, which has the form $\neg[(H) \Rightarrow (K)]$. Therefore, the Provability Theorem calls for proofs of $P', Q' \vdash H$ and $P', Q' \vdash [\neg(K)]$.

Here $P', Q' \vdash [\neg(K)]$ is $[\neg(P)], Q' \vdash \{\neg[\neg(P)]\}$, which follows from the substitution $[\neg(P)] \Rightarrow [\neg(P)]$ in the proof of theorem 1.14.

Also, $P', Q' \vdash H$ is $[\neg(P)], Q' \vdash [(P) \Rightarrow (Q)]$, where *P* is False. Thus the Provability Theorem calls for a proof of $[\neg(P)], Q' \vdash [\neg(P)]$, which again follows from the substitution $[\neg(P)] \Rightarrow [\neg(P)]$ in the proof of theorem 1.14. Hence $[\neg(P)] \Rightarrow [(P) \Rightarrow (Q)]$ by the law of denial of the antecedent (theorem 1.40).

These proofs of $[\neg(P)], Q' \vdash H$ and $[\neg(P)], Q' \vdash [\neg(K)]$ complete the proof of $[\neg(P)], Q' \vdash \{\neg[(H) \Rightarrow (K)]\}$, which is $[\neg(P)], Q' \vdash [\neg(V)]$. Again the law of denial of the antecedent gives a proof of $[\neg(P)], Q' \vdash [(V) \Rightarrow (W)]$, which is

$[\neg(P)], Q' \vdash S$. The proof just obtained gives the following main steps (the final proof replaces every theorem cited by a complete proof of that theorem).

$$\vdash [\neg(P)] \Rightarrow [\neg(K)] \qquad\qquad\qquad \text{theorem 1.14,}$$
$$\vdash [\neg(P)] \Rightarrow [\neg(P)] \qquad\qquad\qquad \text{theorem 1.14,}$$
$$\vdash [\neg(P)] \Rightarrow \{[\neg(Q)] \Rightarrow [\neg(P)]\} \qquad \text{axiom P1,}$$
$$\vdash \{[\neg(Q)] \Rightarrow [\neg(P)]\} \Rightarrow [(P) \Rightarrow (Q)] \ \ \text{axiom P3,}$$
$$\vdash [\neg(P)] \Rightarrow \underbrace{[(P) \Rightarrow (Q)]}_{H} \qquad\qquad \text{theorem 1.16,}$$
$$\vdash [\neg(P)] \Rightarrow (H) \qquad\qquad\qquad\quad \text{substitution;}$$

$$\vdash [\neg(P)] \Rightarrow \{(H) \wedge [\neg(K)]\} \qquad\qquad \text{theorem 1.55,}$$
$$\vdash [\neg(P)] \Rightarrow \{\neg \underbrace{[(H) \Rightarrow (K)]}_{V}\} \qquad\qquad \text{definition of } \wedge,$$
$$\vdash [\neg(P)] \Rightarrow [\neg(V)] \qquad\qquad\qquad \text{substitution;}$$
$$\vdash [\neg(V)] \Rightarrow \{[\neg(W)] \Rightarrow [\neg(V)]\} \qquad \text{axiom P1,}$$
$$\vdash [\neg(P)] \Rightarrow \{[\neg(W)] \Rightarrow [\neg(V)]\} \qquad \text{theorem 1.16,}$$
$$\vdash \{[\neg(W)] \Rightarrow [\neg(V)]\} \Rightarrow [(V) \Rightarrow (W)] \qquad \text{axiom P3,}$$
$$\vdash [\neg(P)] \Rightarrow \underbrace{[(V) \Rightarrow (W)]}_{S} \qquad\qquad\quad \text{theorem 1.16,}$$

$$\vdash [\neg(P)] \Rightarrow \Big[\underbrace{\{[(P) \Rightarrow (Q)] \Rightarrow (P)\}}_{V} \Rightarrow \underbrace{(P)}_{W}\Big] \ \text{substitutions.}$$

P False, Q False   Because P is again False, the preceding reasoning remains valid because it does not use the Truth value of Q.

From the preceding proofs of $(P) \Rightarrow (S)$ and $[\neg(P)] \Rightarrow (S)$, the principle of proofs by cases (theorem 1.85) yields a proof of Peirce's Law (S). Subsequent examinations of the proof produced by the Completeness Theorem can yield simplifications.

$$\vdash (P) \Rightarrow \Big[\{[(P) \Rightarrow (Q)] \Rightarrow (P)\} \Rightarrow (P)\Big] \ \text{axiom P1,}$$
$$\vdash (P) \Rightarrow (S) \qquad\qquad\qquad\qquad \text{substitution;}$$

$$\vdash [\neg(P)] \Rightarrow \{(H) \wedge [\neg(P)]\} \qquad\qquad \text{axiom P1,}$$
$$\vdash [\neg(P)] \Rightarrow \{\neg \underbrace{[(H) \Rightarrow (K)]}_{V}\} \qquad\qquad \text{definition of } \wedge,$$
$$\vdash [\neg(P)] \Rightarrow [\neg(V)] \qquad\qquad\qquad \text{substitution;}$$

$\vdash [\neg(V)] \Rightarrow \{[\neg(W)] \Rightarrow [\neg(V)]\}$ axiom P1,
$\vdash [\neg(P)] \Rightarrow \{[\neg(W)] \Rightarrow [\neg(V)]\}$ theorem 1.16,
$\vdash \{[\neg(W)] \Rightarrow [\neg(V)]\} \Rightarrow [(V) \Rightarrow (W)]$ axiom P3,
$\vdash [\neg(P)] \Rightarrow \underbrace{[(V) \Rightarrow (W)]}_{S}$ theorem 1.16,

$\vdash [\neg(P)] \Rightarrow (S)$ substitution;

$\vdash (P) \Rightarrow (S)$ previous result;
$\vdash S$ rule of inference;
$\vdash \{[(P) \Rightarrow (Q)] \Rightarrow (P)\} \Rightarrow (P)$ substitution.

1.9.4 Exercises on the Deduction Theorem

1.87 . Assume that $[\neg(P)] \Rightarrow (P)$ holds and prove that P holds. In other words, prove that $\{[\neg(P)] \Rightarrow (P)\} \vdash (P)$.

1.88 . Assume that $(P) \Rightarrow [\neg(P)]$ holds and prove that $\neg(P)$ holds. In other words, prove that $\{(P) \Rightarrow [\neg(P)]\} \vdash [\neg(P)]$.

1.89 . Apply the Deduction Theorem to prove $\{[\neg(P)] \Rightarrow (P)\} \Rightarrow (P)$.

1.90 . Apply the Deduction Theorem to prove $\{(P) \Rightarrow [\neg(P)]\} \Rightarrow [\neg(P)]$.

1.91 . Assume $(P) \Rightarrow (Q)$ and prove $\{(P) \Rightarrow [\neg(Q)]\} \Rightarrow [\neg(P)]$. In other words, prove that $[(P) \Rightarrow (Q)] \vdash (\{(P) \Rightarrow [\neg(Q)]\} \vdash [\neg(P)])$.

1.92 . Apply the Deduction Theorem to prove the law of reductio ad absurdum

$$[(P) \Rightarrow (Q)] \Rightarrow (\{(P) \Rightarrow [\neg(Q)]\} \Rightarrow [\neg(P)]).$$

1.93 . Assume $[\neg(P)] \Rightarrow (Q)$ and prove $\{[\neg(P)] \Rightarrow [\neg(Q)]\} \Rightarrow (P)$. In other words, prove that $\{[\neg(P)] \Rightarrow (Q)\} \vdash [\{[\neg(P)] \Rightarrow [\neg(Q)]\} \Rightarrow (P)]$.

1.94 . Apply the Deduction Theorem to prove the law of indirect proof

$$\{[\neg(P)] \Rightarrow [\neg(Q)]\} \Rightarrow [\{[\neg(P)] \Rightarrow [\neg(Q)]\} \Rightarrow (P)].$$

1.95 . Apply the Deduction Theorem to prove the tautology

$$\{[(P) \Rightarrow (Q)] \Rightarrow [(Q) \Rightarrow (R)]\} \Rightarrow [(Q) \Rightarrow (R)].$$

1.96 . Apply the Deduction Theorem to prove the tautology

$$\{[(P) \Rightarrow (Q)] \Rightarrow [(Q) \Rightarrow (R)]\} \Rightarrow \{[(P) \Rightarrow (Q)] \Rightarrow [(P) \Rightarrow (R)]\}.$$

1.97 . Apply the Completeness Theorem to prove $\{[\neg(P)] \Rightarrow (P)\} \Rightarrow (P)$.

1.98 . Apply the Completeness Theorem to prove the special law of reductio ad absurdum: $\{(P) \Rightarrow [\neg(P)]\} \Rightarrow [\neg(P)]$.

1.99 . Apply the Completeness Theorem to prove the law of reductio ad absurdum: $[(P) \Rightarrow (Q)] \Rightarrow (\{(P) \Rightarrow [\neg(Q)]\} \Rightarrow [\neg(P)])$.

1.100 . Apply the Completeness Theorem to prove the law of assertion:

$$(P) \Rightarrow \{[(P) \Rightarrow (Q)] \Rightarrow (Q)\}.$$

1.101 . Apply the Completeness Theorem to prove

$$\{[(P) \Rightarrow (Q)] \Rightarrow (R)\} \Rightarrow \{[(R) \Rightarrow (P)] \Rightarrow (P)\}.$$

1.102 . Apply the Completeness Theorem to prove

$$\{[(P) \Rightarrow (Q)] \Rightarrow (R)\} \Rightarrow \{[(R) \Rightarrow (P)] \Rightarrow [(S) \Rightarrow (P)]\}.$$

1.103 . Apply the Completeness Theorem to prove

$$\big[\{[(P) \Rightarrow (R)] \Rightarrow (Q)\} \Rightarrow (Q)\big] \Rightarrow \{[(Q) \Rightarrow (R)] \Rightarrow [(P) \Rightarrow (R)]\}.$$

1.104 . Apply the Completeness Theorem to prove

$$[(R) \Rightarrow (Q)] \Rightarrow (\{[(R) \Rightarrow (Q)] \Rightarrow (P)\} \Rightarrow [(S) \Rightarrow (P)]).$$

1.105 . Apply the Completeness Theorem to prove

$$\{[(R) \Rightarrow (Q)] \Rightarrow [(S) \Rightarrow (P)]\} \Rightarrow \{[(R) \Rightarrow (P)] \Rightarrow [(S) \Rightarrow (P)]\}.$$

1.106 . Apply the Completeness Theorem to prove

$$(\{[(R) \Rightarrow (P)] \Rightarrow (P)\} \Rightarrow [(S) \Rightarrow (P)]) \Rightarrow (\{[(P) \Rightarrow (Q)] \Rightarrow (R)\} \Rightarrow [(S) \Rightarrow (P)]).$$

Chapter 2
First-Order Logic: Proofs with Quantifiers

2.1 Introduction

This chapter introduces quantifiers and first-order logic. The first few sections demonstrate methods for designing proofs through preliminary versions of the Deduction Theorem for first-order logic, Substitutivity of Equivalences, and transformations into prenex forms. A final section derives features of predicates for equality and inequality, either as primitive predicate constants, or predicates defined from other primitive binary predicate constants. The prerequisite for this chapter is a working knowledge of the Classical Propositional Logic for instance, as in chapter 1.

Pure first-order logic includes *quantifiers* corresponding to phrases such as "for each object" or "there exists an (at least one) object" with templates for functions of objects and relations between objects. Applied first-order logic replaces such templates with functions and relations specific to areas such as algebra, arithmetic, geometry, or set theory.

2.2 The Pure Predicate Calculus of First Order

In grammar, the noun "predicate" designates the verb or verbal phrase that makes a statement about the subject of a clause. In logic, similarly, a predicate is a part of an atomic formula that makes a statement about individual objects in applications.

2.2.1 Logical Predicates

The logical concept of **predicate** depends upon the theory under consideration.

© Springer Science+Business Media New York 2015
Y. Nievergelt, *Logic, Mathematics, and Computer Science*,
DOI 10.1007/978-1-4939-3223-8_2

2.1 Example (predicates in arithmetic). Some versions of arithmetic have only two predicates, which state that a number is the sum or product of two numbers:

$M = K + L$ (read "*M* equals the sum of *K* and *L*"),

$N = K * L$ (read "*N* equals the product of *K* and *L*"),

or equivalent formulae with a different notation [18, p. 318], [72, p. 202–203].

These predicates are called "ternary" because each involves three variables.

2.2 Example (predicates in geometry). In geometry, a predicate may state that a point is on a line, or that a point lies between two other points on the same line, or that two segments, or two angles, are congruent, or that a line lies in a plane:

$P \in L$ (read "the point *P* is on the line *L*"),

$X < Y < Z$ (read "the point *Y* is between the points *X* and *Z*"),

$PQ \equiv RS$ (read "the segment *PQ* is congruent to the segment *RS*"),

$\angle ABC \equiv \angle PQR$ (read "the angle *ABC* is congruent to the angle *PQR*"),

$L \subset E$ (read "the line *L* lies in the plane *E*"),

or equivalent formulations with a different notation [61, Ch. I].

2.3 Example (predicates in set theory). Some versions of set theory have only one predicate, which states that a set is an element of a set:

$X \in Y$ (read "*X* is an element of *Y*"),

$\varnothing \in Y$ (read "the empty set is an element of *Y*"),

$X \in \varnothing$ (read "*X* is an element of the empty set"),

$\varnothing \in \varnothing$ (read 'the empty set is an element of the empty set").

This predicate is called "binary" because it involves two variables, *X* and *Y*.

The formulae in the foregoing examples are called **terms** or **atomic formulae** because they are the simplest formulae in arithmetic, geometry, and set theory. Thus, $X \in Y$ is a term, or, in other words, an atomic formula.

In arithmetic, the symbols 0 and 1 are called **individual constants**, because they always denote the numbers zero and one, respectively. Similarly, in set theory, the symbol $\varnothing$ is an **individual constant**, because it always denotes the empty set. In contrast, the symbols = and $\in$ are called **predicate constants**, because they always denote the relations of equality and set membership, respectively. Logics that include such constants are called **applied** predicate calculi; they may also include other **functional constants** or **relational constants** corresponding to other relations between objects. In contrast, logics that do not include any constants but allow for variables representing arbitrary individuals, predicates, functions, and relations are called **pure** predicate calculi. Thus a pure predicate calculus is a general logic that may later apply to algebra, arithmetic, geometry, and set theory as well.

In applied logics, if an atomic formula contains a variable, then it may, but need not, have a Truth value. For example, the formula $X \in Y$ has no Truth value, because different substitutions for *X* and *Y* can yield different Truth values. However some formulae may contain variables and yet have a Truth value.

2.4 Example. In logics with an "equality" relation, the formula $X = X$ is True for every X.

2.5 Example. In binary arithmetic the formula $0 * X = 0$ is True for every X.

2.6 Example. In binary arithmetic $0 * X = 1$ is False for every X.

2.7 Example. In set theory the formula $X \in \varnothing$ is False for every X.

2.8 Example. In the theory of well-formed sets the formula $X \in X$ is False for every X.

2.2.2 Variables, Quantifiers, and Formulae

The formulae studied here are those specified in definition 2.9.

2.9 Definition (well-formed formulae). Select *three* disjoint lists of symbols.

Every symbol from the *first* list of symbols, which may consist of one or more letter(s) from a specified alphabet, $P, Q, \ldots$, optionally with subscript(s) $P_\flat, P_{\flat\flat}, \ldots$, superscript(s) $P^\sharp, P^{\sharp\sharp}, \ldots$, or "middlescript(s)" $P|, P||, \ldots$, is called a **formulaic letter**. Such formulaic letters are not parts of the predicate calculus, but they help in describing the following rules to define well-formed formulae.

Also, every symbol from the *second* list of symbols, which may consist of one or more letter(s) from a specified alphabet, $A, B, \ldots$, optionally with subscript(s) $A_\flat, A_{\flat\flat}, \ldots$, superscript(s) $A^\sharp, A^{\sharp\sharp}, \ldots$, or "middlescript(s)" $A|, A||, \ldots$, is called a **propositional variable** or a **sentence symbol** [31, p. 17]. (Propositional variables may later be replaced in pure calculi by functional or relational variables, or in applied calculi by atomic formulae, which may include individual variables or constants specific to applications.)

Moreover, every symbol from the *third* list of symbols, which may consist of one or more letter(s) from a specified alphabet, $X, Y, \ldots$, optionally with subscript(s) $X_\flat, X_{\flat\flat}, \ldots$, superscript(s) $X^\sharp, X^{\sharp\sharp}, \ldots$, or "middlescript(s)" $X|, X||, \ldots$, is called an **individual variable**. (Individual variables may later be replaced by items specific to applications, for instance, numbers in arithmetic, or points in geometry.)

Every propositional variable or atomic formula is a **well-formed formula**. For all well-formed formulae P and Q, and for every individual variable X, the following four strings of symbols are also well-formed formulae:

$(W1) \quad \neg(P) \quad$ (read "not P"),
$(W2) \ (P) \Rightarrow (Q) \quad$ (read "P implies Q" or "if P, then Q"),
$(W3) \quad \forall X(P) \quad$ (read "for each X, P"),
$(W4) \quad \exists X(P) \quad$ (read "there exists X such that P").

Furthermore, only strings of symbols built from letters or variables through applications of the rules W1–W4 can be well-formed formulae. Equivalent definitions apply to other connectives and to prefix and postfix notations.

2.2.3 Proper Substitutions of Free or Bound Variables

In the logic presented here, only individual variables may appear immediately after either quantifier, $\forall$ (read "for each") or $\exists$ (read "there exists"). Because of this restriction, this logic is a **first order logic**. Logical systems allowing for propositional variables to appear immediately after a quantifier are of second or higher order.

In Boolean logic, if a formula P is True regardless of X, but if P also contains another variable Z, then substituting Z for X can change the Truth value of P.

2.10 Counterexample. Consider any context with at least two different objects, for instance, two binary numbers in arithmetic, two points in geometry, or two sets in set theory. Thus for each object X there exists a *different* object Z, whence $\forall X\{\exists Z[\neg(X = Z)]\}$ is True. Replacing Z by X in $\exists Z[\neg(X = Z)]$ gives $\exists X[\neg(X = X)]$, which is False, because each object equals itself, by example 2.4. Thus, replacing Z by X in the True formula $\forall X\{\exists Z[\neg(X = Z)]\}$ yields the False formula $\forall X\{\exists X[\neg(X = X)]\}$. Similarly, replacing X by Z in the True formula $\forall X\{\exists Z[\neg(X = Z)]\}$ gives the False formula $\forall Z\{\exists Z[\neg(Z = Z)]\}$.

One way (not pursued here) to avoid the phenomenon exhibited in counterexample 2.10 consists of substituting parameters other than variables [117]. Alternatively, counterexample 2.10 shows that substitutions of a variable by another must obey certain rules, for instance, with the concepts introduced in definition 2.11.

2.11 Definition (free or bound variables). For each individual variable X and for each logical formula P, an occurrence of the variable X is **bound** in the formula P if and only if in P that occurrence of the variable X immediately follows $\forall$ or $\exists$, or if it appears in the **scope** of the quantifier, which is defined to be between either $\forall X($ or $\exists X($ and the corresponding right parenthesis $)$). An occurrence of the variable X is **free** in P if and only if that occurrence of X is not bound in P. A logical formula is **closed** if and only if it does not contain any free occurrence of any variable. A logical **sentence** is a closed logical formula.

2.12 Example. This example focuses on the formula from counterexample 2.10.

In the formula $\exists Z[\neg(X = Z)]$, both occurrences of the variable Z are *bound*.
In the formula $\exists Z[\neg(X = Z)]$, the only occurrence of the variable X is *free*.
The formula $\exists Z[\neg(X = Z)]$ is *not* closed, because it contains a free occurrence of X.

In contrast, the formula $\forall X\{\exists Z[\neg(X = Z)]\}$ is closed, because all occurrences of X and Z are bound.

The following definitions avoid the phenomenon in counterexample 2.10.

2.13 Definition (change of bound variables). The substitution of the variable Z for each *bound* occurrence of the variable X in P is denoted by $\mathrm{Subb}_Z^X(P)$. Such a substitution is called a **change of bound variables** if and only if Z does not occur in P. Such a change of bound variables is **proper** if and only if X does not occur freely in P and Z does not occur in P.

2.14 Example. In the formula $\exists Z[\neg(X = Z)]$ from counterexample 2.10, the substitution of X for the *bound* occurrences of Z is *not* a change of bound variables, because X already occurs in the formula.

Definition 2.13 explicitly applies only to variables. In particular, it does *not* allow for substitutions of constants for bound variables.

2.15 Remark. If $\mathrm{Subb}_Z^X(P)$ is a change of bound variables in a formula P, then $\mathrm{Subb}_X^Z[\mathrm{Subb}_Z^X(P)]$ reproduces P.

Indeed, if $\mathrm{Subb}_Z^X(P)$ is a change of bound variables, then Z does not occur in P. Consequently, the only occurrences of Z in $\mathrm{Subb}_Z^X(P)$ are those replacing bound occurrences of X. Therefore, Z does not occur freely in $\mathrm{Subb}_Z^X(P)$. Thus, $\mathrm{Subb}_X^Z[\mathrm{Subb}_Z^X(P)]$ replaces *all* the occurrences of Z in $\mathrm{Subb}_Z^X(P)$, all of which are bound, by the initially bound occurrences of X, and reverts to the initial formula P.

Depending on the axiomatic system, an axiom, theorem, or inference rule may declare that two formulae that differ from each other only by changes of bound variables are mutually equivalent, so that $(P) \Leftrightarrow [\mathrm{Subb}_Z^X(P)]$ [89, p. 181].

2.16 Definition (change of free variables). A formula P **admits** (the substitution of) Z for an individual variable X, or, in other words, Z is **free** (to be substituted) **for** X in P, if and only if in the substitution of Z for every free occurrence of X

- every free occurrence of X becomes a free occurrence of Z,

or, equivalently, if X, Z, and P satisfy the following condition:

- either Z is an individual constant, or
- Z is an individual variable, and substituting Z for every *free* occurrence of X in P does not convert any free occurrence of X into a bound occurrence of Z.

In the present exposition, the notation $\mathrm{Subf}_Z^X(P)$ states that P admits Z for X, and substitutes Z for each free occurrence of X in P.

2.17 Example. In counterexample 2.10, the formula $\exists Z[\neg(X = Z)]$ does *not* admit Z for X, or, in other words, Z is *not* free for X, because substituting Z for every free occurrence of X converts the free occurrence of X into a bound occurrence of Z in $\exists Z[\neg(Z = Z)]$.

Thus another way to avoid the phenomenon exhibited in counterexample 2.10 consists in allowing only substitutions *admitted* in the sense of definition 2.16 [72, p. 94], [84, p. 48] [108, p. 37], [110, p. 101].

2.18 Remark. If an individual variable Z does not occur in a formula P, then P admits Z for each individual variable X. Indeed, Z does not appear in P after any quantifier ($\forall$ or $\exists$); thus Z is not bound in P, so that every free occurrence of X is replaced by a free occurrence of Z in $\text{Subf}_Z^X(P)$.

2.19 Remark. If an individual variable Z does not occur in a formula P, and if P does not contain any bound occurrences of an individual variable X, then $\text{Subf}_X^Z[\text{Subf}_Z^X(P)]$ is P. Indeed, by hypothesis all occurrences of X are free in P, and by remark 2.18 they all become free occurrences of Z in $\text{Subf}_Z^X(P)$. Also, X does not occur in $\text{Subf}_Z^X(P)$. Again by remark 2.18 but with X and Z swapped, all occurrences of Z in $\text{Subf}_Z^X(P)$ become free occurrences of X in $\text{Subf}_X^Z[\text{Subf}_Z^X(P)]$, which no longer contain any occurrences of Z. Thus $\text{Subf}_X^Z[\text{Subf}_Z^X(P)]$ is P.

2.20 Remark. All well-formed formulae P and Q result from the construction specified in definition 2.9, so that individual variables occur only immediately after quantifiers, or in terms (atomic formulae) that are then combined with connectives. Consequently, the following pairs of formulae are not only mutually equivalent but also mutually identical [117, p. 44]:

- $\text{Subf}_Z^X[\neg(P)]$ and $\neg[\text{Subf}_Z^X(P)]$.
- $\text{Subf}_Z^X[(P) \Rightarrow (Q)]$ and $[\text{Subf}_Z^X(P)] \Rightarrow [\text{Subf}_Z^X(Q)]$.
- $\text{Subb}_Z^X[\neg(P)]$ and $\neg[\text{Subb}_Z^X(P)]$.
- $\text{Subb}_Z^X[(P) \Rightarrow (Q)]$ and $[\text{Subb}_Z^X(P)] \Rightarrow [\text{Subb}_Z^X(Q)]$.

The following abbreviation is convenient.

2.21 Definition (abbreviation). The notation $\exists!X(P)$ (read "there exists a unique X such that P" or "there exists exactly one X such that P") abbreviates the formula

$$\exists X\big[(P) \wedge \big(\forall Z\{[\text{Subf}_Z^X(P)] \Rightarrow (Z = X)\}\big)\big].$$

2.2.4 Axioms and Rules for the Pure Predicate Calculus

As the axioms of the propositional calculus reflect patterns of deductive reasoning with implications and negations, the axioms of the predicate calculus reflect patterns of deductive reasoning with quantifiers. There also exist several choices of initial axioms for use with quantifiers, for instance, the following axioms [18, §30, p. 171–172], [122, p. 170].

2.22 Definition. The following **axioms of the Pure Predicate Calculus** govern the use of the **universal quantifier** $\forall$ and the **existential quantifier** $\exists$.

Axiom Q0 *Axioms of the Propositional Calculus, but here with well-formed formulae as in definition 2.9, are axioms of the predicate calculus.*

Axiom Q1 (specialization) $[\forall X(P)] \Rightarrow [Subf_Z^X(P)]$, *if P admits Z for X.*

Axiom Q2 $\{\forall X[(P) \Rightarrow (Q)]\} \Rightarrow \{(P) \Rightarrow [\forall X(Q)]\}$, *if P contains no free X.*

Axiom Q3 $\{\forall X[\neg(P)]\} \Leftrightarrow \{\neg[\exists X(P)]\}$.

Axiom Q4 $\{\exists X[\neg(P)]\} \Leftrightarrow \{\neg[\forall X(P)]\}$.

The first axiom (schema) of the predicate calculus (Q0) and the rules of inference carry all the theorems from the propositional calculus over to theorems of the predicate calculus. In particular, different propositional calculi, which may result from different axiom systems, may lead to different predicate calculi.

2.23 Example. From definition 1.4, the following formulae (P1) and (P2) form a system of two axioms for the Pure Positive Implicational Propositional Calculus:

> (P1) $(P) \Rightarrow [(Q) \Rightarrow (P)]$,
>
> (P2) $\{(P) \Rightarrow [(Q) \Rightarrow (R)]\} \Rightarrow \{[(P) \Rightarrow (Q)] \Rightarrow [(P) \Rightarrow (R)]\}$.

Formulae (P1), (P2), and

$$\text{(P3)} \quad \{[\neg(Q)] \Rightarrow [\neg(P)]\} \Rightarrow [(P) \Rightarrow (Q)],$$

form a system of three axioms for the Pure Classical Propositional Calculus.

The second axiom (schema) (Q1) corresponds to the notion that if an individual variable X may occur in a formula P, and if P is True regardless of X, in other words, if $\forall X(P)$ is True, then P remains True with X replaced by any individual variable or constant Z. If X and Z are the same variable, then axiom Q1 gives $[\forall X(P)] \Rightarrow (P)$.

The third axiom (schema) (Q2) describes the relation between the universal quantifier ("for each") and the logical connective of the Pure Positive Implicational Propositional Calculus ("if . . . then").

The fourth axiom, for the existential quantifier (Q3), states that a formula P is False for every X if and only if there does *not* exist any X for which P is True.

Similarly, the fifth axiom (schema), for the existential quantifier (Q4), states that there exists some X for which P is False if and only if it is False that P is True for every X. Axiom Q4 asserts the existence of an object. Consequently, axiom Q4 applies neither to "empty" theories where nothing exists, nor to logics that require not only existence but also the determination of which objects satisfy a formula.

Besides the axioms, the predicate calculus allows for proofs of theorems through the following **rules of inference**.

2.24 Definition (rules of inference). The following **rules of inference** hold.

2.25 Rule ("Modus Ponens" (abbreviated by M. P.), or "Detachment").

If P is a theorem, and
if $(P) \Rightarrow (Q)$ is a theorem,
then Q is a theorem.

2.26 Rule (Generalization).

> If P is a theorem,
> then $\forall X(P)$ is also a theorem.

2.27 Definition (theorems and proofs). A **proof** is a sequence of well-formed formulae H, K, L, ... P, Q, R, where each formula is either (a substitution in) an axiom (schema), or results from a previous formula in the sequence by any rule of inference (*Detachment*, *Generalization*, or Substitution).

A formula is a **theorem** if and only if it is a (usually the last) formula in a proof. The notation $\vdash R$ means that R is a theorem.

2.28 Example. Every axiom of the predicate calculus is a theorem.

2.2.5   Exercises on Quantifiers

Each of the following ten exercises lists one formula P. Identify a formula that is logically equivalent to $\neg(P)$ among the same ten exercises.

2.1 . $\forall X[\exists Y(X \in Y)]$

2.2 . $\forall X[\exists Y(Y \in X)]$

2.3 . $\forall X[(X \in A) \Rightarrow (X \in B)]$

2.4 . $\forall X\{(X \in C) \Leftrightarrow [(X \in A) \wedge (X \in B)]\}$

2.5 . $\forall X\{(X \in C) \Leftrightarrow [(X \in A) \vee (X \in B)]\}$

2.6 . $\exists X\big(\{(X \in C) \wedge [\neg(X \in A)] \wedge [\neg(X \in B)]\} \vee \{[\neg(X \in C)] \wedge [(X \in A) \vee (X \in B)]\}\big)$

2.7 . $\exists X\{(X \in A) \wedge [\neg(X \in B)]\}$

2.8 . $\exists X\{\forall Y[\neg(Y \in X)]\}$

2.9 . $\exists X\big([(X \in C) \wedge \{[\neg(X \in A)] \vee [\neg(X \in B)]\}] \vee \{[\neg(X \in C)] \wedge [(X \in A) \wedge (X \in B)]\}\big)$

2.10 . $\exists X\{\forall Y[\neg(X \in Y)]\}$

2.2.6   Examples with Implicational and Predicate Calculi

The examples of theorems and proofs selected for this and the subsequent subsections gradually build up a tool to design proofs by substituting mutually equivalent formulae for one another. As a first step, the following derived rules of inference will simplify proofs by avoiding potentially lengthy instances of axiom Q2.

2.29 Theorem (derived rule). *If X is not free in R, and if $\forall X[(R) \Rightarrow (S)]$ is a theorem, then $(R) \Rightarrow [\forall X(S)]$ is also a theorem.*

Proof. Apply axiom Q2 and *Detachment*:

$\vdash \forall X[(R) \Rightarrow (S)]$ hypothesis,

$\vdash \{\forall X[(R) \Rightarrow (S)]\} \Rightarrow \{(R) \Rightarrow [\forall X(S)]\}$ axiom Q2, no free X in R,

$\vdash (R) \Rightarrow [\forall X(S)]$ *Detachment*.

$\square$

2.30 Theorem (derived rule). *If X does not occur freely in R, and if $(R) \Rightarrow (S)$ is a theorem, then $(R) \Rightarrow [\forall X(S)]$ is also a theorem.*

Proof. Apply *Generalization* and theorem 2.29:

$\vdash (R) \Rightarrow (S)$ hypothesis,

$\vdash \forall X[(R) \Rightarrow (S)]$ *Generalization*,

$\vdash (R) \Rightarrow [\forall X(S)]$ theorem 2.29, no free X in R.

$\square$

Theorem 2.31 reveals a situation where $\forall Y[\text{Subf}_Y^X(U)]$ may be replaced by $\forall X[\text{Subf}_X^X(U)]$, which is $\forall X(U)$.

2.31 Theorem (change of bound variables). *If Y does not occur in U, then $\vdash [\forall X(U)] \Rightarrow \{\forall Y[\text{Subf}_Y^X(U)]\}$.*

If Y does not occur in U, and if U does not contain any bound occurrence of X, then conversely $\vdash [\forall X(U)] \Leftarrow \{\forall Y[\text{Subf}_Y^X(U)]\}$.

Proof. This proof follows Monk's [89, p. 180, thm. 10.55].

$\vdash \underbrace{[\forall X(U)]}_{R} \Rightarrow \underbrace{[\text{Subf}_Y^X(U)]}_{S}$ specialization (axiom Q1), no Y in U,

$\vdash \overbrace{[\forall X(U)]}^{R} \Rightarrow \overbrace{\{\forall Y[\text{Subf}_Y^X(U)]\}}^{\forall Y[S]}$ theorem 2.29, no Y in U, so no Y in R.

For the converse, if Y does not occur in U, and if U does not contain any bound occurrence of X, then each occurrence of X in U is replaced by a free occurrence of Y in $\text{Subf}_Y^X(U)$. Consequently, if V denotes the formula $\text{Subf}_Y^X(U)$, then V contains no occurrences of X and no bound occurrences of Y. Moreover, $\text{Subf}_X^Y[\text{Subf}_Y^X(U)]$ reproduces U, by remark 2.19. Thus, applying the previous result to the formula V and swapping the roles of X and Y give $\vdash [\forall Y(V)] \Rightarrow \{\forall X[\text{Subf}_X^Y(V)]\}$. $\square$

2.32 Remark (change of bound variables). Theorem 2.31 shows that two formulae P and Q that differ from each other only by the names of their bound variables are mutually equivalent. Indeed, if the variable Y does not occur in a formula P, and if U is any atomic formula (for instance, $X \in Z$) that occurs in P, then U does not contain any bound variables; in particular, U does not contain any bound occurrence of X. Consequently $\vdash [\forall X(U)] \Leftrightarrow \{\forall Y[\text{Subf}_Y^X(U)]\}$ by theorem 2.31. This substitution may use different new variables $Y, Y_\flat, Y_{\flat\flat}, \ldots$, for different atomic formulae $U, U_\flat, U_{\flat\flat}, \ldots$ and then proceed to more complicated components of P. Using the same new variables in P and Q results in two identical formulae.

2.33 Example. Let P denote the formula $\forall X\{\exists Z[\neg(X = Z)]\}$, and let Q denote the formula $\forall W\{\exists Y[\neg(W = Y)]\}$. The variables $Y_\flat$ and $Y_{\flat\flat}$ occur in neither P nor Q.

In P, let $U_\flat$ denote the atomic formula $X = Z$. Then $\vdash$ $[\exists Z(U_\flat)]$ $\Leftrightarrow$ $\{\exists Y_\flat[\mathrm{Subf}^Z_{Y_\flat}(U_\flat)]\}$ by theorem 2.31. Let $U_{\flat\flat}$ denote the resulting formula $\exists Y_\flat[\neg(X = Y_\flat)]$. Then theorem 2.31 shows that $\vdash [\forall X(U_{\flat\flat})]$ $\Leftrightarrow$ $\{\forall Y_{\flat\flat}[\mathrm{Subf}^X_{Y_{\flat\flat}}(U_{\flat\flat})]\}$, which is $\forall Y_{\flat\flat}\{\exists Y_\flat[\neg(Y_{\flat\flat} = Y_\flat)]\}$. The same formula results from the same procedure applied to Q.

The following selection of theorems also relates the present axioms to other axiom systems in subsection 2.2.8. Their proofs follow Church's [18, p. 186–188].

2.34 Theorem. *For all P, Q, and X,* $\vdash \{\forall X[(P) \Rightarrow (Q)]\} \Rightarrow \{[\forall X(P)] \Rightarrow (Q)\}$.

Proof. Apply the Implicational Calculus with axiom Q1:

$\vdash \{\forall X[(P) \Rightarrow (Q)]\} \Rightarrow [(P) \Rightarrow (Q)]$ axiom Q1,

$\vdash [\forall X(P)] \Rightarrow (P)$ axiom Q1,

$\vdash \{\forall X[(P) \Rightarrow (Q)]\} \Rightarrow \{[\forall X(P)] \Rightarrow (Q)\}$ derived rule (theorem 1.31). $\qquad\square$

2.35 Theorem. *For all P, Q, and X,* $\vdash \{\forall X[(P) \Rightarrow (Q)]\} \Rightarrow \{[\forall X(P)] \Rightarrow [\forall X(Q)]\}$.

Proof. Apply the Implicational Calculus with *Generalization*, axiom Q2, and theorems 2.30 and 2.34:

$\vdash \underbrace{\{\forall X[(P) \Rightarrow (Q)]\}}_{\substack{R \\ R}} \Rightarrow \underbrace{\{[\forall X(P)] \Rightarrow (Q)\}}_{\substack{S \\ \forall X\{S\}}}$ theorem 2.34,

$\vdash \overbrace{\{\forall X[(P) \Rightarrow (Q)]\}}^{R} \Rightarrow \overbrace{\left(\forall X\{[\forall X(P)] \Rightarrow (Q)\}\right)}^{\forall X\{S\}}$ theorem 2.30,

$\vdash \left(\forall X\{[\forall X(P)] \Rightarrow (Q)\}\right) \Rightarrow \{[\forall X(P)] \Rightarrow [\forall X(Q)]\}$ axiom Q2,

$\vdash \{\forall X[(P) \Rightarrow (Q)]\} \Rightarrow \{[\forall X(P)] \Rightarrow [\forall X(Q)]\}$ transitivity (1.16). $\qquad\square$

2.36 Counterexample. The converse of theorem 2.35, which would be

$$[\forall X(P) \Rightarrow \forall X(Q)] \Rightarrow \{\forall X[(P) \Rightarrow (Q)]\},$$

is *False* in contexts with two *different* objects Y and Z, so that $\neg(Y = Z)$ is True:

- $\forall X[(X = Y) \Rightarrow (X = Z)]$ is False, because substituting Y for X gives $[(Y = Y) \Rightarrow (Y = Z)]$, which is False, because of the True hypothesis $Y = Y$ and the False conclusion $Y = Z$.
- $\forall X(X = Y)$ is False, because substituting Z for X gives $(Z = Y)$, which is False by the assumption that $\neg(Y = Z)$.
- $[\forall X(X = Y)] \Rightarrow [\forall X(X = Z)]$ is True, because of its False hypothesis.
- $\{[\forall X(X = Y)] \Rightarrow [\forall X(X = Z)]\} \Rightarrow \{\forall X[(X = Y) \Rightarrow (X = Z)]\}$ is False, because of the True hypothesis and the False conclusion.

2.37 Theorem (derived rule).   *For all P, Q, and X, if ⊢ (P) ⇒ (Q), then*
⊢ [∀X(P)] ⇒ (Q) *and* ⊢ [∀X(P)] ⇒ [∀X(Q)] .

Proof.   Apply theorems 2.34 and 2.35:

⊢ (P) ⇒ (Q)	hypothesis,
⊢ ∀X[(P) ⇒ (Q)]	*Generalization,*
⊢ {[∀X(P)] ⇒ (Q)}	theorem 2.34 and *Detachment,*
⊢ [∀X(P)] ⇒ [∀X(Q)]	theorem 2.35 and *Detachment.*

$\square$

2.38 Theorem.   *For all P, Q, and X, if P does not contain any free occurrence of X,*
then ⊢ {(P) ⇒ [∀X(Q)]} ⇔ {∀X[(P) ⇒ (Q)]} .

Proof.   Axiom Q2 gives ⊢ {∀X[(P) ⇒ (Q)]} ⇒ {(P) ⇒ [∀X(Q)]}. For the
converse, use the Pure Positive Implicational Propositional Calculus with axioms Q1
and *Generalization*:

⊢ [∀X(Q)]⇒(Q)	axiom Q1,
⊢ {(P)⇒[∀X(Q)]}⇒{(P)⇒[∀X(Q)]}	theorem 1.14,
⊢ {(P)⇒[∀X(Q)]}⇒[(P)⇒(Q)]	theorem 1.32,
⊢ ∀X({(P)⇒[∀X(Q)]}⇒[(P)⇒(Q)])	*Generalization,*

$$\underbrace{\overbrace{\{(P){\Rightarrow}[\forall X(Q)]\}}^{R}}_{R} \Rightarrow \underbrace{(\overbrace{\forall X\{[(P){\Rightarrow}(Q)]\}}^{\forall X\{S\}})}$$

⊢ {(P)⇒[∀X(Q)]}⇒(∀X{[(P)⇒(Q)]})   theorem 2.29, no free X in (P)⇒[∀X(Q)].

$\square$

2.39 Theorem.   *For all P and X, if P does not contain any free occurrence of X,*
then ⊢ [∀X(P)] ⇒ (P) *and* ⊢ [∀X(P)] ⇐ (P) .

Proof.   Axiom Q1 gives ⊢ [∀X(P)] ⇒ (P). For the converse, apply theorems 1.14
and 2.30:

⊢ (P) ⇒ (P)	theorem 1.14,
⊢ (P) ⇒ [∀X(P)]	theorem 2.30.

$\square$

2.40 Remark.   The statement of theorem 2.39 suggests that if *P* contains a free
occurrence of *X*, then the implication (P) ⇒ [∀X(P)] may differ from the
Generalization rule, from ⊢ *P* to infer ⊢ ∀X(P), which applied only if *P* is a
theorem.

2.41 Example.   If *P* denotes the formula *X* = ∅, then (P) ⇒ [∀X(P)] becomes
(X = ∅) ⇒ [∀X(X = ∅)], which is not a theorem. Indeed, if (X = ∅) ⇒
[∀X(X = ∅)] were a theorem, then substituting ∅ for the free occurrences of *X* by
specialization and *Detachment* would yield (∅ = ∅) ⇒ [∀X(X = ∅)], which is
not a theorem, because ∅ = ∅ is True while ∀X(X = ∅) is False in set theory.

Theorem 2.42 provides a converse for theorem 2.35 if *X* is not free in *P*.

2.42 Theorem.   *For all P, Q, and X, if X does not occur freely in P, then* ⊢
{[∀X(P)] ⇒ [∀X(Q)]} ⇒ {∀X[(P) ⇒ (Q)]}.

Proof. Use theorems 2.38 and 2.39, with H, K, L, M as in theorem 1.36:

$$\vdash \{(\underbrace{\ \ P\ \ }_{H}) \Rightarrow \underbrace{[\forall X(Q)]}_{L}\} \Rightarrow \underbrace{\{\forall X[(P) \Rightarrow (Q)]\}}_{M} \qquad \text{theorem 2.38, no free } X \text{ in } P,$$

$$\vdash (\underbrace{\ \ P\ \ }_{H}) \Rightarrow \underbrace{[\forall X(P)]}_{K} \qquad\qquad\qquad\quad \text{theorem 2.39, no free } X \text{ in } P,$$

$$\vdash \{\underbrace{[\forall X(P)]}_{K} \Rightarrow \underbrace{[\forall X(Q)]}_{L}\} \Rightarrow \underbrace{\{\forall X[(P) \Rightarrow (Q)]\}}_{M} \qquad \text{theorem 1.36.}$$

$\square$

2.2.7 Examples with Pure Propositional and Predicate Calculi

The following theorems invoke the full Classical Propositional Calculus, including contraposition and its converse for negations, or Tarski's axioms for equivalences.

2.43 Theorem. *For all P, Q, and X,*
$$\vdash \{\forall X[(P) \Leftrightarrow (Q)]\} \Rightarrow \{[\forall X(P)] \Leftrightarrow [\forall X(Q)]\}.$$

Proof. Apply theorems 2.37 and 2.35 with the transitivity of implication:

$$\begin{aligned}
&\vdash [(P) \Leftrightarrow (Q)] \Rightarrow [(P) \Rightarrow (Q)] && \text{definition of } \Leftrightarrow, \\
&\vdash \{\forall X[(P) \Leftrightarrow (Q)]\} \Rightarrow \{\forall X[(P) \Rightarrow (Q)]\} && \text{theorem 2.37,} \\
&\vdash \{\forall X[(P) \Rightarrow (Q)]\} \Rightarrow \{[\forall X(P)] \Rightarrow [\forall X(Q)]\} && \text{theorem 2.35,} \\
&\vdash \{\forall X[(P) \Leftrightarrow (Q)]\} \Rightarrow \{[\forall X(P)] \Rightarrow [\forall X(Q)]\} && \text{transitivity.}
\end{aligned}$$

The converse conclusion results from the symmetry of $\Leftrightarrow$ and swapping P and Q. The final result then follows from theorem 1.55. $\square$

For the records, theorem 2.44 combines theorems 2.35, 2.42 and 2.43.

2.44 Theorem. *For all P, Q, and X, if X does not occur freely in P, then* $\vdash$ $\{[\forall X(P)] \Rightarrow [\forall X(Q)]\} \Leftrightarrow \{\forall X[(P) \Rightarrow (Q)]\}.$

Proof. Apply theorems 2.35, 2.42 and 2.43. $\square$

2.45 Theorem (derived rule). *For all P, Q, and X, if* $\vdash$ (P) $\Leftrightarrow$ (Q), *then* $\vdash [\forall X(P)] \Leftrightarrow [\forall X(Q)].$

Proof. Apply *Generalization*, theorem 2.43, and *Detachment*:

$$\begin{aligned}
&\vdash (P) \Leftrightarrow (Q) && \text{hypothesis,} \\
&\vdash \forall X[(P) \Leftrightarrow (Q)] && \textit{Generalization,} \\
&\vdash \{\forall X[(P) \Leftrightarrow (Q)]\} \Rightarrow \{[\forall X(P)] \Leftrightarrow [\forall X(Q)]\} && \text{theorem 2.43,} \\
&\vdash [\forall X(P)] \Leftrightarrow [\forall X(Q)] && \textit{Detachment.}
\end{aligned}$$

$\square$

Theorems 2.46 and 2.47 show that $\exists$ could be defined in terms of $\forall$ and double negation, or vice versa, provided that axiom Q0 includes the full Classical Propositional Calculus.

2.46 Theorem.   *For all P and X,* $\vdash [\exists X(P)] \Leftrightarrow (\neg\{[\forall X[\neg(P)]\})$.

Proof. Apply the full propositional calculus and axiom Q3:

$\vdash \{\neg[\exists X(P)]\} \Leftrightarrow \{\forall X[\neg(P)]\}$ axiom Q3,

$\vdash (\neg\{\neg[\exists X(P)]\}) \Leftrightarrow (\neg\{\forall X[\neg(P)]\})$ contraposition and its converse,   □

$\vdash [\exists X(P)] \Leftrightarrow (\neg\{\forall X[\neg(P)]\})$ double negation and transitivity.

2.47 Theorem.   *For all P and X,* $\vdash [\forall X(P)] \Leftrightarrow (\neg\{[\exists X[\neg(P)]\})$.

Proof. Apply the full propositional calculus and axiom Q4:

$\vdash \{\neg[\forall X(P)]\} \Leftrightarrow \{\exists X[\neg(P)]\}$ axiom Q4,

$\vdash (\neg\{\neg[\forall X(P)]\}) \Leftrightarrow (\neg\{\exists X[\neg(P)]\})$ contraposition and its converse,   □

$\vdash [\forall X(P)] \Leftrightarrow (\neg\{\exists X[\neg(P)]\})$ double negation and transitivity.

2.48 Theorem (existential generalization).   *For all X, Y, and P,* $\vdash [Subf_Y^X(P)] \Rightarrow$ $[\exists X(P)]$. *In particular,* $\vdash (P) \Rightarrow [\exists X(P)]$.

Proof. Apply the propositional calculus with axioms Q1 and Q3:

$\vdash \{\forall X[\neg(P)]\} \Rightarrow \{Subf_Y^X[\neg(P)]\}$ axiom Q1,

$\vdash \{\neg[\exists X(P)]\} \Rightarrow \{\forall X[\neg(P)]\}$ axiom Q3,

$\vdash \{Subf_Y^X[\neg(P)]\} \Rightarrow \{\neg[Subf_Y^X(P)]\}$ remark 2.20,

$\vdash \{\neg[\exists X(P)]\} \Rightarrow \{\neg[Subf_Y^X(P)]\}$ transitivity,

$\vdash [Subf_Y^X(P)] \Rightarrow [\exists X(P)]$ converse contraposition & *Detachment*.

                                                           □

Theorem 2.49 provides a converse to theorem 2.48 if X is not free in P.

2.49 Theorem.   *For all P and X, if X is not free in P, then* $\vdash [\exists X(P)] \Leftrightarrow (P)$.

Proof. Apply the propositional calculus with theorems 2.46 and 2.39:

$\vdash [\neg(P)] \Rightarrow \{\forall X[\neg(P)]\}$ theorem 2.39, no free X in P,

$\vdash (\neg\{\forall X[\neg(P)]\}) \Rightarrow \{\neg[\neg(P)]\}$ contraposition,

$\vdash [\exists X(P)] \Rightarrow (\neg\{\forall X[\neg(P)]\})$ theorem 2.46,

$\vdash [\exists X(P)] \Rightarrow \{\neg[\neg(P)]\}$ transitivity,

$\vdash \{\neg[\neg(P)]\} \Rightarrow (P)$ double negation,

$\vdash [\exists X(P)] \Rightarrow (P)$ transitivity.

The converse is theorem 2.48.   □

2.2.8   *Other Axiomatic Systems for the Pure Predicate Calculus*

With the rules of *Detachment* and *Generalization*, and equivalent propositional calculi, Margaris [84, p. 49] and Rosser [110, p. 101] use the following axiom schemata for the predicate calculus:

Axiom A4 $\{\forall X[(P) \Rightarrow (Q)]\} \Rightarrow \{[\forall X(P)] \Rightarrow [\forall X(Q)]\}$.

Axiom A5 $[\forall X(P)] \Rightarrow [Subf_Y^X(P)]$.

Axiom A6 $(P) \Rightarrow [\forall X(P)]$ *if X does not occur freely in P.*

Margaris allows *Generalization*s only of axioms but proves a deduction theorem that then leads to the same rule of *Generalization* [84, p. 49].

In contrast, Kleene [72, p. 107] uses two axiom schemata

∀-schema $[\forall X(P)] \Rightarrow [Subf_Y^X(P)]$,
∃-schema $[Subf_Y^X(P)] \Rightarrow [\exists X(P)]$,

paired with two inference rules, where X does not occur freely in P:

∀-rule   from $(P) \Rightarrow (Q)$ infer $(P) \Rightarrow [\forall X(Q)]$, where X is not free in P,
∃-rule   from $(P) \Rightarrow (Q)$ infer $[\exists X(P)] \Rightarrow (Q)$, where X is not free in Q.

In both systems, $\exists X(P)$ is merely an abbreviation for $\neg\{\forall X[\neg(P)]\}$, as also in the systems of Church [18, p. 171] and Stoll [122, p. 115]. Reversely, in other systems $\forall X(P)$ is merely an abbreviation for $\neg\{\exists X[\neg(P)]\}$, for instance, in Kunen's [74, p. 3]. Axioms such as Q3 and Q4 partially dissociate the quantifiers from the axiom(s) for the negation in the selected propositional calculus.

Also, yet another way to define substitutions of free variables consists in performing substitutions with a different procedure, as outlined in definition 2.50.

2.50 Definition (proper substitution of free variables). A **proper substitution** of a variable Z for each *free* occurrence of a *different* variable X in a logical formula P consists of the following three steps:

(1) Identify a variable that does *not* occur in P, for example, Y.
(2) In P, replace each *bound* occurrence of Z by Y.
(3) Then replace each *free* occurrence of X by Z.

Steps (1) and (2) produce a change of bound variables according to definition 2.13. After step (2), P no longer contains any bound occurrence of Z, and hence no strings of the form $\forall Z$ or $\exists Z$. Consequently, P now admits Z for X in step (3).

Definitions 2.16 and 2.50 thus provide two ways to avoid the phenomenon in counterexample 2.10. Both ways lend themselves to the same notation.

2.51 Definition. The notation $Subf_Z^X(P)$ states that P admits Z for X, and substitutes Z for each free occurrence of X in P. Alternatively, the same notation $Subf_Z^X(P)$ denotes the proper substitution of Z for each free occurrence of X in P. The two alternatives are compatible, by remark 2.15 and definition 2.50. For convenience, $Subf_X^X(P)$ is defined to be P, and if X does *not* occur freely in P, then $Subf_Z^X(P)$ is also defined to be P. The concept and notation for proper substitutions also apply to the substitution of a constant for a free variable. Because constants cannot appear immediately after a quantifier, they are not bound. Consequently, only the last step applies to the proper substitution of constants for free variables. Thus, $Subf_\varnothing^X(P)$ merely substitutes $\varnothing$ for every free occurrence of X in P.

2.52 Example. For P consider the formula $\big(\forall X\{\exists Z[\neg(X = Z)]\}\big) \vee (X = \varnothing)$.

(1) Verify that the variable Y does not occur in P.

(2) Replace the bound occurrences of Z by Y, which gives the formula
$(\forall X\{\exists Y[\neg(X = Y)]\}) \vee (X = \varnothing)$.

(3) Replace the free occurrence of X by Z, which gives the formula
$(\forall X\{\exists Y[\neg(X = Y)]\}) \vee (Z = \varnothing)$ for $\mathrm{Subf}_Z^X(P)$.

In contrast, substituting the constant $\varnothing$ for every free occurrence of X in P yields
$\big(\forall X\{\exists Z[\neg(X = Z)]\}\big) \vee (\varnothing = \varnothing)$ for $\mathrm{Subf}_\varnothing^X(P)$.

2.53 Example. A situation like that in definition 2.50 occurs with computer algorithms to swap two variables X and Z, which typically use a third variable Y distinct from X and Z as a temporary storage. First, the algorithm assigns X to Y, an operation denoted by $Y := X$. Second, the algorithm assigns Z to X, an operation denoted by $X := Z$. Finally, the algorithm assigns Y to Z, an operation denoted by $Z := Y$.

2.2.9 Exercises on Kleene's, Margaris's, and Rosser's Axioms

The following exercises show that Margaris's and Rosser's axioms A4–A6 are derivable from the rules of inference with axioms Q1–Q4 and the Classical Propositional Calculus.

2.11. Prove that the abbreviation $\exists X(P)$ for $\neg\{\forall X[\neg(P)]\}$ is derivable from the rules of inference with axioms Q1–Q4 and the Classical Propositional Calculus.

2.12. Prove that Margaris's and Rosser's axioms A4, A5, and A6 are theorems derivable from the rules of inference with axioms Q1–Q4 and the Classical Propositional Calculus.

The following exercises show that axioms Q1–Q4 are derivable from Margaris's and Rosser's axioms A4–A6 and the Classical Propositional Calculus.

2.13. Prove that axiom Q2 is a theorem derivable from the rules of inference with Margaris's and Rosser's axioms A4–A6 and the Classical Propositional Calculus.

2.14. Prove that axiom Q1 is a theorem derivable from the rules of inference with Margaris's and Rosser's axioms A4–A6 and the Classical Propositional Calculus.

2.15. Prove that axiom Q4 is a theorem derivable from the rules of inference with Margaris's and Rosser's axioms A4–A6 and the Classical Propositional Calculus.

2.16. Prove that axiom Q3 is a theorem derivable from the rules of inference with Margaris's and Rosser's axioms A4–A6 and the Classical Propositional Calculus.

The following exercises show that Kleene's schema and rules are derivable from the rules of inference with axioms Q1–Q4 and the Classical Propositional Calculus.

2.17 . Prove that Kleene's ∃-rule is derivable from the rules of inference with axioms Q1–Q4 and the Classical Propositional Calculus.

2.18 . Prove that Kleene's ∀-rule is derivable from the rules of inference with axioms Q1–Q4 and the Classical Propositional Calculus.

2.19 . Prove that Kleene's ∃-schema is derivable from the rules of inference with axioms Q1–Q4 and the Classical Propositional Calculus.

2.20 . Prove that Kleene's ∀-schema is derivable from the rules of inference with axioms Q1–Q4 and the Classical Propositional Calculus.

2.3 Methods of Proof for the Pure Predicate Calculus

If other considerations guarantee that a well-formed formula P has a proof but do not produce any proof of it, *then* writing down all the proofs of the predicate calculus, for instance, in increasing order of complexity, eventually yields among all such proofs a proof of P [18, p. 99–100, footnote 183]. However, if the shortest proof of P is very long, then this method may take longer than the time available to the user to arrive at any proof of P. Thus for all practical purposes this method may also fail to determine whether a formula is a theorem.

The problem of deciding whether a well-formed formula is a theorem, derivable from specified axioms and inference rules, is called the **decision problem**. For the pure predicate calculus, no algorithms can provide a step-by-step recipe applicable to all well-formed formulae to determine whether any such formula is a theorem, as proved by Church [16, 17]. Nevertheless, methods exist to help in deciding whether a well-formed formula is a theorem.

Trial and error is an option [114, p. 31], sometimes working backward from the particular well-formed formula as a final goal, or forward from the axioms, inference rules, and previous theorems as starting points or intermediate steps [72, p. 54–55]. The methods presented in this section guide this method of designing proofs.

2.3.1 Substituting Equivalent Formulae

One method of proof consists of replacing any occurrence of a formula by an equivalent formula, thanks to theorem 2.54 [18, p. 101, 124, 189], [108, p. 48].

2.54 Theorem (Substitutivity of Equivalence in the Pure Predicate Calculus, preliminary version). *For all well-formed logical formulae U and V, if ⊢ (U) ⇔ (V), and if a formula Q results from substituting any (zero, one, several, or all) occurrence(s) of U by V in a well-formed formula P, then ⊢ (P) ⇔ (Q).*

Proof (Outline). Theorems 1.29 and 1.46 have already established the conclusions for logical implications and negations.

In all cases, if Q is P, which results by substituting none of the occurrences of U by V, then $(P) \Leftrightarrow (Q)$ is $(P) \Leftrightarrow (P)$, which is theorem 1.63.

For the universal quantifier, if P is $\forall X(U)$, then Q is either $\forall X(U)$ or $\forall X(V)$. If Q is $\forall X(V)$, then P with $(U) \Rightarrow (V)$ yield Q, by theorem 2.37, and conversely, Q with $(V) \Rightarrow (U)$ yield P, by theorems 2.37, or also by theorem 2.45.

For the existential quantifier, theorem 2.46 reduces to the previous cases a formula P of the form $\exists X(U)$.

The general case follows by several applications of the previous case and the cases in theorems 1.29 and 1.46, in a way that may be specified more explicitly after the availability of the Principle of Mathematical Induction in chapter 4. $\qquad\square$

2.55 Example. Let P denote the formula $\forall X\big([\exists Y(Y \in X)] \vee \{\forall Z[\neg(Z \in X)]\}\big)$, U the formula $\exists Y(Y \in X)$, and W the formula $\neg\{\forall Y[\neg(Y \in X)]\}$. Then $(U) \Leftrightarrow (W)$ by theorem 2.46. Moreover, let V denote the formula $\neg\{\forall Z[\neg(Z \in X)]\}$. Because Z does not occur in $\neg(Y \in X)$, theorem 2.31 shows that $(W) \Leftrightarrow (V)$. Hence $(U) \Leftrightarrow (V)$ by transitivity. Consequently, $(P) \Leftrightarrow (Q)$ where Q denotes the formula $\forall X\big(\big[\neg\{\forall Z[\neg(Z \in X)]\}\big] \vee \{\forall Z[\neg(Z \in X)]\}\big)$, which is $\forall X\big([\neg\{V\}] \vee \{V\}\big)$. Since $(V) \vee [\neg(V)]$ is a theorem, by *Generalization* Q and hence P is also a theorem.

2.3.2 Discharging Hypotheses

A method to design a proof of an implication $(H) \Rightarrow (C)$ consists of first designing a derivation $H \vdash C$ of C from H, and then transforming the derivation $H \vdash C$ into a proof of $(H) \Rightarrow (C)$ by theorem 2.56 [108, p. 46–47].

2.56 Theorem (Deduction Theorem, preliminary version). *With any axiom system for which axioms P1, P2, and $(P) \Rightarrow (P)$ are axioms or theorems (or schema thereof), and for every derivation $H \vdash C$ of a formula C from a formula H with the propositional calculus, the rules of inference, and axioms Q1–Q4, but without Generalization on any free variable in H, there exists a proof of $(H) \Rightarrow (C)$.*

Proof (Outline). Every step of the derivation $H \vdash C$ is a formula, denoted here by S. If S is H, or an axiom, or results from two previous steps and *Detachment*, then the proof of the deduction theorem 1.22 for the Pure Positive Implicational Propositional Calculus shows how to replace S by $(H) \Rightarrow (S)$.

If S is a *Generalization* $\forall X(R)$ of a previous step R with a variable X that does not occur freely in H, then R has already been replaced by $(H) \Rightarrow (R)$. Hence *Generalization* gives $\forall X[(H) \Rightarrow (R)]$, whence, because X does not occur freely in H, theorem 2.29 yields a proof of $(H) \Rightarrow [\forall X(R)]$, which is $(H) \Rightarrow (S)$.

The general case follows by several applications of the previous cases in a way that may be specified more explicitly after the availability of the Principle of Mathematical Induction in chapter 4. $\qquad\square$

Thus, the selection of axioms P1–P3 and Q1–Q4 leads to the Deduction Theorem (2.56) more directly than would other selections of otherwise equivalent axioms [108, p. 47]. In practice, however, a derivation of $H \vdash C$ of C from H may already suggest other logical steps that shortcut or bypass the entire procedure outlined in the proof of the Deduction Theorem 2.56.

To demonstrate such shortcuts, the following theorems provide means for bringing quantifiers to the "front" of a formula. For example, axioms Q3 and Q4 with theorem 2.54 already allow the replacement of $\neg[\exists X(P)]$ by $\forall X[\neg(P)]$, and of $\neg[\forall X(P)]$ by $\exists X[\neg(P)]$. In an implication $(R) \Rightarrow (S)$, each of R and S can be of the form (P), or $\forall X(P)$, or $\exists X(P)$, starting with $\forall$, or $\exists$, or no quantifiers, which leads to nine different cases. In the case where neither R nor S begins with a quantifier, then no quantifiers need to be brought to the front of $(R) \Rightarrow (S)$. The other eight cases form the object of the following theorems.

Theorem 2.57 handles a case where R is $\exists X(P)$ while S is Q.

2.57 Theorem. *If X does not occur freely in Q, then*
$$\vdash \{\forall X[(P) \Rightarrow (Q)]\} \Leftrightarrow \{[\exists X(P)] \Rightarrow (Q)\}.$$

Proof. Let H denote $\forall X[(P) \Rightarrow (Q)]$, and let C denote $[\exists X(P)] \Rightarrow (Q)$.

$\vdash \forall X[(P) \Rightarrow (Q)]$	hypothesis,
$\vdash \{\forall X[(P) \Rightarrow (Q)]\} \Rightarrow [(P) \Rightarrow (Q)]$	specialization (Q1),
$\vdash (P) \Rightarrow (Q)$	*Detachment*,
$\vdash [\neg(Q)] \Rightarrow [\neg(P)]$	contraposition,
$\vdash \forall X\{[\neg(Q)] \Rightarrow [\neg(P)]\}$	*Generalization*,
$\vdash [\neg(Q)] \Rightarrow \{\forall X[\neg(P)]\}$	theorem 2.29, no free X in Q,
$\vdash (\neg\{\forall X[\neg(P)]\}) \Rightarrow \{\neg[\neg(Q)]\}$	contraposition,
$\vdash [\exists X(P)] \Rightarrow (Q)$	double negation (1.42) and 2.46.

Hence the Deduction Theorem (2.56) leads to a proof of $\{\forall X[(P) \Rightarrow (Q)]\} \Rightarrow \{[\exists X(P)] \Rightarrow (Q)\}$. Yet the foregoing derivation suggests shortcuts:

$\vdash \{\forall X[(P) \Rightarrow (Q)]\} \Rightarrow [(P) \Rightarrow (Q)]$	axiom Q1,
$\vdash [(P) \Rightarrow (Q)] \Rightarrow \{[\neg(Q)] \Rightarrow [\neg(P)]\}$	contraposition,
$\vdash \{\forall X[(P) \Rightarrow (Q)]\} \Rightarrow \{[\neg(Q)] \Rightarrow [\neg(P)]\}$	transitivity,
$\vdash \{\forall X[(P) \Rightarrow (Q)]\} \Rightarrow (\forall X\{[\neg(Q)] \Rightarrow [\neg(P)]\})$	theorem 2.30,
$\vdash (\forall X\{[\neg(Q)] \Rightarrow [\neg(P)]\}) \Rightarrow ([\neg(Q)] \Rightarrow \{\forall X[\neg(P)]\})$	theorem 2.29,
$\vdash ([\neg(Q)] \Rightarrow \{\forall X[\neg(P)]\}) \Rightarrow (\neg\{\forall X[\neg(P)]\}) \Rightarrow \{\neg[\neg(Q)]\}$	contraposition,
$\vdash (\neg\{\forall X[\neg(P)]\}) \Rightarrow \{\neg[\neg(Q)]\} \Rightarrow \{[\exists X(P)] \Rightarrow (Q)\}$	1.42, 2.46,
$\vdash \{\forall X[(P) \Rightarrow (Q)]\} \Rightarrow \{[\exists X(P)] \Rightarrow (Q)\}$	transitivity.

For the converse, let H denote $[\exists X(P)] \Rightarrow (Q)$, and let C denote $\forall X[(P) \Rightarrow (Q)]$.

$\vdash [\exists X(P)] \Rightarrow (Q)$	hypothesis,
$\vdash (P) \Rightarrow [\exists X(P)]$	theorem 2.48,
$\vdash (P) \Rightarrow (Q)$	transitivity,
$\vdash \forall X[(P) \Rightarrow (Q)]$	*Generalization*.

Again the Deduction Theorem (2.56) leads to a proof of $\{[\exists X(P)] \Rightarrow (Q)\} \Rightarrow \{\forall X[(P) \Rightarrow (Q)]\}$ but the foregoing derivation suggests shortcuts:

$\vdash (P) \Rightarrow [\exists X(P)]$ theorem 2.48,

$\vdash (\{[\exists X(P)] \Rightarrow (Q)\}) \Rightarrow [(P) \Rightarrow (Q)]$ theorem 1.34,

$\vdash (\{[\exists X(P)] \Rightarrow (Q)\}) \Rightarrow \{\forall X[(P) \Rightarrow (Q)]\}$ theorem 2.30.

□

The proofs of the following theorems emerge from similar outlines, starting with a derivation $H \vdash C$ of C from H, and transforming it into a proof of $\vdash (H) \Rightarrow (C)$ by shortcuts suggested by the derivation or by the Deduction Theorem (2.56). To this end, the following derived rule proves useful.

2.58 Theorem (derived rule). *If X is not free in Q, and if $\vdash (P) \Rightarrow (Q)$, then $\vdash [\exists X(P)] \Rightarrow (Q)$.*

Proof. Apply theorem 2.57:

$\vdash (P) \Rightarrow (Q)$ *hypothesis,*

$\vdash \forall X[(P) \Rightarrow (Q)]$ *Generalization,*

$\vdash \{\forall X[(P) \Rightarrow (Q)]\} \Rightarrow \{[\exists X(P)] \Rightarrow (Q)\}$ *theorem 2.57, no free X in Q,*

$\vdash [\exists X(P)] \Rightarrow (Q)$ *Detachment.*

□

2.59 Theorem. *If X does not occur freely in P, then $\vdash (P) \Leftrightarrow [\exists X(P)]$.*

Proof. Apply theorems 2.48, 1.12, and 2.58.

$\vdash (P) \Rightarrow [\exists X(P)]$ theorem 2.48,

$\vdash (P) \Rightarrow (P)$ theorem 1.12,

$\vdash [\exists X(P)] \Rightarrow (P)$ theorem 2.58, no free X in P.

□

2.60 Theorem. *For all P, Q, and X, $\vdash \{\forall X[(P) \Rightarrow (Q)]\} \Rightarrow \{[\exists X(P)] \Rightarrow [\exists X(Q)]\}$.*

Proof. Let H denote $\forall X[(P) \Rightarrow (Q)]$, and let C denote $[\exists X(P)] \Rightarrow [\exists X(Q)]$.

$\vdash \{\forall X[(P) \Rightarrow (Q)]\} \Rightarrow [(P) \Rightarrow (Q)]$ axiom Q1,

$\vdash [(P) \Rightarrow (Q)] \Rightarrow \{[\neg(Q)] \Rightarrow [\neg(P)]\}$ contraposition,

$\vdash \{\forall X[(P) \Rightarrow (Q)]\} \Rightarrow \{[\neg(Q)] \Rightarrow [\neg(P)]\}$ transitivity,

$\vdash \{\forall X[(P) \Rightarrow (Q)]\} \Rightarrow (\forall X\{[\neg(Q)] \Rightarrow [\neg(P)]\})$ theorem 2.30,

$\vdash (\forall X\{[\neg(Q)] \Rightarrow [\neg(P)]\})$
$\quad \Rightarrow (\{\forall X[\neg(Q)]\} \Rightarrow \{\forall X[\neg(P)]\})$ theorem 2.35

$\vdash (\{\forall X[\neg(Q)]\} \Rightarrow \{\forall X[\neg(P)]\})$
$\quad \Rightarrow [(\neg\{\forall X[\neg(P)]\}) \Rightarrow (\neg\{\forall X[\neg(Q)]\})]$ contraposition,

$\vdash [(\neg\{\forall X[\neg(P)]\}) \Rightarrow (\neg\{\forall X[\neg(Q)]\})]$
$\quad \Rightarrow \{[\exists X(P)] \Rightarrow [\exists X(Q)]\}$ 1.42, 2.46,

$\vdash \{\forall X[(P) \Rightarrow (Q)]\} \Rightarrow \{[\exists X(P)] \Rightarrow [\exists X(Q)]\}$ transitivity.

□

2.61 Theorem (derived rule). *If $(P) \Leftrightarrow (Q)$ is a theorem, then $[\exists X(P)] \Leftrightarrow [\exists X(Q)]$ is a theorem.*

Proof. Apply *Generalization*, theorem 2.60, and *Detachment*:

$\vdash (P) \Leftrightarrow (Q)$	hypothesis,
$\vdash \forall X[(P) \Leftrightarrow (Q)]$	*Generalization*,
$\vdash \{\forall X[(P) \Leftrightarrow (Q)]\} \Rightarrow \{[\exists X(P)] \Rightarrow [\exists X(Q)]\}$	theorem 2.60,
$\vdash [\exists X(P)] \Rightarrow [\exists X(Q)]$	*Detachment*.

□

2.62 Theorem. $\vdash \{\exists X[(P) \Rightarrow (Q)]\} \Leftrightarrow \{[\forall X(P)] \Rightarrow [\exists X(Q)]\}$.

Proof. This proof follows Church's [18, p. 205]. Apply the law of denial of the antecedent (theorem 1.40) and the law of proof by cases subject to hypotheses (theorem 1.50):

$\vdash [\neg(P)] \Rightarrow [(P) \Rightarrow (Q)]$	theorem 1.40,
$\vdash \forall X\{[\neg(P)] \Rightarrow [(P) \Rightarrow (Q)]\}$	*Generalization*,
$\vdash \{\exists X[\neg(P)]\} \Rightarrow \{\exists X[(P) \Rightarrow (Q)]\}$	theorem 2.60 and *Detachment*,
$\vdash \{\neg[\forall X(P)]\} \Rightarrow \{\exists X[(P) \Rightarrow (Q)]\}$	axiom Q4 and *Detachment*,
$\vdash (Q) \Rightarrow [(P) \Rightarrow (Q)]$	axiom P1,
$\vdash \forall X\{(Q) \Rightarrow [(P) \Rightarrow (Q)]\}$	Generalization,
$\vdash [\exists X(Q)] \Rightarrow \{\exists X[(P) \Rightarrow (Q)]\}$	theorem 2.60 and *Detachment*,
$\vdash \{[\forall X(P)] \Rightarrow [\exists X(Q)]\} \Rightarrow \{\exists X[(P) \Rightarrow (Q)]\}$	theorem 1.50 and *Detachment*.

For the converse, use theorems 2.48 and 2.57:

$\vdash (P) \Rightarrow \{[(P) \Rightarrow (Q)] \Rightarrow (Q)\}$	law of assertion (1.38),
$\vdash [\forall X(P)] \Rightarrow (P)$	axiom Q1,
$\vdash [\forall X(P)] \Rightarrow \{[(P) \Rightarrow (Q)] \Rightarrow (Q)\}$	transitivity,
$\vdash [(P) \Rightarrow (Q)] \Rightarrow \{[\forall X(P)] \Rightarrow (Q)\}$	commutation (1.37),
$\vdash (Q) \Rightarrow [\exists X(Q)]$	theorem 2.48,
$\vdash [(P) \Rightarrow (Q)] \Rightarrow \{[\forall X(P)] \Rightarrow [\exists X(Q)]\}$	derived rule,
$\vdash \forall X([(P) \Rightarrow (Q)] \Rightarrow \{[\forall X(P)] \Rightarrow [\exists X(Q)]\})$	*Generalization*,
$\vdash \{\exists X[(P) \Rightarrow (Q)]\} \Rightarrow \{[\forall X(P)] \Rightarrow [\exists X(Q)]\}$	theorem 2.57 and *Detachment*.

□

2.63 Theorem. *If X does not occur freely in P, then*
$\vdash \{\exists X[(P) \Rightarrow (Q)]\} \Leftrightarrow \{(P) \Rightarrow [\exists X(Q)]\}$.

Proof. This proof follows Church's [18, p. 205]. Apply theorems 2.39 and 2.62:

$\vdash \{\exists X[(P) \Rightarrow (Q)]\} \Leftrightarrow \{[\forall X(P)] \Rightarrow [\exists X(Q)]\}$	theorem 2.62,
$\vdash (P) \Leftrightarrow [\forall X(P)]$	theorem 2.39, no free X in P,
$\vdash \{\exists X[(P) \Rightarrow (Q)]\} \Leftrightarrow \{(P) \Rightarrow [\exists X(Q)]\}$	derived rules.

□

Similar to theorem 2.57, theorem 2.64 handles a case where R is $\forall X(P)$ while S is Q.

2.64 Theorem. *If X does not occur freely in Q, then*
$\vdash \{\forall X[(P) \Rightarrow (Q)]\} \Leftrightarrow \{[\forall X(P)] \Rightarrow (Q)\}$.

Proof. Apply theorems 2.59, 2.62, and 2.54:

$$\vdash (Q) \Leftrightarrow [\exists X(Q)] \qquad\qquad \text{theorem 2.59, no free } X \text{ in } Q,$$
$$\vdash \{\exists X[(P) \Rightarrow (Q)]\} \Leftrightarrow \{[\forall X(P)] \Rightarrow [\exists X(Q)]\} \quad \text{theorem 2.62,}$$
$$\vdash \{\forall X[(P) \Rightarrow (Q)]\} \Leftrightarrow \{[\forall X(P)] \Rightarrow (Q)\} \quad \text{theorem 2.54.}$$

□

2.3.3   Prenex Normal Form

Yet another method of proof consists in transforming a formula into an equivalent formula in which all the quantifiers, if any, are at the beginning.

2.65 Definition. A formula Q is in **prenex normal form** if and only if Q is of the form

$$Q_b X_b Q_{bb} X_{bb} \ldots Q_{b\ldots b} X_{b\ldots b}(R)$$

optionally with brackets and parentheses, where R is a well-formed formula without quantifiers and each string $Q_{b\ldots b}$ is either $\forall$ or $\exists$. The formula R is called the **matrix** of P while the string $Q_b X_b Q_{bb} X_{bb} \ldots Q_{b\ldots b} X_{b\ldots b}$ is called the **prefix** of P.

2.66 Example. The formula $\forall X \exists Z[\neg(Z = X)]$ is in prenex normal form. Its prefix is $\forall X \exists Z$ while its matrix is $\neg(Z = X)$.

Theorem 2.67 reveals that every well-formed formula is equivalent to a formula in prenex normal form [18, §39], [108, p. 49].

2.67 Theorem (prenex normal form, preliminary version). *For every well-formed formula P there exists a well-formed formula Q in prenex normal form such that* $(P) \Leftrightarrow (Q)$.

Proof (Outline). In P, replace bound variables so that different quantifiers bind different variables, which gives a formula equivalent to P, by theorems 2.31 and 2.54.

With different quantified variables, theorem 2.38 then provides a means to bring quantifiers in front of an implication.

Axioms Q3 and Q4 also provide a means to bring any quantifier in front of any negation.

The general case follows by several applications of the previous cases in a way that may be specified more explicitly after the availability of the Principle of Mathematical Induction in chapter 4.   □

2.68 Example. The formula $\neg\{\exists X[\forall Z(Z = X)]\}$ is *not* in prenex normal form. Nevertheless, axiom Q4 gives the equivalent formula $\forall X\{\neg[\forall Z(Z = X)]\}$, whence axiom Q3 and theorem 2.54 yield the equivalent formula $\forall X\{\exists Z[\neg(Z = X)]\}$, which *is* in prenex normal form.

Transforming a logical formula P into an equivalent formula Q in prenex normal form, or partly so, may reveal a proof of Q and hence also of P, as demonstrated in example 2.55. Example 2.69 completes the transformation into prenex normal form, which reveals a propositional theorem in the matrix.

2.69 Example. The formula $\forall X\big([\exists Y(Y \in X)] \vee \{\forall Z[\neg(Z \in X)]\}\big)$ is *not* in prenex normal form. Nevertheless, by the definition of $\vee$ in terms of $\neg$ and $\Rightarrow$ the formula becomes $\forall X\big(\{\neg[\exists Y(Y \in X)]\} \Rightarrow \{\forall Z[\neg(Z \in X)]\}\big)$.

Axiom Q3 gives $\forall X\big(\{\forall Y[\neg(Y \in X)]\} \Rightarrow \{\forall Z[\neg(Z \in X)]\}\big)$.

Theorem 2.38 gives $\forall X\big[\forall Z\big(\{\forall Y[\neg(Y \in X)]\} \Rightarrow [\neg(Z \in X)]\big)\big]$.

Theorem 2.64 yields $\forall X\big[\forall Z\big(\exists Y\{[\neg(Y \in X)] \Rightarrow [\neg(Z \in X)]\}\big)\big]$, which is a theorem: selecting Z for Y gives $[\neg(Z \in X)] \Rightarrow [\neg(Z \in X)]$, which has the pattern $(P) \Rightarrow (P)$ of theorem 1.12.

Besides providing transformations that may facilitate proofs, as in example 2.55, bringing formulae into prenex normal form, in particular, Skolem's normal form with all the existential quantifiers preceding all the universal quantifiers,

$$\exists X_{\flat} \ldots \exists X_{\flat \ldots \flat} \forall Y^{\sharp} \ldots \forall Y^{\sharp \cdots \sharp}(R),$$

leads to Gödel's Completeness Theorem, that a formula is a theorem if and only if it is valid in all applications, even though no mechanical ways to check either may exist [18, §42–§44].

2.3.4 Proofs with More than One Quantifier

The following theorems are examples of theorems involving more than one quantifier. The first theorem allows for the deletion of a redundant universal quantifier.

2.70 Theorem. $\vdash [\forall X(Q)] \Leftrightarrow \{\forall X[\forall X(Q)]\}$.

Proof. Apply theorem 2.39 to $\forall X(Q)$, which has no free X.   □

The second theorem allows for the swap of two consecutive universal quantifiers.

2.71 Theorem. $\vdash \{\forall X[\forall Y(P)]\} \Leftrightarrow \{\forall Y[\forall X(P)]\}$.

Proof. Apply axiom Q1, theorem 2.29, and *Generalization*:

$\vdash \{\forall X[\forall Y(P)]\} \Rightarrow [\forall Y(P)]$ axiom Q1,

$\vdash [\forall Y(P)] \Rightarrow (P)$ axiom Q1,

$\vdash \{\forall X[\forall Y(P)]\} \Rightarrow (P)$ transitivity (theorem 1.16),

$\vdash \{\forall X[\forall Y(P)]\} \Rightarrow [\forall X(P)]$ theorem 2.37,

$\vdash \{\forall X[\forall Y(P)]\} \Rightarrow \{\forall Y[\forall X(P)]\}$ theorem 2.29.

                                                                                        □

The third theorem allows for the swap of two consecutive existential quantifiers.

2.72 Theorem. $\vdash \{\exists X[\exists Y(P)]\} \Leftrightarrow \{\exists Y[\exists X(P)]\}$.

Proof. Apply the complete law of double negation, axiom Q3, and theorem 2.71:

$$\exists X[\exists Y(P)]$$

$\qquad\qquad\quad \updownarrow \qquad$ double negation,

$$\neg\big(\neg\{\exists X[\exists Y(P)]\}\big)$$

$\qquad\qquad\quad \updownarrow \qquad$ axiom Q3 and theorem 2.54,

$$\neg\big(\forall X\{\neg[\exists Y(P)]\}\big)$$

$\qquad\qquad\quad \updownarrow \qquad$ axiom Q3 and theorem 2.54,

$$\neg\big(\forall X\{\forall Y[\neg(P)]\}\big)$$

$\qquad\qquad\quad \updownarrow \qquad$ theorem 2.71,

$$\neg\big(\forall Y\{\forall X[\neg(P)]\}\big)$$

$\qquad\qquad\quad \updownarrow \qquad$ axiom Q3 and theorem 2.54,

$$\neg\big(\forall Y\{\neg[\exists X(P)]\}\big)$$

$\qquad\qquad\quad \updownarrow \qquad$ axiom Q3 and theorem 2.54,

$$\neg\big(\neg\{\exists Y[\exists X(P)]\}\big)$$

$\qquad\qquad\quad \updownarrow \qquad$ double negation (theorems 1.41 and 1.42).

$$\exists Y[\exists X(P)]$$

$\qquad\qquad\qquad\qquad\qquad\qquad\qquad\qquad\qquad\qquad\qquad\qquad\qquad \square$

The fourth theorem allows for the swap of different quantifiers in an implication.

2.73 Theorem. $\vdash \{\exists X[\forall Y(P)]\} \Rightarrow \{\forall Y[\exists X(P)]\}$.

Proof. Apply theorems 2.48, 2.35, and 2.57:

$\vdash (P) \Rightarrow [\exists X(P)]$ $\qquad\qquad\qquad$ theorem 2.48,

$\vdash [\forall Y(P)] \Rightarrow \{\forall Y[\exists X(P)]\}$ $\qquad$ theorem 2.37,

$\vdash \{\exists X[\forall Y(P)]\} \Rightarrow \{\forall Y[\exists X(P)]\}$ $\quad$ theorem 2.57, no free X in $\{\forall Y[\exists X(P)]\}$.

$\qquad\qquad\qquad\qquad\qquad\qquad\qquad\qquad\qquad\qquad\qquad\qquad\qquad \square$

2.74 Counterexample. The converse of theorem 2.73, which would be

$$\{\forall Y[\exists X(P)]\} \Rightarrow \{\exists X[\forall Y(P)]\},$$

can be *False*. For instance, in every context with at least two *different* objects V and W, consider the logical formula $X = Y$ for P.

$\vdash \forall Y[\exists X(X = Y)]$ $\qquad\qquad\qquad$ for each Y, choose $X := Y$;

$\exists X[\forall Y(X = Y)]$ $\qquad\qquad\qquad\qquad$ is False: no X equals V and W;

$\{\forall Y[\exists X(X = Y)]\} \Rightarrow \{\exists X[\forall Y(X = Y)]\}$ $\quad$ is False because $(T) \not\Rightarrow (F)$.

2.3.5   *Exercises on the Substitutivity of Equivalence*

The following exercises focus on details of the proof of theorem 2.54, with the logical equivalence $\Leftrightarrow$ defined either by Tarski's axioms IV, V, VI in example 1.87 on page 55, or with $\Rightarrow$ and $\wedge$ in definition 1.51 on page 38.

2.21 . Prove that if P denotes $(U) \Rightarrow (W)$, if Q denotes $(V) \Rightarrow (W)$, and if $\vdash$ $(V) \Leftrightarrow (U)$, then $\vdash (P) \Rightarrow (Q)$.

2.22 . Prove that if P denotes $(W) \Rightarrow (U)$, if Q denotes $(W) \Rightarrow (V)$, and if $\vdash$ $(V) \Leftrightarrow (U)$, then $\vdash (P) \Rightarrow (Q)$.

2.23 . Prove that if P denotes $(U) \Rightarrow (W)$, if Q denotes $(V) \Rightarrow (W)$, and if $\vdash$ $(V) \Leftrightarrow (U)$, then $\vdash (Q) \Rightarrow (P)$.

2.24 . Prove that if P denotes $(W) \Rightarrow (U)$, if Q denotes $(W) \Rightarrow (V)$, and if $\vdash$ $(V) \Leftrightarrow (U)$, then $\vdash (Q) \Rightarrow (P)$.

2.25 . Prove that if P denotes $\forall X(U)$, if Q denotes $\forall X(V)$, without free occurrences of X in U and V, and if $\vdash (V) \Leftrightarrow (U)$, then $\vdash (P) \Rightarrow (Q)$.

2.26 . Prove that if P denotes $\forall X(U)$, if Q denotes $\forall X(V)$, without free occurrences of X in U and V, and if $\vdash (V) \Leftrightarrow (U)$, then $\vdash (Q) \Rightarrow (P)$.

2.27 . Prove that if P denotes $\neg(U)$, if Q denotes $\neg(V)$, and if $\vdash (V) \Leftrightarrow (U)$, then $\vdash (P) \Rightarrow (Q)$.

2.28 . Prove that if P denotes $\neg(U)$, if Q denotes $\neg(V)$, and if $\vdash (V) \Leftrightarrow (U)$, then $\vdash (Q) \Rightarrow (P)$.

2.29 . Prove that if P denotes either $(U) \Rightarrow (W)$ or $(W) \Rightarrow (U)$, if Q denotes either $(V) \Rightarrow (W)$ or $(W) \Rightarrow (V)$, respectively, and if $\vdash (V) \Leftrightarrow (U)$, then $\vdash (P) \Leftrightarrow (Q)$.

2.30 . Prove that if P denotes $\neg(U)$, if Q denotes $\neg(V)$, and if $\vdash (V) \Leftrightarrow (U)$, then $\vdash (P) \Leftrightarrow (Q)$.

2.4 Predicate Calculus with Other Connectives

This section introduces theorems with quantifiers and conjunctions or disjunctions.

2.4.1 Universal Quantifiers and Conjunctions or Disjunctions

This subsection presents theorems involving the universal quantifier ($\forall$) and a conjunction ($\wedge$) or disjunction ($\vee$), beginning with an equivalence with a conjunction.

2.75 Theorem. $\vdash \{\forall X[(P) \wedge (Q)]\} \Rightarrow \{[\forall X(P)] \wedge [\forall X(Q)]\}$.

Proof. Apply theorems 1.53, 2.37, 1.52, 1.55:

$\vdash [(P) \wedge (Q)] \Rightarrow (P)$	theorem 1.53,
$\vdash \{\forall X[(P) \wedge (Q)]\} \Rightarrow \{\forall X(P)\}$	theorem 2.37,
$\vdash [(P) \wedge (Q)] \Rightarrow (Q)$	theorem 1.52,

$\vdash \{\forall X[(P) \wedge (Q)]\} \Rightarrow \{\forall X(Q)\}$ theorem 2.37,

$\vdash \{\forall X[(P) \wedge (Q)]\} \Rightarrow \{[\forall X(P)] \wedge [\forall X(Q)]\}$ theorem 1.55.

□

The converse implication forms the object of the following theorem.

2.76 Theorem. $\vdash \{[\forall X(P)] \wedge [\forall X(Q)]\} \Rightarrow \{\forall X[(P) \wedge (Q)]\}$.

Proof. Apply axiom Q1 with theorems 1.82 and 2.30:

$\vdash [\forall X(P)] \Rightarrow (P)$ axiom Q1,

$\vdash [\forall X(Q)] \Rightarrow (Q)$ axiom Q1,

$\vdash \{[\forall X(P)] \wedge [\forall X(Q)]\} \Rightarrow [(P) \wedge (Q)]$ theorem 1.82,

$\vdash \{[\forall X(P)] \wedge [\forall X(Q)]\} \Rightarrow \{\forall X[(P) \wedge (Q)]\}$ theorem 2.30.

□

The following theorem gives an implication with a disjunction.

2.77 Theorem. $\vdash \{[\forall X(P)] \vee [\forall X(Q)]\} \Rightarrow \{\forall X[(P) \vee (Q)]\}$.

Proof. Apply axiom Q1, exercise 1.57, theorem 2.29, and *Generalization*:

$\vdash [\forall X(P)] \Rightarrow (P)$ axiom Q1,

$\vdash [\forall X(Q)] \Rightarrow (Q)$ axiom Q1,

$\vdash \{[\forall X(P)] \vee [\forall X(Q)]\} \Rightarrow [(P) \vee (Q)]$ exercise 1.57,

$\vdash \forall X(\{[\forall X(P)] \vee [\forall X(Q)]\} \Rightarrow [(P) \vee (Q)])$ *Generalization*,

$\vdash \{[\forall X(P)] \vee [\forall X(Q)]\} \Rightarrow \{\forall X[(P) \vee (Q)]\}$ theorem 2.29.

□

2.78 Counterexample. The converse of theorem 2.77, which would be

$$\{\forall X[(P) \vee (Q)]\} \Rightarrow \{[\forall X(P)] \vee [\forall X(Q)]\},$$

may be *False*. For instance, in every context with exactly *two different* objects V and W, consider the formulae $X = V$ for P and $X = W$ for Q:

$\vdash \forall X[(X = V) \vee (X = W)]$ because either $(X = V)$ or $(X = W)$;

$\forall X(X = V)$ is False if $X := W$;

$\forall X(X = W)$ is False if $X := V$;

$[\forall X(X = V)] \vee [\forall X(X = W)]$ is False by the preceding two lines;

$\{\forall X[(X = V) \vee (X = W)]\}$

$\quad \Rightarrow \{[\forall X(X = V)] \vee [\forall X(X = W)]\}$ is False because $(T) \not\Rightarrow (F)$.

However, theorem 2.79 shows a converse of theorem 2.77 in a particular case.

2.79 Theorem. *If P has no free X, then* $\vdash \{\forall X[(P) \vee (Q)]\} \Rightarrow \{(P) \vee [\forall X(Q)]\}$ *and*
$\vdash \{\forall X[(P) \vee (Q)]\} \Rightarrow \{[\forall X(P)] \vee [\forall X(Q)]\}$

Proof. Apply the definition of $\vee$:

$$\forall X[(P) \vee (Q)]$$

$\quad\quad\quad\quad\quad\quad\Updownarrow\quad$ definition of $(P) \vee (Q)$,

$$\forall X\{[(P) \Rightarrow (Q)] \Rightarrow (Q)\}$$

$\quad\quad\quad\quad\quad\quad\Downarrow\quad$ theorem 2.35,

$$\{\forall X[(P) \Rightarrow (Q)]\} \Rightarrow [\forall X(Q)]$$

$\quad\quad\quad\quad\quad\quad\Updownarrow\quad$ theorems 2.38, 2.54, no free X in P,

$$\{(P) \Rightarrow [\forall X(Q)]\} \Rightarrow [\forall X(Q)]$$

$\quad\quad\quad\quad\quad\quad\Updownarrow\quad$ definition of $\vee$,

$$(P) \vee [\forall X(Q)]$$

$\quad\quad\quad\quad\quad\quad\Updownarrow\quad$ theorem 2.39, 2.54, no free X in P.

$$[\forall X(P)] \vee [\forall X(Q)]$$

$\hfill\square$

2.4.2 *Existential Quantifiers and Conjunctions or Disjunctions*

This subsection presents theorems involving the existential quantifier ($\exists$) and a conjunction ($\wedge$) or disjunction ($\vee$), beginning with an equivalence with a disjunction.

2.80 Theorem. $\vdash \{[\exists X(P)] \vee [\exists X(Q)]\} \Leftrightarrow \{[\exists X[(P) \vee (Q)]\}$.

Proof. Apply contraposition with theorems 2.75, 2.76, and 2.45:

$$\vdash \big(\forall X\{[\neg(P)] \wedge [\neg(Q)]\}\big) \Leftrightarrow \big(\{\forall X[\neg(P)]\} \wedge \{\forall X[\neg(Q)]\}\big) \quad \text{2.75, 2.76,}$$

$\quad\quad\quad\quad\quad\quad\quad\quad\quad\quad\Updownarrow\quad$ contraposition,

$$\big[\neg\big(\{\forall X[\neg(P)]\} \wedge \{\forall X[\neg(Q)]\}\big)\big] \Leftrightarrow \big[\neg\big(\forall X\{[\neg(P)] \wedge [\neg(Q)]\}\big)\big]$$

$\quad\quad\quad\quad\quad\quad\quad\quad\quad\quad\Updownarrow\quad$ 2.45,

$$\big[\big(\neg\{\forall X[\neg(P)]\}\big) \vee \big(\neg\{\forall X[\neg(Q)]\}\big)\big] \Leftrightarrow \big[\neg\big(\forall X\{\neg[(P) \vee (Q)]\}\big)\big]$$

$\quad\quad\quad\quad\quad\quad\quad\quad\quad\quad\Updownarrow\quad$ axiom Q4,

$$\big(\neg\{\neg[\exists X(P)]\}\big) \vee \big(\neg\{\neg[\exists X(Q)]\}\big) \Leftrightarrow \big[\neg\big(\neg\{\exists X[(P) \vee (Q)]\}\big)\big]$$

$\quad\quad\quad\quad\quad\quad\quad\quad\quad\quad\Updownarrow\quad$ double negations.

$$\{[\exists X(P)] \vee [\exists X(Q)]\} \Leftrightarrow \{\exists X[(P) \vee (Q)]\}$$

$\hfill\square$

A similar equivalence with a conjunction requires that X be not free in P.

2.81 Theorem. *If P has no free X, then* $\vdash \{\exists X[(P) \wedge (Q)]\} \Leftrightarrow \{(P) \wedge [\exists X(Q)]\}$.

Proof. Apply the full propositional calculus with theorems 2.38, 2.61, 2.46, and axiom Q4, and theorem 1.69:

$$\exists X[(P) \wedge (Q)]$$

$\quad\quad\quad\quad\quad\quad\Updownarrow\quad$ double negation,

$$\exists X\big(\neg\{\neg[(P) \wedge (Q)]\}\big)$$

$\quad\quad\quad\quad\quad\quad\Updownarrow\quad$ De Morgan's first law and theorem 2.61,

$$\exists X\big(\neg\{[\neg(P)] \vee [\neg(Q)]\}\big)$$

$\Updownarrow$ definition of $\vee$,

$$\exists X\big(\neg\{(P) \Rightarrow [\neg(Q)]\}\big)$$

$\Updownarrow$ axiom Q4,

$$\neg\big(\forall X\{(P) \Rightarrow [\neg(Q)]\}\big)$$

$\Updownarrow$ theorem 2.38,

$$\neg\big[(P) \Rightarrow \{\forall X[\neg(Q)]\}\big]$$

$\Updownarrow$ definition of $\vee$ by theorem 1.69,

$$\neg\big([\neg(P)] \vee \{\forall X[\neg(Q)]\}\big)$$

$\Updownarrow$ De Morgan's second law and double negation,

$$(P) \wedge \big(\neg\{\forall X[\neg(Q)]\}\big)$$

$\Updownarrow$ theorem 2.46.

$$(P) \wedge [\exists X(Q)]$$

$\square$

2.4.3 Exercises on Quantifiers with Other Connectives

For the following exercises, prove that the stated formulae are theorem schema.

2.31 . $\{\exists X[(P) \vee (Q)]\} \Leftrightarrow \{\exists X[(Q) \vee (P)]\}$.

2.32 . $\{\forall X[(P) \wedge (P)]\} \Leftrightarrow \{\forall X(P)\}$.

2.33 . $\{\exists X[(P) \vee (P)]\} \Leftrightarrow \{\exists X(P)\}$.

2.34 . $(\exists X\{[(P) \vee (Q)] \vee (R)\}) \Leftrightarrow (\exists X\{(P) \vee [(Q) \vee (R)]\})$.

2.35 . $(\forall X\{[(P) \wedge (Q)] \vee (R)\}) \Leftrightarrow (\{\forall X[(P) \vee (R)]\} \wedge \{\forall X[(Q) \vee (R)]\})$.

2.36 . $(\exists X\{[(P) \vee (Q)] \wedge (R)\}) \Leftrightarrow (\{\exists X[(P) \wedge (R)]\} \vee \{\exists[(Q) \wedge (R)]\})$.

2.37 . $[\exists X(Q)] \Leftrightarrow [\exists X (\exists X(Q))]$.

2.38 . If P has no free X, then $\{(P) \wedge [\forall X(Q)]\} \Leftrightarrow \{\forall X[(P) \wedge (Q)]\}$.

2.39 . If P has no free X, then $\{(P) \vee [\forall X(Q)]\} \Leftrightarrow \{\forall X[(P) \vee (Q)]\}$.

2.40 . If P has no free X, then $\{(P) \vee [\exists X(Q)]\} \Leftrightarrow \{\exists X[(P) \vee (Q)]\}$.

2.5 Equality-Predicates

Applications of logic, for instance, algebra, arithmetic, and geometry, may include concepts of "equality" that allow for substitutions of mutually *equal* objects in statements and formulae, which results in mutually *equivalent* statements and formulae.

2.5.1   First-Order Predicate Calculi with an Equality-Predicate

Different applications may define equality differently [8, p. 6–7]. For instance, in some versions of integer arithmetic, the equality $a = b$ means that a and b are two symbols for *one* integer [25, p. 44], [76, p. 1]. In contrast, in some versions of set theory, the equality $A = B$ means that A and B denote sets with identical set-theoretical features: they have the same elements, and they are elements of the same sets [8, 35, p. 6–7]; the question whether A and B denote the same set does not arise in the theory. Nevertheless, such different concepts of equality happen to conform to a logical predicate, denoted by $\mathscr{I}$ to suggest identity, subject to the following axioms (which might also be called *postulates* to distinguish them from logical axioms) [18, § 48].

Axiom $\mathscr{I}$1 (reflexivity of equality) $\vdash \mathscr{I}(X, X)$.

Axiom $\mathscr{I}$2 (substitutivity of equality) $\vdash [\mathscr{I}(X, Y)] \Rightarrow [(P) \Rightarrow (Q)]$ *for all well-formed formula P and Q such that Q results from the substitution of any one free occurrence of X in P by Y, provided that the resulting occurrence of Y is also free, or, in other words, provided that in P this occurrence of X is not within the scope of a quantifier ($\forall X, \forall Y, \exists X, \exists Y$) bounding X or Y.*

The condition stipulated in axiom $\mathscr{I}$2 is similar to the requirement that P admit Y for X, or that Y be free for X, but only for one particular occurrence of X in P.

Using only the Pure Positive Implicational Propositional Calculus, theorems 2.82 and 2.83 show that every predicate $\mathscr{I}$ satisfying axioms $\mathscr{I}$1 and $\mathscr{I}$2 is symmetric and transitive [84, p. 104].

2.82 Theorem (symmetry of equality). $\vdash [\mathscr{I}(X, Y)] \Rightarrow [\mathscr{I}(Y, X)]$.

Proof. In axiom $\mathscr{I}$2, substitute the terms $\mathscr{I}(X, X)$ for P and $\mathscr{I}(Y, X)$ for Q:

$\quad \vdash [\mathscr{I}(X, Y)] \Rightarrow \{[\mathscr{I}(X, X)] \Rightarrow [\mathscr{I}(Y, X)]\}$ axiom $\mathscr{I}$2, ,

$\quad \vdash \mathscr{I}(X, X)$ axiom $\mathscr{I}$1,

$\quad \vdash [\mathscr{I}(X, Y)] \Rightarrow [\mathscr{I}(Y, X)]$ derived rule (theorem 1.15).

$\hfill \square$

2.83 Theorem (transitivity of equality). $\vdash [\mathscr{I}(X, Y)] \Rightarrow \{[\mathscr{I}(Y, Z)] \Rightarrow [\mathscr{I}(X, Z)]\}$. *Hence, if* $\vdash \mathscr{I}(X, Y)$ *and* $\vdash \mathscr{I}(Y, Z)$, *then* $\vdash \mathscr{I}(X, Z)$.

Proof. Use the symmetry of equality (theorem 2.82) and axiom $\mathscr{I}$2:

$\quad \vdash [\mathscr{I}(X, Y)] \Rightarrow [\mathscr{I}(Y, X)]$ theorem 2.82,

$\quad \vdash [\mathscr{I}(Y, X)] \Rightarrow \{[\mathscr{I}(Y, Z)] \Rightarrow [\mathscr{I}(X, Z)]\}$ axiom $\mathscr{I}$2,

$\quad \vdash [\mathscr{I}(X, Y)] \Rightarrow \{[\mathscr{I}(Y, Z)] \Rightarrow [\mathscr{I}(X, Z)]\}$ transitivity (theorem 1.16),

$\quad \vdash \mathscr{I}(X, Y)$ hypothesis,

$\quad \vdash [\mathscr{I}(Y, Z)] \Rightarrow [\mathscr{I}(X, Z)]$ *Detachment*,

$\quad \vdash \mathscr{I}(Y, Z)$ hypothesis,

$\quad \vdash \mathscr{I}(X, Z)$ *Detachment*.

$\hfill \square$

Using only the Pure Positive Implicational Propositional Calculus, theorem 2.84 extends axiom $\mathscr{I}2$ to a converse implication, so that substituting mutually equal objects results in mutually equivalent formulae.

2.84 Theorem (substitutivity of equality). $\vdash [\mathscr{I}(X,Y)] \Rightarrow [(P) \Leftrightarrow (Q)]$ *for all well-formed formula P and Q such that Q results from the substitution of any one free occurrence of X in P provided that in P this occurrence of X is not within the scope of a quantifier* $(\forall X, \forall Y, \exists X, \exists Y)$ *bounding X or Y.*

Proof. The implication $\vdash [\mathscr{I}(X,Y)] \Rightarrow [(P) \Rightarrow (Q)]$ is axiom $\mathscr{I}2$.

For the converse, the hypothesis also states that in Q the resulting occurrence of Y is not within the scope of a quantifier $(\forall X, \forall Y, \exists X, \exists Y)$ bounding Y or X, which allows swapping X and Y, and swapping P and Q, in axiom $\mathscr{I}2$, so that $\vdash [\mathscr{I}(Y,X)] \Rightarrow [(Q) \Rightarrow (P)]$. Hence the conclusion follows from the symmetry $\vdash [\mathscr{I}(X,Y)] \Rightarrow [\mathscr{I}(Y,X)]$ by theorem 2.82 and the transitivity of implication. $\qquad\square$

Repeated applications of theorem 2.84 and the proof of substitutivity of equivalence then show that substituting mutually equal objects in a formula leads to an equivalent formula.

2.5.2 Simple Applied Predicate Calculi with an Equality-Predicate

Some applications of logic might omit all propositional variables and instead have only atomic formulae with a few predicates, or perhaps only one predicate, which might be denoted by some constant $\mathscr{E}$. Such applications are called simple applied predicate calculi. For instance, a version of set theory has no propositional variables and only one predicate, for set membership, so that $\mathscr{E}(X,Y)$ stands for $X \in Y$. In such applications, an additional equality predicate $\mathscr{I}$ allows for substitutions of mutually equivalent objects in statements and formulae if and only if $\mathscr{I}$ is reflexive (a condition that replaces axiom $\mathscr{I}1$), symmetric, transitive, and satisfies the following two conditions, which replace axiom $\mathscr{I}2$ [18, p. 283, exercise 48.3]:

$$[\mathscr{I}(A,B)] \Rightarrow \{[\mathscr{E}(X,A)] \Rightarrow [\mathscr{E}(X,B)]\},$$
$$[\mathscr{I}(A,B)] \Rightarrow \{[\mathscr{E}(A,Y)] \Rightarrow [\mathscr{E}(B,Y)]\}.$$

In applied logics with other predicates, for instance, predicates for the sum and products of integers in arithmetic, two similar conditions must be appended for each predicate to ensure the substitutivity of mutually equal objects. By the postulated symmetry of the equality predicate $\mathscr{I}$ these two conditions are equivalent to

$$[\mathscr{I}(A,B)] \Rightarrow \{[\mathscr{E}(X,A)] \Leftrightarrow [\mathscr{E}(X,B)]\},$$
$$[\mathscr{I}(A,B)] \Rightarrow \{[\mathscr{E}(A,Y)] \Leftrightarrow [\mathscr{E}(B,Y)]\}.$$

These conditions suffice to ensure that if $\mathscr{I}(A, B)$ holds, then substituting any free occurrence of A for any free occurrence of B according to the conditions stipulated by axiom $\mathscr{I}2$ in any formula P produces an equivalent formula Q, because well-formed formulae include only atomic formulae of the form $\mathscr{E}(Z, W)$. The proof follows the pattern of the proof of the substitutivity of equivalence. The resulting theorem is called the *substitutivity of equality*.

In particular, if a simple applied predicate calculus has exactly one predicate, $\mathscr{E}$, which is binary (involving exactly two individual variables), then the same conditions may serve to *define* an equality predicate $\mathscr{I}$ so that $\mathscr{I}(A, B)$ is merely an abbreviation for

$$\big(\forall X\{[\mathscr{E}(X, A)] \Leftrightarrow [\mathscr{E}(X, B)]\}\big) \wedge \big(\forall Y\{[\mathscr{E}(A, Y)] \Leftrightarrow [\mathscr{E}(B, Y)]\}\big). \qquad (2.1)$$

Formula (2.1), abbreviated by $\mathscr{I}(A, B)$, satisfies axiom $\mathscr{I}2$ and is reflexive, symmetric, and transitive, because so is the equivalence $\Leftrightarrow$ in the full propositional calculus. In particular, the theorem on substitutivity of equality holds. An equality predicate $\mathscr{I}$ defined in this manner from another binary predicate $\mathscr{E}$ thus does not add anything to the theory except convenience.

2.85 Example (equality in set theory). In a version of set theory, with the single binary predicate $\mathscr{E}(Z, W)$, the postulate (or axiom) of extensionality states that

$$\big(\forall X\{[\mathscr{E}(X, A)] \Leftrightarrow [\mathscr{E}(X, B)]\}\big) \Leftrightarrow \big(\forall Y\{[\mathscr{E}(A, Y)] \Leftrightarrow [\mathscr{E}(B, Y)]\}\big). \qquad (2.2)$$

The equality $\mathscr{I}(A, B)$ of sets A and B is then an abbreviation of each of the formulae $\forall X\{[\mathscr{E}(X, A)] \Leftrightarrow [\mathscr{E}(X, B)]\}$ and $\forall Y\{[\mathscr{E}(A, Y)] \Leftrightarrow [\mathscr{E}(B, Y)]\}$.

The following theorems confirm that every equality-predicate defined by formula (2.1) is reflexive, symmetric, and transitive.

2.86 Theorem (reflexivity of defined equality-predicates). *Every equality-predicate* $\mathscr{I}(A, B)$ *defined by formula (2.1) is reflexive:*

$$\vdash \forall C[\mathscr{I}(C, C)].$$

Proof. One method to *design* a formal proof transforms the objective, here the *yet unproved* formula $\forall C[\mathscr{I}(C, C)]$, first into its defining formula (2.1), and then into logically equivalent formulae, for instance, in prenex form, until one such equivalent formula appears that is a theorem, thanks to an axiom or to a previously proven theorem. For instance, substituting C for A and also C for B in the defining formula (2.1) gives

$$\mathscr{I}(C, C) \quad \textbf{yet unproved,}$$
$$\Updownarrow \quad \text{definition of } \mathscr{I}$$
$$\big(\forall X\{[\mathscr{E}(X, C)] \Leftrightarrow [\mathscr{E}(X, C)]\}\big) \wedge \big(\forall Y\{[\mathscr{E}(C, Y)] \Leftrightarrow [\mathscr{E}(C, Y)]\}\big)$$
$$\Updownarrow \quad \text{theorem 2.75,}$$
$$\forall X \forall Y\big(\{[\mathscr{E}(X, C)] \Leftrightarrow [\mathscr{E}(X, C)]\} \wedge \{[\mathscr{E}(C, Y)] \Leftrightarrow [\mathscr{E}(C, Y)]\}\big)$$

where each logical formula $[\mathscr{E}(W,Z)] \Leftrightarrow [\mathscr{E}(W,Z)]$ has the pattern of the reflexivity of the logical implication $(P) \Leftrightarrow (P)$ (theorem 1.63). Thus, a complete proof may proceed as follows:

$\vdash (P) \Leftrightarrow (P)$ — theorem 1.63,

$\vdash [\mathscr{E}(X,C)] \Leftrightarrow [\mathscr{E}(X,C)]$ — substitution in $(P) \Leftrightarrow (P)$,

$\vdash [\mathscr{E}(C,Y)] \Leftrightarrow [\mathscr{E}(C,Y)]$ — substitution in $(P) \Leftrightarrow (P)$,

$\vdash \{[\mathscr{E}(X,C)] \Leftrightarrow [\mathscr{E}(X,C)]\} \wedge \{[\mathscr{E}(C,Y)] \Leftrightarrow [\mathscr{E}(C,Y)]\}$ — theorem 1.54,

$\vdash \mathscr{I}(C,C)$ — formula (2.1).

Hence $\vdash \forall C[\mathscr{I}(C,C)]$ results by *Generalization* and theorem 2.75. □

2.87 Theorem (symmetry of defined equality-predicates). *Every equality-predicate defined by formula (2.1) is symmetric: if $\vdash \mathscr{I}(A,B)$, then $\vdash \mathscr{I}(B,A)$; moreover,*

$$\vdash \forall A \forall B\{[\mathscr{I}(A,B)] \Rightarrow [\mathscr{I}(B,A)]\}.$$

Proof. One method to *design* a formal proof transforms the objective, here the *yet unproved* formula $\vdash \forall A \forall B\{[\mathscr{I}(A,B)] \Rightarrow [\mathscr{I}(B,A)]\}$, first into its defining formula (2.1), and then into logically equivalent formulae, for instance, in prenex form, until one such equivalent formula appears that is a theorem, thanks to an axiom or to a previously proven theorem. Here an equivalence will emerge:

$$[\mathscr{I}(A,B)] \Leftrightarrow [\mathscr{I}(B,A)] \quad \textbf{yet unproved,}$$
$$\Updownarrow \quad \text{definition of } \mathscr{I}$$
$$\big(\forall X\{[\mathscr{E}(X,A)] \Leftrightarrow [\mathscr{E}(X,B)]\}\big) \wedge \big(\forall Y\{[\mathscr{E}(A,Y)] \Leftrightarrow [\mathscr{E}(B,Y)]\}\big)$$
$$\Leftrightarrow \big(\forall X\{[\mathscr{E}(X,B)] \Leftrightarrow [\mathscr{E}(X,A)]\}\big) \wedge \big(\forall Y\{[\mathscr{E}(B,Y)] \Leftrightarrow [\mathscr{E}(A,Y)]\}\big),$$

which suggests invoking the symmetry of the logical equivalence $[(P) \Leftrightarrow (Q)] \Leftrightarrow [(Q) \Leftrightarrow (P)]$ (theorem 1.64). Thus, a complete proof may proceed as follows:

$\vdash [(P) \Leftrightarrow (Q)] \Leftrightarrow [(Q) \Leftrightarrow (P)]$ — theorem 1.64,

$\vdash \{[\mathscr{E}(X,A)] \Leftrightarrow [\mathscr{E}(X,B)]\} \Leftrightarrow \{[\mathscr{E}(X,B)] \Leftrightarrow [\mathscr{E}(X,A)]\}$ — substitution,

$\vdash [(R) \Leftrightarrow (S)] \Leftrightarrow [(S) \Leftrightarrow (R)]$ — theorem 1.64,

$\vdash \{[\mathscr{E}(A,Y)] \Leftrightarrow [\mathscr{E}(B,Y)]\} \Leftrightarrow \{[\mathscr{E}(B,Y)] \Leftrightarrow [\mathscr{E}(A,Y)]\}$ — substitution,

$\vdash \{[(P) \Leftrightarrow (Q)] \wedge [(R) \Leftrightarrow (S)]\} \Leftrightarrow \{[(Q) \Leftrightarrow (P)] \wedge [(S) \Leftrightarrow (R)]\}$ — theorem 1.82.

Hence the conclusion results by *Generalization* and theorem 2.75. □

2.88 Theorem (transitivity of defined equality-predicates). *Every equality-predicate defined by formula (2.1) is transitive: if $\vdash \mathscr{I}(A,B)$ and $\vdash \mathscr{I}(B,C)$, then $\vdash \mathscr{I}(A,C)$; moreover,*

$$\vdash \forall A \forall B \forall C\big(\{[\mathscr{I}(A,B)] \wedge [\mathscr{I}(B,C)]\} \Rightarrow [\mathscr{I}(A,C)]\big).$$

Proof. One method to *design* a formal proof transforms the objective, here the *yet unproved* formula $\vdash \forall A \forall B \forall C\big(\{[\mathscr{I}(A,B)] \wedge [\mathscr{I}(B,C)]\} \Rightarrow [\mathscr{I}(A,C)]\big)$, first into its defining formula (2.1), and then into logically equivalent formulae, for instance, in prenex form, until one such equivalent formula appears that is a theorem, thanks to an axiom or to a previously proven theorem.

$$\{[\mathscr{I}(A,B)] \wedge [\mathscr{I}(B,C)]\} \Rightarrow [\mathscr{I}(A,C)] \qquad \textbf{yet unproved,}$$
$$\Updownarrow \quad \text{definition of } \mathscr{I}$$

$$\big[(\forall X\{[\mathscr{E}(X,A)]\Leftrightarrow[\mathscr{E}(X,B)]\}) \wedge (\forall Y\{[\mathscr{E}(A,Y)]\Leftrightarrow[\mathscr{E}(B,Y)]\})$$
$$\wedge(\forall X\{[\mathscr{E}(X,B)]\Leftrightarrow[\mathscr{E}(X,C)]\}) \wedge (\forall Y\{[\mathscr{E}(B,Y)]\Leftrightarrow[\mathscr{E}(C,Y)]\})\big]$$
$$\Rightarrow (\forall X\{[\mathscr{E}(X,A)]\Leftrightarrow[\mathscr{E}(X,C)]\}) \wedge (\forall Y\{[\mathscr{E}(A,Y)]\Leftrightarrow[\mathscr{E}(C,Y)]\}),$$

which suggests invoking the transitivity of the logical equivalence (theorem 1.65):

$$\{[(H) \Leftrightarrow (K)] \wedge [(K) \Leftrightarrow (L)]\} \Rightarrow [(H) \Leftrightarrow (L)],$$

$$\{[(P) \Leftrightarrow (Q)] \wedge [(Q) \Leftrightarrow (R)]\} \Rightarrow [(P) \Leftrightarrow (R)],$$

with the commutativity and associativity of the logical conjunction (theorems 1.57 and 1.66) combined with theorem 1.82:

$$\big(\{[(H) \Leftrightarrow (K)] \wedge [(K) \Leftrightarrow (L)]\} \wedge \{[(P) \Leftrightarrow (Q)] \wedge [(Q) \Leftrightarrow (R)]\}\big)$$
$$\Rightarrow \{[(H) \Leftrightarrow (L)] \wedge [(P) \Leftrightarrow (R)]\}.$$

The conclusion results by *Generalization* and theorem 2.75. □

2.5.3 Other Axiom Systems for the Equality-Predicate

Other axioms systems exist to specify the identity predicate.

Axiom $\mathscr{I}1$ (reflexivity of equality) $\vdash \forall X \, \mathscr{I}(X,X)$.

Axiom $\mathscr{I}2$ (substitutivity of equality) *For every unary predicate variable or predicate constant $\mathscr{F}$, involving only one individual variable,*

$$\vdash \forall X \forall Y \big([\mathscr{I}(X,Y)] \Rightarrow \{[\mathscr{F}(X)] \Rightarrow [\mathscr{F}(Y)]\}\big).$$

For every binary predicate variable or predicate constant $\mathscr{F}$, involving only two individual variables,

$$\vdash \forall X \forall Y \forall W \forall Z\{[\mathscr{I}(X,Y)] \Rightarrow ([\mathscr{I}(W,Z)] \Rightarrow \{[\mathscr{F}(X,W)] \Rightarrow [\mathscr{F}(Y,Z)]\})\}.$$

For every ternary predicate variable or predicate constant $\mathscr{F}$, involving only three individual variables,

$$\vdash \forall U \forall V \forall X \forall Y \forall W \forall Z$$

$$\Big([\mathscr{I}(U,V)] \Rightarrow \{ [\mathscr{I}(X,Y)] \Rightarrow ([\mathscr{I}(W,Z)] \Rightarrow \{ [\mathscr{F}(U,X,W)] \Rightarrow [\mathscr{F}(V,Y,Z)] \}) \} \Big).$$

Similar stipulations also hold for predicate variables or constants involving more than three individual variables.

2.5.4 Defined Ranking-Predicates

Each application with a binary predicate constant $\mathscr{E}$ allows for a corresponding predicate $\mathscr{R}$ of **ranking**, also called **ordering** or **inequality**, defined in terms of $\mathscr{E}$ so that $\mathscr{R}(A,B)$ is an abbreviation of the formula

$$\forall X \{ [\mathscr{E}(X,A)] \Rightarrow [\mathscr{E}(X,B)] \}. \tag{2.3}$$

Formula (2.3) is also denoted by $A \preceq B$ (read "A precedes B") instead of $\mathscr{R}(A,B)$. The resulting predicate $\mathscr{R}$ is reflexive and transitive, but not necessarily symmetric, as verified in the exercises.

2.5.5 Exercises on Equality-Predicates

2.41 . Verify that the ranking-predicate $\mathscr{R}(A,B)$ defined by formula (2.3) is reflexive: prove $\vdash \forall A [\mathscr{R}(A,A)]$.

2.42 . Investigate whether the ranking-predicate $\mathscr{R}(A,B)$ defined by formula (2.3) is symmetric: is $\forall A \forall B \{ [\mathscr{R}(A,B)] \Rightarrow [\mathscr{R}(B,A)] \}$ a theorem?

2.43 . Verify that the ranking-predicate $\mathscr{R}(A,B)$ defined by formula (2.3) is transitive: prove $\vdash \forall A \forall B \forall C ([\mathscr{R}(A,B)] \wedge [\mathscr{R}(B,C)] \} \Rightarrow [\mathscr{R}(A,C)])$.

Exercises 2.45, 2.44, and 2.46 focus on the alternative ranking predicate $\mathscr{A}(A,B)$ defined in terms of the same binary predicate constant $\mathscr{E}$ as an abbreviation of formula (2.4):

$$\forall X \{ [\mathscr{E}(B,Y)] \Rightarrow [\mathscr{E}(A,Y)] \}. \tag{2.4}$$

2.44 . Verify that the alternative ranking-predicate $\mathscr{A}(A,B)$ defined by formula (2.4) is transitive: prove $\vdash \forall A \forall B \forall C ([\mathscr{A}(A,B)] \wedge [\mathscr{A}(B,C)] \} \Rightarrow [\mathscr{A}(A,C)])$.

2.45 . Verify that the alternative ranking-predicate $\mathscr{A}(A, B)$ defined by formula (2.4) is reflexive: prove $\vdash \forall A[\mathscr{A}(A, A)]$.

2.46 . Investigate whether the alternative ranking-predicate $\mathscr{A}(A, B)$ defined by formula (2.4) is symmetric: determine whether $\forall A \forall B\{[\mathscr{A}(A, B)] \Rightarrow [\mathscr{A}(B, A)]\}$ is a theorem.

2.47 . Verify that the equality predicate defined as in example 2.85 for set theory satisfies the alternative axioms $\mathscr{J}1$ and $\mathscr{J}2$ from subsection 2.5.3.

2.48 . Verify that the equality predicate defined as in example 2.85 for set theory satisfies axioms $\mathscr{J}1$ and $\mathscr{J}2$. from subsection 2.5.1.

2.49 . Verify that the equality predicate defined by axioms $\mathscr{J}1$ and $\mathscr{J}2$ from subsection 2.5.1 also satisfies the alternative axioms $\mathscr{J}1$ and $\mathscr{J}2$ from subsection 2.5.3.

2.50 . Verify that the equality predicate defined by axioms $\mathscr{J}1$ and $\mathscr{J}2$ from subsection 2.5.3 also satisfies axioms $\mathscr{J}1$ and $\mathscr{J}2$ from subsection 2.5.1.

Chapter 3
Set Theory: Proofs by Detachment, Contraposition, and Contradiction

3.1 Introduction

This chapter introduces set theory from two parallel perspectives: as an intuitive mathematical theory, and as a simple applied predicate calculus of first order. Starting from first-order logic and some of the Zermelo-Fraenkel axioms (extensionality, empty set, pairing, power set, separation, and union), where all objects under consideration are sets, the chapter first derives relations between sets, subsets, supersets, unions, intersections, and Cartesian products of sets of sets. Subsequent sections introduce relations, functions, injections, surjections, bijections, composite functions, and inverse functions. Another section focuses on the duality between partitions and equivalence relations. The last section deals with pre-orders, partial orders, linear or total orders, and well-orders. Many proofs begin with an informal intuitive proof, then demonstrate how to design a more formal proof, and finally present a detailed outline of such a formal proof in first-order logic. The other Zermelo-Fraenkel axioms (choice and infinity or substitution) are only mentioned here, because they form the topic of subsequent chapters. The prerequisites for this chapter consist of a working knowledge of first-order logic, for instance, as described in chapters 1 and 2, which contain all the logical theorems cited in this chapter.

For some practical problems, features that are essential to their solutions can be specified in terms of **sets** or **collections** of objects.

3.1 Example (Binary arithmetic). The binary arithmetic of computers relies on a *set of two symbols*, 0 and 1, which will be defined with yet other sets in this chapter.

3.2 Example (Geometries). Geometries can be designed entirely with sets. Points are sets (of *sets of coordinates*), while lines, planes, and space are *sets of points*. Points, lines, planes, and space are "primitive" objects that may remain undefined, but relations between them are specified through axioms. For instance, the first axiom of incidence geometry specifies that through any two distinct points passes

© Springer Science+Business Media New York 2015
Y. Nievergelt, *Logic, Mathematics, and Computer Science*,
DOI 10.1007/978-1-4939-3223-8_3

exactly one line [61, p. 3]. Likewise in this chapter, mathematical "sets" are "primitive" objects that remain undefined, while features of sets and relations between sets are specified by axioms.

The foregoing examples already demonstrate a major difficulty in using problems about "real" objects to illustrate logical and mathematical concepts: no exact answers might be available. For an example as elementary as binary arithmetic, electronic digital computers internally do not use anything like the symbols 0 and 1: indeed, they use two electrical potentials confined to two mutually exclusive ranges, neither of which need contain any zero [48, p. 642], with different ranges on different machines [51, p. 60, fig. 3–1; p. 83], [138, p. 4–5, §1], in the reverse order on other machines [103, p. 1–4], sometimes reversing the order within the same machine [23, § 5]. A precise answer would involve more advanced engineering, logic, mathematics, and physics. Therefore, the sets in the present exposition will not contain "real" objects; instead, all the following sets will contain only abstract objects defined by precise rules. The judicious use of such abstract mathematical sets in applied disciplines from astronomy to zoology is precisely the task of such disciplines.

3.2 Sets and Subsets

This section introduces mathematical sets by means of the concept of set membership. The predicate of set membership then allows for the definition of the concepts of subset, superset, and a derived predicate of equality.

3.2.1 Equality and Extensionality

One of the major mathematical achievements around the beginning of the twentieth century was the realization that most of mathematics and computer science consists of logical relations between abstract objects called sets [8, p. 3], [83]. There is no definition of mathematical sets. Indeed, such a definition would have to define sets in terms of yet more foundational objects, but sets *are* the most foundational objects. Henceforth, here and as in other texts [22, p. 50], [141, p. 60], *all mathematical objects are sets, and all quantified variables designate sets*:

> andere Objekte als Mengen existieren für uns überhaupt nicht
> ("for us objects other than sets simply do not exist") [36, p. 271].

Instead of a definition of sets, a few "axioms" specify certain characteristics of sets. Such axioms are also called **postulates** because they are appended to the list of axioms for the underlying logic [18, § 55]. Such axioms of logic describe universal patterns of reasoning applicable in all contexts. In contrast, postulates specify the kind of objects considered in a particular context. The distinction between logical axioms and applied postulates is convenient to distinguish between reasoning and objects. Yet this distinction is somewhat artificial, because changing the logical axioms also changes the provable properties of the applied objects, which amounts to changing the kind of objects under consideration [18, p. 317, footnote 520]. Consequently, following tradition, the postulates of set theory are also labeled here as axioms.

The set theory presented here involves *one* undefined primitive binary relation, denoted by $\in$ and called "membership." The symbol $\in$ is a typographical variation on the lowercase Greek letter ϵ (read "epsilon"), selected here as the first letter of the copula ἐστὶ (pronounced "es-tee"), meaning "is" [36, p. 272]. The notation $X \in Y$ is read in various ways as "X is an element of Y," "X is a member of Y," or "X belongs to Y." For any set X and any set Y, the atomic formula $X \in Y$ has a Truth value, so that $X \in Y$ is either True (if X is an element of Y) or False (if X is not an element of Y). Thus, the following generalization of the tautology $(B) \vee [\neg(B)]$ is universally valid:

$$\vdash \forall X\big(\forall Y\{(X \in Y) \vee [\neg(X \in Y)]\}\big).$$

The relation $\in$ of membership is the only foundational relation between sets. Consequently, the only characteristics of a set are its elements. Because set theory involves only one relation, every set A can relate to other sets X and Y in only four ways:

$$X \in A,$$
$$\neg(X \in A);$$
$$A \in Y,$$
$$\neg(A \in Y).$$

Whether $A \in Y$ or $\neg(A \in Y)$ might also be considered as a characteristic of A. Consequently, rather than involving every element of A (every set X such that $X \in A$), another way to define the characteristics of A might involve every set of which A is an element (every set Y such that $A \in Y$). However, if the only characteristics of a set are its elements, then the two ways ought to be logically equivalent; this equivalence forms the essence of the axiom of extensionality.

Axiom S1 (Axiom of extensionality)

$$\vdash \forall A\big\{\forall B\big[\big(\{\forall X[(X \in A) \Leftrightarrow (X \in B)]\} \Leftrightarrow \{\forall Y[(A \in Y) \Leftrightarrow (B \in Y)]\}\big)\big]\big\}.$$

In the axiom of extensionality (S1), the formula $\forall X[(X \in A) \Leftrightarrow (X \in B)]$ states that each set X is an element of A if and only if X is an element of B. The formula $\forall Y[(A \in Y) \Leftrightarrow (B \in Y)]$ states that A is an element of Y if and only if B is an element of Y. The axiom of extensionality states that these two formulae are logically equivalent.

Yet another way to state that two sets have exactly the same characteristics involves a derived binary relation (derived from $\in$) denoted by $=$ and called "equality." For each set A and each set B, the formula $A = B$ (read "A equals B") means that A and B have exactly the same characteristics. By the axiom of extensionality, the equality of two sets can be stated in two logically equivalent ways:

$$\vdash \forall A\{\forall B[(A = B) \Leftrightarrow \{\forall X[(X \in A) \Leftrightarrow (X \in B)]\}]\},$$

$$\vdash \forall A\{\forall B[(A = B) \Leftrightarrow \{\forall Y[(A \in Y) \Leftrightarrow (B \in Y)]\}]\}.$$

The notation $A = B$ is a shorthand to state that the following two formulae hold:

$$\forall X[(X \in A) \Leftrightarrow (X \in B)],$$

$$\forall Y[(A \in Y) \Leftrightarrow (B \in Y)].$$

The axiom of extensionality states that these two formulae are logically equivalent.

A variation consists in defining $A = B$ as an abbreviation of the first formula, $\forall X[(X \in A) \Leftrightarrow (X \in B)]$, [8, p. 4–5], [36, p. 272–273, Def. 2], and then in adopting the axiom that if $A = B$ then the second formula holds: $\forall Y[(A \in Y) \Leftrightarrow (B \in Y)]$ [36, p. 274, Axiom I].

There is another presentation of set theory with *two* undefined relations, equality ($=$) and membership ($\in$). Then the axiom of extensionality specifies that two sets are the "same" set if and only if they contain the "same" elements. In this exposition the distinction just made does not matter, because equality ($=$) serves only as a shorthand: all operations with sets pertain to elements of those sets. (See also the discussions by Bernays [8, p. 53] and Fraenkel [8, p. 6–8].) For the negations of membership and equality, the following abbreviations prove convenient.

3.3 Definition. The symbols $\notin$ and $\neq$ denote the negations of $\in$ and $=$ so that

$$\vdash \forall X \forall Y\{(X \notin Y) \Leftrightarrow [\neg(X \in Y)]\};$$

$$\vdash \forall A \forall B\{(A \neq B) \Leftrightarrow [\neg(A = B)]\}.$$

At the elementary stage of set theory, most formal logical proofs of relations between sets are straightforward, in the sense that they use only axioms and definitions to establish a sequence of equivalences between the objective of the proof and a theorem or universally valid formula. Such formal logical proofs are usually longer than "informal" proofs. To show a first example of a proof within set theory — a formal version and an informal version — the following theorem states that each set equals itself.

3.4 Remark. In *designing* a proof we may at any stage start from the conclusion — but we may *not* assume it as True — and then search for logically equivalent formulae that connect the conclusion to other formulae that we know how to prove. Smullyan's method of tableaux uses such an approach [117, Ch. II, p. 15–30].

3.5 Theorem. *Each set equals itself: the formula* $\forall S\,(S = S)$ *is universally valid.*

Proof. An informal proof can consist of the following statements.

- Every set X is an element of S if and only if X is an element of S;
- hence $S = S$ by definition of the equality of sets and extensionality (S1).

One method to *design* a formal proof transforms the objective, here the *yet unproved* formula $\forall S\,(S = S)$, into logically equivalent formulae, until one such equivalent formula appears that is a theorem thanks to an axiom or to a previously proven theorem. For instance, substituting S for A and also S for B in the axiom of extensionality gives

$$S = S \quad \textbf{yet unproved,}$$
$$\Updownarrow \quad \text{definition of } =$$
$$\forall X[(X \in S) \Leftrightarrow (X \in S)],$$

which is in prenex form, and where the logical formula $(X \in S) \Leftrightarrow (X \in S)$ has the pattern of the theorem $(P) \Leftrightarrow (P)$. Thus, a complete proof may proceed as follows:

$\vdash (P) \Leftrightarrow (P)$ reflexivity of equivalence (theorem 1.63),
$\vdash (X \in S) \Leftrightarrow (X \in S)$ substitution in the theorem $(P) \Leftrightarrow (P)$,
$\vdash \forall S\{\forall X[(X \in S) \Leftrightarrow (X \in S)]\}$ generalizations, first on X, then on S,
$\vdash \forall S(S = S)$ definition of $=$ and extensionality (S1).

The proof just presented relies on one of the two formulae for the axiom of extensionality: $\forall X[(X \in A) \Leftrightarrow (X \in B)]$. Another proof could rely on the other formula: $\forall Y[(A \in Y) \Leftrightarrow (B \in Y)]$. $\qquad\qquad$ □

More generally, by the properties of a logical equality-predicate derived from a binary predicate, here $\in$, as explained in section 2.5, the equality of sets is

- reflexive: $\vdash \forall S(S = S)$ (every set equals itself), also proved in theorem 3.5,
- symmetric: $\vdash \forall A \forall B[(A = B) \Rightarrow (B = A)]$ (if $A = B$, then $B = A$),
- transitive: $\vdash \forall A \forall B \forall C\{[(A = B) \wedge (B = C)] \Rightarrow (A = C)\}$ (if $A = B$ and $B = C$, then $A = C$),

and equality also allows substitutions of mutually equal sets in theorems.

The axiom of extensionality merely provides two logically equivalent criteria to test whether sets have exactly the same characteristics. However, so far in this theory there is no "set" yet. The "existence" of at least one set — or, more accurately, a convention about an abstract concept of a specific set — requires a second axiom.

3.2.2   The Empty Set

The second axiom, called the **axiom of the empty set,** guarantees the existence of at least one set, denoted by $\varnothing$ or also by $\{\ \ \}$; this set contains no element.

Axiom S2 (Axiom of the empty set) $\vdash \forall X[\neg(X \in \varnothing)]$.

Set theory could dispense with the constant $\varnothing$ and state the axiom of the empty set in the alternative form $\vdash \exists E[\forall X(X \notin E)]$. In the present theory, this alternative form is a consequence of axiom S2. Indeed, with P denoting the formula $\forall X(X \notin E)$, theorem 2.48 becomes

$$[\text{Subf}^E_\varnothing(P)] \Rightarrow [\exists E(P)],$$

$$[\forall X(X \notin E)] \Rightarrow \{\exists E[\forall X(X \notin E)]\}.$$

The alternative form $\vdash \exists E[\forall X(X \notin E)]$ is more cumbersome, because it does not provide a name for any such set.

The determination of the Truth value of an equality $A = B$ requires prior definitions of both sets A and B. In contrast, there exists a different use of the same concept of equality, denoted by $C := D$ (read "let C equal D") to specify a hitherto undefined set C in terms of an already defined set D [59, p. 8], [121, p. 5]. Alternatively, the notation $D =: C$ (also read "let C equal D") can also serve to specify C in terms of D, especially where a derivation leads to a lengthy formula D, which can thus be abbreviated by a shorter variable or string C [121, p. 271, p. 347].

3.6 Example.   The notation $E := \varnothing$ specifies that E stands for $\varnothing$.

3.2.3   Subsets and Supersets

In some circumstances, only some of the elements of a set prove useful; the following definition then allows for the grouping of all such elements into a "subset."

3.7 Definition (Subsets and supersets).   For each set A, for each set B, the set A is a **subset** of the set B if and only if each element of A is also an element of B. Either notation $A \subseteq B$ or $A \subseteqq B$ indicates that "A is a subset of B"; thus,

$$\vdash \forall A\{\forall B[(A \subseteq B) \Leftrightarrow \{\forall X[(X \in A) \Rightarrow (X \in B)]\}]\}.$$

Similarly, a set B is a **superset** of a set A if and only if $A \subseteq B$, a relation also denoted by $B \supseteq A$ or $B \supseteqq A$.

In the definition of subsets, the equivalence ($\Leftrightarrow$) states that the relation of subset ($A \subseteq B$) is logically equivalent to the formula $\forall X[(X \in A) \Rightarrow (X \in B)]$. In this formula, for each set X the logical implication ($\Rightarrow$) states that *if* X is an element of A, *then* X is also an element of B.

The concept of subset is so different from the concept of element as to warrant different terminologies, for instance, reading $A \subseteq B$ as "A is a subset of B" but reading $A \in B$ as "A is an element of B." In contrast, such vague phrases as "A is in B" or "B contains A" do not have any significance, unless they are supplemented with "as an element" or "as a subset" [36, p. 272].

There also exist symbols more specific than $A \subseteq B$. For instance, $A \subset B$ or $A \subsetneq B$ or $A \subsetneqq B$ indicate that A is a subset of B different from B; thus,

$$\vdash \forall A \left(\forall B \left\{ (A \subsetneqq B) \Leftrightarrow (A \subsetneq B) \Leftrightarrow (A \subset B) \Leftrightarrow [(A \subseteq B) \wedge (A \neq B)] \right\} \right).$$

Similarly, $B \supset A$ or $B \supseteq A$ or $B \supsetneqq A$ stand for $(B \supseteq A) \wedge (B \neq A)$.

(However, some authors use $\subset$ to mean $\subseteq$ [140, § 3.12, p. 59–60].)

The following theorems provide further examples and practice with proofs in set theory, and demonstrate how to design a proof. The first three theorems establish features of the concept of subset: reflexivity, anti-symmetry, and transitivity.

3.8 Theorem (reflexivity of $\subseteq$).   *Each set is a subset of itself:* $\vdash \forall S(S \subseteq S)$.

Proof. An informal proof may consist of the single statement that each element of S is also an element of S, whence S is a subset of S, by definition 3.7.

The design of a more formal proof can transform the set-theoretic formula $S \subseteq S$ into a logical formula, and verify that it is a logical theorem:

$$\forall S(S \subseteq S) \quad \textbf{yet unproved,}$$
$$\Updownarrow \quad \text{definition of } \subseteq$$
$$\forall S \left\{ \forall X[(X \in S) \Rightarrow (X \in S)] \right\}$$

which is in the prenex form $\forall S \left\{ \forall X[(P) \Rightarrow (P)] \right\}$, where the matrix is the theorem $(P) \Rightarrow (P)$ (theorem 1.14), here with $X \in S$ instead of P. Thus a complete proof can proceed as follows.

$\vdash (P) \Rightarrow (P)$ theorem from implicational logic (1.14),
$\vdash (X \in S) \Rightarrow (X \in S)$ substitution in theorem 1.14,
$\vdash \forall S \{ \forall X[(X \in S) \Rightarrow (X \in S)] \}$ generalizations, first on X, then on S,
$\vdash \forall S(S \subseteq S)$ definition (3.7) of subsets.

$\square$

3.9 Theorem (transitivity of $\subseteq$).   *For all sets A, B, and C, if $A \subseteq B$ and $B \subseteq C$, then $A \subseteq C$:*

$$\vdash \forall A \forall B \forall C \{ [(A \subseteq B) \wedge (B \subseteq C)] \Rightarrow (A \subseteq C) \}.$$

Proof. An informal proof can proceed as follows.

- If each element of A is an element of B,
- and if each element of B is an element of C,

- then each element of A is an element of B and hence also an element of C,
- whence A is a subset of B, by definition of subset.

The design of a formal proof can unravel the set-theoretic formula $(A \subseteq B) \wedge (B \subseteq C)] \Rightarrow (A \subseteq C)$ into a logical formula by means of the definition of subset, and verify that the resulting formula is a theorem from logic:

$$(A \subseteq B) \wedge (B \subseteq C)] \Rightarrow (A \subseteq C) \qquad \qquad \textbf{yet unproved,}$$

$\updownarrow$ definition of $\subseteq$

$$\big(\{\forall X[(X \in A) \Rightarrow (X \in B)]\} \wedge \{\forall X[(X \in B) \Rightarrow (X \in C)]\}\big)$$
$$\Rightarrow \{\forall X[(X \in A) \Rightarrow (X \in C)]\}$$

$\updownarrow$ theorems 2.54, 2.75, and 2.76,

$$\big[\forall X\big(\{[(X \in A) \Rightarrow (X \in B)]\} \wedge \{[(X \in B) \Rightarrow (X \in C)]\}\big)\big]$$
$$\Rightarrow \{\forall X[(X \in A) \Rightarrow (X \in C)]\}$$

$\Uparrow$ theorem 2.35,

$$\forall X\big\{\big(\{[(X \in A) \Rightarrow (X \in B)]\} \wedge \{[(X \in B) \Rightarrow (X \in C)]\}\big)$$
$$\Rightarrow [(X \in A) \Rightarrow (X \in C)]\big\}$$

which is in prenex normal form, with a matrix of the type

$$\{[(P) \Rightarrow (Q)] \wedge [(Q) \Rightarrow (R)]\} \Rightarrow [(P) \Rightarrow (R)],$$

which is another form of the transitivity of the implication (theorem 1.76). Thus the last line is a theorem. Reversing the order of the steps and inserting $\forall A$ and $\forall B$ before each step (generalizing) then completes the proof. □

3.10 Theorem (anti-symmetry of $\subseteq$). *Two sets are subsets of each other if and only if they equal each other:*

$$\vdash \forall A\big(\forall B\{(A = B) \Leftrightarrow [(A \subseteq B) \wedge (B \subseteq A)]\}\big).$$

Proof. An informal proof can proceed as follows.

- If each element of A is an element of B,
- and if each element of B is an element of A,
- then A and B have exactly the same elements;
- hence $A = B$ by extensionality, and conversely.

A formal proof can establish the following sequence of logical equivalences.

$(A \subseteq B) \wedge (B \subseteq A)$

$\updownarrow$ definition of subset,

$\{\forall X[(X \in A) \Rightarrow (X \in B)]\} \wedge \{\forall X[(X \in B) \Rightarrow (X \in A)]\}$

$\updownarrow$ thanks to $\vdash \{[\forall X(P)] \wedge [\forall X(Q)]\} \Leftrightarrow \{\forall X[(P) \wedge (Q)]\}$, by theorem 2.75,

$\forall X\{[(X \in A) \Rightarrow (X \in B)] \wedge [(X \in B) \Rightarrow (X \in A)]\}$

$\updownarrow$ definition of $\Leftrightarrow$,

$\forall X[(X \in A) \Leftrightarrow (X \in B)]$

$\updownarrow$ axiom of extensionality (S1).

$A = B$

Inserting $\forall A$ and $\forall B$ before each step (generalizing) then completes the proof.   □

The reflexivity, anti-symmetry, and transitivity of the concept of subset also result more generally from the properties of a logical ranking-predicate, here $\subseteq$, derived from a binary predicate, here $\in$, as explained in section 2.5.

The next theorems focus on the subsets and supersets of the empty set.

3.11 Theorem.   *The empty set is a subset of every set:* $\vdash \forall S(\varnothing \subseteq S)$.

Proof.   An informal proof can use the converse law of contraposition (axiom P3):

- Every set not in S is also not in $\varnothing$, because no set belongs to $\varnothing$;
- the contraposition then means that every element in $\varnothing$ is also in S;
- hence $\varnothing$ is a subset of S, by the definition of subsets.

The design of a more formal proof can transform the set-theoretic formula $\varnothing \subseteq S$ into a logical formula, and verify that it is a logical theorem.

$$\forall S(\varnothing \subseteq S) \quad \textbf{yet unproved,}$$
$$\updownarrow \quad \text{definition of } \subseteq$$
$$\forall S\{\forall X[(X \in \varnothing) \Rightarrow (X \in S)]\}$$
$$\updownarrow \quad \text{contraposition and theorem 2.45}$$
$$\forall S\{\forall X[(X \notin S) \Rightarrow (X \notin \varnothing)]\}$$

which is in the prenex form $\forall S \forall X[(P) \Rightarrow (Q)]$, where the matrix $(P) \Rightarrow (Q)$ is a theorem, because so is Q. Thus a complete proof may consist of the following steps.

$\vdash \forall X(X \notin \varnothing)$ axiom of the empty set (S2),

$\vdash X \notin \varnothing$ logical axiom of specialization (Q1),

$\vdash (X \notin S) \Rightarrow (X \notin \varnothing)$ theorem 1.12 from implicational logic,

$\vdash \forall S \{\forall X[(X \notin S) \Rightarrow (X \notin \varnothing)]\}$ generalizations,

$\vdash \forall S \{\forall X[(X \in \varnothing) \Rightarrow (X \in S)]\}$ contraposition and theorem 2.45

$\vdash \forall S(\varnothing \subseteq S)$ definition 3.7 of $\subseteq$.

□

3.12 Theorem. *Every subset of the empty set is the empty set:* $\vdash \forall S[(S \subseteq \varnothing) \Rightarrow (S = \varnothing)]$.

Proof. Theorem 3.11 already guarantees that the empty set is a subset of *every* set: $\vdash \forall S(\varnothing \subseteq S)$, in particular, for every subset S of the empty set, for which the reverse inclusion also holds ($S \subseteq \varnothing$). Hence $S = \varnothing$ by theorem 3.10.

$\vdash \forall S(\varnothing \subseteq S)$ theorem 3.11,

$\vdash \varnothing \subseteq S$ specialization (axiom Q1),

$\vdash (\varnothing \subseteq S) \Rightarrow \{(S \subseteq \varnothing) \Rightarrow [(\varnothing \subseteq S) \wedge (S \subseteq \varnothing)]\}$ theorem 1.54,

$\vdash (S \subseteq \varnothing) \Rightarrow [(\varnothing \subseteq S) \wedge (S \subseteq \varnothing)]$ *Detachment*,

$\vdash (S \subseteq \varnothing) \Rightarrow (S = \varnothing)$ theorem 3.10 and transitivity,

$\vdash \forall S[(S \subseteq \varnothing) \Rightarrow (S = \varnothing)]$ Generalization.

□

The axioms of extensionality (S1) and of the empty set (S2) apply to all sets. Yet so far the theory guarantees the existence of only one set, namely the empty set $\varnothing$, for which $\varnothing \in \varnothing$ is False, but $\varnothing = \varnothing$ and $\varnothing \subseteq \varnothing$ are True. Sets other than the empty set require further axioms, as explained in the next section.

3.2.4 Exercises on Sets and Subsets

3.1 . Write a logical formula stating that a set S is not the empty set.

3.2 . Write a logical formula stating that a set A is not a subset of B.

3.3 . Write a logical formula stating that a set A is not a superset of B.

3.4 . Write a logical formula stating that a set A is not equal to a set B.

3.5 . Prove that $\varnothing \in \varnothing$ is not a theorem.

3.6 . Prove that $\varnothing \subseteq \varnothing$ is a theorem.

3.7 . Prove that $\varnothing = \varnothing$ is a theorem.

3.8 . Use the second formula, $\forall Y[(A \in Y) \Leftrightarrow (B \in Y)]$, in the axiom of extensionality to write a proof that $S = S$ for each set S.

3.9 . Prove that $\varnothing$ is the only subset of $\varnothing$: $\vdash \forall S[(S \subseteq \varnothing) \Rightarrow (S = \varnothing)]$.

3.10 . For each set S, prove that $S \supsetneqq S$ is not a theorem.

3.11 . Prove that if a set S is the only subset of itself, then S is empty.

3.12 . Prove that there does *not* exist a set S such that $S \in Y$ for every set Y.

3.13 . Prove that if a set S is a subset of every set, then S is empty.

3.14 . For all sets A, B, and C, prove that if $A = B$ and $B = C$, then $A = C$.

3.15 . For all sets A, B, and C, prove that if $A \subsetneqq B$ and $B \subsetneqq C$, then $A \subsetneqq C$.

3.16 . For all sets A, B, and C, prove that if $A \subset B$ and $B \subset C$, then $A \subset C$.

3.17 . For all sets A and B, prove that $A = B$ if and only if for every set Z

$$(A \subseteq Z) \Leftrightarrow (B \subseteq Z).$$

3.18 . For all sets A and B, prove that $A \subseteq B$ if and only if for each set Z

$$(B \subseteq Z) \Rightarrow (A \subseteq Z).$$

3.19 . For all sets C and D, prove that $C \supseteq D$ if and only if for each set W

$$(D \supseteq W) \Rightarrow (C \supseteq W).$$

3.20 . For all sets R and S, prove that $R \supseteq S$ if and only if for each set W

$$(W \in S) \Rightarrow (W \in R).$$

3.3 Pairing, Power, and Separation

This section introduces axioms to form sets with one or two elements (pairing), all subsets of a set (power), or selections of specific elements into a subset (separation).

3.3.1 Pairing

A theory allowing for sets other than the empty set requires additional axioms. For instance, the **axiom of pairing** states that for every set H and every set K, there exists a set L, also denoted by $\{H, K\}$, which contains only the elements H and K.

Axiom S3 (Axiom of pairing)

$$\vdash \forall H \{ \forall K [\exists L (\forall X \{ (X \in L) \Leftrightarrow [(X = H) \vee (X = K)] \})] \}.$$

In the axiom of pairing (S3), the equivalence ($\Leftrightarrow$) states that a set X is an element of L if and only if X equals H or X equals K. Because the logical "or" is inclusive, the axiom of pairing thus allows $\{H, K\}$ to contain both H and K. Moreover, because the logical "or" commutes, the order in which H and K appear does not matter.

3.13 Theorem. *For each set H and for each set K, $\vdash \{H, K\} = \{K, H\}$.*

Proof. An informal proof can compare the elements of $\{H, K\}$ and $\{K, H\}$:

* The set $\{H, K\}$ contains the elements H and K, but no other element;
* the set $\{K, H\}$ contains the elements K and H, but no other element;
* thus $\{H, K\}$ and $\{K, H\}$ contain exactly the same elements;
* therefore $\{H, K\} = \{K, H\}$ by the axiom of extensionality (S1).

A formal proof can use $\vdash [(P) \vee (Q)] \Leftrightarrow [(Q) \vee (P)]$ (theorem 1.79) to show that

$$[(X = H) \vee (X = K)] \Leftrightarrow [(X = K) \vee (X = H)]$$

is a theorem, and hence also

$$(X \in \{H, K\}) \Leftrightarrow [(X = H) \vee (X = K)] \Leftrightarrow [(X = K) \vee (X = H)] \Leftrightarrow (X \in \{K, H\}),$$

whence $\vdash \{H, K\} = \{K, H\}$ by the axiom of extensionality (S1).   □

With $H := S$ and $K := S$, the following theorem shows that for each set S, there exists a set, denoted by $\{S\}$, which contains only one element, S.

3.14 Theorem. *For each set S there exists a set L, also denoted by $\{S\}$, which contains only the element S; formally, $\vdash \forall S (\exists L \{\forall X [(X \in L) \Leftrightarrow (X = S)]\})$.*

Proof. An informal proof can use the axiom of pairing with $H := S$ and $K := S$:

* For each set S, the axiom of pairing yields a set $\{S, S\}$;
* by the axiom of pairing, $X \in \{S, S\}$ if and only if $X = S$ or $X = S$;
* yet "$X = S$ or $X = S$" merely repeats the same statement "$X = S$";
* therefore $\{S, S\} = \{S\}$ by the axiom of extensionality.

A formal proof can rely on $\vdash (P) \Leftrightarrow [(P) \vee (P)]$ (theorem 1.68):

$$\vdash \forall H \{\forall K [\exists L (\forall X \{(X \in L) \Leftrightarrow [(X = H) \vee (X = K)]\})]\} \quad \text{axiom S3,}$$
$$\vdash \exists L (\forall X \{(X \in L) \Leftrightarrow [(X = S) \vee (X = S)]\}) \quad \text{Subf}_S^H, \text{Subf}_S^K,$$
$$\vdash \forall S [\exists L (\forall X \{(X \in L) \Leftrightarrow [(X = S) \vee (X = S)]\})] \quad \text{Generalization,}$$
$$\vdash \forall S (\exists L \{\forall X [(X \in L) \Leftrightarrow (X = S)]\}) \quad \vdash (P) \Leftrightarrow [(P) \vee (P)].$$

□

3.15 Definition (singleton). A **singleton** is a set containing exactly one element.

3.16 Example. In theorem 3.14, substituting $\varnothing$ for S gives the set $L = \{\varnothing\}$, which contains the single element $\varnothing$. In particular, $\varnothing \in \{\varnothing\}$, so that $\{\varnothing\}$ is *not* empty.

3.17 Remark. The distinction between $\varnothing$ and $\{\varnothing\}$ is *crucial*. With different notations, to appreciate better the difference between $\varnothing = \{\ \}$ and $\{\varnothing\} = \{\{\ \}\}$, consider $\varnothing$ as an empty bag; then $\{\varnothing\}$ is a bag $\{\ldots\}$ with *another* empty bag $\varnothing$ inside it, also known as a "double bag" in the market place. There, a single empty bag $\{\ \} = \varnothing$ might not be sufficiently strong to hold a six-pack of heavy glass bottles filled with your favorite beverage. (Bottled water, of course, what were you thinking?) That's why you ask for a double bag $\{\varnothing\} = \{\{\ \}\}$ in which to put and then carry the heavy six-pack. If the six-pack also comes with a wrapping, then the combined packaging becomes a "triple bag": $\{\{\varnothing\}\} = \{\{\{\ \}\}\}$, which is *yet another* set.

Theorem 3.18 confirms that the sets $\varnothing$ and $\{\varnothing\}$ have *different* characteristics.

3.18 Theorem. *The sets $\varnothing$ and $\{\varnothing\}$ are two distinct sets: $\varnothing \neq \{\varnothing\}$.*

Proof. An informal proof can utilize substitutions in previous axioms and theorems:

- By definition of the empty set, $\varnothing \notin \varnothing$ (by a substitution in axiom S2);
- moreover, $\varnothing \in \{\varnothing\}$ (by a substitution in theorem 3.14);
- hence the two sets $\varnothing$ and $\{\varnothing\}$ have different elements: $\varnothing \in \{\varnothing\}$ but $\varnothing \notin \varnothing$;
- consequently, $\varnothing \neq \{\varnothing\}$, by the axiom of extensionality.

The following formal proof uses contraposition in the theorems

$$\vdash (P) \Rightarrow \{[(P) \Rightarrow (Q)] \Rightarrow (Q)\} \qquad \text{law of assertion (theorem 1.38),}$$
$$\vdash (P) \Rightarrow ([\neg(Q)] \Rightarrow \{\neg[(P) \Rightarrow (Q)]\}) \qquad \text{theorem 1.54.}$$

whence if P and $\neg(Q)$ are theorems, then so is $\neg[(P) \Rightarrow (Q)]$ by *Detachment*:

$\vdash \neg(\varnothing \in \varnothing)$	$\neg(Q)$: substitution in axiom S2,
$\vdash \varnothing \in \{\varnothing\}$	P: substitution in theorem 3.14,
$\vdash \neg[(\varnothing \in \{\varnothing\}) \Rightarrow (\varnothing \in \varnothing)]$	$\neg[(P) \Rightarrow (Q)]$,
$\vdash \exists X\{\neg[(X \in \{\varnothing\}) \Rightarrow (X \in \varnothing)]\}$	$\text{Subf}^X_\varnothing$ with theorem 2.48,
$\vdash \neg\{\forall X[(X \in \{\varnothing\}) \Rightarrow (X \in \varnothing)]\}$	$\{\exists X[\neg(P)]\} \Leftrightarrow \{\neg[\forall X(P)]\}$ (axiom Q4),
$\vdash \neg(\{\varnothing\} \subseteq \varnothing)$	definition 3.7 of subsets,
$\vdash \varnothing \neq \{\varnothing\}$	contraposition of theorem 3.10 .

$\square$

The distinction between $\varnothing$ and $\{\varnothing\}$ allows for the formation of other sets.

3.19 Example. Substituting $H := \varnothing$ and $K := \{\varnothing\}$ in the axiom of pairing gives the set $L = \{H, K\} = \{\varnothing, \{\varnothing\}\}$. This set L has *two* elements, because $\varnothing \neq \{\varnothing\}$.

3.20 Example. Substituting $S := \{\varnothing\}$ in theorem 3.14 gives the set $L = \{S\} = \{\{\varnothing\}\}$. This set L has *one* element, in effect $\{\varnothing\}$; thus, $\{\varnothing\} \in \{\{\varnothing\}\}$.

3.3.2   Power Sets

For sets with more than two elements, a new axiom becomes necessary. The **axiom of the power set** states that for each set A, the collection of all subsets of A forms a new set, denoted by $\mathscr{P}$ or by $\mathscr{P}(A)$ and called the **power set** of A:

Axiom S4 (Axiom of the power set)

$$\vdash \forall A \big(\exists \mathscr{P} \, \{ \forall S [(S \in \mathscr{P}) \Leftrightarrow (S \subseteq A)] \} \big).$$

In the axiom of the power set (S4), the equivalence ($\Leftrightarrow$) states that a set S is an *element* of the power set $\mathscr{P}(A)$ if and only if S is a *subset* of the set A.

3.21 Example.   The empty set $\varnothing$ is the only subset of itself, by theorem 3.12. Hence its power set has only one element, $\varnothing$, so that $\mathscr{P}(\varnothing) = \{\varnothing\}$.

A set A and its power set $\mathscr{P}(A)$ might have no elements in common. Theorem 3.22 shows that for every element X of A, the singleton $\{X\}$ is an element of $\mathscr{P}(A)$.

3.22 Theorem.   *For all sets A and X, if $X \in A$, then $\{X\} \in \mathscr{P}(A)$, and conversely:*

$$\forall A \forall X \{ (X \in A) \Leftrightarrow [\{X\} \in \mathscr{P}(A)] \}.$$

Proof.   An informal proof can rely on the definitions of subsets and power sets.

- $X \in A$ if and only if $\{X\} \subseteq A$, by definitions of $\{X\}$ and subsets;
- $\{X\} \subseteq A$ if and only if $\{X\} \in \mathscr{P}(A)$ by definition of power sets.

A more formal proof carries out similar verifications from the formal definitions.

$$\forall A \big(\forall X \{ (X \in A) \Leftrightarrow [\{X\} \in \mathscr{P}(A)] \} \big) \quad \textbf{yet unproved},$$
$$\quad \updownarrow \quad \text{definition of power sets,}$$
$$\forall A \big(\forall X \{ (X \in A) \Leftrightarrow [\{X\} \subseteq A] \} \big)$$
$$\quad \updownarrow \quad \text{definition of subsets,}$$
$$\forall A \{ \forall X [(X \in A) \Leftrightarrow \{ \forall Z [(Z \in \{X\}) \Rightarrow (Z \in A)] \}] \}$$
$$\quad \updownarrow \quad \text{definition of } \{X\} \text{ (pairing),}$$
$$\forall A \{ \forall X [(X \in A) \Leftrightarrow \{ \forall Z [(Z = X) \Rightarrow (Z \in A)] \}] \}$$
$$\quad \updownarrow \text{ theorems 2.34 and 2.38,}$$
$$\forall A \{ \forall X \forall Z \{ (X \in A) \Leftrightarrow [(Z = X) \Rightarrow (Z \in A)] \} \}$$

which holds by extensionality and substitutivity of equality.   □

Theorem 3.23 shows that the power set of a singleton has exactly two elements.

3.23 Theorem.   *For each set H, $\mathscr{P}(\{H\}) = \{\varnothing, \{H\}\}$.*

Proof.   An informal proof can list the subsets of $\{H\}$ by cases:

- $\{H\} \subseteq \{H\}$ by theorem 3.8,
- $\varnothing \subseteq \{H\}$ by theorem 3.11,
- $\{\varnothing, \{H\}\} \subseteq \mathscr{P}(\{H\})$ by the previous two lines;
- if $B \subseteq \{H\}$, then B has no element or B has the single element H,
- hence $\{H\}$ has no subsets other than $\varnothing$ and $\{H\}$,
- thus $\mathscr{P}(\{H\}) \subseteq \{\varnothing, \{H\}\}$ by the previous two lines,
- whence $\mathscr{P}(\{H\}) = \{\varnothing, \{H\}\}$.

A formal proof can proceed through two cases, using theorem 1.85:

$$\vdash \big([(P) \Rightarrow (Q)] \wedge \{[\neg(P)] \Rightarrow (Q)\}\big) \Rightarrow (K),$$

with P for $B = \varnothing$ and Q for $B \in \{\varnothing, \{\varnothing\}\}$. In the first case, $(B = \varnothing) \Rightarrow (B \in \{\varnothing, \{\varnothing\}\})$ by substituting $\{\varnothing, \{\varnothing\}\}$ for Y in the axiom of extensionality (S1). In the second case, the definitions of $B \neq \varnothing$ and $B \subseteq \{H\}$ give

$$(B \neq \varnothing) \wedge (B \subseteq \{H\})$$
$$\Updownarrow \qquad \text{definitions of } \varnothing \text{ and } \subseteq,$$
$$[\exists X(X \in B)] \wedge \{\forall Z[(Z \in B) \Rightarrow (Z \in \{H\})]\}$$
$$\Updownarrow \qquad \text{definition of } \{H\} \text{ by theorem 3.14,}$$
$$[\exists X(X \in B)] \wedge \{\forall Z[(Z \in B) \Rightarrow (Z = H)]\}$$
$$\Downarrow \qquad (Z=H) \Rightarrow \{\forall Y[(Z \in Y) \Leftrightarrow (H \in Y)]\}$$
$$\exists X\big[(X \in B) \wedge \{\forall Z[(Z \in B) \Rightarrow (H \in B)]\}\big]$$
$$\Downarrow \qquad \text{specialization Subf}_X^Z.$$
$$\exists X\{(X \in B) \wedge [(X \in B) \Rightarrow (H \in B)]\}$$
$$\Downarrow \qquad \{(P) \wedge [(P) \Rightarrow (Q)]\} \Rightarrow (Q),$$
$$H \in B$$

Thus $B \subseteq \{H\}$, but $H \in B$, whence $\{H\} \subseteq B$, and hence $B = \{H\}$. $\qquad\square$

3.24 Example. The singleton $\{\varnothing\}$ has two subsets: $\{\varnothing\}$ by theorem 3.8, and $\varnothing$ by theorem 3.11. Moreover, $\{\varnothing\}$ has no other subsets. Hence, $\mathscr{P}(\{\varnothing\}) = \{\varnothing, \{\varnothing\}\}$.

3.25 Counterexample. The set $A := \{\{\varnothing\}\}$ has *no* element in common with its power set $\mathscr{P}(\{\{\varnothing\}\}) = \big\{\varnothing, \{\{\varnothing\}\}\big\}$. In particular, A is *not* a subset of $\mathscr{P}(A)$.

The axioms of the empty set (S2), of pairing (S3), and of the power set (S4) allow for increasingly large sets, for instance,

$$\varnothing,$$
$$\mathscr{P}(\varnothing) = \{\varnothing\},$$
$$\mathscr{P}(\{\varnothing\}) = \big\{\varnothing, \{\varnothing\}\big\},$$
$$\mathscr{P}\big(\{\varnothing, \{\varnothing\}\}\big) = \Big\{\varnothing, \{\varnothing\}, \{\{\varnothing\}\}, \{\varnothing, \{\varnothing\}\}\Big\}.$$
$$\vdots$$

However, the "selection" or "separation" of subsets requires yet another axiom.

3.3.3   Separation of Sets

For each set A and for each logical formula P, the **axiom schema of separation** "separates" or "sets aside" from A a subset S consisting of all the elements X of A for which P is True. The logical formula P may *not* involve any free occurrence of the variable S, but it may involve free occurrences of X and of other variables, as indicated in the following axiom by the ellipsis $\ldots$, which may represent other free variables. Such a "separation rule" to form subsets differs from other axioms, because the logic used here allows for the quantification of the elements, with the symbols $\forall X$, but this logic has no provision for the quantification of formulae: it does not allow for expressions like $\forall P$ with P standing for formulae. Thus, the "separation rule" provides a schema for infinitely many axioms, in effect one axiom for each logical formula, whence its name.

Axiom S5 (Axiom schema of separation) For each set A, and for each logical formula P that does *not* contain any free occurrence of the variable S, there exists a subset $S \subseteq A$ that consists of only those elements $X \in A$ for which P is True:

$$\vdash \forall A\big[\exists S\big(\forall X\,\{(X \in S) \Leftrightarrow [(X \in A) \wedge (P)]\}\big)\big].$$

Common notations for the subset S have the formats

$$S = \{X : (X \in A) \wedge (P)\},$$
$$S = \{X \in A : P\}.$$

An alternative notation replaces the colon (:) by a vertical bar (|), but in the context of further mathematics such vertical bars become difficult to recognize against absolute values, norms, and other similar symbols.

3.26 Example. Consider the set

$$A := \mathscr{P}(\{\varnothing, \{\varnothing\}\})$$
$$= \{\,\varnothing,\ \{\varnothing\},\ \{\{\varnothing\}\},\ \{\varnothing, \{\varnothing\}\}\,\},$$

and let P be the formula $\neg(X = \varnothing)$. Then

$$A = \{\,\varnothing,\ \{\varnothing\},\ \{\{\varnothing\}\},\ \{\varnothing, \{\varnothing\}\}\,\},$$
$$(P) \Leftrightarrow [\neg(X = \varnothing)],$$
$$S = \{X : (X \in A) \wedge [\neg(X = \varnothing)]\},$$
$$S = \{X \in A : \neg(X = \varnothing)\},$$
$$S = \{\,\{\varnothing\},\ \{\{\varnothing\}\},\ \{\varnothing, \{\varnothing\}\}\,\}.$$

The existence of this set S would not have followed from the previous axioms.

The following examples demonstrate the axiom schema of separation with various instances of the formula P. Additional examples appear in the next section.

3.27 Example. For each set A and for each set B, let P be $(X \in B)$. Then

$$S = \{X : (X \in A) \land (X \in B)\},$$

which is usually denoted by $S = A \cap B$ and called the **intersection** of A and B.

3.28 Example. For each set A and for each set B, let P be $\neg(X \in B)$. Then

$$S = \{X : (X \in A) \land [\neg(X \in B)]\},$$

which is usually denoted by $S = A \setminus B$ and called the **difference** of A and B.

The symbol $\setminus$ adopted here for the difference of two sets aims at avoiding confusions with the arithmetic difference of sets $A - B$ in such further branches of mathematics as convexity, linear algebra, and functional analysis.

3.29 Example. For each set A, and for all subsets $B \subseteq A$ and $C \subseteq A$, let P be the formula $(X \in B) \lor (X \in C)$. Then

$$S = \{X : (X \in A) \land [(X \in B) \lor (X \in C)]\},$$

which is usually denoted by $S = B \cup C$ and called the **union** of B and C.

Unions and intersections of more than two sets form the topic of the next section.

3.30 Remark. Zermelo introduced the axiom of separation (S5) as the "Axiom der Aussonderung" [145, p. 263, Axiom III], which also translates as the "axiom of triage" or "sifting" [8, p. 11, footnote 2]. With the axiom of separation, the theory presented here can be called Zermelo's set theory. Fraenkel and Skolem substituted for the axiom of separation a more general axiom called the "axiom of replacement" ("Axiom der Erzetzung") [36, p. 309, Axiom VIII], [116], which can be stated with the notation from definition 2.21 [22, p. 52]:

$$\forall Z \ldots \forall W (\{\forall X [\exists! Y(Q)]\} \Rightarrow \forall A \exists B \forall C [(C \in B) \Leftrightarrow (\exists D\{(D \in A) \land [\mathrm{Subf}_D^X \mathrm{Subf}_C^Y(Q)]\})])$$

or, alternatively [128, p. 202],

$$\forall A \{[\forall W \forall X \forall Y \forall Z \{(X \in A) \land [\mathrm{Subf}_Y^W(Q)] \land [\mathrm{Subf}_Z^W(Q)]\} \Rightarrow (Y = Z)]$$
$$\Rightarrow (\exists B \forall Y [(Y \in B) \Leftrightarrow \{\exists X(X \in A) \land [\mathrm{Subf}_Y^W(Q)]\}])\}.$$

This formula can be better described with the concept of "mathematical function" defined in section 3.6. The axiom of separation follows from the axiom of replacement by substituting $(P) \land (W = X)$ for Q.

3.3.4 *Exercises on Pairing, Power, and Separation of Sets*

3.21 . Give examples of sets X, Y, Z with $X \in Y$ and $Y \in Z$ but $X \notin Z$.

3.22 . Provide examples of sets X, Y such that $X \in Y$ but $X \subsetneq Y$.

3.23 . Provide examples of sets X, Y such that $X \subseteq Y$ but $X \notin Y$.

3.24 . Provide examples of sets X, Y such that $X \in Y$ and $X \subseteq Y$.

3.25 . Provide examples of sets A, X such that $X \in A$ but $X \notin \mathscr{P}(A)$.

3.26 . Provide examples of sets A, X such that $X \subseteq A$ but $X \subsetneq \mathscr{P}(A)$.

3.27 . Prove that $\{\varnothing\} \neq \{\{\varnothing\}\}$.

3.28 . Prove that $\{\varnothing\} \neq \{\varnothing, \{\varnothing\}\}$.

3.29 . Prove that $\{\{\varnothing\}\} \neq \{\varnothing, \{\varnothing\}\}$.

3.30 . Determine whether $\{\varnothing\} \subseteq \{\{\varnothing\}\}$.

3.31 . Determine whether $\{\{\varnothing\}\} \subseteq \{\{\varnothing\}\}$.

3.32 . Determine whether $\{\varnothing\} \subseteq \{\varnothing, \{\varnothing\}\}$.

3.33 . Determine whether $\{\{\varnothing\}\} \subseteq \{\varnothing, \{\varnothing\}\}$.

3.34 . Prove that replacing $\vee$ by $\wedge$ in axiom S3 yields the *False* formula

$$\forall H\{\forall K[\exists L(\forall X \{(X \in L) \Leftrightarrow [(X = H) \wedge (X = K)]\})]\}.$$

3.35 . For theorem 3.13, explain how the word "and" in the informal proof corresponds to the logical connective $\vee$ in the formal proof.

3.36 . For each set S, prove that $\{\varnothing, S\} \subseteq \mathscr{P}(S)$.

3.37 . Prove that two sets equal each other if and only if they have the same power sets: $\vdash \forall A(\forall B \{(A = B) \Leftrightarrow [\mathscr{P}(A) = \mathscr{P}(B)]\})$.

3.38 . Identify the set $\{\varnothing\} \cap \{\varnothing, \{\varnothing\}\}$.

3.39 . Identify the set $\{\{\varnothing\}\} \cap \{\varnothing, \{\varnothing\}\}$.

3.40 . Identify the set $\{\varnothing\} \cap \{\{\varnothing\}\}$.

3.41 . Identify the set $\{\varnothing\} \cup \{\varnothing, \{\varnothing\}\}$ and one of its supersets.

3.42 . Identify the set $\{\{\varnothing\}\} \cup \{\varnothing, \{\varnothing\}\}$ and one of its supersets.

3.43 . For each set S, identify the set $S \cap \varnothing$.

3.44 . For each set S, identify the set $S \cup \varnothing$.

3.45 . Prove that $\vdash \forall A[(A \setminus \varnothing) = A]$.

3.46 . Prove that $\vdash \forall A[(A \setminus A) = \varnothing]$.

3.47 . Prove that $\vdash \forall A\big(\forall B\{[(A \setminus B) = \varnothing] \Leftrightarrow (A \subseteq B)\}\big)$.

3.48 . Prove that $\vdash \forall A\big(\forall B\{[(A \setminus B) = A] \Leftrightarrow [(A \cap B) = \varnothing)]\}\big)$.

3.49 . Prove or disprove $\forall A\{\forall B[(A \setminus B) = (B \setminus A)]\}$.

3.50 . Prove or disprove $\forall A\big[\forall B\big(\forall C\{[(A \setminus B) \setminus C] = [A \setminus (B \setminus C)]\}\big)\big]$.

3.4   Unions and Intersections of Sets

3.4.1   Unions of Sets

Many mathematical situations involve unions and intersections of any "collection" or "family" of sets, or, in other words, of any set of sets. For instance, the **Venn diagram** in figure 3.1 (adapted from Hamburger and Pippert's [52, 53]) illustrates all the possible unions and intersections for a set of five sets. Scientists, for instance, biologists, also use such Venn diagrams [7, p. 402, Fig. 2(B)], [90, p. 1472, Fig. 3]. For the existence of such general unions and intersections, however, new axioms become necessary. Thus, the **axiom of union** asserts that for each set $\mathscr{F}$ there exists a new set U, denoted by $\bigcup \mathscr{F}$ and called the **union** of $\mathscr{F}$, which consists of all the elements X of all the elements S of $\mathscr{F}$:

Axiom S6 (Axiom of union)   For each set $\mathscr{F}$, there exists a set U, also denoted by $\bigcup \mathscr{F}$, which consists of all the elements that belong to any element of $\mathscr{F}$:

$$\vdash \forall \mathscr{F}\big(\exists U\{\forall X[(X \in U) \Leftrightarrow \{\exists S[(S \in \mathscr{F}) \wedge (X \in S)]\}]\}\big).$$

Fig. 3.1 Venn diagram for the unions and intersections of five sets [52, 53]. Scientists, for instance, biologists, also use such Venn diagrams [7, p. 402, Fig. 2(B)], [90, p. 1472, Fig. 3].

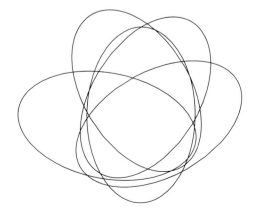

In the axiom of union (S6), the equivalence ($\Leftrightarrow$) states that a set X is an element of the union $U = \bigcup \mathscr{F}$ if and only if there exists an element S of $\mathscr{F}$ such that $X \in S$.

3.31 Example. If $\mathscr{F} = \varnothing$, then $(\bigcup \varnothing) = \varnothing$, because in the axiom of union (S6) the condition $(S \in \mathscr{F})$ is False for every set S, and then the equivalent condition $X \in (\bigcup \varnothing)$ is False for every set X. Thus $\bigcup \varnothing$ contains no element.

For the union of two sets, a special notation proves convenient.

3.32 Definition. The notation $A \cup B$ stands for $\bigcup \{A, B\}$.

With only two sets in $\mathscr{F}$, the definition of $\bigcup \mathscr{F}$ simplifies considerably.

3.33 Theorem. *For all sets A and B, if $\mathscr{F} = \{A, B\}$, then*

$$\forall X \left\{ \left[X \in \bigcup \{A, B\} \right] \Leftrightarrow [(X \in A) \vee (X \in B)] \right\}.$$

Proof. For $\mathscr{F} = \{A, B\}$, the axiom of pairing (S3) shows that

$$(S \in \mathscr{F}) \Leftrightarrow [(S = A) \vee (S = B)],$$

whence the axiom of union (S6) gives the following condition for $X \in (A \cup B)$:

$$\begin{aligned}
[X \in (A \cup B)] &\Leftrightarrow \{\exists S[(S \in \{A, B\}) \wedge (X \in S)]\} \\
&\Leftrightarrow \left(\exists S\{[(S = A) \vee (S = B)] \wedge (X \in S)\} \right) \\
&\Leftrightarrow \left(\exists S\{[(S = A) \wedge (X \in S)] \vee [(S = B) \wedge (X \in S)]\} \right) \\
&\Leftrightarrow \{\exists S[(X \in A) \vee (X \in B)]\} \\
&\Leftrightarrow [(X \in A) \vee (X \in B)].
\end{aligned}$$

The last equivalences result from extensionality, which gives

$$\begin{aligned}
[(S = A) \wedge (X \in S)] &\Leftrightarrow (X \in A), \\
[(S = B) \wedge (X \in S)] &\Leftrightarrow (X \in B).
\end{aligned}$$

Moreover, S does not occur in $(X \in A) \vee (X \in B)$, which makes $\exists S$ superfluous, so that $\{\exists S[(X \in A) \vee (X \in B)]\} \Leftrightarrow [(X \in A) \vee (X \in B)]$ by theorem 2.49. $\square$

3.34 Example. For $A := \{\varnothing, \{\varnothing\}\}$ and $B := \left\{ \{\varnothing, \{\varnothing\}\} \right\}$, the union $A \cup B$ consists of the elements that belong to A (these are $\varnothing$ and $\{\varnothing\}$) or that belong to B (where there is only one element, $\{\varnothing, \{\varnothing\}\}$). Therefore, $A \cup B = \left\{ \varnothing, \{\varnothing\}, \{\varnothing, \{\varnothing\}\} \right\}$.

3.35 Example. Abbreviations are common [81, p. 98], [83, p. 453], [128, p. 129]:

$$0 := \varnothing,$$

$$1 := 0 \cup \{0\} = \varnothing \cup \{\varnothing\} = \{\varnothing\},$$

$$2 := 1 \cup \{1\} = \{\varnothing\} \cup \{\{\varnothing\}\} = \{\varnothing, \{\varnothing\}\},$$

$$3 := 2 \cup \{2\} = \{\varnothing, \{\varnothing\}\} \cup \{\{\varnothing, \{\varnothing\}\}\} = \{\varnothing, \{\varnothing\}, \{\varnothing, \{\varnothing\}\}\},$$

$$4 := 3 \cup \{3\} = \Big\{ \varnothing, \{\varnothing\}, \{\varnothing, \{\varnothing\}\}, \{\varnothing, \{\varnothing\}, \{\varnothing, \{\varnothing\}\}\} \Big\},$$

$$5 := 4 \cup \{4\},$$
$$6 := 5 \cup \{5\},$$
$$7 := 6 \cup \{6\},$$
$$8 := 7 \cup \{7\},$$
$$9 := 8 \cup \{8\}.$$

The sets $0, 1$ are the **binary digits**; $0, 1, 2, 3, 4, 5, 6, 7, 8, 9$ are the **decimal digits**.

3.36 Example.   For all sets A, B, and C, there exists a set V whose only elements are A, B, and C. Indeed, the axiom of pairing yields two sets $\{A, B\}$ and $\{B, C\}$. With

$$\mathscr{F} = \{\, \{A, B\}, \{B, C\} \,\},$$

the axiom of union produces a set $\bigcup \mathscr{F} = \{A, B\} \cup \{B, C\}$ such that for each set X,

$$
\begin{aligned}
[X \in (\{A, B\} \cup \{B, C\})] &\Leftrightarrow [(X \in \{A, B\}) \vee (X \in \{B, C\})] \\
&\Leftrightarrow \{[(X = A) \vee (X = B)] \vee [(X = B) \vee (X = C)]\} \\
&\Leftrightarrow [(X = A) \vee (X = B) \vee (X = C)].
\end{aligned}
$$

The common notation $\{A, B, C\}$ can replace $\{A, B\} \cup \{B, C\}$, so that

$$\{A, B, C\} = \{A, B\} \cup \{B, C\}.$$

3.37 Example.   If the set $\mathscr{F}$ contains only three sets A, B, and C, so that $\mathscr{F} = \{A, B, C\}$, then $\bigcup\{A, B, C\}$ is also denoted by $A \cup B \cup C$, so that

$$A \cup B \cup C = \bigcup\{A, B, C\} = \bigcup \mathscr{F}$$

consists of all the elements that are elements of at least one of A, B, or C.

The following theorems show that unions and intersections of sets have features similar to — and derived from — those of logical disjunctions and conjunctions.

3.38 Theorem. *The union of sets commutes:* $\vdash \forall A \{\forall B[(A \cup B) = (B \cup A)]\}$.

Proof. An informal proof may consist of the statements that a set belongs to $A \cup B$ if and only if it belongs to A or B, which means that it belongs B or A, and hence to $B \cup A$, whence $A \cup B = B \cup A$ by the axiom of extensionality. A formal proof reveals that the commutativity of the union $\cup$ corresponds to the commutativity of the logical connective $\vee$ in the proof of theorem 3.13, which states that $\{A, B\} = \{B, A\}$:

$$A \cup B = \bigcup\{A, B\} = \bigcup\{B, A\} = B \cup A.$$

$\square$

3.39 Theorem. *The union of sets is associative:*

$$\vdash \forall A \forall B \big[\forall C \big(\{[(A \cup B) \cup C] = [A \cup (B \cup C)]\}\big)\big].$$

Proof. An informal proof can merely point out that a set "A or B, or C" is equivalent to "A, or B or C." A formal proof reveals that the associativity of the union $\cup$ corresponds to the associativity of the logical connective $\vee$ in a translation of $[(A \cup B) \cup C] = [A \cup (B \cup C)]$ into a theorem with atomic formulae and connectives:

$$[(A \cup B) \cup C] = [A \cup (B \cup C)] \quad \textbf{yet unproved,}$$
$$\Updownarrow \quad \text{axiom of extensionality (S1),}$$
$$\forall X \,(\{X \in [(A \cup B) \cup C]\}$$
$$\Leftrightarrow \{X \in [A \cup (B \cup C)]\})$$
$$\Updownarrow \quad \text{theorem 3.33 twice,}$$
$$\forall X \,(\{[(X \in A) \vee (X \in B)] \vee (X \in C)\}$$
$$\Leftrightarrow \{(X \in A) \vee [(X \in B) \vee (X \in C)]\})$$

which is in prenex form with a matrix that is a theorem:

$$\{[(P) \vee (Q)] \vee (R)\} \Leftrightarrow \{(P) \vee [(Q) \vee (R)]\},$$

in other words, by associativity of the logical disjunction $\vee$, by theorem 1.71. $\quad\square$

The following theorem shows that if a set B is an *element* of a set $\mathscr{F}$, then B is a *subset* of the union $\bigcup \mathscr{F}$.

3.40 Theorem. *For each set $\mathscr{F}$ and for each set B, if $B \in \mathscr{F}$, then $B \subseteq (\bigcup \mathscr{F})$:*

$$\vdash \forall \mathscr{F} \left(\forall B \left\{(B \in \mathscr{F}) \Rightarrow \left[B \subseteq \left(\bigcup \mathscr{F}\right)\right]\right\}\right).$$

Proof. An informal proof can substitute B for S in the axiom of union (S6):

- If $B \in \mathscr{F}$, then for each $X \in B$ there exists $S \in \mathscr{F}$ with $X \in S$, namely $S = B$;
- hence $X \in \bigcup \mathscr{F}$ for every $X \in B$, by definition of $\bigcup \mathscr{F}$ (axiom S6);
- consequently $B \subseteq \bigcup \mathscr{F}$ by definition of subsets (definition 3.7).

A formal proof consists in transforming the proposed formula until a theorem appears. Because two steps involve not an equivalence but an implication, however, the final proof reorders the investigative five steps in their reverse order:

(5) $\forall \mathcal{F}\big(\forall B\,\{[B \in \mathcal{F}] \Rightarrow [B \subseteq (\bigcup \mathcal{F})]\}\big)$ **yet unproved**,

$\updownarrow$ definitions of $\subseteq$,

(4) $\forall \mathcal{F}\Big[\forall B\big([B \in \mathcal{F}] \Rightarrow \{\forall X\,\{(X \in B) \Rightarrow [X \in (\bigcup \mathcal{F})]\}\}\big)\Big]$

$\updownarrow$ definitions of $\bigcup$,

(3) $\forall \mathcal{F}\Big[\forall B\big([B \in \mathcal{F}] \Rightarrow \{\forall X\big[(X \in B) \Rightarrow \{\exists A[(A \in \mathcal{F}) \wedge (X \in A)]\}\big]\}\big)\Big]$

$\Uparrow$ no X in $(B \in \mathcal{F})$, hence $\{\forall X[(P) \Rightarrow (Q)]\} \Rightarrow \{(P) \Rightarrow [\forall X(Q)]\}$ (theorem 2.38),

(2) $\forall \mathcal{F}\Big\{\forall B\big(\forall X\{[B \in \mathcal{F}] \Rightarrow \big[(X \in B) \Rightarrow \{\exists A[(A \in \mathcal{F}) \wedge (X \in A)]\}\big]\}\big)\Big\}$

$\Uparrow \vdash [\mathrm{Subf}_B^A(P)] \Rightarrow [\exists A(P)]$,

(1) $\forall \mathcal{F}\big(\forall B\{\forall X\big[(B \in \mathcal{F}) \Rightarrow \{(X \in B) \Rightarrow [(B \in \mathcal{F}) \wedge (X \in B)]\}\big]\}\big)$

which holds thanks to $\vdash (P) \Rightarrow \{(Q) \Rightarrow [(P) \wedge (Q)]\}$ (theorem 1.54). $\square$

Theorem 3.41 shows that if $\mathcal{F}$ is a set of subsets of a set A, then its union $\bigcup \mathcal{F}$ is a subset of A.

3.41 Theorem.   *For all sets A and $\mathcal{F}$, if $\mathcal{F} \subseteq \mathcal{P}(A)$, then $\bigcup \mathcal{F} \subseteq A$.*

Proof. By the axiom of union (S6)

$$\forall X \Big\{\Big(X \in \bigcup \mathcal{F}\Big) \Leftrightarrow \{\exists B[(X \in B) \wedge (B \in \mathcal{F})]\}\Big\}.$$

Yet $(B \in \mathcal{F}) \Rightarrow (B \subseteq A)$ by the hypothesis that $\mathcal{F} \subseteq \mathcal{P}(A)$, whence

$$[(X \in B) \wedge (B \in \mathcal{F})] \Rightarrow [(X \in B) \wedge (B \subseteq A)]$$

by theorem 1.82, and hence $X \in A$ by definition 3.7 of subset. Consequently, because $X \in A$ has no free occurrences of B, theorem 2.57 gives

$$\forall X \Big[\Big(X \in \bigcup \mathcal{F}\Big) \Rightarrow (X \in A)\Big],$$

so that $\bigcup \mathcal{F} \subseteq A$ again by definition 3.7 of subset. $\square$

3.4.2 Intersections of Sets

Based on the axiom of union and the axiom of separation, definition 3.42 specifies for each *nonempty* set $\mathscr{F}$ a new set, denoted by $\bigcap \mathscr{F}$ and called the **intersection** of $\mathscr{F}$, which consists of every element X that is an element of every element of $\mathscr{F}$.

3.42 Definition (intersection of sets). For each set $\mathscr{F}$, apply the axiom of union to define $\mathscr{A} := \bigcup \mathscr{F}$, and apply the axiom of separation to the set $\mathscr{A}$ and to the formula

$$\forall Y[(Y \in \mathscr{F}) \Rightarrow (X \in Y)].$$

Then define the **intersection** of $\mathscr{F}$, a set denoted by $\bigcap \mathscr{F}$, through the formula

$$\bigcap \mathscr{F} := \left\{ X \in \bigcup \mathscr{F} : \forall Y[(Y \in \mathscr{F}) \Rightarrow (X \in Y)] \right\},$$

so that

$$\forall X \{ (X \in \bigcap \mathscr{F}) \Leftrightarrow [(X \in \bigcup \mathscr{F}) \wedge \{ \forall Y[(Y \in \mathscr{F}) \Rightarrow (X \in Y)] \}] \}.$$

The definition of the intersection $\bigcap \mathscr{F}$ of a set of sets $\mathscr{F}$ states that a set X is an element of $\bigcap \mathscr{F}$ if and only if $\forall Y[(Y \in \mathscr{F}) \Rightarrow (X \in Y)]$ holds, which occurs if and only if X is an element of every element Y of $\mathscr{F}$. This definition also holds if $\mathscr{F}$ is empty because of the requirement that $\bigcap \mathscr{F}$ first be a subset of the union $\bigcup \mathscr{F}$. (This definition of $\bigcap$ in terms of $\bigcup$ conforms to Bernays's [8, p. 14].) If the set $\mathscr{F}$ contains only two elements, then the definition of $\bigcap \mathscr{F}$ simplifies considerably.

3.43 Theorem. *For all sets A and B, if* $\mathscr{F} = \{A, B\}$ *then*

$$\bigcap \{A, B\} = \{X : [X \in (A \cup B)] \wedge [(X \in A) \wedge (X \in B)]\}.$$

Proof. Apply theorem 1.55, $[(H) \Rightarrow (K)] \Rightarrow ([(H) \Rightarrow (L)] \Rightarrow \{(H) \Rightarrow [(K) \wedge (L)]\})$:

$\vdash \left(\forall Y \{ (Y \in \{A, B\}) \Rightarrow (X \in Y) \} \right)$
$\qquad \Rightarrow [(A \in \{A, B\}) \Rightarrow (X \in A)]$ specialization Subf_A^Y,
$\vdash A \in \{A, B\}$ axiom of pairing (S3),
$\vdash X \in A$ *Detachment*;
$\vdash \left(\forall Y \{ (Y \in \{A, B\}) \Rightarrow (X \in Y) \} \right)$
$\qquad \Rightarrow [(B \in \{A, B\}) \Rightarrow (X \in B)]$ specialization Subf_B^Y,
$\vdash B \in \{A, B\}$ axiom of pairing (S3),
$\vdash X \in B$ *Detachment*;

$$\vdash \big(\forall Y\{(Y \in \{A, B\}) \Rightarrow (X \in Y)\}\big)$$
$$\Rightarrow [(X \in A) \wedge (X \in B)] \qquad \text{theorem 1.55;}$$

$$\vdash (X \in A) \Rightarrow [(A \in \{A, B\}) \Rightarrow (X \in A)] \quad \text{from axiom P1,}$$
$$\vdash (X \in B) \Rightarrow [(B \in \{A, B\}) \Rightarrow (X \in B)] \quad \text{from axiom P1,}$$
$$\vdash [(X \in A) \wedge (X \in B)]$$
$$\Rightarrow \big(\forall Y\{(Y \in \{A, B\}) \Rightarrow (X \in Y)\}\big)$$

which holds thanks to theorem 1.82:

$$\vdash \{[(P) \Rightarrow (Q)] \wedge [(R) \Rightarrow (S)]\} \Rightarrow \{[(P) \wedge (R)] \Rightarrow [(Q) \wedge (S)]\}. \qquad \square$$

For the intersection of two sets, a special notation proves convenient.

3.44 Definition. The notation $A \cap B$ stands for $\bigcap\{A, B\}$.

3.45 Example. For the sets

$$A := \big\{\, \varnothing, \{\varnothing\}, \{\{\varnothing\}\} \,\big\},$$

$$B := \big\{\, \{\varnothing\}, \{\{\varnothing\}\}, \{\{\{\varnothing\}\}\} \,\big\},$$

$A \cap B$ contains only the elements belonging "simultaneously" to both A and B:

$$A \cap B = \big\{\, \varnothing, \{\varnothing\}, \{\{\varnothing\}\} \,\big\} \cap \big\{\, \{\varnothing\}, \{\{\varnothing\}\}, \{\{\{\varnothing\}\}\} \,\big\}$$

$$= \big\{\, \{\varnothing\}, \{\{\varnothing\}\} \,\big\}.$$

Because the definition of the intersection of sets relies upon the conjunction $\wedge$, the intersection has formal features similar to the logical features of the conjunction, for instance, commutativity and associativity, as demonstrated in the following theorems.

3.46 Theorem. *The intersection of sets commutes:*

$$\vdash \forall A \{\forall B[(A \cap B) = (B \cap A)]\}.$$

Proof. An informal proof can state that a set X is an element of A and B if and only if X is an element of B and A. A formal proof shows that the commutativity of the intersection $\cap$ corresponds to the commutativity of the logical connective $\wedge$:

$$(A \cap B) = (B \cap A) \qquad \textbf{yet unproved,}$$
$$\Updownarrow \qquad \text{axiom S1,}$$
$$\forall X \{[X \in (A \cap B)] \Leftrightarrow [X \in (B \cap A)]\}$$
$$\Updownarrow \qquad \text{theorem 3.43,}$$
$$\forall X \{[(X \in A) \wedge (X \in B)] \Leftrightarrow [(X \in B) \wedge (X \in A)]\}$$

which holds by the commutativity $\vdash [(P) \wedge (Q)] \Leftrightarrow [(Q) \wedge (P)]$ (theorem 1.57).   □

3.47 Theorem. *The intersection of sets is associative:*

$$\vdash \forall A \big[\forall B \big(\forall C \{ [(A \cap B) \cap C] = [A \cap (B \cap C)] \} \big) \big].$$

Proof. An informal proof can rely on the equivalence of "*A* and *B*, and *C*" with "*A*, and *B* and *C*." A formal proof shows that the associativity of the intersection ∩ corresponds to the associativity of the logical connective ∧.

$$[(A \cap B) \cap C] = [A \cap (B \cap C)] \quad \textbf{yet unproved,}$$
$$\Updownarrow \quad \text{axiom of extensionality (S1),}$$
$$\forall X \big(\{ X \in [(A \cap B) \cap C] \}$$
$$\Leftrightarrow \{ X \in [[A \cap (B \cap C)]] \} \big)$$
$$\Updownarrow \quad \text{theorem 3.43 twice,}$$
$$\forall X \big(\{ [(X \in A) \wedge (X \in B)] \wedge (X \in C) \}$$
$$\Leftrightarrow \{ (X \in A) \wedge [(X \in B) \wedge (X \in C)] \} \big)$$

which holds thanks to the associativity of ∧ (theorem 1.66):

$$\{ [(P) \wedge (Q)] \wedge (R) \} \Leftrightarrow \{ (P) \wedge [(Q) \wedge (R)] \}.$$ □

Theorem 3.48 shows that the intersection $\bigcap \mathscr{F}$ of a set of sets $\mathscr{F}$ is a subset of every element of $\mathscr{F}$. In other terms, for each set B, if $B \in \mathscr{F}$, then $(\bigcap \mathscr{F}) \subseteq B$.

3.48 Theorem. *For each (nonempty) set $\mathscr{F}$ and for each $B \in \mathscr{F}$, $(\bigcap \mathscr{F}) \subseteq B$:*

$$\vdash \forall \mathscr{F} \left(\forall B \left\{ (B \in \mathscr{F}) \Rightarrow \left[\left(\bigcap \mathscr{F} \right) \subseteq B \right] \right\} \right).$$

Proof. An informal proof can consist of the following statements:

- If $X \in \bigcap \mathscr{F}$, then X is an element of every $Y \in \mathscr{F}$, and hence of $B \in \mathscr{F}$;
- consequently, $(\bigcap \mathscr{F}) \subseteq B$, by definition of subsets.

A formal proof can consist in transforming the proposed formula into its prenex form until a theorem appears in its matrix.

$$\forall \mathscr{F} \left(\forall B \{ (B \in \mathscr{F}) \Rightarrow [(\bigcap \mathscr{F}) \subseteq B] \} \right) \quad \textbf{yet unproved,}$$
$$\Updownarrow \quad \text{definition of } \subseteq,$$
$$\forall \mathscr{F} \left\{ \forall B \big[(B \in \mathscr{F}) \Rightarrow \{ \forall X [(X \in \bigcap \mathscr{F}) \Rightarrow (X \in B)] \} \big] \right\}$$
$$\Updownarrow \quad \text{definition of } \bigcap,$$
$$\forall \mathscr{F} \left\{ \forall B \big[(B \in \mathscr{F}) \Rightarrow$$
$$\{ \forall X [\{ \forall Y [(Y \in \mathscr{F}) \Rightarrow (X \in Y)] \} \Rightarrow (X \in B)] \} \big] \right\}$$
$$\Uparrow \quad \text{no } X \text{ in } B \in \mathscr{F} \text{ (theorem 2.38),}$$

$$\forall \mathscr{F} \Big\{ \forall B \Big[\forall X \big\{ (B \in \mathscr{F}) \Rightarrow$$
$$\big[\{ \forall Y [(Y \in \mathscr{F}) \Rightarrow (X \in Y)] \} \Rightarrow (X \in B) \big] \big\} \Big] \Big\}$$

$\Updownarrow$ theorem 1.37,

$$\forall \mathscr{F} \Big\{ \forall B \Big[\forall X (\{ \overbrace{\forall Y [(Y \in \mathscr{F}) \Rightarrow (X \in Y)]}^{P} \}$$
$$\Rightarrow \big[\underbrace{(B \in \mathscr{F})}_{Q} \Rightarrow \underbrace{(X \in B)}_{R} \big] \big) \Big] \Big\}$$

which holds by specialization (Subf$_B^Y$) and the law of commutation (theorem 1.37):

$$\vdash \big[(P) \Rightarrow \{ (Q) \Rightarrow (R) \} \big] \Leftrightarrow \big[(Q) \Rightarrow \{ (P) \Rightarrow (R) \} \big].$$

□

3.49 Definition (disjoint sets). Two sets A and B are **disjoint** if and only if $A \cap B = \varnothing$. Similarly, a set of sets $\mathscr{F}$ is **pairwise disjoint** if and only if either $A = B$ or $A \cap B = \varnothing$ for all elements A and B of $\mathscr{F}$.

For the union of disjoint sets, a special notation proves convenient.

3.50 Definition (disjoint unions). A union $A \cup B$ is **disjoint** if and only if $A \cap B = \varnothing$; only for disjoint sets, the notation $A \dot{\cup} B$ stands for $A \cup B$. Similarly, a union $\bigcup \mathscr{F}$ is **pairwise disjoint** if and only if either $A = B$ or $A \cap B = \varnothing$ for all elements A and B of $\mathscr{F}$; only for pairwise disjoint sets, the notation $\dot{\bigcup} \mathscr{F}$ stands for $\bigcup \mathscr{F}$.

3.4.3   Unions and Intersections of Sets

Another notation for unions and intersections of a set $\mathscr{G}$ proves convenient, especially where the set $\mathscr{G}$ relates in a specific way to a second set $\mathscr{F}$. The new notation may then "index" the elements of $\mathscr{G}$ with the corresponding elements of $\mathscr{F}$.

3.51 Example. For each set $\mathscr{F} \subseteq \mathscr{P}(U)$ of subsets of a set U, consider the set $\mathscr{G}$ of all the complements $U \setminus S$ of all the elements $S \in \mathscr{F}$; thus,

$$\mathscr{G} := \{ B \in \mathscr{P}(U) : \exists S [(S \in \mathscr{F}) \wedge (B = U \setminus S)] \}.$$

The indexed notation then denotes the union and the intersection of $\mathscr{G}$ as follows:

$$\bigcup_{S \in \mathscr{F}} (U \setminus S) := \bigcup \mathscr{G},$$

$$\bigcap_{S \in \mathscr{F}} (U \setminus S) := \bigcap \mathscr{G}.$$

The notation on the left-hand sides avoids the need to write a formula for the set $\mathscr{G}$.

Because the definitions of the intersection and union of sets rely on conjunction and disjunction, the union and intersection have formal features similar to the logical features of the conjunction and disjunction, for instance, distributivity.

3.52 Theorem (De Morgan's Laws). *For each set U, and for each set $\mathscr{F} \subseteq \mathscr{P}(U)$ of subsets of U, the complement of the union equals the intersection of the complements,*

$$U \setminus \left(\bigcup \mathscr{F}\right) = \bigcap_{A \in \mathscr{F}} (U \setminus A),$$

whereas for each set U, and for each nonempty family $\mathscr{F} \subseteq \mathscr{P}(U)$ of subsets of U, the complement of the intersection equals the union of the complements,

$$U \setminus \left(\bigcap \mathscr{F}\right) = \bigcup_{A \in \mathscr{F}} (U \setminus A).$$

Proof. For the complement of the intersection, an informal proof can proceed as follows.

- For each set X, $X \in U \setminus (\bigcap \mathscr{F})$ if and only if $X \in U$ but $X \notin (\bigcap \mathscr{F})$;
- by definition, $X \notin (\bigcap \mathscr{F})$ if and only if there exists $A \in \mathscr{F}$ with $X \notin A$;
- hence $X \in U \setminus (\bigcap \mathscr{F})$ if and only if $X \in U$ and there exists $A \in \mathscr{F}$ with $X \notin A$;
- equivalently, $X \in U \setminus (\bigcap \mathscr{F})$ if and only if there exists $A \in \mathscr{F}$ with $X \in (U \setminus A)$;
- hence $X \in U \setminus (\bigcap \mathscr{F})$ if and only if $X \in \bigcup_{A \in \mathscr{F}} (U \setminus A)$.

The foregoing informal proof does not justify the permutation of the two statements "$X \in U$" and "there exists $A \in \mathscr{F}$" but the following formal proof justifies such a permutation by the absence of any free occurrence of A in $X \in U$.

$$U \setminus (\bigcap \mathscr{F}) = \bigcup_{A \in \mathscr{F}} (U \setminus A)$$
$$\Updownarrow \quad \text{(axiom S1)},$$
$$\forall X \left\{ [X \in \{U \setminus (\bigcap \mathscr{F})\}] \Leftrightarrow [X \in \bigcup_{A \in \mathscr{F}} (U \setminus A)] \right\}$$
$$\Updownarrow \quad (\setminus),$$
$$\forall X \left\{ [(X \in U) \wedge \{\neg (X \in \bigcap \mathscr{F})\}] \Leftrightarrow [X \in \bigcup_{A \in \mathscr{F}} (U \setminus A)] \right\}$$
$$\Updownarrow \quad \bigcup, \bigcap,$$
$$\forall X \{ [(X \in U) \wedge (\neg \{\forall A [(A \in \mathscr{F}) \Rightarrow (X \in A)]\})]$$
$$\Leftrightarrow \{\exists A [(A \in \mathscr{F}) \wedge \{(X \in U) \wedge [\neg (X \in A)]\}]\}\}$$
$$\Updownarrow \quad \begin{array}{l} \neg [\forall A(P)], \\ \exists A[\neg (P)], \end{array}$$
$$\forall X \{ [(\exists A \{(X \in U) \wedge \{\neg \{[\neg (A \in \mathscr{F})\} \vee (X \in A)]\}\})]$$
$$\Leftrightarrow [\exists A \{(A \in \mathscr{F}) \wedge [(X \in U) \wedge [\neg (X \in A)]]\}]\}$$
$$\Updownarrow \quad \text{De Morgan},$$
$$\forall X \{ [(\exists A \{(X \in U) \wedge \{(A \in \mathscr{F}) \wedge [\neg (X \in A)]\}\})]$$
$$\Leftrightarrow \{\exists A [(A \in \mathscr{F}) \wedge \{(X \in U) \wedge [\neg (X \in A)]\}]\}\}$$

which holds by associativity $\vdash \{(P) \wedge [(Q) \wedge (R)] \Leftrightarrow \{[(P) \wedge (Q)] \wedge (R)\}$ (theorem 1.66).

For the complement of the union, an informal proof can proceed as follows.

- For each set X, $X \in U \setminus (\bigcup \mathscr{F})$ if and only if $X \in U$ but $X \notin (\bigcup \mathscr{F})$;
- by definition, $X \notin (\bigcup \mathscr{F})$ if and only if $X \notin A$ for every $A \in \mathscr{F}$;
- hence $X \in U \setminus (\bigcup \mathscr{F})$ if and only if $X \in U$ and $X \notin A$ for every $A \in \mathscr{F}$;
- equivalently, $X \in U \setminus (\bigcup \mathscr{F})$ if and only if $X \in (U \setminus A)$ for every $A \in \mathscr{F}$;
- hence $X \in U \setminus (\bigcup \mathscr{F})$ if and only if $X \in \bigcap_{A \in \mathscr{F}} (U \setminus A)$.

The foregoing informal proof does not justify the permutation of the two statements "$X \in U$" and "for every $A \in \mathscr{F}$" because it hides the permutation by placing the quantifier at the end of the statements. The following formal proof justifies such a permutation by the absence of any free occurrence of A in $X \in U$.

$$U \setminus (\bigcup \mathscr{F}) = \bigcap_{A \in \mathscr{F}} (U \setminus A)$$
$$\Updownarrow \text{(S1)},$$
$$\forall X \left\{ [X \in U \setminus (\bigcup \mathscr{F})] \Leftrightarrow [X \in \bigcap_{A \in \mathscr{F}} (U \setminus A)] \right\}$$
$$\Updownarrow (\setminus),$$
$$\forall X \left\{ [(X \in U) \wedge \{\neg (X \in \bigcup \mathscr{F})\}] \Leftrightarrow [X \in \bigcap_{A \in \mathscr{F}} (U \setminus A)] \right\}$$
$$\Updownarrow \bigcup, \bigcap,$$
$$\forall X \{[(X \in U) \wedge \{\neg (\exists A \{(A \in \mathscr{F}) \wedge (X \in A)\})\}]$$
$$\Leftrightarrow [(X \in \bigcup_{A \in \mathscr{F}} (U \setminus A)) \wedge \forall A \{(A \in \mathscr{F}) \Rightarrow \{(X \in U) \wedge [\neg (X \in A)]\}\}\}$$
$$\Updownarrow$$
$$\forall X \{[(\forall A \{(X \in U) \wedge [\neg ((A \in \mathscr{F}) \wedge (X \in A))]\})]$$
$$\Leftrightarrow [(X \in U \setminus \bigcap \mathscr{F}) \wedge \forall A \{[\neg (A \in \mathscr{F})] \vee \{(X \in U) \wedge [\neg (X \in A)]\}\}]\}$$
$$\Updownarrow$$
$$\forall X \{[(\forall A \{(X \in U) \wedge [([\neg (A \in \mathscr{F})] \vee [\neg (X \in A)])]\})]$$
$$\Leftrightarrow ([X \in U] \wedge \exists B [(B \in \mathscr{F}) \wedge \{\neg (X \in B)\}])$$
$$\wedge (\forall A \{[\neg (A \in \mathscr{F})] \vee \{(X \in U) \wedge [\neg (X \in A)]\}\})\}$$
$$\Uparrow$$
$$\forall X \{[(\forall A \{(X \in U) \wedge [([\neg (A \in \mathscr{F})] \vee [\neg (X \in A)])]\})]$$
$$\Leftrightarrow [(X \in U) \wedge \forall A \{[\neg (A \in \mathscr{F})] \vee \{(X \in U) \wedge [\neg (X \in A)]\}\}]\}$$

where the first implication, $\Uparrow$, used the assumption that $\mathscr{F} \neq \varnothing$ for the formula beginning with $\exists B \ldots$; the last line follows from the distributivity of $\wedge$ over $\vee$.   □

3.53 Theorem. *For all sets $\mathscr{F}$ and $\mathscr{G}$, intersection distributes over their unions:*

$$\vdash \left[\left(\bigcup \mathscr{F} \right) \cap \left(\bigcup \mathscr{G} \right) \right] = \left[\bigcup \{A \cap B : (A \in \mathscr{F}) \wedge (B \in \mathscr{G})\} \right].$$

Proof. An informal proof can proceed as follows.

- For each set X, $X \in [(\bigcup \mathscr{F}) \cap (\bigcup \mathscr{G})]$ if and only if $X \in (\bigcup \mathscr{F})$ and $X \in (\bigcup \mathscr{G})$;
- equivalently, X belongs to some $A \in \mathscr{F}$ and X belongs to some $B \in \mathscr{G}$,
- equivalently, $X \in (A \cap B)$ for some elements $A \in \mathscr{F}$ and $B \in \mathscr{G}$;
- equivalently, $X \in [\bigcup \{A \cap B : (A \in \mathscr{F}) \wedge (B \in \mathscr{G})\}]$.

A formal proof can proceed as follows.

$$[(\bigcup \mathscr{F}) \cap (\bigcup \mathscr{G})]$$
$$= [\bigcup \{A \cap B : (A \in \mathscr{F}) \wedge (B \in \mathscr{G})\}] \quad \textbf{yet unproved,}$$
$$\Updownarrow \quad \text{extensionality,}$$
$$\forall X (\{X \in [(\bigcup \mathscr{F}) \cap (\bigcup \mathscr{G})]\}$$
$$\Leftrightarrow \{X \in [\bigcup \{A \cap B : (A \in \mathscr{F}) \wedge (B \in \mathscr{G})\}]\})$$
$$\Updownarrow \quad \text{definitions: } \cap, \bigcup,$$
$$\forall X (\{[X \in (\bigcup \mathscr{F})] \wedge [X \in (\bigcup \mathscr{G})]\}$$
$$\Leftrightarrow \left[\exists A (\exists B \{(A \in \mathscr{F}) \wedge (B \in \mathscr{G}) \wedge [X \in (A \cap B)]\})\right])$$
$$\Updownarrow \quad \text{definition of } \bigcup,$$
$$[\forall X (\{\exists A[(A \in \mathscr{F}) \wedge (X \in A)]\} \wedge \{\exists B[(B \in \mathscr{G}) \wedge (X \in B)]\})]$$
$$\Leftrightarrow \left[\exists A (\exists B \{(A \in \mathscr{F}) \wedge (B \in \mathscr{G}) \wedge [X \in (A \cap B)]\})\right]$$
$$\Updownarrow \quad \text{no free } A \text{ or } B,$$
$$[\forall X (\exists A \{\exists B[(A \in \mathscr{F}) \wedge (B \in \mathscr{G}) \wedge (X \in A) \wedge (X \in B)]\})]$$
$$\Leftrightarrow \left[\exists A (\exists B \{(A \in \mathscr{F}) \wedge (B \in \mathscr{G}) \wedge [X \in (A \cap B)]\})\right]$$

which holds by definition of $A \cap B$ (theorem 3.43).   □

3.54 Theorem. *For all sets $\mathscr{F}$ and $\mathscr{G}$, union distributes over their intersections:*

$$\vdash \left[\left(\bigcap \mathscr{F}\right) \cup \left(\bigcap \mathscr{G}\right)\right] = \left[\bigcap \{A \cup B : (A \in \mathscr{F}) \wedge (B \in \mathscr{G})\}\right].$$

Proof. An informal proof can proceed as follows.

- For each set X, $X \in [(\bigcap \mathscr{F}) \cup (\bigcap \mathscr{G})]$ if and only if $X \in (\bigcap \mathscr{F})$ or $X \in (\bigcap \mathscr{G})$;
- equivalently, X belongs to every $A \in \mathscr{F}$ or X belongs to every $B \in \mathscr{G}$,
- equivalently, $X \in (A \cup B)$ for all elements $A \in \mathscr{F}$ and $B \in \mathscr{G}$;
- equivalently, $X \in [\bigcap \{A \cup B : (A \in \mathscr{F}) \wedge (B \in \mathscr{G})\}]$.

A formal proof can proceed as follows.

$$[(\bigcap \mathscr{F}) \cup (\bigcap \mathscr{G})]$$
$$= [\bigcap \{A \cup B : (A \in \mathscr{F}) \wedge (B \in \mathscr{G})\}] \qquad \textbf{unproved,}$$
$$\Updownarrow \qquad \text{(S1)},$$
$$\forall X \, (\{X \in [(\bigcap \mathscr{F}) \cup (\bigcap \mathscr{G})]\}$$
$$\Leftrightarrow \{X \in [\bigcap \{A \cup B : (A \in \mathscr{F}) \wedge (B \in \mathscr{G})\}]\})$$
$$\Updownarrow \qquad \text{def. } \cup, \bigcap,$$
$$\forall X \, (\{[X \in (\bigcap \mathscr{F})] \vee [X \in (\bigcap \mathscr{G})]\}$$
$$\Leftrightarrow \big[\forall A \big(\forall B \{[(A \in \mathscr{F}) \wedge (B \in \mathscr{G})] \Rightarrow [X \in (A \cup B)]\}\big)\big])$$
$$\Updownarrow \qquad \text{definition: } \bigcap,$$
$$[\forall X \, (\{\forall A[(A \in \mathscr{F}) \Rightarrow (X \in A)]\} \vee \{\forall B[(B \in \mathscr{G}) \Rightarrow (X \in B)]\})]$$
$$\Leftrightarrow \big[\forall A \big(\forall B \{[(A \in \mathscr{F}) \wedge (B \in \mathscr{G})] \Rightarrow [X \in (A \cup B)]\}\big)\big]$$
$$\Updownarrow \qquad \text{no free } A, B,$$
$$\{\forall X \, [\forall A \, (\forall B \{[(A \in \mathscr{F}) \Rightarrow (X \in A)] \vee [(B \in \mathscr{G}) \Rightarrow (X \in B)]\})]\}$$
$$\Leftrightarrow \big[\forall A \big(\forall B \{(A \in \mathscr{F}) \wedge (B \in \mathscr{G}) \wedge [X \in (A \cup B)]\}\big)\big]$$
$$\Updownarrow \qquad [(P) \wedge (Q)] \Rightarrow (P),$$
$$\{\forall X \, [\forall A \, (\forall B \{[(A \in \mathscr{F}) \Rightarrow (X \in A)] \vee [(B \in \mathscr{G}) \Rightarrow (X \in B)]\})]\}$$
$$\Leftrightarrow \big[\forall A \big(\forall B \{[(A \in \mathscr{F}) \wedge (B \in \mathscr{G})] \Rightarrow [(X \in A) \vee (X \in B)]\}\big)\big]$$
$$\Updownarrow \qquad \text{definition: } \cup,$$
$$\{\forall X \, [\forall A \, (\forall B \{[(A \in \mathscr{F}) \Rightarrow (X \in A)] \vee [(B \in \mathscr{G}) \Rightarrow (X \in B)]\})]\}$$
$$\Leftrightarrow \big[\forall A \big(\forall B \{[(A \in \mathscr{F}) \wedge (B \in \mathscr{G})] \Rightarrow [X \in (A \cup B)]\}\big)\big]$$

which holds by definition of $A \cup B$ (theorem 3.33).   □

Besides union, intersection, and difference, another operation with sets is useful.

3.55 Definition (symmetric difference). The **symmetric difference** of any sets A and B is denoted by $A \triangle B$ and defined by

$$A \triangle B := (A \cup B) \setminus (A \cap B).$$

3.4.4 Exercises on Unions and Intersections of Sets

3.51 . List all the elements of $\{2, 3, 7\} \cup \{3, 5, 7\}$.

3.52 . List all the elements of $\{4, 6, 8\} \cup \{4, 8, 9\}$.

3.53 . List all the elements of $\{2, 3, 7\} \cap \{3, 5, 7\}$.

3.54 . List all the elements of $\{4, 6, 8\} \cap \{4, 8, 9\}$.

3.55 . List all the elements of $\{2, 3, 7\} \triangle \{3, 5, 7\}$.

3.56 . List all the elements of $\{4, 6, 8\} \triangle \{4, 8, 9\}$.

3.57 . Provide an example of a set $\mathscr{F}$ such that $\bigcup \mathscr{F} = \mathscr{F}$.

3.58 . Provide an example of a set $\mathscr{G}$ such that $\bigcap \mathscr{G} = \mathscr{G}$.

3.59 . Provide an example of a set $\mathscr{F}$ such that $\bigcup \mathscr{F} \neq \mathscr{F}$.

3.60 . Provide an example of a set $\mathscr{G}$ such that $\bigcap \mathscr{G} \neq \mathscr{G}$.

3.61 . Provide examples of sets B and $\mathscr{F}$ such that $B \in \mathscr{F}$ but $B \notin \bigcup \mathscr{F}$.

3.62 . Provide examples of sets B and $\mathscr{G}$ such that $B \in \mathscr{G}$ but $\bigcap \mathscr{G} \notin B$.

3.63 . Provide examples of sets A, B, X, Y such that $X \in A$ and $Y \in B$ but $(X \cup Y) \notin (A \cup B)$.

3.64 . Provide examples of sets A, B, X, Y such that $X \in A$ and $Y \in B$ but $(X \cap Y) \notin (A \cap B)$.

3.65 . For each set A, prove that $\bigcup \{A\} = A$.

3.66 . For each set A, prove that $\bigcup \mathscr{P}(A) = A$.

3.67 . Prove that $\forall A \forall B \forall C \{[(A \cup B) \cap C] = [(A \cap C) \cup (B \cap C)]\}$.

3.68 . Prove that $\forall A \forall B \forall C \{[(A \cap B) \cup C] = [(A \cup C) \cap (B \cup C)]\}$.

3.69 . Prove that $\forall A[(A \cup \varnothing) = A]$.

3.70 . Prove that $\forall A[(A \cap \varnothing) = \varnothing]$.

3.71 . Prove that $\forall A[(A \cup A) = A]$.

3.72 . Prove that $\forall A[(A \cap A) = A]$.

Prove the following formulae for all subsets A and B of each set U.

3.73 . $[U \setminus (A \cap B)] = (U \setminus A) \cup (U \setminus B)$

3.74 . $[U \setminus (A \cup B)] = (U \setminus A) \cap (U \setminus B)$

3.75 . $[(A \setminus B) \setminus U] = (A \setminus U) \setminus (B \setminus U)$

3.76 . $[U \setminus (A \setminus B)] = (U \setminus A) \cup (U \cap B)$

3.77 . $[(A \cup B) \setminus U] = (A \setminus U) \cup (B \setminus U)$

3.78 . $[(A \cap B) \setminus U] = (A \setminus U) \cap (B \setminus U)$

3.79 . $[U \cap (A \setminus B)] = (U \setminus B) \setminus (U \setminus A)$

3.80 . Prove or disprove that $([\mathscr{P}(A)] \cup [\mathscr{P}(B)]) \subseteq \mathscr{P}(A \cup B)$.

3.81 . Prove or disprove that $([\mathscr{P}(A)] \cap [\mathscr{P}(B)]) \supseteq \mathscr{P}(A \cap B)$.

3.82 . Prove or disprove that $([\mathscr{P}(A)] \cap [\mathscr{P}(B)]) \subseteq \mathscr{P}(A \cap B)$.

3.83. Prove or disprove that $([\mathscr{P}(A)] \cup [\mathscr{P}(B)]) \supseteq \mathscr{P}(A \cup B)$.

3.84. Prove or disprove that $([\mathscr{P}(A)] \setminus [\mathscr{P}(B)]) \subseteq \mathscr{P}(A \setminus B)$.

3.85. Prove or disprove that $([\mathscr{P}(A)] \setminus [\mathscr{P}(B)]) \supseteq \mathscr{P}(A \setminus B)$.

3.86. For each nonempty set $\mathscr{F} \subseteq \mathscr{P}(U)$, and each $B \subseteq U$, prove that

$$\left(\bigcap \mathscr{F} \right) \cup B = \bigcap_{A \in \mathscr{F}} (A \cup B).$$

3.87. For each set $\mathscr{F} \subseteq \mathscr{P}(U)$, and for each subset $B \subseteq U$, prove that

$$\left(\bigcup \mathscr{F} \right) \cap B = \bigcup_{A \in \mathscr{F}} (A \cap B).$$

3.88. Prove that $A \triangle \varnothing = A$.

3.89. Prove that $A \triangle A = \varnothing$.

3.90. Prove that the symmetric difference is associative: $(A \triangle B) \triangle C = A \triangle (B \triangle C)$.

3.91. Prove that the symmetric difference commutes: $A \triangle B = B \triangle A$.

3.92. Prove or disprove that $[(A \triangle C) \cup (B \triangle C)] \subseteq [(A \cup B) \triangle C]$.

3.93. Prove or disprove that $[(A \triangle C) \cup (B \triangle C)] \supseteq [(A \cup B) \triangle C]$.

3.94. Prove or disprove that $[(A \triangle C) \cap (B \triangle C)] \subseteq [(A \cap B) \triangle C]$.

3.95. Prove or disprove that $[(A \triangle C) \cap (B \triangle C)] \supseteq [(A \cap B) \triangle C]$.

3.96. Prove or disprove that $[(A \triangle C) \setminus (B \triangle C)] \subseteq [(A \setminus B) \triangle C]$.

3.97. Prove or disprove that $[(A \triangle C) \setminus (B \triangle C)] \supseteq [(A \setminus B) \triangle C]$.

3.98. Prove or disprove that $[(A \cup C) \triangle (B \cup C)] \subseteq [(A \triangle B) \cup C]$.

3.99. Prove or disprove that $[(A \cup C) \triangle (B \cup C)] \supseteq [(A \triangle B) \cup C]$.

3.100. Prove or disprove that $[(A \cap C) \triangle (B \cap C)] \subseteq [(A \triangle B) \cap C]$.

3.101. Prove or disprove that $[(A \cap C) \triangle (B \cap C)] \supseteq [(A \triangle B) \cap C]$.

3.102. Prove or disprove that $[(A \setminus C) \triangle (B \setminus C)] \subseteq [(A \triangle B) \setminus C]$.

3.103. Prove or disprove that $[(A \setminus C) \triangle (B \setminus C)] \supseteq [(A \triangle B) \setminus C]$.

3.104. Prove or disprove that $[(C \setminus A) \triangle (C \setminus B)] \subseteq [C \setminus (A \triangle B)]$.

3.105. Prove or disprove that $[(C \setminus A) \triangle (C \setminus B)] \supseteq [C \setminus (A \triangle B)]$.

3.106. Prove or disprove that $([\mathscr{P}(A)] \triangle [\mathscr{P}(B)]) \subseteq \mathscr{P}(A \triangle B)$.

3.107. Prove or disprove that $([\mathscr{P}(A)] \triangle [\mathscr{P}(B)]) \supseteq \mathscr{P}(A \triangle B)$.

3.108 . Prove that if $A \cap B = \varnothing$, then $A \cup B = (A \setminus B) \dot{\cup} (B \setminus A)$.

3.109 . Prove that $A \cup B = (A \bigtriangleup B) \dot{\cup} (A \cap B)$.

3.110 . Prove that $A \bigtriangleup B = (A \setminus B) \dot{\cup} (B \setminus A)$.

3.5 Cartesian Products and Relations

Beyond logic and sets, much of mathematics consists of connections between types of sets called Cartesian products, mathematical functions, and mathematical relations. These types of sets allow for mathematical specifications, analysis, synthesis, and processing of such concepts as graphs, maps, algorithms, and rankings.

3.5.1 Cartesian Products of Sets

Cartesian products contain certain sets with two elements. Whereas $\{X, Y\} = \{Y, X\}$ for all sets X and Y, however, some situations require a method for listing the elements of a set in a specific order by means of "ordered pairs" or otherwise, for example, in geography and in navigation as in figure 3.2. The following definition (attributed to Wiener and Kuratowski [83, p. 455]) and theorem 3.58 derive such

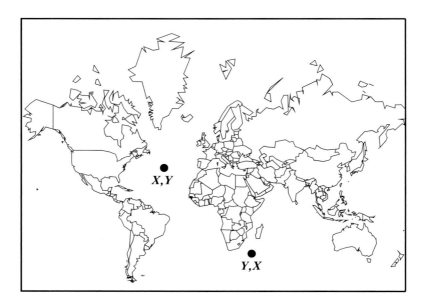

Fig. 3.2 If $X \neq Y$, then $(X, Y) \neq (Y, X)$.

ordered pairs from sets, which shows that the concept of ordered pairs does not require any additional axiom.

3.56 Definition (ordered pairs). For all sets X and Y, the **ordered pair** (X, Y) is the set defined by three applications of the axiom of pairing as

$$(X, Y) = \{\{X\}, \{X, Y\}\}.$$

X is the **first coordinate** of (X, Y), and Y is the **second coordinate** of (X, Y).

3.57 Example.

- If $X = 0$ and $Y = 1$, then $(X, Y) = \{\{0\}, \{0, 1\}\}$.
- If $X = 1$ and $Y = 0$, then $(X, Y) = \{\{1\}, \{1, 0\}\} = \{\{1\}, \{0, 1\}\}$.
- If $X = 0$ and $Y = 0$, then $(X, Y) = \{\{0\}, \{0, 0\}\} = \{\{0\}, \{0\}\} = \{\{0\}\}$.

The following theorem confirms that, in contrast to sets with two elements, ordered pairs record the order of their coordinates.

3.58 Theorem. *For all sets X and Y, if $X \neq Y$, then $(X, Y) \neq (Y, X)$.*

Proof. An informal proof can consist in showing that the two sets (X, Y) and (Y, X) contain different elements, whence $(X, Y) \neq (Y, X)$.

- For all sets X and Y, $X \in \{X\}$ and $Y \in \{Y\}$ by pairing (S3);
- if $X \neq Y$, then $X \notin \{Y\}$ and $Y \notin \{X\}$, by pairing (S3);
- hence if $X \neq Y$, then $\{X\} \neq \{Y\}$ by extensionality (S1);
- from $X \notin \{Y\}$ and $X \in \{Y, X\}$ it follows that $\{X\} \neq \{Y, X\}$ by (S1);
- from $\{X\} \neq \{Y\}$, $\{X\} \neq \{Y, X\}$ follows $\{X\} \notin \{\{Y\}, \{Y, X\}\} = (Y, X)$;
- yet $\{X\} \in \{\{X\}, \{X, Y\}\} = (X, Y)$;
- from $\{X\} \in (X, Y)$ and $\{X\} \notin (Y, X)$ follows $(X, Y) \neq (Y, X)$, by (S1).

A formal proof can parallel the same reasoning.

(1)	$\vdash \forall Z[(Z \in \{X\}) \Leftrightarrow (Z = X)]$	axiom of pairing (S3),
(2)	$\vdash \forall Z[(Z \neq X) \Leftrightarrow (Z \notin \{X\})]$	contraposition and transposition,
(3)	$\vdash (Y \neq X) \Leftrightarrow (Y \notin \{X\})$	specialization Subf_{Y}^{Z},
(4)	$\vdash Y \neq X$	hypothesis,
(5)	$\vdash Y \notin \{X\}$	*Detachment*;
(6)	$\vdash Y \in \{X, Y\}$	axiom of pairing (S3),
(7)	$\vdash \{X\} \neq \{X, Y\}$	(5), (6), and extensionality (S1);
(8)	$\vdash Y \in \{Y\}$	axiom of pairing (S3),
(9)	$\vdash \{X\} \neq \{Y\}$	(5), (8), and extensionality (S1);
(10)	$\vdash \{X\} \notin \{\{Y\}, \{Y, X\}\} = (Y, X)$	(7), (9), and extensionality (S1);
(11)	$\vdash \{X\} \in \{\{X\}, \{X, Y\}\} = (X, Y)$	pairing (S3),
(12)	$\vdash (X, Y) \neq (Y, X)$	(10), (11), and extensionality (S1).

□

The definition of the ordered pair (X, Y) holds for all sets X and Y, in particular, for all elements X and Y of two sets A and B. The following theorem shows that all these ordered pairs (X, Y) are themselves elements of a set.

3.59 Theorem. *For all sets A and B, for each $X \in A$ and for each $Y \in B$, the pair (X, Y) belongs to $\mathscr{P}[\mathscr{P}(A \cup B)]$.*

Proof. An informal proof can trace back the definition of the ordered pair (X, Y):

- From $X \in A$ follows $X \in A \cup B$, because $A \subseteq A \cup B$, whence $\{X\} \in \mathscr{P}(A \cup B)$;
- from $Y \in B$ it follows that $Y \in A \cup B$, because $B \subseteq A \cup B$;
- from $X \in A \cup B$ and $Y \in A \cup B$ it follows that $\{X, Y\} \in \mathscr{P}(A \cup B)$;
- from $\{X\} \in \mathscr{P}(A \cup B)$ and $\{X, Y\} \in \mathscr{P}(A \cup B)$, it follows that $\{\{X\}, \{X, Y\}\} \in \mathscr{P}[\mathscr{P}(A \cup B)]$.

A formal proof can parallel the foregoing argument.

$\vdash X \in A$	hypothesis,
$\vdash A \subseteq (A \cup B)$	theorem 3.40 ,
$\vdash X \in (A \cup B)$	definition of subsets and specialization,
$\vdash \{X\} \subseteq (A \cup B)$	definitions of subsets and singletons,
$\vdash \{X\} \in \mathscr{P}(A \cup B)$	definition of power sets.
$\vdash Y \in B$	hypothesis,
$\vdash Y \in (A \cup B)$	as for $X \in (A \cup B)$,
$\vdash \{X, Y\} \in \mathscr{P}(A \cup B)$	$X \in (A \cup B)$ and $Y \in (A \cup B)$,
$\vdash \{\{X\}, \{X, Y\}\} \subseteq \mathscr{P}(A \cup B)$	$\{X\} \in \mathscr{P}(A \cup B)$, $\{X, Y\} \in \mathscr{P}(A \cup B)$,
$\vdash \{\{X\}, \{X, Y\}\} \in \mathscr{P}[\mathscr{P}(A \cup B)]$	definition of power set.

$\square$

Because every ordered pair (X, Y) belongs to the set $\mathscr{P}[\mathscr{P}(A \cup B)]$, the axiom of separation guarantees that the collection of all such ordered pairs is a set.

3.60 Definition (Cartesian product). For all sets A and B, the **Cartesian product** of A and B is the set $A \times B$ (read "A cross B"), consisting of all ordered pairs (X, Y) with $X \in A$ and $Y \in B$. Thus,

$$A \times B$$

$$= \left\{ C \in \mathscr{P}\left[\mathscr{P}(A \cup B)\right] : \exists X \big(\exists Y \{(X \in A) \wedge (Y \in B) \wedge [C = (X, Y)]\}\big) \right\}$$

$$= \{(X, Y) : (X \in A) \wedge (Y \in B)\}$$

with $\exists X \big(\exists Y \{(X \in A) \wedge (Y \in B) \wedge [C = (X, Y)]\}\big)$ for P in the axiom of separation (S5). The sets A and B are the **factors** of the Cartesian product $A \times B$.

A common graphical representation of a Cartesian product $A \times B$ lists all the elements of A along a horizontal axis, and all the elements of B along a vertical axis, so that the element (X, Y) of $A \times B$ appears directly above X and across from Y.

3.61 Example. For the sets

$$A := \{0, 1, 2\},$$
$$B := \{0, 1\},$$

the Cartesian product $A \times B$ takes the form

$$A \times B = \{(0, 0), (1, 0), (2, 0), (0, 1), (1, 1), (2, 1)\},$$

with the following graphical representation:

B $\qquad\qquad\qquad\qquad\qquad A \times B$

1 $\qquad (0, 1)\ (1, 1)\ (2, 1)$

0 $\qquad (0, 0)\ (1, 0)\ (2, 0)$

$\qquad\quad 0 \quad\ 1 \quad 2 \qquad A$

For practice and for future use, the following theorems establish relations between Cartesian products and other operations with sets.

3.62 Theorem. *For all sets A, B, C, and D,*

$$[(A \cap C) \times (B \cap D)] = [(A \times B) \cap (C \times D)].$$

Proof. An informal proof can establish that $[(A \cap C) \times (B \cap D)]$ and $[(A \times B) \cap (C \times D)]$ have exactly the same elements. These two sets are Cartesian products, and consequently their elements are ordered pairs.

- An ordered pair (X, Y) is an element of $(A \cap C) \times (B \cap D)$ if and only if $X \in (A \cap C)$ and $Y \in (B \cap D)$;
- hence $(X, Y) \in [(A \cap C) \times (B \cap D)]$ if and only if $X \in A$ and $X \in C$, and $Y \in B$ and $Y \in D$,
- which is equivalent to $X \in A$ and $Y \in B$, and $X \in C$ and $Y \in D$,
- which is equivalent to $(X, Y) \in (A \times B)$ and $(X, Y) \in (C \times D)$,
- which is thus equivalent to $(X, Y) \in [(A \times B) \cap (C \times D)]$.

As the preceding informal proof swapped $X \in C$ and $Y \in B$, a formal proof can rely on the commutativity $[(P) \wedge (Q)] \Leftrightarrow [(Q) \wedge (P)]$ of $\wedge$ by theorem 1.57:

$$(X, Y) \in [(A \cap C) \times (B \cap D)]$$

$\Updownarrow$ definition of Cartesian products,

$$[X \in (A \cap C)] \wedge [Y \in (B \cap D)]$$

$\Updownarrow$ definition of intersection,

$$[(X \in A) \wedge (X \in C)] \wedge [(Y \in B) \wedge (Y \in D)]$$

$\Updownarrow$ commutativity and associativity of $\wedge$,

$$[(X \in A) \wedge (Y \in B)] \wedge [(X \in C) \wedge (Y \in D)]$$

$\Updownarrow$ definition of Cartesian products,

$$[(X, Y) \in (A \times B)] \wedge [(X, Y) \in (C \times D)]$$

$\Updownarrow$ definition of intersection.

$$(X, Y) \in [(A \times B) \cap (C \times D)]$$

$\square$

3.63 Theorem. *For all sets A, B, C, and D,*

$$[(A \times B) \cup (C \times D) \cup (A \times D) \cup (C \times B)] = [(A \cup C) \times (B \cup D)].$$

Proof. An informal proof can establish that $[(A \times B) \cup (C \times D) \cup (A \times D) \cup (C \times B)]$ and $[(A \cup C) \times (B \cup D)]$ have exactly the same elements.

- An ordered pair (X, Y) is an element of $(A \times B) \cup (C \times D) \cup (A \times D) \cup (C \times B)$ if and only if $(X, Y) \in (A \times B)$, or $(X, Y) \in (C \times D)$, or $(X, Y) \in (A \times D)$, or $(X, Y) \in (C \times B)$,
- which is equivalent to $X \in A$ and $Y \in B$, or $X \in C$ and $Y \in D$, or $X \in A$ and $Y \in D$, or $X \in C$ and $Y \in B$,
- which is equivalent to $X \in A$ or $X \in C$, and $Y \in B$ or $Y \in D$,
- which is equivalent to $(X, Y) \in [(A \cup C) \times (B \cup D)]$.

A formal proof translates the alleged equivalences into the logical theorem

$$\{[(P) \wedge (Q)] \vee [(P) \wedge (S)] \vee [(R) \wedge (Q)] \vee [(R) \wedge (S)]\} \Leftrightarrow \{[(P) \vee (R)] \wedge [(Q) \vee (S)]\},$$

which follows from the distributivity of $\wedge$ over $\vee$ (theorem 1.75), with

$$(P) \Leftrightarrow (X \in A), \quad (Q) \Leftrightarrow (Y \in B), \quad (R) \Leftrightarrow (X \in C), \quad (S) \Leftrightarrow (Y \in D).$$

Thus,

$$(X, Y) \in [(A \times B) \cup (C \times D) \cup (A \times D) \cup (C \times B)]$$

$\Updownarrow$ definition of $\bigcup$,

$$[(X, Y) \in (A \times B)] \vee [(X, Y) \in (C \times D)]$$
$$\vee [(X, Y) \in (A \times D)] \vee [(X, Y) \in (C \times B)]$$

$\Updownarrow$ definition of $\times$,

$$[(X \in A) \wedge (Y \in B)] \vee [(X \in C) \wedge (Y \in D)]$$
$$\vee [(X \in A) \wedge (Y \in D)] \vee [(X \in C) \wedge (Y \in B)] \vee$$

$\Updownarrow$ theorem 1.75,

$$[(X \in A) \vee (X \in C)] \wedge [(Y \in B) \vee (Y \in D)]$$

$\Updownarrow$ definition of $\bigcup$,

$$[X \in (A \cup C)] \wedge [Y \in (B \cup D)]$$

$\Updownarrow$ definition of $\times$.

$$(X, Y) \in [(A \cup C) \times (B \cup D)]$$

$\square$

3.5.2   *Cartesian Products of Unions and Intersections*

The foregoing theorems generalize to Cartesian products of unions or intersections of any sets of sets.

3.64 Theorem. *For all sets of sets $\mathscr{F}$ and $\mathscr{G}$,*

$$\left(\bigcup \mathscr{F}\right) \times \left(\bigcup \mathscr{G}\right) = \bigcup_{(A,B) \in \mathscr{F} \times \mathscr{G}} (A \times B).$$

Proof. An informal proof can show that $(\bigcup \mathscr{F}) \times (\bigcup \mathscr{G})$ and $\bigcup_{(A,B) \in \mathscr{F} \times \mathscr{G}} (A \times B)$ have exactly the same elements.

- A pair (X, Y) is an element of $(\bigcup \mathscr{F}) \times (\bigcup \mathscr{G})$ if and only if $X \in (\bigcup \mathscr{F})$ and $Y \in (\bigcup \mathscr{G})$,
- which is equivalent to $X \in A$ for some $A \in \mathscr{F}$ and $Y \in B$ for some $B \in \mathscr{G}$,
- which is equivalent to $(X, Y) \in (A \times B)$ for some $(A, B) \in (\mathscr{F} \times \mathscr{G})$.

A formal proof can parallel the foregoing reasoning.

$$(X, Y) \in (\bigcup \mathscr{F}) \times (\bigcup \mathscr{G})$$

$\Updownarrow$ definition of $\times$,

$$[X \in (\bigcup \mathscr{F})] \wedge [Y \in (\bigcup \mathscr{G})]$$

$\Updownarrow$ definition of $\bigcup$,

$$\{\exists A[(A \in \mathscr{F}) \wedge (X \in A)]\} \wedge \{\exists B[(B \in \mathscr{G}) \wedge (Y \in B)]\}$$

$\Updownarrow$ no B, no A,

$$\exists A (\exists B \{[(A \in \mathscr{F}) \wedge (X \in A)] \wedge [(B \in \mathscr{G}) \wedge (Y \in B)]\})$$

$\Updownarrow$ properties of $\wedge$,

$$\exists A (\exists B \{[(A \in \mathscr{F}) \wedge (B \in \mathscr{G})] \wedge [(X \in A) \wedge (Y \in B)]\})$$

$\Updownarrow$ definition of $\times$,

$$\exists A (\exists B \{[(A \times B) \in (\mathscr{F} \times \mathscr{G})] \wedge [(X, Y) \in (A \times B)]\})$$

$\Updownarrow$ definition of $\bigcup$.

$$(X, Y) \in \bigcup_{(A \times B) \in (\mathscr{F} \times \mathscr{G})} (A \times B)$$

$\square$

3.65 Theorem. *For all sets of sets $\mathscr{F}$ and $\mathscr{G}$,*

$$\left(\bigcap \mathscr{F}\right) \times \left(\bigcap \mathscr{G}\right) = \bigcap_{(A,B)\in\mathscr{F}\times\mathscr{G}} (A \times B).$$

Proof. An informal proof can show that $\left(\bigcap \mathscr{F}\right) \times \left(\bigcap \mathscr{G}\right)$ and $\bigcap_{(A,B)\in\mathscr{F}\times\mathscr{G}}(A \times B)$ have exactly the same elements.

- A pair (X, Y) is an element of $\left(\bigcap \mathscr{F}\right) \times \left(\bigcap \mathscr{G}\right)$ if and only if $X \in \left(\bigcap \mathscr{F}\right)$ and $Y \in \left(\bigcap \mathscr{G}\right)$,
- which is equivalent to $X \in A$ for every $A \in \mathscr{F}$ and $Y \in B$ for every $B \in \mathscr{G}$,
- which is equivalent to $(X, Y) \in (A \times B)$ for every $(A, B) \in (\mathscr{F} \times \mathscr{G})$.

The foregoing informal proof glosses over the case in which at least one of $\mathscr{F}$ or $\mathscr{G}$ is empty, which would require invoking the corresponding unions as supersets of the intersections, because of the definition of intersections. One remedy could consist in proving such cases separately. Indeed, $\mathscr{F} = \varnothing$ or $\mathscr{G} = \varnothing$ if and only if $(\mathscr{F} \times \mathscr{G}) = \varnothing$. A formal proof can parallel the foregoing reasoning with theorem 1.82,

$$\vdash \{[(P) \Rightarrow (Q)] \wedge [(R) \Rightarrow (S)]\} \Rightarrow \{[(P) \wedge (R)] \Rightarrow [(Q) \wedge (S)]\},$$

to prove that

$$\left(\bigcap \mathscr{F}\right) \times \left(\bigcap \mathscr{G}\right) \subseteq \bigcap_{(A,B)\in\mathscr{F}\times\mathscr{G}} (A \times B),$$

with the hypotheses $C \in \mathscr{F}$ for P and $D \in \mathscr{G}$ for R for the converse, so that

$$\vdash \{[(P) \Rightarrow (Q)] \wedge [(R) \Rightarrow (S)]\} \Leftarrow \left([(P) \wedge (R)] \wedge \{[(P) \wedge (R)] \Rightarrow [(Q) \wedge (S)]\}\right).$$

Thus,

$$(X, Y) \in \left(\bigcap \mathscr{F}\right) \times \left(\bigcap \mathscr{G}\right)$$

$\Updownarrow$ definition of $\times$,

$$[X \in \left(\bigcap \mathscr{F}\right)] \wedge [Y \in \left(\bigcap \mathscr{G}\right)]$$

$\Updownarrow$ $\bigcap \mathscr{H} \subseteq \bigcup \mathscr{H}$,

$$\left(\{\forall A[(A \in \mathscr{F}) \Rightarrow (X \in A)]\} \wedge \{\forall B[(B \in \mathscr{G}) \Rightarrow (Y \in B)]\}\right)$$
$$\wedge \left(\{\exists C[(C \in \mathscr{F}) \wedge (X \in C)]\} \wedge \{\exists D[(D \in \mathscr{G}) \wedge (Y \in D)]\}\right)$$

$\Updownarrow$ no B, no A,

$$\left[\forall A\left(\forall B\{[(A \in \mathscr{F}) \Rightarrow (X \in A)] \wedge [(B \in \mathscr{G}) \Rightarrow (Y \in B)]\}\right)\right]$$
$$\wedge \left[\exists C\left(\exists D\{[(C \in \mathscr{F}) \wedge (X \in C) \wedge (D \in \mathscr{G}) \wedge (Y \in D)]\}\right)\right]$$

$\Updownarrow$ theorems,

$$\big[\forall A\big(\forall B\{[(A \in \mathscr{F}) \wedge (B \in \mathscr{G})] \Rightarrow [(X \in A) \wedge (Y \in B)]\}\big)\big]$$
$$\wedge\big[\exists C\big(\exists D\{[(C \in \mathscr{F}) \wedge (X \in C) \wedge (D \in \mathscr{G}) \wedge (Y \in D)]\}\big)\big]$$

$\Updownarrow$ definition of $\times$,

$$\big[\forall A\big(\forall B\{[(A, B) \in (\mathscr{F} \times \mathscr{G})] \Rightarrow [(X, Y) \in (A \times B)]\}\big)\big]$$
$$\wedge\big[\exists C\big(\exists D\{[(C, D) \in (\mathscr{F} \times \mathscr{G})] \wedge [(X, Y) \in (C \times D)]\}\big)\big]$$

$\Updownarrow$ definition of $\bigcap$.

$$(X, Y) \in \bigcap\nolimits_{(A \times B) \in (\mathscr{F} \times \mathscr{G})} (A \times B)\}$$

$\square$

3.5.3 *Mathematical Relations and Directed Graphs*

The Cartesian product $A \times B$ provides a means to draw connections, or, in other words, to specify relations, between elements of the two sets A and B.

3.66 Definition (relation). A **relation** between elements of sets A and B is a subset $R \subseteq A \times B$ of their Cartesian product.

- Two elements $X \in A$ and $Y \in B$ are **related** with respect to the relation R if and only if $(X, Y) \in R$.
- For each relation $R \subseteq A \times B$, the **domain** $\mathscr{D}(R) \subseteq A$ of the relation R consists of those elements of A related by R to at least one element of B:

$$\mathscr{D}(R) = \{X \in A : \exists Y\{(Y \in B) \wedge [(X, Y) \in R]\}\}.$$

- Similarly, the **range** $\mathscr{R}(R) \subseteq B$ of the relation R consists of those elements of B related by R to at least one element of A:

$$\mathscr{R}(R) := \{Y \in B : \exists X\{(X \in A) \wedge [(X, Y) \in R]\}\}.$$

Another common notation for $(X, Y) \in R$ is XRY, especially if such a special symbol as $\subseteq$, $\subsetneqq$, or $=$ denotes the relation. Relations may also be represented by graphs.

3.67 Definition (directed graph). A **directed graph** is an ordered pair $G := (V, E)$ with a relation $E \subseteq V \times V$ between elements of the same set $A = V = B$. Elements of V are called **vertices**. A pair $(X, Y) \in E$ is called an **edge from X to Y**:

3.68 Example. For each set A, the **diagonal** $\varDelta_A \subseteq A \times A$ is the subset

$$\varDelta_A = \{(X, X) \in A \times A : X \in A\}.$$

Thus, the diagonal corresponds to the relation called "equality," or, equivalently, "identity": a pair of elements (X, Y) lies on the diagonal Δ_A if and only if $X = Y$. Because $X = X$, and hence $(X, X) \in \Delta_A$, for every $X \in A$, it follows that the diagonal Δ_A relates every element of A to itself, whence $\mathscr{D}(R) = A = \mathscr{R}(R)$. For example, if

$$A := \{ \varnothing, \{\varnothing\} \},$$

then

$$A \times A = \left\{ \begin{array}{cc} (\varnothing, \{\varnothing\}) \ (\{\varnothing\}, \{\varnothing\}) \\ (\varnothing, \varnothing) \quad (\{\varnothing\}, \varnothing) \end{array} \right\},$$

$$\Delta_A = \left\{ \begin{array}{cc} (\{\varnothing\}, \{\varnothing\}) \\ (\varnothing, \varnothing) \end{array} \right\},$$

because the only pairs (X, Y) with $X = Y$ are the pairs $(\varnothing, \varnothing)$ and $(\{\varnothing\}, \{\varnothing\})$. As a graph, the diagonal consists of a single loop at each vertex:

3.69 Example. For all sets H and K, consider their respective power sets $A := \mathscr{P}(H)$ and $B := \mathscr{P}(K)$. The relation $R \subseteq \mathscr{P}(H) \times \mathscr{P}(K)$ of **inclusion** consists of all, but only those, pairs (V, W) of subsets $V \subseteq H$ and $W \subseteq K$ such that $V \subseteq W$:

$$R = \{(V, W) \in \mathscr{P}(H) \times \mathscr{P}(K) : (V \subseteq H) \wedge (W \subseteq K) \wedge (V \subseteq W)\}.$$

The domain of the relation $\subseteq$ consists of all subsets of A included as a subset in at least one subset of B. Similarly, the range of the relation $\subseteq$ consists of all subsets of B that contain as a subset at least one subset of A.

3.70 Example. For all sets H and K, let $A := \mathscr{P}(H)$ and $B := \mathscr{P}(K)$. The relation $S \subseteq \mathscr{P}(H) \times \mathscr{P}(K)$ of **strict inclusion** consists of all, but only those, pairs (V, W) of subsets $V \subseteq H$ and $W \subseteq K$ such that $V \subseteq W$ but $V \neq W$:

$$S = \{(V, W) \in \mathscr{P}(H) \times \mathscr{P}(K) : (V \subseteq H) \wedge (W \subseteq K) \wedge (V \subseteq W) \wedge (V \neq W)\}.$$

The strict inclusion $(V \subseteq W) \wedge (V \neq W)$, is also denoted by $V \subset W$ or by $V \subsetneqq W$:

$$\vdash [(V \subseteq W) \wedge (V \neq W)] \Leftrightarrow (V \subset W) \Leftrightarrow (V \subsetneqq W).$$

The domain of the relation of strict inclusion consists of all subsets of A included in, but not equal to, at least one subset of B; its range consists of all subsets of B that contain, but do not coincide with, at least one subset of A.

3.71 Example. For all sets A and B, the relational constant $\in$ relates every element $X \in A$ that is an element of some $Y \in B$. Denote this relation on $A \times B$ by E. (The symbol $\in$ cannot denote a subset E of $A \times B$ because $\in$ is neither an individual variable nor an individual constant.) Thus, $E \subseteq (A \times B)$ is defined by

$$[(X, Y) \in E] \Leftrightarrow [(X \in A) \wedge (X \in Y) \wedge (Y \in B)].$$

For example, if

$$A := \{\, \varnothing, \ \{\varnothing\} \,\},$$
$$B := \{\, \varnothing, \ \{\varnothing\} \,\},$$

then

$$A \times B = \left\{ \begin{array}{c} (\varnothing, \{\varnothing\}) \ (\{\varnothing\}, \{\varnothing\}) \\ (\varnothing, \varnothing) \quad (\{\varnothing\}, \varnothing) \end{array} \right\},$$

$$E = \left\{ \begin{array}{c} (\varnothing, \{\varnothing\}) \end{array} \right\},$$

because in $A \times B$ the only pair (X, Y) such that $X \in Y$ is the pair $(\varnothing, \{\varnothing\})$.

3.72 Example. For the sets

$$A := \{\, \varnothing, \ \{\varnothing\} \,\},$$
$$B := \{\, \varnothing, \ \{\varnothing\}, \ \{\, \varnothing, \{\varnothing\} \,\} \,\},$$

the Cartesian product $A \times B$ and the relation E take the forms

$$A \times B = \left\{ \begin{array}{c} \big(\varnothing, \{\, \varnothing, \{\varnothing\} \,\}\big) \ \big(\{\varnothing\}, \{\, \varnothing, \{\varnothing\} \,\}\big) \\ (\varnothing, \{\varnothing\}) \qquad (\{\varnothing\}, \{\varnothing\}) \\ (\varnothing, \varnothing) \qquad (\{\varnothing\}, \varnothing) \end{array} \right\},$$

$$E = \left\{ \begin{array}{c} \big(\varnothing, \{\, \varnothing, \{\varnothing\} \,\}\big) \ \big(\{\varnothing\}, \{\, \varnothing, \{\varnothing\} \,\}\big) \\ (\varnothing, \{\varnothing\}) \end{array} \right\},$$

because in $A \times B$ the only pairs (X, Y) such that $X \in Y$ are those just displayed.

In some contexts a relation $R \subseteq A \times B$ may also prove useful with B and A listed in the reverse order. Then the "inverse" relation $R^{\circ -1} \subseteq B \times A$ contains similar pairs as R does but with coordinates listed in the reverse order. Some texts denote the inverse relation by R^{-1}, which can cause confusion because the same notation also represents reciprocals in arithmetic and algebra. The notation adopted here for the inverse relation, $R^{\circ -1}$, conforms to [70].

3.73 Definition (inverse relation). The **inverse** of a relation $R \subseteq A \times B$ between two sets A and B is the relation $R^{\circ -1} \subseteq B \times A$ defined by

$$R^{\circ -1} = \{(Y, X) \in B \times A : (X, Y) \in R\}.$$

3.74 Example. The inverse of the diagonal is the same diagonal.

3.75 Example. The inverse of the subset relation $\subseteq$ is the superset relation $\supseteq$.

3.76 Definition (composite relation). The **composition** of two relations $R \subseteq A \times B$ and $S \subseteq B \times C$ is the relation $S \circ R$ on $A \times C$ defined as follows. The **composite** relation $S \circ R$ relates an element $U \in A$ to an element $W \in C$ if and only if R relates U to some element $V \in B$ and S relates the same element V to $W \in C$:

$$(S \circ R) = \{(U, W) \in (A \times C) : \exists V([V \in B] \wedge [(U, V) \in R] \wedge [(V, W) \in S])\}.$$

3.77 Example. For each relation $R \subseteq A \times B$ between any sets A and B, the composition $R \circ R^{\circ -1}$ contains the diagonal $\triangle_{\mathscr{R}(R)}$ for the range of R. Indeed, by definition of its range, R relates every $Y \in \mathscr{R}(R)$ to an element $X \in A$ so that $(X, Y) \in R$; then the inverse $R^{\circ -1}$ relates Y to X so that $(Y, X) \in R^{\circ -1}$. From $(Y, X) \in R^{\circ -1}$ and $(X, Y) \in R$ it follows that $(Y, Y) \in (R \circ R^{\circ -1})$ for every $(Y, Y) \in \triangle_{\mathscr{R}(R)}$. Therefore $\triangle_{\mathscr{R}(R)} \subseteq (R \circ R^{\circ -1})$. Similarly, $R^{\circ -1} \circ R$ contains the diagonal $\triangle_{\mathscr{D}(R)}$ for the domain of R. Indeed, by definition of its domain, R relates every $X \in \mathscr{D}(R)$ to an element $Y \in B$ so that $(X, Y) \in R$; then the inverse $R^{\circ -1}$ relates Y to X so that $(Y, X) \in R^{\circ -1}$. From $(X, Y) \in R$ and $(Y, X) \in R^{\circ -1}$ it follows that $(X, X) \in (R^{\circ -1} \circ R)$ for every $(X, X) \in \triangle_{\mathscr{D}(R)}$. Therefore $\triangle_{\mathscr{D}(R)} \subseteq (R^{\circ -1} \circ R)$.

Because every relation R between sets A and B is a subset $R \subseteq (A \times B)$, operations with sets apply to all relations.

3.78 Definition (unions and intersections of relations). For all sets A, B, C, E, and for all relations $R \subseteq A \times B$ and $T \subseteq C \times E$, the **union** of the relations R and T is the relation $R \cup T$ between $A \cup C$ and $B \cup E$, so that

$$(R \cup T) := \{(X, Y) \in [(A \cup C) \times (B \cup E)] : [(X, Y) \in R] \vee [(X, Y) \in T]\}.$$

Similarly, the **intersection** of the relations R and T is the relation $R \cap T$ between $A \cap C$ and $B \cap E$, so that

$$(R \cap T) = \{(X, Y) \in [(A \cap C) \times (B \cap E)] : [(X, Y) \in R] \wedge [(X, Y) \in T]\}.$$

A particular instance of intersections of relations $R \subseteq (A \times B)$ and $T \subseteq (A \times B)$ consists of the intersection of R with a subset $S \subseteq A$ and its Cartesian product with B. The "restriction" of R to S is then the relation $R \cap T$ with $T = (S \times B)$. The concept of the "restriction" of a relation to a subset is useful if the subset has characteristics that are useful for the purpose at hand while the complement of the subset does not.

3.79 Definition (restricted relation). For each relation $R \subseteq (A \times B)$, and for each subset $S \subseteq A$ of A, the **restriction of R to S** is the relation $R|_S \subseteq (S \times B)$ defined by

$$R|_S = R \cap (S \times B)$$
$$= \{(X, Y) \in R : X \in S\}.$$

Thus, $R|_S$ restricts its first coordinates only to those elements of S.

There also exists a similar instance of intersections of relations $R \subseteq (A \times B)$ and $T \subseteq (A \times B)$ as the intersection of R with a subset $V \subseteq B$ and its Cartesian product with A. The "restriction" of R to V is then the relation $R \cap T$ with $T = (A \times V)$. Similarly, the restriction of R to $S \subseteq A$ and $V \subseteq B$ is the relation $R \cap (S \times V)$.

3.5.4 Exercises on Cartesian Products of Sets

3.111 . Determine whether the Cartesian product is associative.

3.112 . For each set A, prove that $A \times \varnothing = \varnothing$, and that $\varnothing \times A = \varnothing$.

3.113 . Prove that $A \times B = \varnothing$ if and only if $A = \varnothing$ or $B = \varnothing$.

3.114 . Prove that $\times$ distributes over $\cap$: $[(A \times B) \cap (C \times B)] = [(A \cap C) \times B]$.

3.115 . Prove that $\times$ distributes over $\cup$: $[(A \times B) \cup (C \times B)] = [(A \cup C) \times B]$.

3.116 . Prove that $[(A \setminus C) \times B] = [(A \times B) \setminus (C \times B)]$.

3.117 . Prove that $[A \times (B \setminus D)] = [(A \times B) \setminus (A \times D)]$.

3.118 . Prove or disprove that $[(A \setminus C) \times (B \setminus D)] \subseteq [(A \times B) \setminus (C \times D)]$.

3.119 . Prove or disprove that $[(A \setminus C) \times (B \setminus D)] \supseteq [(A \times B) \setminus (C \times D)]$.

3.120 . Prove or disprove that $[(A \Delta C) \times (B \Delta D)] \subseteq [(A \times B) \Delta (C \times D)]$.

3.121 . Prove or disprove that $[(A \Delta C) \times (B \Delta D)] \supseteq [(A \times B) \Delta (C \times D)]$.

3.122 . Prove or disprove that $([\mathscr{P}(A)] \times [\mathscr{P}(B)]) \subseteq \mathscr{P}(A \times B)$.

3.123 . Prove or disprove that $([\mathscr{P}(A)] \times [\mathscr{P}(B)]) \supseteq \mathscr{P}(A \times B)$.

3.124 . Provide a formula for the inverse of the relation of inclusion.

3.125 . Give a formula for the inverse of the relation of strict inclusion.

3.126 . For each relation $R \subseteq A \times B$, and for each subset $S \subseteq A$ of A, prove that $R|_S = R \cap (S \times B)$.

3.127 . For all sets A and B, prove that $\varnothing$ is an element of the domain $\mathscr{D}$ of the relation $\subseteq$ on $\mathscr{P}(A) \times \mathscr{P}(B)$.

3.128 . For all sets A and B, prove that $\mathscr{P}(B)$ is the range $\mathscr{R}$ of the relation $\subseteq$ on $\mathscr{P}(A) \times \mathscr{P}(B)$.

3.129 . Provide examples of sets A and B, such that $\mathscr{P}(A)$ is the domain $\mathscr{D}$ of the relation $\subseteq$ on $\mathscr{P}(A) \times \mathscr{P}(B)$.

3.130 . Provide examples of sets A and B, such that $\mathscr{P}(A)$ is *not* the domain $\mathscr{D}$ of the relation $\subseteq$ on $\mathscr{P}(A) \times \mathscr{P}(B)$, so that $\mathscr{D} \subsetneq \mathscr{P}(A)$.

3.6   Mathematical Functions

Functions are relations relating exactly one element of their domain to each element of their range.

3.6.1   *Mathematical Functions*

In some applications of relations, the domain and the range can contain measurements.

3.80 Example. Results from astronomical observations can consist of a relation between two coordinates of position, with ordered pairs (X, Y) where X is the observed ascension (elevation) and Y is the observed declination (azimuth) of an asteroid. For example, the following pairs (X, Y) record the ascension X and the declination Y of the asteroid Pallas, measured by Baron von Zach about 1800 A.D. [14, p. 5]:

$$(0, 408) \quad (30, 89) \quad (60, -66) \quad (90, 10) \quad (120, 338) \quad (150, 807)$$
$$(180, 1238) \quad (210, 1511) \quad (240, 1583) \quad (270, 1462) \quad (300, 1183) \quad (330, 804)$$

with the corresponding graphical representation in figure 3.3.

Results from astronomical observations can also consist of a relation between time and position, with ordered pairs (T, Y) where Y is the observed position (declination or azimuth) of a planet at time T. Such a relation has the following properties.

- If no observation was made at some time T, then the results do not contain any ordered pair with T as their first coordinate.
- Each observation yields only one position Y at any time T.
- Observations can yield the same position Y at several times, for instance, if the motion is periodic.

Mathematical "functions" are relations corresponding to such applications.

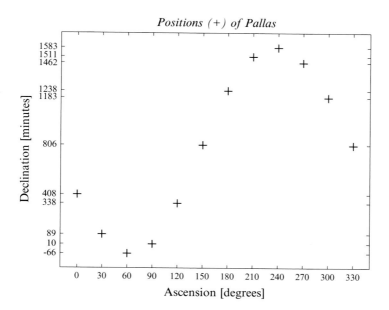

Fig. 3.3 Positions include at most one declination for each ascension.

3.81 Definition. A **function** from a set A to a set B is a relation $F \subseteq (A \times B)$ such that for each $X \in A$ there exists *at most one* $Y \in B$ for which $(X, Y) \in F$.

The **domain** of F is the subset $\mathscr{D}(F) \subseteq A$ consisting of every $X \in A$ for which there exists some $Y \in B$ such that $(X, Y) \in F$.

$$\mathscr{D}(F) = \{X \in A : \exists Y[(Y \in B) \wedge \{(X, Y) \in F\}]\}.$$

The **range** of F, denoted by $\mathscr{R}(F)$, consists of every $Y \in B$ such that there exists some $X \in A$ such that $(X, Y) \in F$:

$$\mathscr{R}(F) = \{Y \in B : \exists X[(X \in A) \wedge \{(X, Y) \in F\}]\}.$$

Moreover, B is called the **co-domain** of F.

For each $X \in \mathscr{D}(F)$, the unique Y such that $(X, Y) \in F$ is called the **value of F at X**, or also the **image of X by F**. The same Y is denoted by $F(X)$ (read "F of X"); thus, $Y = F(X)$ if and only if $(X, Y) \in F$. The element X is called the **argument** of F in the expression $F(X)$. The notation

$$F : A \to B$$

(read "F maps A to B") means that F is a function from A to B; the notation $X \mapsto F(X)$ (read "F maps X to $F(X)$") may specify $F(X)$ by a formula or otherwise.

3.82 Remark. A variant of the concept of a function from a set A to a set B is a subset $F \subseteq (A \times B)$ such that for each $X \in A$ there exists *exactly one* $Y \in B$ for which $(X, Y) \in F$, and then the notation $F : A \to B$ implies that $\mathscr{D}(F) = A$.

The requirement that $\mathscr{D}(F) = A$ remains harmless with simple examples of functions, but it presents unnecessary obstacles with more realistic examples of functions, whose complexity can make the domain difficult or impossible to identify, especially if the identification of the domain is irrelevant to the task at hand.

The specification with *"at most"* in definition 3.81 is common in set theory [128, p. 58 & 86], algebraic geometry (especially with "rational" functions) [136, p. 34–35], complex analysis (especially with "meromorphic" functions) [1, p. 128], and functional analysis [130, p. 4 & 18], in instances where several functions called "operators" have different domains but arise from a relation common to all of them.

3.83 Example. The pairs in example 3.80 define a function $F : A \to B$ from the set of ascensions $A := \{0, 30, 60, 90, 120, 150, 180, 210, 240, 270, 300, 330\}$ to the set of declinations $B := \{-66, 10, 89, 338, 408, 804, 807, 1183, 1238, 1462, 1511, 1583\}$. This function $F : A \to B$ has domain A, codomain B, and range B.

3.84 Example. With $A := \{6, 7, 8, 9\}$ and $B := \{1, 2, 3, 4\}$, let

$$F := \{(6, 2), (7, 1), (8, 2), (9, 3)\}.$$

Then $F : A \to B$ is a function with domain A, codomain B, and range $\{1, 2, 3\}$.

3.85 Example. For all nonempty sets A and B, and for any element $Z \in B$, there exists a **constant function**

$$\begin{aligned} C_Z : A &\to B, \\ X &\mapsto Z. \end{aligned}$$

The constant function C_Z maps every element $X \in A$ to the same value Z, so that

$$C_Z := \{(X, Z) : X \in A\}.$$

In particular, $\mathscr{D}(C_Z) = A$ for the domain of C_Z, and $\mathscr{R}(C_Z) = \{Z\}$ for its range.

3.86 Example. For each set A, there exists an **identity function**

$$\begin{aligned} I_A : A &\to A, \\ X &\mapsto X. \end{aligned}$$

Thus the identity function I_A maps every element to itself, so that

$$I_A := \{(X, X) : X \in A\}.$$

In particular, $\mathscr{D}(I_A) = A$ and $\mathscr{R}(I_A) = A$, because $(X, X) \in I_A$ for *every* $X \in A$. The function I_A is also denoted by Δ_A and is called the the "diagonal" of $A \times A$.

3.87 Example. For all sets A and B, the **canonical projection functions** from the Cartesian product $A \times B$ into its factors A and B are the functions P_A and P_B with

$$P_A : (A \times B) \to A,$$
$$(X, Y) \mapsto X;$$

$$P_B : (A \times B) \to B,$$
$$(X, Y) \mapsto Y.$$

Thus, P_A maps (X, Y) to its first coordinate X in A, whereas P_B maps (X, Y) to its second coordinate Y in B. The domain of P_A and P_B is $A \times B$, but the range of P_A is A, whereas the range of P_B is B.

3.88 Example. For all sets A and B, for any subset $V \subseteq A$ and any element $Z \in B$, the **slice function** $S_{V,Z}$ maps each element $X \in V$ to $(X, Z) \in (A \times B)$:

$$S_{V,Z} : V \to (A \times B),$$
$$X \mapsto (X, Z).$$

Thus, $S_{V,Z}$ maps its domain V to the **slice** $V \times \{Z\}$ in $A \times B$.

3.89 Example. For each set A and each subset $S \subseteq A$, the **characteristic function** χ_S, with the Greek letter χ (read "chi"), maps every element of the subset S to 1, and every element outside the subset S to 0, (with $0 = \varnothing$, $1 = \{\varnothing\}$, and $2 = \{0, 1\}$):

$$\chi_S : A \to 2,$$
$$X \mapsto \begin{cases} 1 \text{ if } X \in S, \\ 0 \text{ if } X \notin S. \end{cases}$$

The following theorem provides a means to compare two functions to each other.

3.90 Theorem. *Two functions $F : A \to B$ and $G : A \to B$ are equal, $F = G$, if and only if they have the same domain $D \subseteq A$ and $F(X) = G(X)$ for every $X \in D$.*

Proof. This proof rewrites $F(X) = G(X)$ in terms of the definition of functions:

$$F = G$$
$$\Updownarrow \quad \text{extensionality},$$
$$\forall X \big(\forall Y \{ [(X, Y) \in F] \Leftrightarrow [(X, Y) \in G] \} \big)$$
$$\Updownarrow \quad \text{functional notation}.$$
$$\forall X \big(\forall Y \{ [Y = F(X)] \Leftrightarrow [Y = G(X)] \} \big)$$

$\square$

Some situations involve only parts of a function, or combinations of several functions, for instance, as defined by the concepts introduced here.

3.91 Definition (restriction of functions). For each function $F : A \to B$, and for each subset $S \subseteq A$ of A, the **restriction of F to S** is the function $F|_S \subseteq S \times B$ defined by

$$F|_S : S \to B,$$
$$X \mapsto F(X).$$

Thus, $F|_S(X) = F(X)$, but $F|_S$ maps only the elements of $S \cap \mathscr{D}(F)$.

3.92 Example. With $A := \{6, 7, 8, 9\}$ and $B := \{1, 2, 3, 4\}$, let

$$F := \{(6, 2), (7, 1), (8, 2), (9, 3)\}.$$

Then the restriction of F to the subset $S := \{6, 8\}$ is $F|_S = \{(6, 2), (8, 2)\}$.

3.93 Example. For each set B, and for each subset $W \subseteq B$, the **inclusion function**, denoted by the Greek letter ι ("iota"), is the restriction of the identity function to W:

$$\iota_W : W \to B,$$
$$X \mapsto X.$$

Thus $\iota_W : W \to B$ is the restriction of $I_B : B \to B$ to the subset $W \subseteq B$.

3.94 Definition (union of functions). For all *disjoint* sets A and C, so that $A \cap C = \varnothing$, for all sets B and E, and for all functions $F : A \to B$ and $G : C \to E$, the **union** of the functions F and G is the union $F \dot\cup G$ of the two sets $F \subseteq A \times B$ and $G \subseteq C \times E$:

$$F \dot\cup G : (A \dot\cup C) \to (B \cup E),$$

$$X \mapsto \begin{cases} F(X) \text{ if } X \in \mathscr{D}(F) \subseteq A, \\ G(X) \text{ if } X \in \mathscr{D}(G) \subseteq C. \end{cases}$$

3.95 Example (overloading operators). In a computer language such as C++ "overloading" the addition $(+)$ from numbers to pairs of numbers corresponds to forming the union of the addition function F defined on a set A of numbers and an addition function G defined on a set C of pairs of numbers.

3.96 Definition (intersection of functions). For all sets A, B, C, E, and all functions $F : A \to B$ and $G : C \to E$, the **intersection** of the functions F and G is the intersection $F \cap G$ of the two sets $F \subseteq A \times B$ and $G \subseteq C \times E$, so that

$$F \cap G : D \to (B \cap E),$$
$$X \mapsto F(X) = G(X),$$

with domain $D := \mathscr{D}(F \cap G) = \{X \in A \cap C : F(X) = G(X)\}$.

3.6.2   *Images and Inverse Images of Sets by Functions*

Some situations involve the images by functions of not only single elements in the domain, but also subsets of the domain, as defined here.

3.97 Definition (image of a set). For each function $F : A \to B$ and each subset $V \subseteq A$, the **image of V by F** is the subset $F"(V)$ consisting of all images of each $X \in V \cap \mathscr{D}(F)$ by F:

$$F"(V) = \{Y \in B : \exists X\,[(X \in V) \wedge (F(X) = Y)]\}$$
$$= \{F(X) : X \in V \cap \mathscr{D}(F)\}.$$

For each function $F : A \to B$, $F"$ is a function of subsets: $F" : \mathscr{P}(A) \to \mathscr{P}(B)$.

3.98 Remark. The notation $F"(V)$ adopted in definition 3.97 is common in set theory [74, p. 14], [128, p. 65]. Informal usage employs the notation $F(V)$ for the image of a subset $V \subseteq A$ by a function $F : A \to B$, but this usage is ambiguous. For example, consider the set $A := \{\,\varnothing,\, \{\varnothing\}\,\}$ and the constant function

$$C_\varnothing : \{\,\varnothing,\, \{\varnothing\}\,\} \to \{\,\varnothing,\, \{\varnothing\}\,\},$$
$$X \mapsto \varnothing.$$

The set $\{\varnothing\}$ is an element of A, whence $C_\varnothing(\{\varnothing\}) = \varnothing$ because $C_\varnothing(X) = \varnothing$ for every $X \in A$. Yet $\{\varnothing\}$ is also a subset of A, containing the single element $\varnothing \in \{\varnothing\}$; because $C_\varnothing(\varnothing) = \varnothing$ it follows that $C"_\varnothing(\{\varnothing\}) = \{\varnothing\}$ as the image of a subset:

$$C_\varnothing(\{\varnothing\}) = \varnothing,$$
$$C"_\varnothing(\{\varnothing\}) = \{\varnothing\}.$$

The common informal notation leads to the contradiction

$$C_\varnothing(\{\varnothing\}) = \varnothing,$$
$$C_\varnothing(\{\varnothing\}) = \{\varnothing\}.$$

In a formal theory containing this contradiction, every proposition would be True.

3.99 Example. If $C_Z : A \to B$ is a constant function that maps each $X \in A$ to the same $Z \in B$, then $C_Z"(V) = \{Z\}$ for each *nonempty* subset $V \subseteq A$.

3.100 Example. If $I_A : A \to A$, $X \mapsto X$ is an identity function, so that $I_A(X) = X$. for each $X \in A$ then $I_A"(V) = V$ for each subset $V \subseteq A$. Thus, $I_A"$ is the identity function $I_A" = I_{\mathscr{P}(A)} : \mathscr{P}(A) \to \mathscr{P}(A)$, with $V \mapsto V$ for every $V \in \mathscr{P}(A)$.

Besides images of subsets, such problems as the solution of equations involve the identification of a subset, called a pre-image, mapped to a specified image.

3.101 Definition (pre-image). For each function $F : A \to B$, and for each element $Y \in B$, the **pre-image of** Y **by** F is the *subset* of A denoted by $F^{\circ-1}{}''(\{Y\})$ and consisting of all elements $X \in A$ such that $F(X) = Y$:

$$F^{\circ-1}{}''(\{Y\}) = \{X \in A : F(X) = Y\}.$$

For each subset $W \subseteq B$, the **inverse image**, or **pre-image, of** W **by** F is the subset of $F^{\circ-1}{}''(W)$ in A and consisting of all pre-images of all elements of V by F:

$$F^{\circ-1}{}''(W) = \{X \in A : F(X) \in W\}.$$

3.102 Example. If $C_Z : A \to B$ is a constant function that maps each $X \in A$ to the same $Z \in B$, then $C_Z^{\circ-1}{}''(\{Z\}) = A$.

Also, $C_Z^{\circ-1}{}''(W) = A$ for each subset $W \subseteq B$ for which $Z \in W$.

In contrast, $C_Z^{\circ-1}{}''(S) = \varnothing$ for each subset $S \subseteq B$ for which $Z \notin S$. Thus,

$$C_Z^{\circ-1}{}''(W) = \begin{cases} A & \text{if } Z \in W, \\ \varnothing & \text{if } Z \notin W. \end{cases}$$

3.103 Example. If $I_A : A \to A$ is an identity function, with $I_A(X) = X$ for each $X \in A$, then $I_A^{\circ-1}{}''(W) = W$ for each subset $W \subseteq A$.

Theorem 3.104 relates images and pre-images to unions and intersections.

3.104 Theorem. *For each function $F : A \to B$, for each set $\mathscr{F}$ of subsets of A, and for each set $\mathscr{G}$ of subsets of B, the following relations hold.*

$$F^{\circ-1}{}''\left(\bigcup \mathscr{G}\right) = \bigcup_{W \in \mathscr{G}} F^{\circ-1}{}''(W)$$

$$F^{\circ-1}{}''\left(\bigcap \mathscr{G}\right) = \bigcap_{W \in \mathscr{G}} F^{\circ-1}{}''(W)$$

$$F''\left(\bigcup \mathscr{F}\right) = \bigcup_{V \in \mathscr{F}} F''(V)$$

$$F''\left(\bigcap \mathscr{F}\right) \subseteq \bigcap_{V \in \mathscr{F}} F''(V).$$

Proof. Apply the definitions of union and intersection. For $F^{\circ-1}{}''(\bigcup \mathscr{G})$,

$$F^{\circ-1}{}''\left(\bigcup \mathscr{G}\right) = \bigcup_{W \in \mathscr{G}} F^{\circ-1}{}''(W) \quad \textbf{yet unproved,}$$

$$\Updownarrow \quad \text{extensionality (S1),}$$

$$\forall X \left\{\left[X \in F^{\circ-1}{}''\left(\bigcup \mathscr{G}\right)\right]\right.$$
$$\left.\Leftrightarrow \left[X \in \bigcup_{W \in \mathscr{G}} F^{\circ-1}{}''(W)\right]\right\}$$

$$\Updownarrow \quad \text{definitions of } F^{\circ-1} \text{" and } \bigcup,$$

$$\forall X \left\{ [F(X) \in (\bigcup \mathscr{G})] \right.$$
$$\left. \Leftrightarrow \left(\exists W \left\{ (W \in \mathscr{G}) \wedge [F(X) \in W] \right\} \right) \right\}$$

$$\Updownarrow \quad \text{definition of } \bigcup,$$

$$\forall X \left\{ \left(\exists W \left\{ (W \in \mathscr{G}) \wedge [F(X) \in W] \right\} \right) \right.$$
$$\left. \Leftrightarrow \left(\exists W \left\{ (W \in \mathscr{G}) \wedge [F(X) \in W] \right\} \right) \right\}$$

which holds thanks to theorem 1.63: $(P) \Leftrightarrow (P)$. For $F^{\circ-1"}(\bigcap \mathscr{G})$,

$$F^{\circ-1"}(\bigcap \mathscr{G}) = \bigcap_{W \in \mathscr{G}} F^{\circ-1"}(W) \quad \textbf{yet unproved},$$

$$\Updownarrow \quad \text{extensionality (S1)},$$

$$\forall X \left\{ [X \in F^{\circ-1"}(\bigcap \mathscr{G})] \right.$$
$$\left. \Leftrightarrow [X \in \bigcap_{W \in \mathscr{G}} F^{\circ-1"}(W)] \right\}$$

$$\Updownarrow \quad \text{definitions of } F^{\circ-1} \text{ and } \bigcap,$$

$$\forall X \left\{ [F(X) \in (\bigcap \mathscr{G})] \Leftrightarrow \right.$$
$$\left. \left(\forall W \left\{ (W \in \mathscr{G}) \Rightarrow [F(X) \in W] \right\} \right) \right\}$$

$$\Updownarrow \quad \text{definitions of } \bigcap,$$

$$\forall X \left\{ \left(\forall W \left\{ (W \in \mathscr{G}) \Rightarrow [F(X) \in W] \right\} \right) \right.$$
$$\left. \Leftrightarrow \left(\forall W \left\{ (W \in \mathscr{G}) \Rightarrow [F(X) \in W] \right\} \right) \right\}$$

which holds thanks to theorem 1.63: $(P) \Leftrightarrow (P)$. For $F"(\bigcup \mathscr{F})$,

$$F"(\bigcup \mathscr{F}) = \bigcup_{V \in \mathscr{F}} F"(V) \quad \textbf{yet unproved},$$

$$\Updownarrow \quad \text{extensionality (S1)},$$

$$\forall Y \left\{ [Y \in F"(\bigcup \mathscr{F})] \Leftrightarrow [Y \in \bigcup_{V \in \mathscr{F}} F"(V)] \right\}$$

$$\Updownarrow \quad \text{definitions: } F", \bigcup,$$

$$\forall Y \left[\left(\exists X \left\{ [X \in (\bigcup \mathscr{F})] \wedge [Y = F(X)] \right\} \right) \right.$$
$$\left. \Leftrightarrow \left(\exists V \left\{ [V \in \mathscr{F}] \wedge [Y \in F"(V)] \right\} \right) \right]$$

$$\Updownarrow \quad \text{definitions: } \bigcup, F",$$

$$\forall Y \left[\left(\exists X \left\{ \exists V [(V \in \mathscr{F}) \wedge (X \in V)] \right\} \wedge [Y = F(X)] \right) \right.$$
$$\left. \Leftrightarrow \left\{ \exists V \left[(V \in \mathscr{F}) \wedge \left(\exists X \{ (X \in V) \wedge [Y = F(X)] \} \right) \right] \right\} \right]$$

$$\Updownarrow \quad \text{no } X \text{ in } V \in \mathscr{F},$$

$$\forall Y \left[\left(\exists X \left\{ \exists V [(V \in \mathscr{F}) \wedge (X \in V)] \right\} \wedge [Y = F(X)] \right) \right.$$
$$\left. \Leftrightarrow \left(\exists V \{ \exists X [(V \in \mathscr{F}) \wedge \{ (X \in V) \wedge [Y = F(X)] \}] \} \right) \right]$$

which holds thanks to the associativity of $\wedge$ (theorem 1.66). For $F"(\bigcap \mathscr{F})$,

$$F"\left(\bigcap \mathscr{F}\right) \subseteq \bigcap_{V\in\mathscr{F}} F"(V) \qquad \textbf{yet unproved,}$$

$$\Updownarrow \qquad \text{definition 3.7,}$$

$$\forall Y \left\{[Y \in F"\left(\bigcap \mathscr{F}\right)] \Rightarrow \left[Y \in \bigcap_{V\in\mathscr{F}} F"(V)\right]\right\}$$

$$\Updownarrow \qquad \text{definitions: } F", \bigcap,$$

$$\forall Y \big[\{\exists X\, [X \in \left(\bigcap \mathscr{F}\right)] \wedge [Y = F(X)]\}$$
$$\Rightarrow \left(\forall V \{(V \in \mathscr{F}) \Rightarrow [Y \in F"(V)]\}\right)\big]$$

$$\Updownarrow \qquad \text{definitions: } \bigcap, F",$$

$$\forall Y \Big(\{\exists X\, [\forall V\, ([V \in \mathscr{F}] \Rightarrow [X \in V])] \wedge [Y = F(X)]\}$$
$$\Rightarrow \{\forall V [(V \in \mathscr{F}) \Rightarrow (\exists X\{(X \in V) \wedge [Y = F(X)]\})]\}\Big)$$

$$\text{theorem 2.63: } \Updownarrow \quad \text{no } X \text{ in } (V \in \mathscr{F}),$$

$$\forall Y \Big(\{\exists X\, [\forall V\, ([V \in \mathscr{F}] \Rightarrow [X \in V])] \wedge [Y = F(X)]\}$$
$$\Rightarrow \{\forall V \exists X [\underbrace{(V \in \mathscr{F})}_{P} \Rightarrow (\{\underbrace{(X \in V)}_{Q} \wedge \underbrace{[Y = F(X)]}_{R}\})]\}\Big)$$

which holds thanks to $\vdash \{[(P) \Rightarrow (Q)] \wedge (R)\} \Rightarrow \{(P) \Rightarrow [(Q) \wedge (R)]\}$ by theorem 1.84, and $\vdash \{\exists X[\forall V(W)]\} \Rightarrow \{\forall V[\exists X(W)]\}$ by theorem 2.73. $\qquad\square$

3.6.3 Exercises on Mathematical Functions

3.131 . Determine whether $F := \{(0,1), (2,3), (1,2), (0,4)\}$ is a function.

3.132 . Determine whether $G := \{(9,2), (7,3), (8,2), (6,1)\}$ is a function.

3.133 . Determine whether the following relation is a function:

$$R := \{(0,1), (1,2), (2,4), (3,8), (4,8), (5,4), (6,2), (7,1), (8,0)\}.$$

3.134 . Determine whether the following relation is a function:

$$S := \{(0,2), (1,5), (2,7), (3,5), (4,2), (5,5), (6,7), (7,5), (8,2)\}.$$

3.135 . Determine whether the following relation is a function:

$$Z := \{(0,0), (1,0), (2,0), (3,0), (4,0), (5,0), (6,0), (7,0), (8,0), (9,0)\}.$$

3.136 . Prove that exactly one function exists from $A := \varnothing$ to $B := \varnothing$.

3.137 . Investigate whether a function $F : \varnothing \to B$ exists from $\varnothing$ to any set B.

3.138 . Investigate whether a function $F : A \to \varnothing$ exists from any set A to $\varnothing$.

3.139 . For each set A let $\mathbf{1}_A$ denote the constant function 1:

$$\mathbf{1}_A : A \to \{1\}, \quad \mathbf{1}_A(a) = 1.$$

For each subset $B \subseteq A$, prove that $\mathbf{1}_A|_B$ coincides with χ_B.

3.140 . For all subsets V, W of each set A, investigate whether the characteristic function of the intersection, $\chi_{V \cap W} : A \to 2$, is the intersection of the two characteristic functions $\chi_V : A \to 2$ and $\chi_W : A \to 2$, so that $\chi_{V \cap W} = \chi_V \cap \chi_W$.

3.141 . For all subsets V, W of each set A, investigate whether the characteristic function of the union, $\chi_{V \cup W} : A \to 2$, is the union of the two characteristic functions $\chi_V : A \to 2$ and $\chi_W : A \to 2$, so that $\chi_{V \cup W} = \chi_V \cup \chi_W$.

3.142 . Prove that for each function $F : A \to B$, for all subsets $R, S \subseteq A$, and for all subsets $V, W \subseteq B$, the following relations hold.

$$F^{\circ-1"}(V \cup W) = F^{\circ-1"}(V) \cup F^{\circ-1"}(W)$$
$$F^{\circ-1"}(V \cap W) = F^{\circ-1"}(V) \cap F^{\circ-1"}(W)$$
$$F^{"}(R \cup S) = F^{"}(R) \cup F^{"}(S)$$
$$F^{"}(R \cap S) \subseteq F^{"}(R) \cap F^{"}(S).$$

3.143 . Provide an example for which $F^{"}(R \cap S) \subsetneqq F^{"}(R) \cap F^{"}(S)$.

3.144 . For each function $F : A \to B$ and for all subsets $V, W \subseteq B$, prove that $F^{\circ-1"}(W \setminus V) = [F^{\circ-1"}(W)] \setminus [F^{\circ-1"}(V)]$.

3.145 . For each function $F : A \to B$ and for all subsets $H, K \subseteq A$, investigate whether inclusion or equality holds for $F^{"}(K \setminus H)$ and $[F^{"}(K)] \setminus [F^{"}(H)]$.

3.146 . For each function $F : A \to B$, prove that $F^{"}(V) = \mathscr{R}(F|_V)$.

3.147 . For each function $F : A \to B$, prove that $F^{"} : \mathscr{P}(A) \to \mathscr{P}(B)$ contains all the information about F, in the sense that $\{F(X)\} = F^{"}(\{X\})$ for every $X \in A$.

3.148 . For each function $\mathscr{F} : \mathscr{P}(A) \to \mathscr{P}(B)$, investigate whether there exists a function $F : A \to B$ such that $F^{"} = \mathscr{F}$.

3.149 . Consider the function $F : A \to B$ defined by

$$A := 3 = \{0, 1, 2\} = \Big\{ \varnothing, \{\varnothing\}, \{\varnothing, \{\varnothing\}\} \Big\},$$
$$B := A,$$
$$F := \{(0, 2), (1, 2), (2, 0)\}.$$

Recall that the superscript " indicates images of *subsets* (rather than of elements).

(1) $\{\varnothing, \{\varnothing\}\}$ is a *subset* of A. Find its image: $F^{"}(\{\varnothing, \{\varnothing\}\})$.
(2) $\{\varnothing, \{\varnothing\}\}$ is a *subset* of B. Find its inverse image $F^{\circ-1"}(\{\varnothing, \{\varnothing\}\})$.

(3) $\{\varnothing, \{\varnothing\}\}$ is an *element* of A. Find its image $F(\{\varnothing, \{\varnothing\}\})$.

(4) $\{\varnothing, \{\varnothing\}\}$ is an *element* of B. Find $F^{\circ-1"}(\{\{\varnothing, \{\varnothing\}\}\})$.

3.150 . Consider the function $G : C \to D$ defined by

$$C := 4 = \{0, 1, 2, 3\}$$
$$= \left\{ \varnothing, \{\varnothing\}, \{\varnothing, \{\varnothing\}\}, \left\{ \varnothing, \{\varnothing\}, \{\varnothing, \{\varnothing\}\} \right\} \right\},$$
$$D := C,$$
$$G := \{(0, 3), (1, 0), (2, 3), (3, 1)\}.$$

Recall that the superscript " indicates images of *subsets* (rather than of elements).

(1) 3 is a *subset* of C. Find its image: $G^{"}(3)$.

(2) 3 is a *subset* of D. Find its inverse image $G^{\circ-1"}(3)$.

(3) 3 is an *element* of C. Find its image $G(3)$.

(4) 3 is an *element* of D. Find $G^{\circ-1"}(\{3\})$.

3.7   Composite and Inverse Functions

3.7.1   *Compositions of Functions*

Some situations involve sequences of operations corresponding to sequences of functions. For instance, if a first function consists of pairs (T, X) with the ascension (elevation) X of a planet at time T, and if a second function consists of pairs (X, Y) with the declination (azimuth) Y of the planet at ascension X, then the composition of the two functions consists of pairs (T, Y) with the declination Y of the planet at time T.

3.105 Definition (Composition of functions). For all functions $F : A \to B$ and $G : B \to C$, the **composite function** $G \circ F$ (read "G preceded by F" or "F followed by G" or "the composition of G and F") is the function $G \circ F : A \to C$ defined by

$$(G \circ F)(X) := G[F(X)]$$

for each $X \in F^{\circ-1"}[\mathscr{D}(G)]$. Thus,

$$[(X, Z) \in (G \circ F)] \Leftrightarrow \{[\exists Y(Y \in B)] \wedge [(X, Y) \in F] \wedge [(Y, Z) \in G]\}.$$

3.106 Example.   Consider the following functions F and G:

$$
\begin{aligned}
A &= \{0, 1\}, & B &= \{0, 1, 2\}, \\
B &= \{0, 1, 2\}, & C &= \{0, 1\}, \\
F&:A \rightarrow B, & G&:B \rightarrow C, \\
F &= \{(0, 0), (1, 2)\}; & G &= \{(0, 1), (1, 0), (2, 1)\}.
\end{aligned}
$$

Their composition

$$(G \circ F): A \rightarrow C,$$

has values

$$
\begin{aligned}
(G \circ F)(0) &= G[F(0)] = G[0] = 1, \\
(G \circ F)(1) &= G[F(1)] = G[2] = 1,
\end{aligned}
$$

so that

$$(G \circ F) = \{(0, 1), (1, 1)\}.$$

3.107 Theorem.   *The composition of functions is associative: For all functions* $F : A \rightarrow B, G : B \rightarrow C,$ *and* $H : C \rightarrow D,$

$$[H \circ (G \circ F)] = [(H \circ G) \circ F].$$

Proof.   For each $X \in F^{\circ -1}"\{G^{\circ -1}"[\mathscr{D}(H)]\}$, apply the definition of $\circ$ repeatedly:

$$
\begin{aligned}
[H \circ (G \circ F)](X) &= H\{(G \circ F)(X)\} \\
&= H\{G[F(X)]\} \\
&= [H \circ G][F(X)] \\
&= ([H \circ G] \circ F)(X).
\end{aligned}
$$

$\square$

   In contrast to its associativity, the composition of functions is *not* commutative.

3.108 Counterexample.   Consider the following functions F and G:

$$
\begin{aligned}
A &= \{0, 1\}, & B &= \{0, 1, 2\}, \\
B &= \{0, 1, 2\}, & C &= \{0, 1\}, \\
F&:A \rightarrow B, & G&:B \rightarrow C, \\
F &= \{(0, 0), (1, 2)\}; & G &= \{(0, 1), (1, 0), (2, 1)\}.
\end{aligned}
$$

Their composition

$$(F \circ G): C \rightarrow A,$$

has values

$$(F \circ G)(0) = F[G(0)] = F[1] = 2,$$
$$(F \circ G)(1) = F[G(1)] = F[0] = 0,$$

so that

$$(F \circ G) = \{(0, 2), (1, 0)\}.$$

In contrast,

$$(G \circ F) = \{(0, 1), (1, 1)\}$$

from example 3.106, which confirms that $(F \circ G) \neq (G \circ F)$.

3.7.2 Injective, Surjective, Bijective, and Inverse Functions

Such problems as the solution of equations involve the determination of whether an equation has no solution, exactly one solution, or more than one solution, which correspond to the features of functions introduced here.

3.109 Definition (injectivity). A function $F : A \to B$ is **injective** if and only if for all $W \in \mathscr{D}(F)$ and $X \in \mathscr{D}(F)$, if $W \neq X$, then $F(W) \neq F(X)$:

$$\forall W \big[\forall X \big(\{[W \in \mathscr{D}(F)] \wedge [X \in \mathscr{D}(F)]\} \Rightarrow \{[(W \neq X) \Rightarrow [F(X) \neq F(W)]\}\big) \big].$$

By contraposition, the condition just stated is equivalent to the following alternative condition: for all $W \in \mathscr{D}(F)$ and $X \in \mathscr{D}(F)$, if $F(W) = F(X)$, then $W = X$:

$$\forall W \big[\forall X \big(\{[W \in \mathscr{D}(F)] \wedge [X \in \mathscr{D}(F)]\} \Rightarrow \{[F(X) = F(W)] \Rightarrow (W = X)\}\big) \big].$$

The notation $F : A \hookrightarrow B$ indicates that F is an injection.

Another common usage consists of saying that F maps only **one** X to each **one** Y; yet this alternative terminology fails to indicate which "one" it emphasizes (the *first* "one"), which leads to confusion, and, therefore, will not be used here.

3.110 Example. With $A := \{0, 1, 2\}$ and $B := \{0, 1, 2, 3, 4, 5, 6, 7, 8, 9\}$, the function $G := \{(0, 1), (1, 3), (2, 9)\}$ is injective.

3.111 Example. With $A := \{4, 6, 8, 9\}$ and $B := \{2, 3\}$, the function

$$H := \{(4, 2), (6, 3), (8, 2), (9, 3)\}$$

is *not* injective, because $4 \neq 8$ but $H(4) = 2 = H(8)$.

3.112 Definition (surjectivity). A function $F : A \to B$ is **surjective** if and only if for each $Y \in B$, there exists some $X \in A$ for which $Y = F(X)$. In other words, the condition just stated means that the range of F consists of all of the co-domain B:

$$\forall Y\{(Y \in B) \Rightarrow (\exists X\{(X \in A) \wedge [F(X) = Y]\})\}.$$

The notation $F : A \twoheadrightarrow B$ indicates that F is a surjection.

Another common mathematical usage consists of saying that F maps A **onto** B.

3.113 Example. With $A := \{0, 1, 2\}$ and $B := \{0, 1, 2, 3, 4, 5, 6, 7, 8, 9\}$, the function $G := \{(0, 1), (1, 3), (2, 9)\}$ is *not* surjective: there is *no* $X \in A$ with $G(X) = 6$.

3.114 Example. With $A := \{4, 6, 8, 9\}$ and $B := \{2, 3\}$, the function

$$H := \{(4, 2), (6, 3), (8, 2), (9, 3)\}$$

is surjective. Indeed, $H(4) = 2$ for $Y := 2$, and $H(6) = 3$ for $Y := 3$,

3.115 Example. With $A := \{6, 7, 8, 9\}$ and $B := \{1, 2, 3, 4\}$, the function

$$F := \{(6, 2), (7, 1), (8, 2), (9, 3)\}$$

is *neither* injective, because $F(6) = 2 = F(8)$ with $6 \neq 8$, *nor* surjective, because there does not exist any $X \in A$ with $F(X) = 4$.

3.116 Definition. For all sets A, B, C, D, and for all functions $F : A \to B$ and $G : C \to D$, define the function

$$F \boxtimes G : A \times C \to B \times D,$$
$$(W, Z) \mapsto (F(W), G(Z)).$$

Thus $\left((W, Z), (F(W), G(Z))\right) \in F \boxtimes G$ if and only if $(W, F(W)) \in F$ and $(Z, G(Z)) \in G$, so that $\left((W, F(W)), (Z, G(Z))\right) \in F \times G$.

3.117 Theorem. *If $F : A \to B$ and $G : C \to D$ are both injective, or both surjective, then $F \boxtimes G$ is injective, or surjective, respectively.*

Proof. If $F : A \to B$ and $G : C \to D$ are both surjective, then the function $F \boxtimes G$ is surjective: indeed, for each $(U, V) \in B \times D$, or, equivalently, for all $U \in B$ and $V \in D$, there exist $X \in A$ and $Y \in C$ such that $F(X) = U$ and $G(Y) = V$, by surjectivity of F and G. Hence

$$(F \boxtimes G)(X, Y) = (F(X), G(Y)) = (U, V).$$

If $F : A \to B$ and $G : C \to D$ are both injective, then the function $F \boxtimes G$ is injective: indeed, if $(F \boxtimes G)(W, Z) = (F \boxtimes G)(R, S)$, then

$$\big(F(W), G(Z)\big) = (F \boxtimes G)(W, Z) = (F \boxtimes G)(R, S) = \big(F(R), G(S)\big).$$

Hence $F(W) = F(R)$ and $G(Z) = G(S)$, by equality of ordered pairs. Consequently, $W = R$ and $Z = S$, by injectivity of F and G.   □

3.118 Definition (Bijectivity). A function $F : A \to B$ is **bijective** if and only if F is both injective and surjective, which can be denoted by $F : A \rightleftarrows B$ or $F : A \rightleftharpoons B$.

3.119 Example. With $A := \{0, 1, 2, 3\}$ and $B := \{1, 2, 4, 8\}$, the function $P : A \to B$ defined by $P := \{(0, 1), (1, 2), (2, 4), (3, 8)\}$ is bijective.

3.120 Example. For each set A the identity function $I_A : A \to A, X \mapsto X$ is bijective. Indeed, I_A is injective, because if $I_A(W) = I_A(X)$ then $W = I_A(W) = I_A(X) = X$. Similarly, I_A is surjective, because for each $Y \in A$ there exists $X \in A$, in effect $X := Y$, with $Y = I_A(Y)$.

3.121 Example. For each *proper* subset $S \subsetneqq A$, the inclusion function $\iota : S \to A$, $X \mapsto X$, is injective but not surjective. Indeed, ι is injective, for if $W, X \in S$ and $W \neq X$, then $\iota(W) = W \neq X = \iota(X)$, whence $\iota(W) \neq \iota(X)$. However, ι is not surjective: because S is a proper subset of A, there exists some element Z in $A \setminus S$; in particular, $Z \neq X$ for each $X \in S$, and, consequently, $\iota(X) = X \neq Z$, which means that ι is not surjective. Thus, ι is not bijective either.

3.122 Example. For each nonempty set A and for each set B containing more than one element, the canonical projection $P_A : (A \times B) \to A$ is surjective but not injective. Indeed, B contains at least one element $Y \in B$, because $B \neq \varnothing$; hence, $X = P_A(X, Y)$ for each $X \in A$. However, B also contains some $Z \in B$ such that $Y \neq Z$. Consequently, $(X, Y) \neq (X, Z)$, and yet $P_A(X, Y) = X = P_A(X, Z)$, which means that P_A is not injective. Thus, P_A is not bijective either.

3.123 Theorem. *For all functions $F : A \to B$ and $G : B \to C$,*

* *if F and G are both injective, then $G \circ F$ is also injective;*
* *if F and G are both surjective, then $G \circ F$ is also surjective;*
* *if F and G are both bijective, then $G \circ F$ is also bijective;*
* *if $G \circ F$ is injective, then F is injective;*
* *if $G \circ F$ is surjective, then G is surjective;*
* *if $G \circ F$ is bijective, then F is injective and G is surjective.*

Proof. Assume that F and G are both injective. For all distinct elements $W \neq X$ in $F^{\circ -1}[\mathscr{D}(G)]$, the injectivity of F ensures that $F(W) \neq F(X)$. Hence $G(F(W)) \neq G(F(X))$ by the injectivity of G. Hence, $(G \circ F)(W) = G(F(W)) \neq G(F(X)) = (G \circ F)(X)$, whence $(G \circ F)(W) \neq (G \circ F)(X)$, so that $G \circ F$ is injective.

Assume that F and G are both surjective. For each $Z \in C$, the surjectivity of G ensures the existence of an element $Y \in B$ such that $G(Y) = Z$. Hence, by the surjectivity of F, there exists an element $X \in A$ for which $F(X) = Y$. Therefore, $(G \circ F)(X) = G(F(X)) = G(Y) = Z$, which means that $(G \circ F)$ is surjective.

In particular, if F and G are both bijective, then $G \circ F$ is also bijective.

Assume that $G \circ F$ is injective. For all distinct elements $W \neq X$ in $F^{\circ-1}{}''[\mathscr{D}(G)]$, the injectivity of $G \circ F$ ensures that $(G \circ F)(W) \neq (G \circ F)(X)$, whence $G(F(W)) \neq G(F(X))$. Because G is a function, G cannot take different values at the same argument, whence it follows that $F(W) \neq F(X)$, so that F is injective.

Assume that $G \circ F$ is surjective. For each element $Z \in C$, the surjectivity of $G \circ F$ ensures the existence of an element $X \in A$ such that $(G \circ F)(X) = Z$. Hence, letting $Y := F(X)$ demonstrates the existence of an element $Y \in B$ for which $G(Y) = G(F(X)) = (G \circ F)(X) = Z$, which means that G is surjective.

In particular, if $G \circ F$ is bijective, then F is injective and G is surjective. □

3.124 Definition (invertibility). A function $F : A \to B$ is **invertible** if and only if there exists a function $G : B \to A$ for which $G \circ F = I_{\mathscr{D}(F)}$ and $F \circ G = I_{\mathscr{D}(G)}$. Such a function G is denoted by $F^{\circ-1}$ and called the **inverse function** of F. Thus,

$$F^{\circ-1} \circ F = I_{\mathscr{D}(F)},$$
$$F \circ F^{\circ-1} = I_{\mathscr{D}(G)}.$$

3.125 Example. With $A := \{0, 1, 2, 3\}$ and $B := \{1, 2, 4, 8\}$, the function

$$F := \{(0, 1), (1, 2), (2, 4), (3, 8)\}$$

is invertible, with inverse $F^{\circ-1} := \{(1, 0), (2, 1), (4, 2), (8, 3)\}$.

3.126 Theorem. *Each function $F : A \to B$ admits at most one inverse function. Moreover, if $G : B \to A$ is an inverse function for F, then G consists of all pairs obtained by swapping the coordinates in each pair of F.*

Proof. Assume that G is an inverse function for F, which means that $G \circ F = I_{\mathscr{D}(F)}$ and $F \circ G = I_{\mathscr{D}(G)}$, and consider the set

$$H := \{(Y, X) : (X, Y) \in F\}.$$

If $(X, Y) \in F$, then $X \in \mathscr{D}(F)$. Because $G \circ F = I_{\mathscr{D}(F)}$, it follows that there exists some $Z \in B$ such that $(X, Z) \in F$ and $(Z, X) \in G$. With $(X, Y) \in F$ and $(X, Z) \in F$, it follows that $Y = Z$, because F is a function. This shows that if $(Y, X) \in H$, so that $(X, Y) \in F$, then $(Y, X) = (Z, X) \in G$, whence $G \subseteq H$.

Conversely, If $(Z, X) \in G$, then $Z \in \mathscr{D}(G)$. Because $F \circ G = I_{\mathscr{D}(G)}$ it follows that there exists some $W \in \mathscr{D}(F)$ such that $(Z, W) \in G$ and $(W, Z) \in F$. With $(Z, X) \in G$ and $(Z, W) \in G$, it follows that $W = X$, because G is a function. This shows that if $(Z, X) \in G$, then $(X, Z) \in F$, whence $(Z, X) \in H$, so that $G \subseteq H$.

Finally, $G = H$, which shows that if F has an inverse function G, then the only possibility is $G = H$. Thus, $G = H = F^{\circ-1} = \{(Y, X) : (X, Y) \in F\}$. □

3.127 Theorem. *For each function $F : A \to B$ with $\mathscr{D}(F) = A$, the function F is invertible if and only if F is bijective.*

Proof. Assume that F is invertible, with inverse $G := F^{\circ -1}$. Because $F \circ F^{\circ -1} = I_{\mathscr{D}(G)}$ is surjective, it follows from theorem 3.123 that F is surjective. Similarly, because $F^{\circ -1} \circ F = I_{\mathscr{D}(F)}$ is injective, it follows from theorem 3.123 that F is injective.

Conversely, assume that F is bijective. Construct an inverse function by means of the set G defined by swapping both coordinates in each pair of the function F:

$$G := \{(Y, X) : (X, Y) \in F\}.$$

Then verify that G is the inverse function of F.

First, the injectivity of F ensures that G is a function: if $(Y, X) \in G$ and $(Y, W) \in G$, then $(X, Y) \in F$ and $(W, Y) \in F$, whence $X = W$ by injectivity.

Second, $G \circ F = I_{\mathscr{D}(F)}$. Indeed, if $(X, Z) \in (G \circ F)$, then there exists some $Y \in B$ for which $(X, Y) \in F$ and $(Y, Z) \in G$. Consequently, $(Z, Y) \in F$, and again the injectivity of F shows that $X = Z$, whence $(X, Z) = (X, X)$. Thus, $(G \circ F) \subseteq I_{\mathscr{D}(F)}$. Because the foregoing reasoning holds for each $X \in \mathscr{D}(F)$, however, it also follows that $I_{\mathscr{D}(F)} \subseteq (G \circ F)$, and thus $(G \circ F) = I_{\mathscr{D}(F)}$.

Finally, $F \circ G = I_{\mathscr{D}(G)}$. Indeed, if $(Y, W) \in (F \circ G)$, then there exists some $X \in \mathscr{D}(G)$ with $(Y, X) \in G$ and $(X, W) \in F$. From $(Y, X) \in G$, it follows that $(X, Y) \in F$. Because F is a function, it also follows that $Y = W$, whence $(Y, W) = (Y, Y)$. Thus, $(F \circ G) \subseteq I_{\mathscr{D}(G)}$. Then the surjectivity of F guarantees that for each $Y \in \mathscr{D}(G)$ there exists some $X \in A$ with $(X, Y) \in F$. Hence, $(Y, X) \in G$ and then $(Y, Y) \in (F \circ G)$, so that $I_B \subseteq (F \circ G)$. Therefore, $(F \circ G) = I_B$. □

Some situations involve a concept more general than invertibility.

3.128 Definition (left or right invertibility). A function $F : A \to B$ is **invertible on its left**, or **left invertible**, if and only if there exists a function $G : B \to A$ for which $G \circ F = I_{\mathscr{D}(F)}$. Such a function G is called a **left inverse function** for F.

Similarly, a function $F : A \to B$ is **invertible on its right**, or **right invertible**, if and only if there exists a function $G : B \to A$ for which $F \circ G = I_{\mathscr{D}(G)}$. Such a function G is called a **right inverse function** for F.

3.129 Example. With $A := \{0, 1, 2\}$ and $B := \{0, 1, 2, 3, 4, 5, 6, 7, 8, 9\}$, the function $F : A \to B$ defined by $F := \{(0, 1), (1, 3), (2, 9)\}$ has a left inverse function

$$G := \{(1, 0), (2, 0), (3, 1), (4, 0), (5, 0), (6, 0), (7, 0), (8, 0), (9, 2)\}.$$

3.130 Example. For $A := \{4, 6, 8, 9\}$ and $B := \{2, 3\}$, the function $F : A \to B$ with $F := \{(4, 2), (6, 3), (8, 2), (9, 3)\}$ has a right inverse $G := \{(2, 4), (3, 6)\}$

3.131 Example. Perspectives to draw a picture of space A on a flat screen $B \subset A$ can be represented by a function $F : A \to B$, mapping each point X in space A to its image $F(X)$ on the screen B. For such perspectives, each point $Y \in B$ on the screen B is its own image, so that $F(Y) = Y$. Thus the inclusion function $\iota_B : B \to A$ is a right inverse for F, because $(F \circ \iota_B)(Y) = F[\iota_B(Y)] = F(Y) = Y$ for every $Y \in B$, so that $F \circ \iota_B = I_B$. In contrast, such a perspective F has no left inverse, because F maps many points in space to the same image on the screen.

The existence of a *left*-inverse $G : B \to A$ for a function $F : A \to B$ indicates that for each $Y \in B$ the equation $F(X) = Y$ has *at most* one solution. In contrast, the existence of a *right*-inverse $G : B \to A$ for a function $F : A \to B$ indicates that for each $Y \in B$ the equation $F(X) = Y$ has *at least* one solution.

3.7.3 The Set *of all Functions from a Set to a Set*

This subsection shows that all the functions from any fixed set into any fixed co-domain form a set.

Theorem 3.132 shows that all the functions between two fixed sets form a set.

3.132 Theorem. *For all sets A and B, all the functions from any subset of A to B are the elements of a set, of which all such functions defined on all of A form a subset.*

Proof. By definition 3.66, the power set $\mathscr{P}(A \times B)$ is the *set* of all the relations between A and B. By definition 3.81, every function $F : A \to B$ defined on any subset of A is also a relation, so that $F \in \mathscr{P}(A \times B)$. By the axiom of separation (page 124), with a formula stating that a relation $F \in \mathscr{P}(A \times B)$ is a function, all such functions form a subset of $\mathscr{P}(A \times B)$, denoted here by $\mathscr{F}_{A \to B}$:

$$\mathscr{F}_{A \to B}$$
$$:= \{F \in \mathscr{P}(A \times B) : \forall W \forall X \forall Y [([(X, Y) \in F] \wedge [(W, Y) \in F]) \Rightarrow (X = W)]\}.$$

For each subset $D \subseteq A$, again by the axiom of separation (page 124), with a formula stating that the domain of F is D, the functions $F : D \to B$ defined on all of D form a subset of $\mathscr{F}_{D \to B}$, denoted here by B^D:

$$B^D := \{F \in \mathscr{F}_{D \to B} : \forall X [(X \in D) \Rightarrow (\exists Y [(X, Y) \in F])]\}.$$

The case $D := A$ corresponds to the set B^A of all functions from all of A to B. □

Definition 3.133 sets the notation from the proof of theorem 3.132.

3.133 Definition. For all sets A and B, the set of all functions $F : A \to B$ defined on any subset of A is denoted by $\mathscr{F}_{A \to B}$. The set of all functions $F : A \to B$ defined on all of A is denoted by B^A.

3.134 Example. If $A = \varnothing$, then $A \times B = \varnothing \times B = \varnothing$ for every set B regardless of B. The only subset $F := \varnothing \subseteq \varnothing = A \times B$ is a function from A to B with domain A. Hence if $K = \varnothing = 0$, then $B^K = B^0 = \{\varnothing\} = \{0\}$. Thus there exists a function of zero variable with zero value in every set B, and this function is the natural number zero. This function of zero variable allows for a minimal set of starting functions in some contexts, for instance, with primitive recursive functions [109, p. 926].

3.135 Theorem. *For all sets A and B,* $\bigcup_{D \subseteq A} B^D = \mathscr{F}_{A \to B}$.

Proof. For each subset $D \subseteq A$, the set B^D consists of all the functions $F : D \to B$ defined on all of $D \subseteq A$, whence $B^D \subseteq \mathscr{F}_{A \to B}$, or, equivalently, $B^D \in \mathscr{P}(\mathscr{F}_{A \to B})$ and hence $\{B^D \in \mathscr{P}(\mathscr{F}_{A \to B}) : D \in \mathscr{P}(A)\}$ is also a set. Consequently so is its union by the axiom of union (page 127): $\bigcup_{D \subseteq A} B^D \subseteq \mathscr{F}_{A \to B}$. Conversely, each function $F \in \mathscr{F}_{A \to B}$ is defined on all of its domain, which is a subset $D \subseteq A$, so that $F \in B^D$. Hence also $\mathscr{F}_{A \to B} \subseteq \bigcup_{D \subseteq A} B^D$.   □

3.136 Theorem. *For all sets A and B, for each set* $\mathscr{D} \subseteq \mathscr{P}(A)$ *of subsets of A, all the sets* B^D *for every* $D \in \mathscr{D}$ *also form a set* $\mathscr{F}_{\mathscr{D}, B} \subseteq \mathscr{P}(\mathscr{F}_{A \to B})$. *Hence the union* $\bigcup \mathscr{F}_{\mathscr{D}, B} = \bigcup_{D \in \mathscr{D}} B^D$ *is also a set.*

Proof. For each set $\mathscr{D} \subseteq \mathscr{P}(A)$ of subsets of A, all the sets B^D for every $D \in \mathscr{D}$ also form a subset of $\mathscr{P}(\mathscr{F}_{A \to B})$, denoted here by $\mathscr{F}_{\mathscr{D}, B}$:

$$\mathscr{F}_{\mathscr{D}, B} := \{E \in \mathscr{P}[\mathscr{P}(A \times B)] : \exists D[(D \in \mathscr{D}) \wedge (E = B^D)]\},$$

where $E = B^D$ is an abbreviation of the formula resulting from the axiom of extensionality (page 111) and the preceding two applications of the axiom of separation (page 124). Hence the union $\bigcup \mathscr{F}_{\mathscr{D}, B} = \bigcup_{D \in \mathscr{D}} B^D$ is also a set,

$$\bigcup \mathscr{F}_{\mathscr{D}, B} = \{F \in \mathscr{F}_{A \to B} : \exists D[(D \in \mathscr{D}) \wedge (F \in B^D)]\},$$

by the axiom of union (page 127).   □

Theorem 3.137 reveals a bijection between the set $C^{A \times B}$ of all functions defined on a Cartesian product $A \times B$ into a set C and the set $(C^B)^A$ of all functions from A into the set C^B of all functions defined on B into C.

3.137 Theorem. *For all sets A, B, C, there is a bijection from* $C^{A \times B}$ *onto* $(C^B)^A$.

Proof. Define a function $H : C^{A \times B} \to (C^B)^A$ as follows. For each element $F \in C^{A \times B}$, which is a function $F : A \times B \to C$ defined on all of $A \times B$, define an element $H(F) \in (C^B)^A$, which is a function $H(F) : A \to C^B$ defined on all of A, so that for each $X \in A$ the image $[H(F)](X) \in C^B$ is a function $[H(F)](X) : B \to C$ defined on all of B, defined for each $Y \in B$ by

$$H : C^{A \times B} \to (C^B)^A,$$

$$F \mapsto H(F),$$

$$\{[H(F)](X)\}(Y) := F(X, Y).$$

Similarly, define a function $L : (C^B)^A \to C^{A \times B}$ as follows. For each element $G \in (C^B)^A$, define an element $L(G) \in C^{A \times B}$, which is a function $L(G) : A \times B \to C$, defined by $[L(G)](X, Y) := [G(X)](Y)$.

$$L : (C^B)^A \to C^{A \times B},$$

$$G \mapsto L(G),$$

$$[L(G)](X, Y) := [G(X)](Y).$$

Then $H \circ L = I_{(C^B)^A}$ and $L \circ H = I_{C^{A \times B}}$, whence H and L are inverses of each other:

$$\begin{aligned} [(L \circ H)(F)](X, Y) &= \{L[H(F)]\}(X, Y) \\ &= \{[H(F)](X)\}(Y) \\ &= F(X, Y); \\ \{[(H \circ L)(G)](X)\}(Y) &= [\{H[L(G)]\}(X)](Y) \\ &= [L(G)](X, Y) \\ &= [G(X)](Y). \end{aligned}$$

Thus $(L \circ H)(F) = F$ and $(H \circ L)(G) = G$. Hence H and L are bijections. $\qquad\square$

3.7.4 Exercises on Injective, Surjective, and Inverse Functions

3.151 . For each $F : A \to B$ and $I_B : B \to B$, prove that $I_B \circ F = F$.

3.152 . For each $F : A \to B$ and $I_A : A \to A$, prove that $F \circ I_A = F$.

3.153 . For each $F : A \to B$ and $\varnothing : \varnothing \to A$, prove that $F \circ \varnothing = \varnothing$.

3.154 . Prove that if a function has a left inverse, then it is injective.

3.155 . Prove that if a function is injective, then it has a left inverse.

3.156 . Prove that if a function has a right inverse, then it is surjective.

3.157 . Provide an example of a function with a right inverse but no left inverse.

3.158 . Provide a function with a left inverse but no right inverse.

3.159 . Provide a function that has more than one left inverse.

3.160 . Provide a function that has more than one right inverse.

3.8 Equivalence Relations

3.8.1 *Reflexive, Symmetric, Transitive, or Anti-Symmetric Relations*

Besides functions, mathematics contains several other types of relations. For instance, ordering relations define orders or rankings, whereas equivalence relations define equivalences relative to certain criteria. Such various types of relations can be defined by combinations of several features called reflexivity, symmetry, and transitivity.

3.138 Definition (Reflexivity). For each set A, a relation $R \subseteq A \times A$ is **reflexive** if and only if $(X, X) \in R$ for each $X \in A$:

$$\forall X\{(X \in A) \Rightarrow [(X, X) \in R]\}.$$

As a graph, a reflexive relation contains at least a single loop at each vertex:

3.139 Theorem. *A relation $\mathscr{R} \subseteq A \times A$ is reflexive if and only if $\Delta_A \subseteq \mathscr{R}$.*

Proof. A relation $\mathscr{R} \subseteq A \times A$ is reflexive if and only if $(X, X) \in \mathscr{R}$ for every $X \in A$, in other words, if and only if $\mathscr{R}$ contains the set $\{(X, X) : X \in A\} = \Delta_A$.   □

3.140 Example. For each set A, the diagonal Δ_A is a reflexive relation.

3.141 Example. For each set A and its power set $\mathscr{P}(A)$, the relation $\subseteq$ on $\mathscr{P}(A)$ is reflexive. Indeed, $B \subseteq B$ for each $B \in \mathscr{P}(A)$.

3.142 Counterexample. The relation of membership $\in$ is *not* reflexive. For instance, if $A = \{\varnothing\}$, then $\varnothing \in A$, but $\varnothing \notin \varnothing$. Thus $\in$ is *not* reflexive on $A \times A$.

3.143 Definition (symmetry). For each set A, a relation $R \subseteq A \times A$ is **symmetric** if and only if $(X, Y) \in R$ is equivalent to $(Y, X) \in R$ for all $X \in A$ and $Y \in A$:

$$\forall X \forall Y([(X, Y) \in R] \Leftrightarrow [(Y, X) \in R]).$$

As a graph, a symmetric relation contains edges in either both directions or neither:

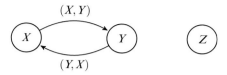

3.144 Example. For each set A, the diagonal Δ_A is a symmetric relation.

3.145 Counterexample. The relation of membership $\in$ is *not* symmetric in general: if $A := \big\{ \varnothing, \{\varnothing\} \big\}$, then $\varnothing \in \{\varnothing\}$ but $\{\varnothing\} \notin \varnothing$; thus $\in$ is *not* symmetric.

3.146 Definition (transitivity). For each set A, a relation $R \subseteq A \times A$ is **transitive** if and only if $(X, Y) \in R$ and $(Y, Z) \in R$ imply $(X, Z) \in R$ for all $X, Y, Z \in A$:

$$\forall X \forall Y \forall Z \big(\{[(X, Y) \in R] \wedge [(Y, Z) \in R]\} \Rightarrow [(X, Z) \in R]\big)$$

As a graph, a transitive relation completes two *consecutive* edges into a triangle:

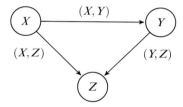

3.147 Example. For each set A, the diagonal Δ_A is a transitive relation.

3.148 Example. For each set A, the relation $\subseteq$ on $\mathscr{P}(A)$ is transitive: for all $U \in \mathscr{P}(A)$, $V \in \mathscr{P}(A)$, $W \in \mathscr{P}(A)$, if $U \subseteq V$ and $V \subseteq W$, then $U \subseteq W$.

3.149 Counterexample. The relation of membership $\in$ is *not* transitive in general: if $A = \big\{ \varnothing, \{\varnothing\}, \{\{\varnothing\}\} \big\}$, then $\varnothing \in \{\varnothing\}$ and $\{\varnothing\} \in \{\{\varnothing\}\}$, but $\varnothing \notin \{\{\varnothing\}\}$.

3.8.2   Partitions and Equivalence Relations

The concept of an equivalence relation on a set corresponds to a "partition" of that set into a union of disjoint subsets called "equivalence" classes.

3.150 Definition (equivalence relations). For each set A, a relation $R \subseteq A \times A$ is an **equivalence relation** if and only if R is *reflexive, symmetric, and transitive*.

3.151 Example. For each set A, the diagonal Δ_A is an equivalence relation.

3.152 Example. The set $A := \{0, 1, 2, 3\}$ admits an equivalence relation

$$\begin{aligned} \mathscr{R} &:= \{(0, 0), (1, 1), (2, 2), (3, 3), (0, 2), (2, 0), (1, 3), (3, 1)\} \\ &= \Delta_A \cup \{(0, 2), (2, 0), (1, 3), (3, 1)\}. \end{aligned}$$

- The relation $\mathscr{R}$ is reflexive because it contains the diagonal

$$\Delta_A = \{(0,0), (1,1), (2,2), (3,3)\}.$$

- The relation $\mathscr{R}$ is symmetric: it contains $(0,2)$, $(2,0)$, as well as $(1,3)$, $(3,1)$.
- The relation $\mathscr{R}$ is transitive because if it contains (X,Y) and (Y,Z), then it contains (X,Z), for instance, $(0,2)$ and $(2,0)$ hence $(0,0)$; $(2,0)$ and $(0,2)$ hence $(2,2)$, as well as $(1,3)$ and $(3,1)$ hence $(1,1)$; $(3,1)$ and $(1,3)$ hence $(3,3)$.

3.153 Counterexample. The relation of membership $\in$ is *not* an equivalence relation in general, because it is *not* symmetric and *not* transitive.

3.154 Definition (partition). A **partition** of a set A is a set $\mathscr{F} \subseteq \mathscr{P}(A)$ of subsets of A with all of the following properties.

- No member of $\mathscr{F}$ is empty: $\forall V\{(V \in \mathscr{F}) \Rightarrow (V \neq \varnothing)\}$.
- The union of all the members of $\mathscr{F}$ "covers" A, which means that $A = (\bigcup \mathscr{F})$.
- All pairs of distinct members of $\mathscr{F}$ are disjoint:

$$\forall V\big(\forall W\{[(V \in \mathscr{F}) \wedge (W \in \mathscr{F}) \wedge (V \neq W)] \Rightarrow [(V \cap W) = \varnothing]\}\big).$$

3.155 Example. The empty set $\varnothing$ admits only one partition: the empty set $\mathscr{F} = \varnothing$.

3.156 Example. Each nonempty set S has a partition: the singleton $\mathscr{F} = \{S\}$.

3.157 Example. The set $A := \{0,1,2,3\}$ admits a partition $\mathscr{F}$ into two disjoint sets, $\mathscr{F} := \big\{ \{0,2\}, \{1,3\} \big\}$:

- No member of $\mathscr{F}$ is empty: $\{0,2\} \neq \varnothing$ and $\{1,3\} \neq \varnothing$.
- The unions of the members of $\mathscr{F}$ covers A, so that $\{0,1,2,3\} = \{0,2\} \cup \{1,3\}$.
- Distinct members of $\mathscr{F}$ are disjoint: $\{0,2\} \cap \{1,3\} = \varnothing$.

3.158 Definition (relations from partitions). For each partition $\mathscr{F} \subseteq \mathscr{P}(A)$ of each set A, define a relation $R_{\mathscr{F}}$ on A so that R relates elements $X \in A$ and $Y \in A$ if and only if X and Y belong to the same element of the partition $\mathscr{F}$:

$$[(X,Y) \in R_{\mathscr{F}}] \Leftrightarrow \{\exists B[(B \in \mathscr{F}) \wedge (X \in B) \wedge (Y \in B)]\}.$$

3.159 Theorem (equivalence relations from partitions). *For each set A and for each partition $\mathscr{F}$ of A, the relation $R_{\mathscr{F}}$ is an equivalence relation.*

Proof. The relation $R_{\mathscr{F}}$ is reflexive: for each $X \in A$, the partition $\mathscr{F}$ has an element $B \in \mathscr{F}$ that contains X, because $\mathscr{F}$ covers A. Thus X and X both belong to the same $B \in \mathscr{F}$, whence $(X,X) \in R_{\mathscr{F}}$. Symbolically,

$$\forall X\{(X \in A) \Rightarrow [(X,X) \in R_{\mathscr{F}}]\}$$ **yet unproved,**

$\quad\quad\quad\quad\quad\quad\quad\quad\quad \Updownarrow$ definition 3.158 for $R_{\mathscr{F}}$,

$$\forall X(\{X \in A\} \Rightarrow$$
$$\{\exists B[(B \in \mathscr{F}) \wedge (X \in B) \wedge (X \in B)]\})$$

$\quad\quad\quad\quad\quad\quad\quad\quad\quad \Updownarrow$ tautology $[(P) \wedge (P)] \Leftrightarrow (P)$,

$$\forall X(\{X \in A\} \Rightarrow$$
$$\{\exists B[(B \in \mathscr{F}) \wedge (X \in B)]\})$$

$\quad\quad\quad\quad\quad\quad\quad\quad\quad \Updownarrow$ definition of $\bigcup$.

$$A \subseteq (\bigcup \mathscr{F})$$

which is universally valid by definition of a partition.

The relation $R_{\mathscr{F}}$ is symmetric: $(X,Y) \in R_{\mathscr{F}}$ if and only if the partition $\mathscr{F}$ has an element $B \in \mathscr{F}$ that contains X and Y, whence B contains Y and X, which means that $(Y,X) \in R_{\mathscr{F}}$. Symbolically, from the tautology $[(P) \wedge (Q)] \Leftrightarrow [(Q) \wedge (P)]$,

$$\vdash \forall X \forall Y \Big\{[(X \in A) \wedge (Y \in A)] \Rightarrow$$
$$\big(\{\exists B[(B \in \mathscr{F}) \wedge (X \in B) \wedge (Y \in B)]\} \Leftrightarrow$$
$$\{\exists B[(B \in \mathscr{F}) \wedge (Y \in B) \wedge (X \in B)]\}\big)\Big\}$$

$\quad\quad\quad\quad\quad\quad\quad\quad\quad \Updownarrow$ definition 3.158 for $R_{\mathscr{F}}$.

$$\forall X \big[\forall Y([(X \in A) \wedge (Y \in A)] \Rightarrow$$
$$\{[(X,Y) \in R_{\mathscr{F}}] \Leftrightarrow [(Y,X) \in R_{\mathscr{F}}]\})\big]$$

The relation $R_{\mathscr{F}}$ is transitive: if $(X,Y) \in R_{\mathscr{F}}$, then the partition $\mathscr{F}$ has an element $B \in \mathscr{F}$ that contains X and Y. If also $(Y,Z) \in R_{\mathscr{F}}$, then the partition $\mathscr{F}$ has an element $C \in \mathscr{F}$ that contains Y and Z. However, from $X \in B$ and $Y \in C$ it follows that $X \in (B \cap C)$, whence B and C are not disjoint and hence $B = C$, by definition of a partition. Consequently, $X \in B$ and $Z \in B$, and, therefore, $(X,Z) \in R_{\mathscr{F}}$. $\square$

3.160 Definition (equivalence classes).   For each equivalence relation $R \subseteq (A \times A)$, and for each element $X \in A$, define the subset $[X]_R$ of all the elements $Y \in A$ equivalent to X with respect to R, called the **equivalence class of** X:

$$[X]_R := \{Y \in A : (X,Y) \in R\}.$$

Then let $\mathscr{F}_R$ consist of all such equivalence classes:

$$\mathscr{F}_R := \{[X]_R \in \mathscr{P}(A) : X \in A\}.$$

Another common notation for $\mathscr{F}_R$ is A/R, so that $\mathscr{F}_R = A/R = \{[X]_R : X \in A\}$. The set $\mathscr{F}_R = A/R$ of all equivalence classes is also called the **quotient** of the set A by the relation R.

3.161 Example.   For each set A, the equivalence classes of the "diagonal" equivalence relation Δ_A consist of every singleton $[X]_{\Delta_A} = \{X\}$ for every $X \in A$. Thus $A/\Delta_A = \big\{\{X\} : X \in A\big\}$.

3.162 Example. For the set $A := \{0, 1, 2, 3\}$, the equivalence relation

$$\mathscr{R} := \{(0,0), (1,1), (2,2), (3,3), (0,2), (2,0), (1,3), (3,1)\}$$
$$= \Delta_A \cup \{(0,2), (2,0), (1,3), (3,1)\}$$

corresponds to the equivalence classes

$$[0]_{\mathscr{R}} = \{0, 2\},$$
$$[1]_{\mathscr{R}} = \{1, 3\}.$$

Thus $A/\mathscr{R} = \{[0]_{\mathscr{R}}, [1]_{\mathscr{R}}\}$.

3.163 Theorem (partitions from equivalence relations). *For each equivalence relation $R \subseteq (A \times A)$, the set of subsets $\mathscr{F}_R \subseteq \mathscr{P}(A)$ is a partition of A.*

Proof. The partition $\mathscr{F}_R$ covers A: the reflexivity of R guarantees that $(X, X) \in R$ for each $X \in A$, whence $X \in [X]_R$, and hence $X \in (\bigcup_{Z \in A} B_Z) = (\bigcup \mathscr{F}_R)$. Thus $A \subseteq (\bigcup \mathscr{F})$. The reverse inclusion follows from $(\bigcup \mathscr{F}) \subseteq [\bigcup \mathscr{P}(A)] = A$:

$$\vdash \forall X\{(X \in A) \Rightarrow [(X, X) \in R]\} \qquad \text{reflexivity of } R,$$
$$\qquad \qquad \Updownarrow \qquad \text{definition of } [X]_R,$$
$$\forall X\{(X \in A) \Rightarrow (X \in [X]_R)\} \qquad$$
$$\qquad \qquad \Updownarrow \qquad \text{definition of } \mathscr{F}_R,$$
$$\forall X\{(X \in A) \Rightarrow [([X]_R \in \mathscr{F}_R) \wedge (X \in [X]_R)]\} \qquad$$
$$\qquad \qquad \Updownarrow \qquad \text{definition of } \exists,$$
$$\forall X[(X \in A) \Rightarrow \{\exists B[(B \in \mathscr{F}_R) \wedge (X \in B)]\}] \qquad$$
$$\qquad \qquad \Updownarrow \qquad \text{definition of } \bigcup,$$
$$\forall X\{(X \in A) \Rightarrow [X \in (\bigcup \mathscr{F})]\} \qquad$$
$$\qquad \qquad \Updownarrow \qquad \text{definition of } \subseteq.$$
$$A \subseteq (\bigcup \mathscr{F})]$$

Any two distinct elements of $\mathscr{F}_R$ are disjoint: if two members B and C of $\mathscr{F}_R$ are not disjoint, then their intersection $B \cap C$ contains an element $X \in A$; by definition of $\mathscr{F}_R$, however, every element of B is equivalent to X, and so is every element of C, whence $B = [X]_R = C$, which is a negation of the distinctness of B and C. Finally, $\mathscr{F}_R$ does not contain any empty element. Indeed, if $A = \varnothing$, then $R \subseteq A \times A$, whence $R = \varnothing$ and $\mathscr{F}_R = \varnothing$, which does not contain any element, and hence no empty element. If $A \neq \varnothing$, then $\mathscr{F}_R = \{[X]_R : X \in A\}$ where $X \in [X]_R$, whence $[X]_R \neq \varnothing$. ☐

3.164 Example. For each equivalence relation $R \subseteq A \times A$ on a set A, the **canonical map,** also called the **quotient map,** is the function $P : A \to A/R$ that maps each element $X \in A$ to its equivalence class $[X]_R = \{Y \in A : (X, Y) \in R\}$:

$$P: A \to A/R,$$
$$X \mapsto [X]_R.$$

3.8.3 *Exercises on Equivalence Relations*

3.161 . Prove that the empty relation $\varnothing \subseteq A \times A$ is reflexive, symmetric, and transitive.

3.162 . Prove that for each set A the relation of strict inclusion $\subsetneqq$ on $\mathscr{P}(A)$ is *not* reflexive.

3.163 . Prove that for each set A the relation of strict inclusion $\subsetneqq$ on $\mathscr{P}(A)$ is anti-symmetric.

3.164 . Prove that for each set A the relation of strict inclusion $\subsetneqq$ on $\mathscr{P}(A)$ is transitive.

3.165 . For the set $A := \{0, 1, 2, 3, 4, 5\}$, verify that the following relation $\mathscr{R}$ is an equivalence relation, and list all its equivalence classes:

$$\mathscr{R} := \left\{ \begin{array}{ccc} & (1,5) \quad (3,5) \quad (5,5) \\ (0,4) \quad (2,4) \quad (4,4) \\ & (1,3) \quad (3,3) \quad (5,3) \\ (0,2) \quad (2,2) \quad (4,2) \\ & (1,1) \quad (3,1) \quad (5,1) \\ (0,0) \quad (2,0) \quad (4,0) \end{array} \right\}.$$

3.166 . For the set $A := \{0, 1, 2, 3, 4, 5\}$, verify that the following relation $\mathscr{S}$ is an equivalence relation, and list all its equivalence classes:

$$\mathscr{S} := \left\{ \begin{array}{ccc} & (5,2) \quad\quad (5,5) \\ (1,4) \quad\quad (4,4) \\ (0,3) \quad\quad (3,3) \\ & (2,2) \quad\quad (2,5) \\ (1,1) \quad\quad (4,1) \\ (0,0) \quad\quad (3,0) \end{array} \right\}.$$

3.167 . For the set $B := \{0, 1, 2, 3, 4, 5, 6, 7\}$, verify that the following set $\mathscr{F}$ of subsets is a partition, and list the corresponding equivalence relation:

$$\mathscr{F} := \big\{ \{0, 2, 4, 6\}, \ \{1, 3, 5, 7\} \big\}.$$

3.168 . For the set $B := \{0, 1, 2, 3, 4, 5, 6, 7\}$, verify that the following set $\mathscr{G}$ of subsets is a partition, and list the corresponding equivalence relation:

$$\mathscr{G} := \big\{ \{0, 4\}, \ \{1, 5\}, \ \{2, 6\}, \ \{3, 7\} \big\}.$$

3.169 . Prove that $R_{(\mathscr{F}_R)} = R$ for each equivalence relation R.

3.170 . Prove that $\mathscr{F}_{(R_{\mathscr{F}})} = \mathscr{F}$ for each partition $\mathscr{F}$.

3.9   Ordering Relations

3.9.1   Preorders and Partial Orders

Besides the reflexivity, symmetry, and transitivity used to define equivalence relations, such other types of relations as rankings also require different variations of these concepts, as introduced here (with the terminology of Suppes [128, §3.2, p. 69]).

3.165 Definition (strict, irreflexivity). For each set A, a relation $R \subseteq A \times A$ is **irreflexive**, or, equivalently, **strict**, if and only if R does *not* relate any element of A to itself:

$$\forall X \{(X \in A) \Rightarrow [(X, X) \notin R]\}.$$

3.166 Example. The empty relation $\varnothing$ is strict, because the conclusion $(X, X) \notin \varnothing$ is universally valid.

3.167 Example. For each set A, the relation of strict inclusion $\subset$ is strict on $\mathscr{P}(A)$. Indeed, for all subsets $V \subseteq A$ and $W \subseteq A$, the definition of $V \subset W$ includes the requirement that $V \neq W$, so that $(V \subset W) \Rightarrow (V \neq W)$. Contraposition then confirms that $(V = W) \Rightarrow (V \not\subset W)$, so that $V \not\subset V$.

One method to specify a ranking or direction on a set removes the requirement of symmetry from the concept of equivalence, which gives the following concept of preorder.

3.168 Definition (preorder or quasi-order). For each set A, a relation $R \subseteq A \times A$ is a **preorder** or a **quasi-order**, if and only if R is *reflexive and transitive*. It is a **strict preorder** if and only if R is *irreflexive and transitive*.

3.169 Example. Consider the set $A := \{0, 1, 2\}$.
The relation $\mathscr{Q} := \{(0, 0), (0, 1), (1, 0), (1, 1), (2, 2)\}$ is a preorder.
The relation $\mathscr{R} := \{(0, 0), (0, 1), (1, 1), (2, 2)\}$ is a preorder.
The relation $\mathscr{S} := \{(0, 1)\}$ is a strict preorder.

3.170 Example. For each set A, the diagonal Δ_A is a preorder. If $A \neq \varnothing$, then Δ_A is *not* strict, because there exists $X \in A$ and then $(X, X) \in \Delta_A$.

3.171 Example. For each set A the relation $\subseteq$ is a preorder on $\mathscr{P}(A)$. The relation $\subseteq$ is *not* strict because $\varnothing \in \mathscr{P}(A)$ and $\varnothing \subseteq \varnothing$.

3.172 Example. For each set A, the relation $\subset$ is a *strict* preorder on $\mathscr{P}(A)$.

The concept of a preorder R allows for "circular" rankings, with $(X, Y) \in R$ and $(Y, X) \in R$ even though $X \neq Y$. To specify different types of rankings, a different concept — anti-symmetry — becomes necessary.

3.173 Definition (anti-Symmetry). For each set A, a relation $R \subseteq A \times A$ is **anti-symmetric** if and only if $(X, Y) \in R$ and $(Y, X) \in R$ imply $X = Y$:

$$\forall X\big(\forall Y\big[\{[(X, Y) \in R] \wedge [(Y, X) \in R]\} \Rightarrow (X = Y)\big]\big).$$

3.174 Example. For each set A, the diagonal Δ_A is an anti-symmetric relation.

3.175 Example. For each set A, the relation $\subseteq$ on $\mathscr{P}(A)$ is anti-symmetric. Indeed, for each $B \in \mathscr{P}(A)$ and for each $C \in \mathscr{P}(A)$, if $B \subseteq C$ and $C \subseteq B$, then $B = C$.

Similar features can also be defined through the following concept of asymmetry.

3.176 Definition (asymmetry). For each set A, a relation $R \subseteq A \times A$ is **asymmetric** if and only if $(X, Y) \in R$ implies $(Y, X) \notin R$ for each $X \in A$ and each $Y \in A$:

$$\forall X\big(\forall Y\{[(X, Y) \in R] \Rightarrow [(Y, X) \notin R]\}\big).$$

3.177 Example. Consider the set $A := \{0, 1, 2\}$.
The relation $\mathscr{Q} := \{(0,0), (0,1), (1,0), (1,1), (2,2)\}$ is *not* asymmetric: $\mathscr{Q}$ contains $(0, 1)$ and $(1, 0)$.
The relation $\mathscr{R} := \{(0,0), (0,1), (1,1), (2,2)\}$ is *not* asymmetric: $\mathscr{R}$ contains $(0, 0)$, $(1, 1)$, and $(2, 2)$.
The relation $\mathscr{S} := \{(0, 1)\}$ is asymmetric.

3.178 Example. The following relation is not asymmetric, not anti-symmetric, not irreflexive, not reflexive, not symmetric, and not transitive:

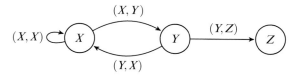

In contrast to preorders, "partial orders" do *not* allow for circular rankings.

3.179 Definition (partial order). For each set A, a relation $R \subseteq A \times A$ is a **partial order** if and only if R is *reflexive, anti-symmetric, and transitive*. It is a **strict partial order** if and only if R is *irreflexive and transitive*.

3.180 Example. Consider the set $A := \{0, 1, 2\}$.
The relation $\mathscr{Q} := \{(0,0), (0,1), (1,0), (1,1), (2,2)\}$ is *not* a partial order, because it is *not* anti-symmetric: $\mathscr{Q}$ contains $(0, 1)$ and $(1, 0)$ but $0 \neq 1$.
The relation $\mathscr{R} := \{(0,0), (0,1), (1,1), (2,2)\}$ is a partial order.
The relation $\mathscr{S} := \{(0, 1)\}$ is a strict partial order.

3.181 Example. For each set A, the diagonal Δ_A is a partial order.

3.182 Example. For each set A the relation $\subseteq$ is a partial order on $\mathscr{P}(A)$.

3.183 Example. For each set A the relation $\subsetneqq$ is a *strict* partial order on $\mathscr{P}(A)$.

As their names might suggest, neither preorders nor partial orders need relate all elements to one another. Relations that do so are called strongly connected.

3.184 Definition (Strong connectivity). For each set A, a relation $R \subseteq A \times A$ is **strongly connected** if and only if $(X, Y) \in R$ or $(Y, X) \in R$ (including both possibilities) for each $X \in A$ and for each $Y \in A$:

$$\forall X \big[\forall Y \big([(X \in A) \wedge (Y \in A)] \Rightarrow \{[(X, Y) \in R] \vee [(Y, X) \in R]\} \big) \big].$$

Strict rankings replace strong connectivity by connectivity.

3.185 Definition (connectivity). For each set A, a relation $R \subseteq A \times A$ is **connected** if and only if $(X, Y) \in R$ or $(Y, X) \in R$ (including both possibilities) for all *distinct* $X \in A$ and $Y \in A$ (such that $X \neq Y$):

$$\forall X \big[\forall Y \big([(X \in A) \wedge (Y \in A) \wedge (X \neq Y)] \Rightarrow \{[(X, Y) \in R] \vee [(Y, X) \in R]\} \big) \big].$$

3.186 Example. Consider the set $A := \{0, 1, 2\}$.

The relation $\mathcal{Q} := \{(0, 0), (0, 1), (1, 0), (1, 1), (2, 2)\}$ is neither connected nor strongly connected, because it contains neither $(0, 2)$ nor $(2, 0)$.

The relation $\mathcal{T} := \{(0, 0), (0, 1), (0, 2), (1, 1), (1, 2), (2, 2)\}$ is strongly connected.

3.9.2   Total Orders and Well-Orderings

The geometric "direction along a line" corresponds to a **total order**, also called **linear order, complete order** [71, p. 14], or **simple order** [71, p. 14], [128, p. 69].

3.187 Definition (total order). For each set A, a relation $R \subseteq A \times A$ is a **total order,** or **total ordering,** if and only if R is a *strongly connected partial order* (strongly connected, reflexive, anti-symmetric, and transitive). It is a **strict total order** if and only if R is *connected, irreflexive, and transitive*.

3.188 Example. The empty relation on the empty set is a *strict* total order.

3.189 Example. For each set S and the singleton $A = \{S\}$, the relation $\subseteq$ on $\mathcal{P}(A) = \{\varnothing, \{S\}\}$, is a total order. Indeed, $\varnothing \subseteq \varnothing$, $\varnothing \subseteq \{S\}$, and $\{S\} \subseteq \{S\}$.

3.190 Counterexample. The relation $\subset$ is *not* a total order in general. Thus, if

$$A := \big\{ \varnothing, \{\varnothing\} \big\},$$

then $\subset$ is *not* a total order on the power set

$$\mathcal{P}(A) = \Big\{ \varnothing, \{\varnothing\}, \{\{\varnothing\}\}, \{\varnothing, \{\varnothing\}\} \Big\},$$

because

$$\{\varnothing\} \not\subset \{\{\varnothing\}\},$$
$$\{\{\varnothing\}\} \not\subset \{\varnothing\}.$$

Thus the relation $\subset$ contains neither the pair $(\{\varnothing\}, \{\{\varnothing\}\})$ nor the pair $(\{\{\varnothing\}\}, \{\varnothing\})$. Instead, for this set A the relation $\subset$ takes the following form:

$$\{\{\varnothing\}\} \subset \{\varnothing, \{\varnothing\}\}$$

$$\cup \qquad\qquad \cup$$

$$\varnothing \quad \subset \quad \{\varnothing\}$$

Some relations that are not total orders can restrict to total orders on subsets.

3.191 Definition (chain). For each set A, and for each relation $R \subseteq A \times A$, a subset $B \subseteq A$ is a **chain** if and only if the restriction $R|_B$ is a total order on B.

In particular, for each total order R on each set A, the set A is a chain relative to R.

In partially ordered sets, a subset might have a first element, which precedes every element of that subset, or a last element, which follows every element of that subset.

3.192 Definition (first or last element). For each set A, for each subset $B \subseteq A$, and for each partial order $R \subseteq A \times A$, an element $X \in B$ is a **minimum**, or **first, or smallest, element in** B if and only if $(X, Y) \in R$ for each $Y \in B$. A first element of B is also denoted by $\min(B)$. Similarly, an element $Z \in B$ is a **maximum**, or **last, or largest, element in** B if and only if $(Y, Z) \in R$ for each $Y \in B$. A last element of B is also denoted by $\max(B)$.

3.193 Example. For each set A and for the relation $\subseteq$ on $\mathscr{P}(A)$, the element $\varnothing \in \mathscr{P}(A)$ is a first element of $\mathscr{P}(A)$. Indeed, $\varnothing \subseteq C$ for each $C \in \mathscr{P}(A)$. Also, the element $A \in \mathscr{P}(A)$ is a last element of $\mathscr{P}(A)$. Indeed, $C \subseteq A$ for each $C \in \mathscr{P}(A)$.

3.194 Definition (well-ordering). For each set A and for each partial order $R \subseteq A \times A$, the set A is **well-ordered** by R if and only if each nonempty subset $B \subseteq A$ has a first element. A relation R is called a **well-ordering** (for the lack of a grammatical and logically equivalent terminology) if and only if A is well-ordered by R.

3.195 Remark. Instead of requiring that a well-ordering be a partial order [30, p. 31], some texts impose the stronger requirement that a well-ordering be a total order [71, p. 29]. Yet the insistence on a total order is redundant; indeed, if each nonempty subset has a first element, then a partial order is automatically a total order [128, p. 74–76].

3.196 Example. The set $\{\varnothing\} = \mathscr{P}(\varnothing)$ is well-ordered by the relation of inclusion $\subseteq$. Indeed, the only nonempty subset of $\{\varnothing\} = \mathscr{P}(\varnothing)$ is $\{\varnothing\}$, and it has a first element, namely $\varnothing$, because $\varnothing \subseteq C$ for each $C \in \{\varnothing\}$.

3.197 Theorem. *Every subset of a well-ordered set is well-ordered.*

Proof. Each nonempty subset $E \subseteq B$ of a subset $B \subseteq A$ of a set A with a well-order $\preceq$ is also a subset $E \subseteq A$ and hence has a first element. Thus $\preceq$ induces a well-order on B.   □

3.198 Definition (upper or lower bound). For each set A, for each subset $B \subseteq A$, and for each partial order $R \subseteq A \times A$, an element $Z \in A$ is an **upper bound for** B if and only if $(Y, Z) \in R$ for each $Y \in B$. Similarly an element $X \in A$ is a **lower bound for** B if and only if $(X, Y) \in R$ for each $Y \in B$.

Thus a last element can differ from an upper bound because an upper bound need not belong to the same subset. Similarly a first element can differ from a lower bound because a lower bound need not belong to the same subset.

3.199 Definition (maximal element). For each set A, for each subset $B \subseteq A$, and for each partial order $R \subseteq A \times A$, an element $Z \in B$ is a **maximal element of** B if and only if $[(Z, Y) \in R] \Rightarrow [(Y, Z) \in R]$ for each $Y \in B$. In other words, $Z \in B$ is a maximal element of B if and only if every element $Y \in B$ that follows Z also precedes Z (which allows for the possibility that no $Y \in B$ follows Z).

Similarly, an element $Z \in B$ is a **minimal element of** B if and only if $[(Y, Z) \in R] \Rightarrow [(Z, Y) \in R]$ for each $Y \in B$.

Thus, either a maximal element precedes no element, or it follows every element that it precedes, but in either case it belongs to the same subset. For instance, a last element is a maximal element. However, a maximal element need not be a last element.

3.200 Example. Consider the set $A := \{0, 1, 2\}$ and the relation

$$\mathscr{Q} := \{(0, 0), (0, 1), (1, 0), (1, 1), (2, 2)\}.$$

The element $Z := 2$ is maximal in A, because the hypothesis $(Z, Y) \in \mathscr{Q}$ in $[(Z, Y) \in \mathscr{Q}] \Rightarrow [(Y, Z) \in \mathscr{Q}]$ is False for each $Y \neq 2$ in A. Yet 2 is not a last element in A, because 2 does not follow every element: $(0, 2) \notin \mathscr{Q}$ and $(1, 2) \notin \mathscr{Q}$.

Such notions as the geometric direction on an "infinite line, plane, or space" require another axiom, the axiom of infinity, which forms the subject of chapter 4.

Axiom S7 (Axiom of infinity) There exists a set $\mathbb{I}$, such that $\varnothing \in \mathbb{I}$, and for each element $C \in \mathbb{I}$ its successor $C \cup \{C\}$ is also an element of $\mathbb{I}$:

$$\vdash \exists \mathbb{I}\big[(\varnothing \in \mathbb{I}) \wedge \big(\forall C \{(C \in \mathbb{I}) \Rightarrow [(C \cup \{C\}) \in \mathbb{I}]\}\big)\big].$$

The concept of well-ordered sets allows for many logically equivalent statements of the last axiom of Zermelo-Frænkel set theory [30, Ch. 1, §9, p. 23, #9.2(3)].

Axiom S8 (Axiom of choice) For each set $\mathscr{F}$ of *nonempty* sets ($A \neq \varnothing$ for each $A \in \mathscr{F}$), there is a "choice" function $F : \mathscr{F} \to \bigcup \mathscr{F}$ with $F(A) \in A$ for each $A \in \mathscr{F}$.

The axiom of choice is logically equivalent to each of the following two theorems, as proved in chapter 6.

3.201 Theorem (Zermelo's Theorem). Every set is well-ordered by some relation.

3.202 Theorem (Zorn's Lemma). In a pre-ordered set A, if every chain in A has an upper bound, then A has a maximal element.

3.9.3 *Exercises on Ordering Relations*

For the following exercises, determine all of the characteristics — among connectivity, strong connectivity, reflexivity, irreflexivity, symmetry, anti-symmetry, asymmetry, transitivity — of the given relation on the set $A := \{0, 1, 2, 3, 4, 5, 6, 7, 8, 9\}$.

3.171 .

$$
\mathscr{Q} := \left\{
\begin{array}{l}
(0, 9)\ (1, 9) \qquad\ (3, 9) \qquad\quad (6, 9) \\
(0, 8)\ (1, 8)\ (2, 8) \qquad (4, 8)\ (6, 8) \\
(0, 7)\ (1, 7) \\
(0, 6)\ (1, 6)\ (2, 6)\ (3, 6)\ (4, 6) \\
(0, 5)\ (1, 5) \\
(0, 4)\ (1, 4)\ (2, 4) \\
(0, 3)\ (1, 3) \\
(0, 2)\ (1, 2) \\
(0, 1)
\end{array}
\right\}.
$$

3.172 .

$$
\mathscr{R} := \left\{
\begin{array}{l}
(0, 9)\ (1, 9) \qquad\ (3, 9) \qquad\qquad (6, 9) \qquad\qquad (9, 9) \\
(0, 8)\ (1, 8)\ (2, 8) \qquad (4, 8) \qquad (6, 8) \qquad (8, 8) \\
(0, 7)\ (1, 7) \qquad\qquad\qquad\qquad\qquad (7, 7) \\
(0, 6)\ (1, 6)\ (2, 6)\ (3, 6)\ (4, 6) \qquad (6, 6) \\
(0, 5)\ (1, 5) \qquad\qquad\quad (5, 5) \\
(0, 4)\ (1, 4)\ (2, 4) \qquad (4, 4) \\
(0, 3)\ (1, 3) \qquad (3, 3) \\
(0, 2)\ (1, 2)\ (2, 2) \\
(0, 1)\ (1, 1) \\
(0, 0)
\end{array}
\right\}.
$$

3.173 .

$$
\mathscr{S} := \left\{
\begin{array}{l}
(0,9)\ (1,9)\quad\ (3,9)\qquad\qquad (6,9) \\
(0,8)\ (1,8)\ (2,8)\quad (4,8)\qquad (6,8) \\
(0,7)\ (1,7) \\
(0,6)\ (1,6)\ (2,6)\ (3,6)\ (4,6)\qquad\qquad\qquad (8,6)\ (9,6) \\
(0,5)\ (1,5) \\
(0,4)\ (1,4)\ (2,4)\qquad\qquad (6,4)\qquad (8,4) \\
(0,3)\ (1,3)\qquad\qquad\qquad (6,3)\qquad\qquad (9,3) \\
(0,2)\ (1,2)\qquad (4,2)\qquad (6,2)\qquad (8,2) \\
(0,1)\qquad (2,1)\ (3,1)\ (4,1)\ (5,1)\ (6,1)\ (7,1)\ (8,1)\ (9,1) \\
\qquad\ (1,0)\ (2,0)\ (3,0)\ (4,0)\ (5,0)\ (6,0)\ (7,0)\ (8,0)\ (9,0)
\end{array}
\right\}.
$$

3.174 .

$$
\mathscr{U} := \left\{
\begin{array}{l}
(0,9)\ (1,9)\quad\ (3,9)\qquad\qquad (6,9)\qquad\qquad (9,9) \\
(0,8)\ (1,8)\ (2,8)\quad (4,8)\qquad (6,8)\qquad (8,8) \\
(0,7)\ (1,7)\qquad\qquad\qquad\qquad\qquad (7,7) \\
(0,6)\ (1,6)\ (2,6)\ (3,6)\ (4,6)\qquad (6,6)\qquad (8,6)\ (9,6) \\
(0,5)\ (1,5)\qquad\qquad\qquad (5,5) \\
(0,4)\ (1,4)\ (2,4)\qquad (4,4)\qquad (6,4)\qquad (8,4) \\
(0,3)\ (1,3)\qquad (3,3)\qquad\qquad (6,3)\qquad\qquad (9,3) \\
(0,2)\ (1,2)\ (2,2)\qquad (4,2)\qquad (6,2)\qquad (8,2) \\
(0,1)\ (1,1)\ (2,1)\ (3,1)\ (4,1)\ (5,1)\ (6,1)\ (7,1)\ (8,1)\ (9,1) \\
(0,0)\ (1,0)\ (2,0)\ (3,0)\ (4,0)\ (5,0)\ (6,0)\ (7,0)\ (8,0)\ (9,0)
\end{array}
\right\}.
$$

3.175 .

$$
\mathscr{V} := \left\{
\begin{array}{l}
\qquad (1,9)\qquad\qquad\qquad\qquad\qquad (9,9) \\
\qquad (1,8)\qquad\qquad\qquad\qquad (8,8) \\
\qquad (1,7)\qquad\qquad\qquad (7,7) \\
\qquad (1,6)\qquad\qquad (6,6) \\
\qquad (1,5)\qquad (5,5) \\
\qquad (1,4)\qquad (4,4) \\
\qquad (1,3)\quad (3,3)\qquad\qquad\qquad (9,3) \\
\qquad (1,2)\ (2,2)\quad (4,2)\qquad\qquad (8,2) \\
(0,1)\ (1,1) \\
\qquad (1,0)
\end{array}
\right\}.
$$

3.176 .

$$\mathscr{W} := \left\{ \begin{array}{lllllllll} (1,9) & (3,9) & & & & & & & (9,9) \\ (1,8) & (2,8) & & & & & & (8,8) & \\ (1,7) & & & & & & (7,7) & & \\ (1,6) & & & & & (6,6) & & & \\ (1,5) & & & & (5,5) & & & & \\ (1,4) & (2,4) & & (4,4) & & & & & \\ (1,3) & & (3,3) & & & & & & (9,3) \\ (1,2) & (2,2) & & (4,2) & & & & (8,2) & \\ (0,1) & (1,1) & (2,1) & (3,1) & (4,1) & (5,1) & (6,1) & (7,1) & (8,1) & (9,1) \\ (1,0) & & & & & & & & \end{array} \right\} .$$

3.177 . Prove that for each set A the empty relation $\varnothing \subseteq A \times A$ is a partial order.

3.178 . Prove that a relation $\mathscr{R}$ is *symmetric* if and only if $\mathscr{R} = \mathscr{R}^{\circ -1}$.

3.179 . Prove that a relation $\mathscr{R}$ on A is *irreflexive* if and only if $\mathscr{R} \cap \Delta_A = \varnothing$.

3.180 . Prove that a relation $\mathscr{R}$ on a set A is *anti-symmetric* if and only if $\mathscr{R} \cap \mathscr{R}^{\circ -1} \subseteq \Delta_A$.

3.181 . Prove that every strict partial order is asymmetric

3.182 . Prove that if a relation is irreflexive and anti-symmetric, then it is also asymmetric

3.183 . Prove that every asymmetric relation is also anti-symmetric.

3.184 . Prove that every asymmetric relation is also irreflexive.

3.185 . Prove that a relation is a strict partial order if and only if it is asymmetric and transitive

3.186 . Provide an example of an anti-symmetric but not asymmetric relation.

3.187 . Provide an example of an asymmetric and strongly connected relation.

3.188 . Exhibit an example of a set A for which the relation $\subseteq$ on $\mathscr{P}(A)$ is *not* a total order: supply subsets B and C such that neither $B \subseteq C$ nor $C \subseteq B$.

3.189 . Exhibit an example of a set A and a subset $S \subseteq \mathscr{P}(A)$ that is a chain with respect to the relation $\subseteq$ on $\mathscr{P}(A)$.

3.190 . Prove that if a function is surjective, then it has a right inverse.

Chapter 4
Mathematical Induction: Definitions and Proofs by Induction

4.1 Introduction

This chapter introduces the concepts of mathematical induction and recursion. Starting from first-order logic and the Zermelo-Fraenkel axioms of set theory, including the axiom of infinity, the chapter establishes the theoretical basis for proofs by the Principle of Mathematical Induction, followed by definitions through mathematical induction, which forms the basis for the concept of primitive-recursive functions. As an extended example, one section defines integer addition and multiplication as primitive-recursive functions, and hence derives the basic properties of integer arithmetic (associativity, commutativity, and distributivity) by induction. Subsequent sections define ordering relations and derive cancellation laws and exponential laws for integer and rational arithmetic. Another section then applies arithmetic to the cardinalities of unions, intersections, differences, and Cartesian products of finite sets. Yet another section focuses on denumerable and other not-necessarily finite sets, leading to J. M. Whitaker's proof of the Bernstein-Cantor-Schröder Theorem within Zermelo-Fraenkel set theory. The prerequisites for this chapter consist of a working knowledge of first-order logic and Zermelo-Fraenkel set theory, for instance, as described in chapters 1, 2, and 3, which contain all the logical and set-theoretical theorems cited in this chapter.

The concepts of "numbers" and "counting" allow for the determination and comparison of the "sizes" of various objects. The same concepts also lead to precise definitions of "infinite" sets, which then reveals an "infinite" variety of "infinite" sets. For example, the number of particles in the universe is related to its mass, volume, and density, which quantities might affect the fate of the universe, in particular, whether the universe will remain finite or will expand to "infinity" [69, p. 445].

Mathematically, this chapter shows how the concepts of "numbers," "arithmetic," "counting," "computing," and "infinity" fit within set theory, without any new axiomatic system.

© Springer Science+Business Media New York 2015
Y. Nievergelt, *Logic, Mathematics, and Computer Science*,
DOI 10.1007/978-1-4939-3223-8_4

4.2 Mathematical Induction

The principle of "mathematical induction" provides a method for proving theorems, and also a method for specifying and analyzing many practical numerical calculations.

4.2.1 The Axiom of Infinity

The Principle of Mathematical Induction forms the theoretical basis for such algorithms as counting and arithmetic, by means of *sequences* of definitions, computations, and verifications, where the length of the sequence depends on the situation. One framework to allow for sequences of yet unspecified lengths consists in embedding all such sequences into one set $\mathbb{N}$ that already allows for sequences of all lengths. To this end, with a construction attributed to John von Neumann [8, p. 22], [128, p. 129], [135], for each element $X \in \mathbb{N}$, the set $\mathbb{N}$ also contains a "next" element $X \cup \{X\}$.

4.1 Definition. For each set X the **successor** of X is $X \cup \{X\}$.

4.2 Example. The successor of the empty set $\varnothing$ is the set $\varnothing \cup \{\varnothing\} = \{\varnothing\}$.

A subsequent theorem will verify that for each set $X \in \mathbb{N}$ the successor $X \cup \{X\}$ is strictly larger than X, so that $X \subsetneq (X \cup \{X\})$. Within the theory presented so far, however, nothing guarantees the existence of a set containing the successor of each of its elements. To this end, a new axiom becomes necessary [8, p. 21]. The following version is identical to that of axiom S7 already mentioned in chapter 3.

Axiom S7 (Axiom of infinity) There exists a set $\mathbb{I}$, such that $\varnothing \in \mathbb{I}$, and for each element $C \in \mathbb{I}$ its successor $C \cup \{C\}$ is also an element of $\mathbb{I}$:

$$\vdash \exists \mathbb{I} \big[(\varnothing \in \mathbb{I}) \wedge \big(\forall C \{ (C \in \mathbb{I}) \Rightarrow [(C \cup \{C\}) \in \mathbb{I}] \} \big) \big].$$

Digits or other symbols can abbreviate the set notation for the elements of $\mathbb{I}$.

4.3 Example. The set $\mathbb{I}$ described in axiom S7 contains the following elements.

$$0 := \varnothing,$$

$$1 := 0 \cup \{0\} = \varnothing \cup \{\varnothing\} = \{\varnothing\},$$

$$2 := 1 \cup \{1\} = \{\varnothing\} \cup \{\{\varnothing\}\} = \{\varnothing, \{\varnothing\}\},$$

$$3 := 2 \cup \{2\} = \{\varnothing, \{\varnothing\}\} \cup \left\{\{\varnothing, \{\varnothing\}\}\right\} = \left\{\varnothing, \{\varnothing\}, \{\varnothing, \{\varnothing\}\}\right\},$$

$$4 := 3 \cup \{3\} = \left\{ \varnothing, \{\varnothing\}, \{\varnothing, \{\varnothing\}\}, \left\{\varnothing, \{\varnothing\}, \{\varnothing, \{\varnothing\}\}\right\} \right\},$$

$$\vdots$$

Such elements are called "natural numbers" (see remark 4.7 about 0).

The axiom of infinity does *not* restrict $\mathbb{I}$ to contain only the elements in example 4.3, so that the set $\mathbb{I}$ might also contain other elements. Therefore, a further construction with the preceding axioms becomes necessary to form a set containing *only* the natural numbers. The construction selected here forms the intersection of subsets of $\mathbb{I}$, using a method common to several parts of mathematics [30, p. 66], [34, p. 21], [60, p. 132]. Specifically, denote by P the logical formula defined by

$$(P) \Leftrightarrow \left[(\varnothing \in A) \wedge \left(\forall X \{(X \in A) \Rightarrow [(X \cup \{X\}) \in A]\}\right)\right].$$

Thus the formula P asserts that a set A contains among its elements the empty set and the successor of every one of its elements. The Axiom of Infinity asserts that there exists a set $\mathbb{I}$ for which $\mathrm{Subf}_{\mathbb{I}}^A(P)$ is True:

$$\vdash \exists \mathbb{I} \left[(\varnothing \in \mathbb{I}) \wedge \left(\forall X \{(X \in \mathbb{I}) \Rightarrow [(X \cup \{X\}) \in \mathbb{I}]\}\right)\right].$$

The following construction forms a subset $\mathbb{N} \subseteq \mathbb{I}$ containing only elements such as those in example 4.3. To this end, consider the set $\mathscr{F} \subseteq \mathscr{P}(\mathbb{I})$ of all subsets $B \subseteq \mathbb{I}$ for which $\mathrm{Subf}_B^A(P)$ is also True:

$$\mathscr{F} := \{B \in \mathscr{P}(\mathbb{I}) : \mathrm{Subf}_B^A(P)\}$$

$$= \left\{ B \in \mathscr{P}(\mathbb{I}) : \right.$$
$$\left. \left[(\varnothing \in B) \wedge \left(\forall X \{(X \in B) \Rightarrow [(X \cup \{X\}) \in B]\}\right)\right]\right\}.$$

The following definition and theorems will confirm that $\mathbb{N}$ can be defined as $\bigcap \mathscr{F}$.

4.4 Theorem. *The set* $\mathscr{F} = \{B \in \mathscr{P}(\mathbb{I}) : \mathrm{Subf}_B^A(P)\}$ *is not empty:* $\mathbb{I} \in \mathscr{F}$.

Proof. By the Axiom of Infinity (S7), the formula P is True for $\mathbb{I}$, so that $\mathrm{Subf}_{\mathbb{I}}^A(P)$ is True. Moreover, $\mathbb{I} \subseteq \mathbb{I}$. Therefore, $\mathbb{I}$ is a subset of $\mathbb{I}$ for which P is True, so that $(\mathbb{I} \subseteq \mathbb{I}) \wedge [\mathrm{Subf}_{\mathbb{I}}^A(P)]$ holds, which means that $\mathbb{I} \in \mathscr{F}$. □

From $\mathbb{I}$ the following definition extracts $\mathbb{N}$ as the smallest subset for which P holds.

4.5 Definition (natural numbers). The set of **natural numbers**, denoted by $\mathbb{N}$, is the intersection of all the elements of $\mathcal{F}$:

$$\mathbb{N} := \bigcap \mathcal{F}$$

A **natural number** is an element of $\mathbb{N}$. Also, define $\mathbb{N}^* := \mathbb{N} \setminus \{\varnothing\}$.

4.6 Remark. In the present theory, every natural number is a *set*.

4.7 Remark. Some texts exclude the element $\varnothing$ from the set $\mathbb{N}$. The two definitions differ only in their terminology and lead to the same theory. Yet the definition of $\mathbb{N}$ with $\varnothing \in \mathbb{N}$ has proved more convenient than without it in the context of set theory [128, p. 121], ordinals [74], and in such situations as Kurt Gödel's work [46] on logic and mathematics. Therefore, the definition adopted here includes $\varnothing \in \mathbb{N}$.

The following theorems verify that P also holds for $\mathbb{N}$.

4.8 Theorem. *The set* $\mathbb{N} = \bigcap \mathcal{F}$ *is not empty; indeed,* $\varnothing \in \mathbb{N}$.

Proof. First, $\varnothing \in \mathbb{I}$ and $\mathbb{I} \in \mathcal{F}$ by theorem 4.4, whence $\varnothing \in \bigcup \mathcal{F}$. Second, for each $B \in \mathcal{F}$ the formula $\mathrm{Subf}_B^A(P)$ holds, hence $\varnothing \in B$; consequently, $\varnothing \in \bigcap \mathcal{F} = \mathbb{N}$. $\qquad\square$

4.9 Theorem. *The formula P is True for* $\mathbb{N}$: $\forall X \{(X \in \mathbb{N}) \Rightarrow [(X \cup \{X\}) \in \mathbb{N}]\}$.

Proof. For each X, if $X \in \mathbb{N} = \bigcap \mathcal{F}$, then $X \in B$ for each $B \in \mathcal{F}$, but then $(X \cup \{X\}) \in B$ for each $B \in \mathcal{F}$, whence $(X \cup \{X\}) \in \bigcap \mathcal{F} = \mathbb{N}$. $\qquad\square$

4.10 Theorem. *The set* $\mathbb{N}$ *is an element of* $\mathcal{F}$.

Proof. Theorems 4.8 and 4.9 show that $\varnothing \in \mathbb{N}$ and $\mathrm{Subf}_{\mathbb{N}}^A(P)$ are True. Moreover, $\mathbb{N} = \bigcap \mathcal{F} \subseteq \mathbb{I}$. Thus, $(\mathbb{N} \subseteq \mathbb{I}) \wedge [\mathrm{Subf}_{\mathbb{N}}^A(P)]$ holds, whence $\mathbb{N} \in \mathcal{F}$. $\qquad\square$

The concept of "successor" can serve to specify functions defined on $\mathbb{N}$.

4.11 Example. The "successor" function is defined by

$$G := \{(X, Y) \in (\mathbb{N} \times \mathbb{N}) : Y = (X \cup \{X\})\},$$

so that

$$G : \mathbb{N} \to \mathbb{N},$$
$$X \mapsto X \cup \{X\}.$$

4.12 Definition (sequence). For each set E, a **sequence** in E is a function $F : \mathbb{N} \to E$. For each $N \in \mathbb{N}$, the value $F(N)$ can also be denoted by F_N, and then the function F can also be denoted by (F_N) or $(F_N)_{N \in \mathbb{N}}$. Also, a **finite sequence** of **length** L in E is a function $S : L \to E$ defined on some $L \in \mathbb{N}$. The value $S(K)$ is also denoted by S_K; then the function S is also denoted by (S_K) or $(S_K)_{K \in L}$.

4.13 Example. The "successor" function in example 4.11 is a sequence in $\mathbb{N}$.

For convenience, mathematical usage abbreviates the successor $N \cup \{N\}$ as $N + 1$.

4.14 Definition. For each natural number N, let $N + 1 := N \cup \{N\}$.

In general, the specification of functions defined on $\mathbb{N}$ requires a method known as the Principle of Mathematical Induction, as described in the next subsections.

4.2.2 The Principle of Mathematical Induction

The following theorem forms the theoretical basis for the methods of proof by induction and recursive computation. Specifically, the "Principle of Mathematical Induction" shows that if a subset $S \subseteq \mathbb{N}$ contains $\varnothing$, and if S also contains the successor $X \cup \{X\}$ of each of its elements X, then S contains all the natural numbers: $S = \mathbb{N}$.

4.15 Theorem (Principle of Mathematical Induction). *For each subset $S \subseteq \mathbb{N}$, if S contains the empty set and the successor of every one of its elements, then $S = \mathbb{N}$. Thus, $\vdash \forall S \left\{ (S \subseteq \mathbb{N}) \wedge \left[\text{Subf}_S^A(P) \right] \Rightarrow (S = \mathbb{N}) \right\}$; or, with $\text{Subf}_S^A(P)$ spelled out:*

$$\vdash \forall S \left[(S \subseteq \mathbb{N}) \wedge (\varnothing \in S) \wedge \left(\forall X \left\{ (X \in S) \Rightarrow [(X \cup \{X\}) \in S] \right\} \right) \Rightarrow (S = \mathbb{N}) \right].$$

Proof. If $S \subseteq \mathbb{N}$ then $S \subseteq \mathbb{N} = \bigcap \mathscr{F} \subseteq \mathbb{I}$. If moreover $\text{Subf}_S^A(P)$ is True, then $S \in \mathscr{F}$ by definition of $\mathscr{F}$. From $S \in \mathscr{F}$, it then follows that $\bigcap \mathscr{F} \subseteq S$. Thus, $\bigcap \mathscr{F} \subseteq S \subseteq \bigcap \mathscr{F}$, whence (by theorem 3.10) equality holds: $S = \bigcap \mathscr{F} = \mathbb{N}$. □

4.16 Remark. The proof of theorem 4.15 derives, or, equivalently, *deduces* the Principle of Mathematical Induction from the preceding axioms and inference rules of logic and axioms or postulates of set theory. Therefore, in mathematics, all inductive proofs are deductive:

> In mathematical English, the words "inductive" and "deductive" may be used interchangeably for any kind of reasoning, except that the principle listed in [theorem 4.15] is usually called the Induction Principle [114, p. 195, Remark].

The following example demonstrates a pattern amenable to induction.

4.17 Example. The elements of $\mathbb{N}$ in example 4.3 reveal the following pattern:

- Every *element* of 0 is also a *subset* of 0, so that $\vdash \forall X[(X \in 0) \Rightarrow (X \subseteq 0)]$. The hypothesis $X \in 0$ is False, whence the implication is universally valid.
- Every *element* of $1 = \{\varnothing\}$ is also a *subset* of 1, because $\varnothing \in 1$ and $\varnothing \subseteq 1$.
- Every *element* of $2 = \{\varnothing, \{\varnothing\}\}$ is also a *subset* of 2. Indeed $\varnothing \in 2$ and $\varnothing \subseteq 2$; also, $\{\varnothing\} \in 2$ and $\{\varnothing\} \subseteq 2$.
- Similarly, every *element* of $3 = \{\varnothing, \{\varnothing\}, \{\varnothing, \{\varnothing\}\}\}$ is also a *subset* of 3. Indeed $\varnothing \in 3$ and $\varnothing \subseteq 3$; moreover, $\{\varnothing\} \in 3$ and $\{\varnothing\} \subseteq 3$; furthermore, $\{\varnothing, \{\varnothing\}\} \in 3$ and $\{\varnothing, \{\varnothing\}\} \subseteq 3$.

The proof of the following theorem shows the use of the Principle of Mathematical Induction to verify the pattern of example 4.17 for all natural numbers.

4.18 Theorem. *For all $L \in \mathbb{N}$ and $N \in \mathbb{N}$, if $L \in N$ then $L \subseteq N$.*

Proof. This proof proceeds by induction with N. Define a set $S \subseteq \mathbb{N}$ by

$$S := \{N \in \mathbb{N} : \forall L[(L \in N) \Rightarrow (L \subseteq N)]\}.$$

Initial step

If $N := \varnothing$, then $L \in \varnothing$ is False, whence $(L \in \varnothing) \Rightarrow (L \subseteq \varnothing)$ is universally valid; thus $\varnothing \in S$, by definition of S.

Inductive hypothesis

Assume $K \in S$; thus $\forall L[(L \in K) \Rightarrow (L \subseteq K)]$ is True.

Inductive step

To verify the theorem for $N := K + 1$, assume that $L \in (K + 1) = (K \cup \{K\})$; then $L \in K$ or $L \in \{K\}$, by definition of the union $K \cup \{K\}$.
  In the case $L \in K$, it follows from the inductive hypothesis that $L \subseteq K$, whence $L \subseteq K \subseteq (K \cup \{K\}) = (K + 1)$ and then $L \subseteq (K + 1)$.
  In the case $L \in \{K\}$, it follows that $L = K$, whence $L = K \subseteq (K \cup \{K\}) = (K + 1)$ and then $L \subseteq (K + 1)$.
  Consequently $L \subseteq (K + 1)$ in either case. Because $L \subseteq (K + 1)$ for every $L \in K$, it follows that $(K + 1) \in S$.

Completion of the proof by induction

From the initial step $\varnothing \in S$, and from the inductive step, if $K \in S$ then $(K + 1) \in S$; it follows from the Principle of Mathematical Induction (theorem 4.15) that $S = \mathbb{N}$. Thus $N \in S$ for every $N \in \mathbb{N}$: for every $N \in \mathbb{N}$, if $L \in N$ then $L \subseteq N$.   □

Theorem 4.18 holds for all natural numbers, but it can fail for other sets, as demonstrated in counterexample 4.19.

4.19 Counterexample. If $X = \{\varnothing\}$ and $Y = \{\{\varnothing\}\}$, then $X \in Y$ but $X \nsubseteq Y$, because $\varnothing \in X$ but $\varnothing \notin Y$. Theorem 4.18 does not apply to Y because $Y \notin \mathbb{N}$.

Similarly, theorem 4.20 shows that every natural number is also a subset of $\mathbb{N}$.

4.20 Theorem. *For every set N, if $N \in \mathbb{N}$, then $N \subset \mathbb{N}$.*

Proof. This proof proceeds by induction with N.

For $N = 0 = \varnothing$, $\varnothing \subset \mathbb{N}$ by theorem 3.11.

As an induction hypothesis, assume that there exists $K \in \mathbb{N}$ such that $K \subset \mathbb{N}$.

For the induction step, from $K \in \mathbb{N}$ follows $\{K\} \subset \mathbb{N}$ by theorem 3.22. Also, $K \subset \mathbb{N}$ by induction hypothesis. Hence $K + 1 = K \cup \{K\} \subset \mathbb{N}$. □

Theorem 4.20 holds for $\mathbb{N}$ but fails for other sets, as already demonstrated in counterexample 4.19.

The following subsection demonstrates how to use the Principle of Mathematical Induction to prove the "existence" of certain functions.

4.2.3   Definitions by Mathematical Induction

The preceding discussion has introduced the Principle of Mathematical Induction as a method to *prove* theorems and verify formulae involving natural numbers. In addition, the following theorem shows how the Principle of Mathematical Induction, and the concept of unions of functions, also form the foundation of a method to *define* by induction the values of some functions. The same method is also known as **recursion** [18, p. 322, n. 526], [63, p. 10]; the resulting functions are also called **recursive functions**, with a terminology attributed [49, p. 167] to Kurt Gödel [43], [46, p. 46].

4.21 Theorem (Definition by Induction or Recursion). *For each nonempty set C, for each element $A \in C$, and for each function $G : C \to C$ with domain $\mathscr{D}(G) = C$, there exists* exactly one *function $F : \mathbb{N} \to C$ that satisfies the following two conditions:*

(DMI.0) $F(0) = A$.
(DMI.1) $F(N + 1) = G[F(N)]$ *for each $N \in \mathbb{N}$.*

Proof. This proof verifies by induction that there exists a sequence of functions (F_N), with each function $F_N \subseteq (\mathbb{N} \times C)$ defined on the set $S_N := (N \cup \{N\})$ so that the following two conditions hold:

(N.0) $F_N(0) = A$.
(N.1) $F_N(I + 1) = G[F_N(I)]$ for each $I \in S_N \setminus \{N\} = N$.

Then the proof ends by verifying that the function F defined by $F := \bigcup_{N \in \mathbb{N}} F_N$, so that $F(N) = F_N(N)$, satisfies all the requirements.

Initial step

For $N = 0$, consider the set $S_0 = \{0\}$, and consider the function

$$F_0 : \{0\} \to C,$$
$$0 \mapsto A.$$

Then the function F_0 just defined satisfies the following two conditions:

(0.0) $F_0(0) = A$.
(0.1) $F_0(I + 1) = G[F_0(I)]$ for each $I \in S_0 \setminus \{0\} = \varnothing$.

Moreover, there exists only one function of singletons $F_0 : \{0\} \to \{A\}$.

Induction hypothesis

Assume that there exists a natural number $J \in \mathbb{N}$ such that the theorem holds for $N := J$, so that for each $L \in S_J = J \cup \{J\}$ there exists exactly one function F_L satisfying the two conditions $(N.0)$ and $(N.1)$ for $N := L$.

Induction step

Let $S \subseteq \mathbb{N}$ consist of every $N \in \mathbb{N}$ for which there exists exactly one function F_N satisfying $(N.0)$ and $(N.1)$.

There exists exactly one function of singletons $H_J : \{J + 1\} \to \{G[F_J(J)]\}$, because there exists exactly one function F_J and hence exactly one value $F_J(J)$.

Define $F_{J+1} := F_J \dot\cup H_J$, which is a function because of the disjoint domains $S_J \cap \{J + 1\} = \varnothing$ (by definition 3.94 in chapter 3).

For $I := J + 1$, the definition of F_{J+1} gives $F_{J+1}(J + 1) = H_J(J + 1) = G[F_J(J)] = G[F_{J+1}(J)]$. For each $L \in S_J \setminus \{J\} = J$, the definition of F_{J+1} gives $F_{J+1}(L + 1) = F_J(L + 1) = G[F_J(L)] = G[F_{J+1}(L)]$ so that the following two conditions hold:

(J + 1.0) $F_{J+1}(0) = F_J(0) = A$,
(J + 1.1) $F_{J+1}(L + 1) = G[F_{J+1}(L)]$ for every $L \in S_{J+1} \setminus \{J + 1\}$.

Moreover, there exists exactly one such function F_{J+1} because there exists exactly one restriction $F_J|_{S_J}$, namely F_J, and exactly one value for $F_{J+1}(J + 1) = G[F_J(J)]$.

Completion of the proof of the theorem

Having constructed the sequence of functions (F_N), define a function $F : \mathbb{N} \to C$ by $F(N) := F_N(N)$. Specifically, let

$$F := \bigcup_{N \in \mathbb{N}} F_N.$$

Then verify that

(DMI.1) $F(0) = F_0(0) = 0$.
(DMI.2) $F(N+1) = F_{N+1}(N+1) = G[F_{N+1}(N)] = G[F_N(N)] = G[F(N)]$.

The uniqueness of F follows from the uniqueness of each restriction $F|_{N+1} = F_N$.

$\square$

Because the proof of the validity of the method of recursion (theorem 4.21) relies on mathematical induction (theorem 4.15), recursion is a logical consequence of induction. The following sections show how recursion suffices to define arithmetic with natural numbers.

4.2.4 Exercises on Mathematical Induction

4.1 . With $5 := 4 \cup \{4\}$, write 5 in terms of sets, as in example 4.3.

4.2 . With $6 := 5 \cup \{5\}$, write 6 in terms of sets, as in example 4.3.

4.3 . Prove that every set is an *element* and a *subset* of its successor.

4.4 . Provide an example of a nonempty proper subset $S \subsetneqq \mathbb{N}$ that contains the successor of every one of its elements.

Prove or disprove each of the following statements for all sets A and B.

4.5 . Prove or disprove that if $A \subseteq B$, then $(A \cup \{A\}) \subseteq (B \cup \{B\})$.

4.6 . Prove or disprove that $(A \cup \{A\}) \cup (B \cup \{B\}) = (A \cup B) \cup \{A \cup B\}$.

4.7 . Prove or disprove that $(A \cup \{A\}) \cap (B \cup \{B\}) = (A \cap B) \cup \{A \cap B\}$.

4.8 . Prove or disprove that $(A \cup \{A\}) \setminus (B \cup \{B\}) = (A \setminus B) \cup \{A \setminus B\}$.

Prove or disprove each of the following statements for all $K, L, M, \in \mathbb{N}$.

4.9 . Prove that if $K \in M$ and $L \in M$, then $K \cup L \in M$.

4.10 . Prove that if $K \in M$ and $L \in M$, then $K \cap L \in M$.

4.11 . Outline a formal proof of theorem 4.4.

4.12 . Outline a formal proof of theorem 4.8.

4.13 . Outline a formal proof of theorem 4.9.

4.14 . Outline a formal proof of theorem 4.10.

4.15 . Outline a formal proof of theorem 4.15.

4.16 . Prove that $\bigcup \mathbb{N} = \mathbb{N}$.

4.17 . Prove that $\bigcap \mathbb{N} = \varnothing$.

4.18 . For the successor function, prove that $N = G^{\circ N}(\varnothing)$ for each $N \in \mathbb{N}^*$.

4.19 . Prove that if a subset $V \subseteq \mathbb{N}$ contains 2, and if V also contains the "successor" $X \cup \{X\}$ of each of its element X, then V contains all the natural numbers except 0 and 1: $V = \mathbb{N} \setminus \{ \varnothing, \{\varnothing\} \}$. In other words, replace $\varnothing$ by $\{ \varnothing, \{\varnothing\} \}$ in P to obtain the logical formula R defined by

$$(R) \Leftrightarrow \Big[(\{ \varnothing, \{\varnothing\} \} \in A) \wedge \forall X \{(X \in A) \Rightarrow [(X \cup \{X\}) \in A]\}\Big].$$

Prove that $\vdash \forall V \Big[\{(V \subseteq \mathbb{N}) \wedge [\mathrm{Subf}_V^A(R)]\} \Rightarrow (V = \mathbb{N} \setminus \{ \varnothing, \{\varnothing\} \})\Big].$

4.20 . Prove that if a subset $U \subseteq \mathbb{N}$ contains 1, and if U also contains the "successor" $X \cup \{X\}$ of each of its elements X, then U contains all the *positive* natural numbers: $U = \mathbb{N} \setminus \{\varnothing\}$. In other words, replace $\varnothing$ by $\{\varnothing\}$ in P to obtain the logical formula Q defined by

$$(Q) \Leftrightarrow \Big[(\{\varnothing\} \in A) \wedge \forall X \{(X \in A) \Rightarrow [(X \cup \{X\}) \in A]\}\Big].$$

Prove that $\vdash \forall U \Big[\{(U \subseteq \mathbb{N}) \wedge [\mathrm{Subf}_U^A(Q)]\} \Rightarrow (U = \mathbb{N} \setminus \{\varnothing\})\Big].$

4.3   Arithmetic with Natural Numbers

4.3.1   Addition with Natural Numbers

This subsection defines the addition of natural numbers and establishes some of its properties, all by induction. Besides explaining the foundations of integer arithmetic, the following considerations also provide examples of proofs by induction.

4.22 Definition (Addition).   For every $M \in \mathbb{N}$, define

$$M + 0 := M, \qquad (A0)$$
$$M + 1 := M \cup \{M\}. \quad (A1)$$

Then for every $M \in \mathbb{N}$, define an addition function by induction, so that for every $N \in \mathbb{N}$,

$$M + (N + 1) := (M + N) + 1. \ (A2)$$

4.23 Remark.   According to definition 4.22, adding N amounts to adding 1 repeatedly N times. With an alternative but logically equivalent notation, for each $M \in \mathbb{N}$ definition 4.22 uses theorem 4.21 and the successor function

$$G : \mathbb{N} \to \mathbb{N},$$
$$N \mapsto N + 1,$$

to specify by induction (recursively) an addition function $F^{(M)} : \mathbb{N} \to \mathbb{N}$ such that

$$F^{(M)}(0) := M, \qquad (A0)$$
$$G(L) := L + 1, \qquad (A1)$$
$$F^{(M)}(N + 1) := G\big[F^{(M)}(N)\big]. \quad (A2)$$

The following theorem shows that addition is associative.

4.24 Theorem. *For all* $P, Q, R \in \mathbb{N}$, $(P + Q) + R = P + (Q + R)$.

Proof. For each P and each Q, proceed by induction with R. For $R := 0$,

$$\begin{aligned}
(P + Q) + 0 &= P + Q && \text{by (A0) with } M := (P + Q), \\
&= P + (Q + 0) && \text{by (A0) with } M := Q.
\end{aligned}$$

Second, *assume* that there exists some $K \in \mathbb{N}$ such that the theorem holds for $R := K$, so that $(P + Q) + K = P + (Q + K)$ for each P and each Q; then

$$\begin{aligned}
(P + Q) + (K + 1) &= [(P + Q) + K] + 1 \ (A2), M := (P + Q), N := K, \\
&= [P + (Q + K)] + 1 \text{ induction hypothesis}, \\
&= P + [(Q + K) + 1] \ (A2), M := P, N := (Q + K), \\
&= P + [Q + (K + 1)] \ (A2), M := Q, N := K.
\end{aligned}$$

$\square$

The following three theorems show that addition commutes. The first theorem shows that adding 0 commutes.

4.25 Theorem. *For each* $N \in \mathbb{N}$, $0 + N = N$.

Proof. Proceed by induction with N. First, establish the conclusion for $N := 0$:

$$0 + 0 = 0 \text{ by (A0) with } M := 0.$$

Second, *assume* that $0 + K = K$ for some $K \in \mathbb{N}$, so that the theorem holds for $N := K$; then

$$\begin{aligned}
0 + (K + 1) &= (0 + K) + 1 \text{ by (A2)}, \\
&= K + 1 && \text{induction hypothesis}.
\end{aligned}$$

$\square$

The second theorem shows that adding 1 commutes.

4.26 Theorem. *For each* $P \in \mathbb{N}$, $1 + P = P + 1$.

Proof. Proceed by induction with P. First, for $P := 0$,

$$1 + 0 = 1 \qquad \text{by (A0) with } M := 1,$$
$$= 0 + 1 \text{ by theorem 4.25.}$$

Second, *assume* that $1 + K = K + 1$ for some $K \in \mathbb{N}$, so that the theorem holds for $P := K$; then

$$1 + (K + 1) = (1 + K) + 1 \text{ (A2) with } M := 1 \text{ and } N := K,$$
$$= (K + 1) + 1 \text{ induction hypothesis.}$$

$\square$

Finally, the third theorem shows that addition commutes.

4.27 Theorem. *For all natural numbers P and Q, $P + Q = Q + P$.*

Proof. For each $P \in \mathbb{N}$, proceed by induction with Q. First,

$$P + 0 = P \qquad \text{by (A0),}$$
$$= 0 + P \text{ by theorem 4.25.}$$

Second, *assume* that there exists $K \in \mathbb{N}$ such that the theorem holds for $Q := K$, so that $R + K = K + R$ for each $R \in \mathbb{N}$; then for each $P \in \mathbb{N}$,

$$P + (K + 1) = P + (1 + K) \text{ induction hypothesis with } R := 1,$$
$$= (P + 1) + K \text{ theorem 4.24,}$$
$$= K + (P + 1) \text{ induction hypothesis with } R := (P + 1),$$
$$= K + (1 + P) \text{ theorem 4.26,}$$
$$= (K + 1) + P \text{ theorem 4.24.}$$

$\square$

4.3.2 Multiplication with Natural Numbers

This subsection defines by induction the multiplication of natural numbers and establishes some of its properties, as well as some relations between addition and multiplication. Besides a continuation of the foundations of integer arithmetic, the following considerations also provide further examples of proofs by induction.

4.28 Definition (Multiplication). For every nonnegative integer $M \in \mathbb{N}$, define a multiplication function by induction based on the following specifications:

$$M * 0 := 0, \qquad\qquad (M0)$$
$$M * (N + 1) := (M * N) + M. \ (M1)$$

4.29 Remark. By definition 4.28, multiplying M and N amounts to starting from 0 and adding M repeatedly N times. With an alternative but logically equivalent notation, for each $M \in \mathbb{N}$, definition 4.28 uses a recursive definition with G (in theorem 4.21) replaced by the addition function $F^{(M)}$ from remark 4.23,

$$G := F^{(M)} : \mathbb{N} \to \mathbb{N},$$
$$N \mapsto N + M,$$

to specify by induction a multiplication function $H^{(M)} : \mathbb{N} \to \mathbb{N}$ such that

$$G(L) := F^{(M)}(L) = L + M,$$
$$H^{(M)}(0) := 0, \qquad (M0)$$
$$H^{(M)}(N+1) := G\left[H^{(M)}(N)\right]. \ (M1)$$

4.30 Example. Multiplication by repeated addition, as in definition 4.28, was used in the IBM® 604® [64, p. 23 & 71] and IBM 650® [65, p. 35 & 36, §2].

The following theorem shows that multiplication distributes over addition on the right-hand side.

4.31 Theorem. *For all $P, Q, R \in \mathbb{N}$, $(P + Q) * R = (P * R) + (Q * R)$.*

Proof. For each P and each Q, proceed by induction with R. For $R := 0$,

$$(P + Q) * 0 = 0 \qquad \qquad (M0) \text{ with } M := (P + Q),$$
$$= 0 + 0 \qquad \qquad (A0) \text{ with } M := 0,$$
$$= (P * 0) + (Q * 0) \ (M0) \text{ with } M := P \text{ and } (M0) \text{ with } M := Q.$$

Second, *assume* that there exists $K \in \mathbb{N}$ such that the theorem holds for $R := K$, so that $(P + Q) * K = (P * K) + (Q * K)$ for each P and each Q; then

$$
\begin{aligned}
(P + Q) * (K + 1) &= [(P + Q) * K] + (P + Q) & (M1), M := P + Q, \\
&= [(P * K) + (Q * K)] + (P + Q) & \text{induction hypothesis}, \\
&= (P * K) + [(Q * K) + (P + Q)] & \text{theorem 4.24}, \\
&= (P * K) + [\{(Q * K) + P\} + Q] & \text{theorem 4.24}, \\
&= (P * K) + [\{P + (Q * K)\} + Q] & \text{theorem 4.27}, \\
&= (P * K) + [P + \{(Q * K) + Q\}] & \text{theorem 4.24}, \\
&= [(P * K) + P] + [(Q * K) + Q] & \text{theorem 4.24}, \\
&= [P * (K + 1)] + [Q * (K + 1)] & (M1) \text{ twice}.
\end{aligned}
$$

□

The following three theorems show that multiplication commutes. The first theorem shows that multiplication by 0 commutes.

4.32 Theorem. *For each natural number $N \in \mathbb{N}$, $0 * N = 0$.*

Proof. Proceed by induction with N. First, establish the conclusion for $N := 0$:

$$0 * 0 = 0 \ (M0).$$

Second, *assume* that there exists $K \in \mathbb{N}$ such that the theorem holds for $N := K$, so that $0 * K = 0$; then

$$
\begin{aligned}
0 * (K + 1) &= (0 * K) + 0 \ (M1), \\
&= 0 + 0 && \text{induction hypothesis,} \\
&= 0 && (A0).
\end{aligned}
$$

□

The second theorem shows that multiplication by 1 commutes.

4.33 Theorem. *For each nonnegative integer $N \in \mathbb{N}$, $1 * N = N$.*

Proof. Proceed by induction with N. First, establish the conclusion for $N := 0$:

$$1 * 0 = 0 \ (M0).$$

Second, *assume* that there exists $K \in \mathbb{N}$ such that the theorem holds for $N := K$, so that $1 * K = K$; then

$$
\begin{aligned}
1 * (K + 1) &= (1 * K) + 1 \ (M1), \\
&= K + 1 && \text{induction hypothesis.}
\end{aligned}
$$

□

Finally, the third theorem shows that multiplication commutes.

4.34 Theorem. *For all nonnegative integers $P, Q \in \mathbb{N}$, $P * Q = Q * P$.*

Proof. For each $P \in \mathbb{N}$, proceed by induction with Q. First, for $Q := 0$,

$$
\begin{aligned}
P * 0 &= 0 && (M0), \\
&= 0 * P && \text{theorem 4.32.}
\end{aligned}
$$

Second, *assume* that there exists $K \in \mathbb{N}$ such that the theorem holds for $Q := K$, so that $P * K = K * P$ for each $P \in \mathbb{N}$; then

$$
\begin{aligned}
P * (K + 1) &= (P * K) + P && (M1), \\
&= (K * P) + P && \text{induction hypothesis,} \\
&= (K * P) + (1 * P) && \text{theorem 4.33,} \\
&= (K + 1) * P && \text{theorem 4.31.}
\end{aligned}
$$

□

The next theorem shows that multiplication distributes over addition also on the left-hand side.

4.35 Theorem. *For all* $P, Q, R \in \mathbb{N}$, $P * (Q + R) = (P * Q) + (P * R)$.

Proof. Use commutativity and distributivity on the right-hand side (theorems 4.34, 4.31):

$$
\begin{aligned}
P * (Q + R) &= (Q + R) * P & \text{theorem 4.34,} \\
&= (Q * P) + (R * P) & \text{theorem 4.31,} \\
&= (P * Q) + (P * R) & \text{theorem 4.34.}
\end{aligned}
$$

$\square$

The following theorem shows that multiplication is associative.

4.36 Theorem. *For all* $P, Q, R \in \mathbb{N}$, $P * (Q * R) = (P * Q) * R$.

Proof. For all $P, Q \in \mathbb{N}$, proceed by induction with R. For $R := 0$,

$$
\begin{aligned}
P * (Q * 0) &= P * 0 & (M0), \\
&= 0 & (M0), \\
&= (P * Q) * 0 & \text{(M0) with } M := (P * Q).
\end{aligned}
$$

Second, *assume* that there exists $K \in \mathbb{N}$ such that the theorem holds for $R := K$, so that $P * (Q * K) = (P * Q) * K$ for all $P, Q \in \mathbb{N}$; then

$$
\begin{aligned}
(P * Q) * (K + 1)] &= [(P * Q) * K] + (P * Q) & \text{(M1) with } M := (P * Q), \\
&= [P * (Q * K)] + (P * Q) & \text{induction hypothesis,} \\
&= P * [(Q * K) + Q] & \text{theorem 4.35,} \\
&= P * [Q * (K + 1)] & (M1).
\end{aligned}
$$

$\square$

There are other arithmetic operations with natural numbers, such as the factorial.

4.37 Definition. For each $N \in \mathbb{N}$, define $N!$ (read "N **factorial**") recursively by

$$
0! := 1,
$$
$$
(N + 1)! := (N!) * (N + 1).
$$

4.3.3   Exercises on Arithmetic by Induction

The following exercises involve the following sets (also defined in example 3.35):

$$
\begin{aligned}
0 &:= \emptyset, \\
1 &:= \{0\}, \\
2 &:= \{0, 1\},
\end{aligned}
$$

$$3 := \{0, 1, 2\},$$
$$4 := \{0, 1, 2, 3\},$$
$$5 := \{0, 1, 2, 3, 4\},$$
$$6 := \{0, 1, 2, 3, 4, 5\},$$
$$7 := \{0, 1, 2, 3, 4, 5, 6\},$$
$$8 := \{0, 1, 2, 3, 4, 5, 6, 7\},$$
$$9 := \{0, 1, 2, 3, 4, 5, 6, 7, 8\}.$$

4.21 . Prove that $1 + 1 = 2$.

4.22 . Prove that $2 + 1 = 3$.

4.23 . Prove that $3 + 1 = 4$.

4.24 . Prove that $4 + 1 = 5$.

4.25 . Prove that $5 + 1 = 6$.

4.26 . Prove that $6 + 1 = 7$.

4.27 . Prove that $7 + 1 = 8$.

4.28 . Prove that $8 + 1 = 9$.

4.29 . Prove that $2 + 2 = 4$.

4.30 . Prove that $3 + 2 = 5$.

4.31 . Prove that $4 + 2 = 6$.

4.32 . Prove that $5 + 2 = 7$.

4.33 . Prove that $6 + 2 = 8$.

4.34 . Prove that $7 + 2 = 9$.

4.35 . Prove that $3 + 3 = 6$.

4.36 . Prove that $4 + 3 = 7$.

4.37 . Prove that $5 + 3 = 8$.

4.38 . Prove that $6 + 3 = 9$.

4.39 . Prove that $4 + 4 = 8$.

4.40 . Prove that $5 + 4 = 9$.

4.41 . Prove that $2 * 2 = 4$.

4.42 . Prove that $3 * 2 = 6$.

4.43 . Prove that $4 * 2 = 8$.

4.44 . Prove that $3 * 3 = 9$.

4.45 . Prove or disprove that addition distributes over multiplication on the left.

4.46 . Prove or disprove that addition distributes over multiplication on the right.

The following exercises refer to the factorial specified by definition 4.37.

4.47 . Identify a set C and functions F and G that fit theorem 4.21 to justify definition 4.37.

4.48 . Calculate 0!, 1!, 2!, 3!.

4.49 . Prove or disprove that $(P + Q)! = (P!) + (Q!)$ for all $P, Q \in \mathbb{N}$.

4.50 . Prove or disprove that $(P * Q)! = (P!) * (Q!)$ for all $P, Q \in \mathbb{N}$.

4.4   Orders and Cancellations

4.4.1   Orders on the Natural Numbers

The set of natural numbers can model geometric concepts, for instance, a direction from left to right, and algebraic concepts, for instance, increasing magnitudes:

$$0 < 1 < 2 < 3 < 4 < 5 < 6 < 7 < 8 < 9 < \ldots$$

For both types of concepts — geometric and algebraic — it suffices to introduce an ordering relation on the natural numbers. To this end, this subsection shows that the natural numbers are well-ordered by a relation $\leq$ defined by

$$(M \leq N) \Leftrightarrow [(M = N) \vee (M \in N)].$$

From this well-ordering relation will result the laws of arithmetic cancellations, which will also allow for the solutions of certain equations. The first results define a strict order $<$ in terms of the foundational relation $\in$ of set membership.

4.38 Definition. For all $M, N \in \mathbb{N}$, define $M < N$ (read "M is less than N") by

$$(M < N) \Leftrightarrow (M \in N).$$

4.39 Example. $0 < 1$ because $0 \in 1$: $0 = \varnothing$ and $1 = \{\varnothing\}$, so $\varnothing \in \{\varnothing\}$.
    $1 < 2$ because $1 \in 2$; $1 = \{\varnothing\}, 2 = \{\varnothing, \{\varnothing\}\}$, whence $\{\varnothing\} \in \{\varnothing, \{\varnothing\}\}$.

The following theorem establishes the *transitivity* of the relation $\in$ on $\mathbb{N}$.

4.40 Theorem. *For all $L, M, N \in \mathbb{N}$, if $L \in M$ and $M \in N$, then $L \in N$:*

$$\forall L \forall M \forall N\{[(L \in \mathbb{N}) \wedge (M \in \mathbb{N}) \wedge (N \in \mathbb{N})]$$

$$\Rightarrow [\{(L \in M) \wedge (M \in N)\} \Rightarrow (L \in N)]\}.$$

Proof. Proceed by induction with N.

Initial step

If $N := 0 = \varnothing$, then for all nonnegative integers $L, M \in \mathbb{N}$ the proposition $M \in \varnothing$ is False, whence so is the conjunction $(L \in M) \wedge (M \in 0)$, and hence the implication $[(L \in M) \wedge (M \in 0)] \Rightarrow (L \in 0)$ is True.

Induction hypothesis

Assume that there exists some $K \in \mathbb{N}$ such that the theorem holds for $N := K$, so that $[(L \in M) \wedge (M \in K)] \Rightarrow (L \in K)$ for all $L, M \in \mathbb{N}$.

Induction step

If $L \in M$ and $M \in K + 1 = K \cup \{K\}$, then two cases arise: $M \in \{K\}$ or $M \in K$.

In the first case, $M \in \{K\}$, whence $M = K$, and the hypothesis $L \in M$ then yields $L \in M = K \subseteq K \cup \{K\}$, so that $L \in K + 1$.

In the second case, $M \in K$, whence the induction hypothesis yields $L \in K \subset K \cup \{K\} = K + 1$, and then again $L \in K + 1$.   □

The transitivity of the relation $\in$ holds on the set $\mathbb{N}$, but it can fail on other sets.

4.41 Counterexample. If $X := \varnothing$, $Y := \{\varnothing\}$, and $Z := \{\{\varnothing\}\}$, then $X \in Y$ and $Y \in Z$, but $X \notin Z$. Theorem 4.40 fails for $Z = \{\{\varnothing\}\}$ because $\{\{\varnothing\}\} \notin \mathbb{N}$.

The following theorem shows that on the natural numbers the relation $\in$ is *neither* reflexive *nor* symmetric, but, instead, $\in$ is both *irreflexive* and *asymmetric*.

4.42 Theorem. *For all $I, L, M \in \mathbb{N}$, $M \notin M$, and $(I \notin L) \vee (L \notin I)$.*

Proof. This proof of the first result $(M \notin M)$ proceeds by induction with M.

Initial step

If $M = 0$, then $M = \varnothing$. From $\varnothing \notin \varnothing$ it follows that $M \notin M$.

Induction hypothesis

Assume that there exists some $K \in \mathbb{N}$ such that the theorem holds for $M := K$, so that $K \notin K$.

Induction step

This step of the proof proceeds by contraposition. If $(K \cup \{K\}) \in (K \cup \{K\})$, then two cases can arise: $(K \cup \{K\}) \in K$ or $(K \cup \{K\}) \in \{K\}$.

First, if $(K \cup \{K\}) \in K$, then the transitivity of the relation $\in$ (theorem 4.40) and $K \in (K \cup \{K\})$ give $K \in K$, which contradicts the induction hypothesis.

Second, if $(K \cup \{K\}) \in \{K\}$, then $(K \cup \{K\}) = K$, whence $K \in \{K\} \subseteq (K \cup \{K\}) = K$ gives $K \in K$, which contradicts the induction hypothesis.

The proof of the second result proceeds by contraposition. For all $I \in \mathbb{N}$ and $L \in \mathbb{N}$, the conjunction $(I \in L) \wedge (L \in I)$ and the transitivity of $\in$ (theorem 4.40) give $I \in I$, which contradicts the result just proved. Therefore, $\neg[(I \in L) \wedge (L \in I)]$. $\qquad\square$

The following theorem shows that each natural number differs from its successor.

4.43 Theorem. *For each natural number $N \in \mathbb{N}$, $N \neq N + 1$.*

Proof. From $N \in \{N\} \subseteq (N \cup \{N\}) = N + 1$ it follows that $N \in (N + 1)$. Yet $N \notin N$ by theorem 4.42. Therefore $N \neq N + 1$ by the axiom of extensionality (S1). $\qquad\square$

The following theorem shows that adding 1 to both sides of a valid inequality gives a valid inequality, so that if $M < N$ then $(M + 1) < (N + 1)$.

4.44 Theorem. *For all $M, N \in \mathbb{N}$, if $M \in N$, then $(M + 1) \in (N + 1)$:*

$$\vdash \forall M \forall N \left([(M \in \mathbb{N}) \wedge (N \in \mathbb{N})] \Rightarrow \{(M \in N) \Rightarrow [(M + 1) \in (N + 1)]\} \right).$$

Proof. Apply induction with N. For $N := 0 = \varnothing$, and for each $M \in \mathbb{N}$, the proposition $M \in \varnothing$ is False; hence the implication $(M \in \varnothing) \Rightarrow [(M + 1) \in (N + 1)]$ is True.

Next, *assume* that there exists $K \in \mathbb{N}$ such that the theorem holds for $N := K$, so that $(M \in K) \Rightarrow [(M + 1) \in (N + 1)]$ holds for every $M \in \mathbb{N}$. If $M \in K + 1 = K \cup \{K\}$, then two cases occur.

In the first case, $M \in \{K\}$, and then $M = K$, whence $M + 1 = K + 1 \in (K + 1) \cup \{K + 1\} = (K + 1) + 1$.

In the second case, $M \in K$, whence $M + 1 \in K + 1$ by induction hypothesis; with $K + 1 \subseteq (K + 1) \cup \{K + 1\} = (K + 1) + 1$, it follows that $M + 1 \in (K + 1) + 1$. $\qquad\square$

The next theorem shows that $0 = \varnothing$ is the smallest natural number.

4.45 Theorem. *For each natural number $M \in \mathbb{N}$, if $M \neq 0$, then $0 \in M$.*

Proof. This proof proceeds by induction with M.

For $M := 0$, the proposition $M \neq 0$ is False, whence the implication $(M \neq 0) \Rightarrow (0 \in M)$ is True.

Next, *assume* that there exists $K \in \mathbb{N}$ such that the theorem holds for $M := K$, so that $(K \neq 0) \Rightarrow (0 \in K)$. Two cases arise: $K = 0$ or $K \neq 0$.

In the first case, if $K = 0$, then $0 = \varnothing \in (\varnothing \cup \{\varnothing\}) = K \cup \{K\} = K + 1$; thus, $0 \in K + 1$ is True.

In the second case, if $K \neq 0$, then $0 \in K$, by the induction hypothesis, whence $0 \in K \subseteq K \cup \{K\} = K + 1$, and hence $0 \in K + 1$.   □

The following theorem shows that the relation $\in$ is *connected* on the natural numbers: if $M \neq N$, then either $M \in N$ or $N \in M$.

4.46 Theorem. *For all natural numbers $M, N \in \mathbb{N}$, exactly one of the following three formulae is True, while the other two are False:*

$$M \in N,$$
$$M = N,$$
$$N \in M.$$

Proof. First, observe that at most one of the three formulae may hold. Indeed, $M \notin M$ by theorem 4.42, whence $(M \in N) \wedge (M = N)$ is False, and $(M = N) \wedge (N \in M)$ is False. Similarly, also by theorem 4.42, $(M \in N) \wedge (N \in M)$ is also False. Second, at least one of the three formulae must hold. This proof uses induction with N.

For $N := 0$, and for each $M \in \mathbb{N}$ either $M = 0 = N$ is True, or $M \neq 0$, and then $0 \in M$ is True by theorem 4.45. To complete the induction, assume that there exists $K \in \mathbb{N}$ such that the theorem holds for $N := K$, so that $(M \in K) \vee (M = K) \vee (K \in M)$ for each $M \in \mathbb{N}$, and examine all three formulae.

If $M \in K$, then $M \in K \subseteq K + 1$, whence $M \in K + 1$.

If $M = K$, then $M = K \in \{K\} \subseteq K \cup \{K\}$, whence $M \in K + 1$.

If $K \in M$, then $K + 1 \in M + 1 = M \cup \{M\}$, (theorem 4.44); two cases occur.

In the first case, $K + 1 \in \{M\}$, whence $K + 1 = M$. In the second case, $K + 1 \in M$ already. Thus, $(M \in [K + 1]) \vee (M = [K + 1]) \vee ([K + 1] \in M)$ is True.   □

The foregoing result completes the proof that $<$ is a strict total order (irreflexive, asymmetric, and transitive) on the natural numbers. The following theorem shows that the natural numbers are well-ordered by the relation $\leq$.

4.47 Theorem (Well-Ordering Principle). *Each nonempty subset of the natural numbers has a smallest element.*

Proof. By contraposition, this proof establishes that every subset of the natural numbers without a smallest element (definition 3.192) is empty. To this end, assume that a subset $E \subseteq \mathbb{N}$ has no smallest element. Thus every $N \in \mathbb{N}$ is *not* the smallest element of E, which means that $N \notin E$ or that E contains an element M such that $M < N$.

To prove that $E = \emptyset$, the proof proceeds by induction with the set $S \subseteq \mathbb{N}$ consisting of every $N \in \mathbb{N}$ such that the complement $\mathbb{N} \setminus E$ contains every $M \leq N$:

$$S := \{N \in \mathbb{N} : (N \cup \{N\}) \subseteq (\mathbb{N} \setminus E)\}.$$

Initial step

The set $S = \mathbb{N} \setminus E$ contains 0; indeed, contraposition shows that if E contained 0, then 0 would be the smallest element of E, because $0 \leq N$ for every $N \in \mathbb{N}$ by theorem 4.45 and hence also for every $N \in E$.

Induction step

Assume that $K \in S$ for some $K \in \mathbb{N}$; thus, every $M \leq K$ belongs to the complement of E. Contraposition again confirms that if E contained $K + 1$, then $K + 1$ would be the smallest element of E; consequently, $K + 1 \notin E$, whence $K + 1 \in S$. Therefore, $S = \mathbb{N}$, whence $E = \mathbb{N} \setminus S = \emptyset$. □

4.48 Definition (minimum). For each nonempty subset $E \subseteq \mathbb{N}$, the **minimum** of E is the smallest element of E and it is denoted by $\min(E)$.

In contrast to the Well-Ordering Principle, some nonempty subsets of the natural numbers have no maximum. Yet every nonempty subset of the natural numbers that has an upper bound (definition 3.198) also has a largest element (definition 3.192).

4.49 Theorem. *Each nonempty subset of the natural numbers with an upper bound in the natural numbers has a largest element.*

Proof. If there exists an upper bound $M \in \mathbb{N}$ for a nonempty subset $E \subseteq \mathbb{N}$, then $I \leq M$ for every $I \in E$. Let $S \subseteq \mathbb{N}$ be the set of all the upper bounds for E:

$$S := \{N \in \mathbb{N} : \forall I[(I \in E) \Rightarrow (I \leq N)]\}.$$

Then $M \in S$ by hypothesis on E; in particular, $S \neq \emptyset$. Consequently, by the Well-Ordering Principle (theorem 4.47) S has a smallest element $K := \min(S)$. Because $K \in S$, it follows that $I \leq K$ for every $I \in E$.

Moreover, $K \in E$, as proved by the following induction. If $K = 0$, then $E = \{0\}$ because $K = 0 < I$ for every $I \in \mathbb{N} \setminus \{0\}$; hence $K = 0 \in E$.

Suppose that there exists $L \in \mathbb{N}$ such that the theorem holds for $K := L$, so that for each $E \subset \mathbb{N}$ if $L = \min(S)$, then $L \in E$. If $\min(S) = L + 1$, then $L + 1 \in E$, for otherwise $I \leq L$ for every $I \in E$ and then $L \in S$; thus the theorem holds for $L + 1$. □

4.50 Definition (maximum). For each nonempty subset $E \subseteq \mathbb{N}$ with an upper bound in $\mathbb{N}$, the **maximum** of E is the largest element of E and it is denoted by $\max(E)$.

The following abbreviations occasionally prove convenient.

4.51 Definition. For all $I \in \mathbb{N}$ and $N \in \mathbb{N}$ define a set $\{I, \dots, N\}$ by

$$\{I, \dots, N\} := \{K \in \mathbb{N} : (I \leq K) \wedge (K \leq N)\},$$

which thus consists of all the natural numbers from I through N . Similarly,

$$\{I, \dots\} := \{I, I + 1, I + 2, \dots\} := \{K \in \mathbb{N} : I \leq K\},$$

which thus consists of all the natural numbers larger than or equal to I .

4.52 Example. If $E := \{2, \dots, 7\}$, then $E = \{2, 3, 4, 5, 6, 7\}$. Also, $\min(E) = 2$ and $\max(E) = 7$.

4.53 Example. If $E := \{2, 3, \dots\}$, then $\min(E) = 2$ but E has no maximum.

4.4.2   Laws of Arithmetic Cancellations

This subsection establishes rules to cancel terms in equations with additions or multiplications. The material also provides more practice with proofs by mathematical induction. The first rule shows how to solve equations of the form $M + 1 = L + 1$.

4.54 Theorem. *For all $L, M \in \mathbb{N}$, if $M + 1 = L + 1$, then $M = L$.*

Proof. If $M + 1 = L + 1$, then $M \cup \{M\} = L \cup \{L\}$, and the sets on both sides have the same elements. In particular, $L \in (L \cup \{L\})$ and thus $L \in (M \cup \{M\})$, whence two cases arise: $L \in \{M\}$, or $L \in M$.

In the first case, $L \in \{M\}$, and then $L = M$ indeed.

In the second case, $L \in M$; then $M \notin \{L\}$, for otherwise $M = L$ and $L \in L$ would contradict theorem 4.42. However, $M \in L \cup \{L\}$, whence $M \in L$, but that would also contradict theorem 4.42. Therefore, this second case does not occur.   □

The following theorem allows for the cancellation of an additive term N common to both sides of an equation of the type $M + N = L + N$.

4.55 Theorem. *For all $M, N, L \in \mathbb{N}$, if $M + N = L + N$, then $M = L$.*

Proof. For all natural numbers $L, M \in \mathbb{N}$, proceed by induction with N. For $N := 0$, if $M + 0 = L + 0$, then $M = M + 0 = L + 0 = L$, by hypothesis and by (A0).

Second, *assume* that there exists $K \in \mathbb{N}$ such that the theorem holds for $N := K$, so that for all natural numbers $L, M \in \mathbb{N}$, if $M + K = L + K$ then $M = L$.

If $M + (K + 1) = L + (K + 1)$, then the associativity and the commutativity of addition lead to $(M + 1) + K = (L + 1) + K$, whence $M + 1 = L + 1$ by induction hypothesis, whence finally $M = L$ by theorem 4.54.   □

The following theorem shows that addition preserves the ordering.

4.56 Theorem. *For all $L, M, N \in \mathbb{N}$, if $L < M$ then $L + N < M + N$.*

Proof. For all L and M, proceed by induction with N.

For $N := 0$, and for all L and M, if $L + 0 = M + 0$, then $L = L + 0 = M + 0 = M$.

To complete the induction, assume that there exists $K \in \mathbb{N}$ such that the theorem holds for $N := K$, so that for all L and M, if $L < M$ then $L + K < M + K$. Hence, if $L < M$ then $L + 1 < M + 1$ by theorem 4.44, whence

$$L + (K + 1) = L + (1 + K)$$
$$= (L + 1) + K$$
$$< (M + 1) + K$$
$$= M + (1 + K)$$
$$= M + (K + 1).$$

□

4.57 Theorem. *For all $M, N \in \mathbb{N}$, if $M \neq 0$, then $N < N + M$.*

Proof. Set $L := 0$ in theorem 4.56. If $M \neq 0$ then $0 < M$ by theorem 4.45, whence $L < M$. Hence $N = 0 + N < M + N$ by theorem 4.56.   □

The following theorem forms the basis for the concept of subtraction of a natural number from a larger natural number.

4.58 Theorem. *For all $M, N \in \mathbb{N}$, if $M < N$, then there exists* exactly one *natural number $L \in \mathbb{N}$ such that $M + L = N$.*

Proof. This proof establishes the existence and the uniqueness separately.

Existence

First, establish the existence of such a number L. For each $M \in \mathbb{N}$, proceed by induction with N.

If $N := 0$, then $(M < 0)$ is False, because $(M < 0) \Leftrightarrow (M \in \varnothing)$. Therefore

$$\vdash \forall M \big([(M \in \mathbb{N}) \wedge (M < 0)] \Rightarrow \{\exists L[(L \in \mathbb{N}) \wedge (M + L = N)]\} \big)$$

is universally valid, and hence the theorem holds for $N = 0$.

For each $M \in \mathbb{N}$, assume that there exists $K \in \mathbb{N}$ such that the theorem holds for $N := K$, so that for each $M \in \mathbb{N}$, if $M < K$ then there exists $L \in \mathbb{N}$ such that $M + L = K$. If $M < (K + 1) = K \cup \{K\}$, then two cases arise.

In the first case, $M \in \{K\}$, whence $M = K$, and then $M + 1 = K + 1$, so that $L = 1$. In the second case, $M \in K$, that is, $M < K$, and the induction hypothesis yields some $L \in \mathbb{N}$ for which $M + L = K$; hence $M + (L+1) = (M+L)+1 = K+1$.

Uniqueness

Second, verify the uniqueness of L, which results from the theorem that if $M + L = N = M + K$, then $L = K$ (theorem 4.55).   □

The following definition specifies the concept of subtraction of a natural number from a larger natural number.

4.59 Definition (Subtraction). For all $L, M, N, \in \mathbb{N}$, if $M < N$, then $N - M := L$ is the natural number L defined in theorem 4.58 such that $M + L = N$

4.60 Example. $8 - 5 = 3$ because $5 + 3 = 8$ by exercise 4.37.

The following theorem shows that multiplication by a *nonzero* natural number preserves the ordering.

4.61 Theorem. *For all $L, M, N \in \mathbb{N}$, if $M < N$ and $0 < L$, then $L * M < L * N$.*

Proof. For all nonnegative integers $M, N \in \mathbb{N}$, use induction with L, the smallest nonzero value of L being 1. For $L := 1$, if $M < N$, then $1 * M = M < N = 1 * N$.

Assume that there exists $K \in \mathbb{N}$ with $0 < K$ such that the theorem holds for $L := K$, so that for all natural numbers $M, N \in \mathbb{N}$, if $M < N$, then $K * M < K * N$. Thus, if $M < N$, then apply theorem 4.56 twice:

$$(K + 1) * M = (K * M) + M < (K * M) + N < (K * N) + N = (K + 1) * N.$$

□

The following theorem allows for the cancellation of a nonzero multiplicative term on both sides of an equation, which forms the basis for the division of a natural number by a nonzero natural number, and the solution of equations of the type $L * M = L * N$.

4.62 Theorem. *For all $L, M, N \in \mathbb{N}$, if $0 < L$ and $L * M = L * N$, then $M = N$.*

Proof. Proceed by contraposition.
If $M \neq N$, then either $M < N$ or $N < M$.
If $M < N$, then $L * M < L * N$ and $L * M \neq L * N$.
Similarly, if $N < M$, then $L * N < L * M$ and $L * M \neq L * N$.   □

4.63 Definition (division). For all $K, M, N \in \mathbb{N}$, if $0 < L$ and $K * L = N$, then define N/L by $N/L := K$, as defined uniquely by theorem 4.62.

4.64 Example. $8/4 = 2$ because $2 * 4 = 8$ by exercise 4.43.

4.65 Example. Division by repeated subtraction was used in the Whirlwind I computer [82, p. 3-7–3-9].

4.4.3   Exercises on Orders and Cancellations

The following exercises involve the natural numbers (sets) $0, 1, 2, 3, 4, 5, 6, 7, 8, 9$ defined in example 3.35, page 128, and reviewed in subsection 4.3.3.

4.51 . Prove that $2 < 5$.

4.52 . Prove that $5 < 7$.

4.53 . Prove that $2 < 7$.

4.54 . Prove that $3 < 5$.

4.55 . Prove that $3 < 7$.

4.56 . Prove that $4 < 9$.

4.57 . Prove that $2 < 8$.

4.58 . Prove: there are no natural numbers $K > 1, L > 1$ with $K * L = 2$.

4.59 . Prove: there are no natural numbers $K > 1, L > 1$ with $K * L = 3$.

4.60 . Prove: there are no natural numbers $K > 1, L > 1$ with $K * L = 5$.

4.61 . Prove: there are no natural numbers $K > 1, L > 1$ with $K * L = 7$.

4.62 . Determine all the natural numbers $K > 1$ and $L > 1$ such that $K * L = 6$ and prove that there are no other such natural numbers.

4.63 . Determine all the natural numbers $K > 1$ and $L > 1$ such that $K * L = 9$ and prove that there are no other such natural numbers.

4.64 . Prove that the relation $\leq$ is a total ordering on $\mathbb{N}$, with

$$(M \leq N) \Leftrightarrow [(M = N) \vee (M \in N)].$$

4.65 . Prove: for each nonempty $S \subseteq \mathbb{N}$ there exists $I \in S$ with $I \cap S = \emptyset$.

4.66 . Prove that for each $L \in \mathbb{N}^*$ there exists $I \in L$ such that $I \cap L = \emptyset$.

4.67 . Prove that $(I \notin K) \vee (K \notin L) \vee (L \notin I)$ for all $I, K, L \in \mathbb{N}$.

4.68 . Prove that $1 < N$ for each $N \in (\mathbb{N} \setminus \{0, 1\})$.

4.69 . Prove that $\mathbb{N} \notin \mathbb{N}$.

4.70 . Prove that $\mathbb{N} \neq (\mathbb{N} \cup \{\mathbb{N}\})$.

4.5   Integers

4.5.1   Negative Integers

Several operations remain undefined with natural numbers. For instance, the ordering relation does not include elements smaller than the empty set, which precludes the use of natural numbers to model relations extending in two opposite directions. Also, the arithmetic of natural numbers does not contain provisions for the "difference" from a larger natural number to a smaller one. Finally, the arithmetic of natural numbers does not include any concept of division other than special cases.

There exist several methods to extend the ordering and the arithmetic of natural numbers, to allow for elements "smaller" than zero, and for differences of any two elements. Such methods begin with the specification of a larger set of "integers" $\mathbb{Z}$ [from the German "Zahl(en)" for "number(s)"].

One method of defining a larger set, outlined by Kunen [74, p. 35], is sufficiently general to produce not only the integers, but also the rational numbers and the real numbers. Essentially, the method consists in defining the new numbers in terms of equivalence classes. For the integers, the strategy consists in introducing the concept of the "difference" between two natural numbers M and N by means of the pair (M, N). Then relate (M, N) to every other pair (P, Q) with the same "difference" between P and Q. Because the concept of "difference" has not yet been defined for all pairs of natural numbers, however, a precise definition uses sums instead.

Two cases arise: either $N \leq M$, or $N > M$. In the first case, if $N \leq M$, then definition 4.59 specifies their difference $J := M - N$. If also (M, N) and (P, Q) represent the same difference, then $P - Q = J = M - N$. An equivalent statement without subtractions results from adding J to both extremes:

$$M - N = J, \quad M = N + J;$$
$$P - Q = J, \quad P = Q + J.$$

In the second case, if $N > M$ with $J := N - M$ and also $Q - P = J = N - M$, then

$$N - M = J, \quad N = M + J;$$
$$Q - P = J, \quad Q = P + J.$$

In either case, the statement that (M, N) and (P, Q) represent the same "difference" can be reworded without subtractions but with sums instead: there exists $J \in \mathbb{N}$ with

- either $M = N + J$ and $P = Q + J$ (if $M > N$ and $P > Q$),

$$M \qquad N$$

$$\cdots \; \bullet \; \bullet \; \bullet \; \bullet \; \bullet \; \bullet \; \bullet \; \bullet \; \cdots$$

$$P \qquad Q$$

- or $N = M + J$ and $Q = P + J$ (if $M < N$ and $P < Q$),

$$N \qquad M$$

$$\cdots \; \bullet \; \bullet \; \bullet \; \bullet \; \bullet \; \bullet \; \bullet \; \bullet \; \cdots$$

$$Q \qquad P$$

Swapping coordinates, from (M, N) to (N, M), will amount geometrically to reversing an order or a direction, which will amount algebraically to passing from a positive to a negative integer. Such considerations lead to the following definition.

4.66 Definition. Define a relation $\simeq$ on the set $A := \mathbb{N} \times \mathbb{N}$ for all $M, N, P, Q \in \mathbb{N}$ by

$$(M, N) \simeq (P, Q)$$
$$\Updownarrow$$
$$\exists J \big[(J \in \mathbb{N}) \wedge \{ [(M = N + J) \wedge (P = Q + J)] \vee [(N = M + J) \wedge (Q = P + J)] \} \big]:$$

there is a $J \in \mathbb{N}$ with $M = N + J$ and $P = Q + J$, or $N = M + J$ and $Q = P + J$.

4.67 Example. $(3, 6) \simeq (5, 8)$ because $J := \mathbf{3}$ confirms that $6 = 3 + \mathbf{3}$ and $8 = 5 + \mathbf{3}$, by exercises 4.35 and 4.37.

The following theorem provides an equivalent formulation of this relation.

4.68 Theorem. *For all pairs of natural numbers,*

$$(M, N) \simeq (P, Q)$$
$$\Updownarrow$$
$$M + Q = N + P.$$

Proof. This proof establishes the two implications ($\Rightarrow$ and $\Leftarrow$) separately.
   Assume first that $(M, N) \simeq (P, Q)$. Then two cases can arise.
   If there exists $I \in \mathbb{N}$ with $M = N + I$ and $P = Q + I$, then

$$M + Q = (N + I) + Q = N + (I + Q) = N + (Q + I) = N + P.$$

If there exists $I \in \mathbb{N}$ with $N = M + I$ and $Q = P + I$, then

$$M + Q = M + (P + I) = (M + (I + P) = (M + I) + P = N + P.$$

For the converse, assume that $M + Q = N + P$. Then two cases can arise.
If $M \leq N$, then there exists $I \in \mathbb{N}$ such that $N = M + I$. Hence

$$M + Q = N + P = (M + I) + P = M + (P + I)$$

whence (by theorem 4.55) cancelling M yields $Q = P + I$.

If $M > N$, then there exists $I \in \mathbb{N}^*$ such that $M = N + I$. Hence

$$N + P = M + Q = (N + I) + Q = N + (Q + I)$$

whence (by theorem 4.55) cancelling N yields $P = Q + I$.   □

4.69 Example. $(5, 4) \simeq (2, 1)$ because $5 + 1 = 4 + 2$, by exercises 4.25 and 4.31.

Instead of definition 4.66 the derivations can hence utilize theorem 4.68. Thus the following theorem shows that the relation $\simeq$ is an equivalence relation (definition 3.150).

4.70 Theorem. *The relation $\simeq$ is an equivalence relation on $\mathbb{N} \times \mathbb{N}$.*

Proof. This proof verifies that the relation $\simeq$ is reflexive, symmetric, and transitive.

Reflexivity

For each pair $(M, N) \in (\mathbb{N} \times \mathbb{N})$, the equality $M + N = N + M$ holds by the commutativity of addition, whence $(M, N) \simeq (M, N)$.

Symmetry

If $(M, N) \simeq (P, Q)$, then $M + Q = N + P$, whence $Q + M = P + N$, which means that $(P, Q) \simeq (M, N)$.

Transitivity

If $(K, L) \simeq (M, N)$, then $K + N = L + M$. If also $(M, N) \simeq (P, Q)$, then $M + Q = N + P$. Consequently,

$$K + (M + Q) = K + (N + P) = (K + N) + P = (L + M) + P = L + (M + P)$$

whence (theorem 4.55) cancelling M yields $K + Q = L + P$, and $(K, L) \simeq (P, Q)$.
                                                                                               □

4.71 Definition (Kunen's definition of the integers). The set of integers is the set $\mathbb{Z} := (\mathbb{N} \times \mathbb{N}) / \simeq$ of all equivalence classes $[(M, N)]_{\simeq}$ for the relation $\simeq$ [74, p. 35].

4.72 Example. Setting $I := 3$ shows that $(0, 3) \simeq (1, 4) \simeq (2, 5) \simeq (3, 6) \simeq (4, 7)$. Thus the pairs $(0, 3)$, $(1, 4)$, $(2, 5)$, $(3, 6)$, $(4, 7)$ are elements of the equivalence class $[(0, 3)]_{\simeq}$. Similarly, $(3, 0) \simeq (4, 1) \simeq (5, 2) \simeq (6, 3) \simeq (7, 4)$.

Thus the pairs $(3, 0), (4, 1), (5, 2), (6, 3), (7, 4)$ are elements of the equivalence class $[(3, 0)]_\simeq$.

Each pair $(I, J) \in \mathbb{N} \times \mathbb{N}$ is equivalent to a pair of the type $(K, 0)$ or $(0, K)$.

4.73 Theorem. *If $I > J$, then $(I, J) \simeq (I - J, 0)$.*
If $I \leq J$, then $(I, J) \simeq (0, J - I)$.

Proof. If $I > J$, then $I = (I - J) + J$ and $J = 0 + J$ whence $(I, J) \simeq (I - J, 0)$.
If $I \leq J$, then $I = 0 + I$ and $J = (J - I) + I$ whence $(I, J) \simeq (0, J - I)$.   □

The following diagram shows a few elements from three equivalence classes: $[(0, 2)]_\simeq$, $[(0, 0)]_\simeq$, and $[(3, 0)]_\simeq$ relative to the relation $\simeq$ on $A := \mathbb{N} \times \mathbb{N}$:

$$\begin{array}{ccccc}
(1, 3) & (3, 3) & & \mathbf{(6, 3)} \\
(0, 2) & (2, 2) & & \mathbf{(5, 2)} \\
(1, 1) & & \mathbf{(4, 1)} \\
(0, 0) & & \mathbf{(3, 0)}
\end{array}$$

4.5.2 Arithmetic with Integers

From the preceding definition of integers in terms of equivalent pairs of natural numbers — which represent equivalent differences — follows a definition of arithmetic in terms of such pairs.

4.74 Definition (Kunen's definition of integer arithmetic). Define an arithmetic with pairs of natural numbers as follows:

$$(M, N) + (P, Q) := (M + P, N + Q),$$

$$(M, N) * (P, Q) := ([M * P] + [N * Q], [M * Q] + [N * P]).$$

The following theorem verifies that addition and multiplication of pairs commute.

4.75 Theorem. *For all $M, N, P, Q \in \mathbb{N}$,*

$$(M, N) + (P, Q) = (P, Q) + (M, N),$$

$$(M, N) * (P, Q) = (P, Q) * (M, N).$$

Proof. Apply the commutativity of addition and multiplication with natural numbers:

$$(M, N) + (P, Q) = (M + P, N + Q)$$
$$= (P + M, Q + N)$$
$$= (P, Q) + (M, N);$$

$$(M, N) * (P, Q) = ([M * P] + [N * Q], [M * Q] + [N * P])$$
$$= ([P * M] + [Q * N], [Q * M] + [P * N])$$
$$= ([P * M] + [Q * N], [P * N] + [Q * M])$$
$$= (P, Q) * (M, N).$$

□

The next theorem checks that arithmetic with equivalent pairs yields equivalent results: different pairs representing the same difference yield the same sum or product.

4.76 Theorem. *For all $I, J, K, L, M, N, P, Q \in \mathbb{N}$, if*

$$(I, J) \simeq (K, L),$$

$$(M, N) \simeq (P, Q),$$

then

$$(I, J) + (M, N) \simeq (K, L) + (P, Q),$$

$$(I, J) * (M, N) \simeq (K, L) * (P, Q).$$

Proof. If $(I, J) \simeq (K, L)$, then $I + L = J + K$. If also $(M, N) \simeq (P, Q)$, then $M + Q = N + P$. Moreover,

$$(I, J) + (M, N) = (I + M, J + N),$$

$$(K, L) + (P, Q) = (K + P, L + Q),$$

whence

$$(I + M) + (L + Q) = (I + L) + (M + Q)$$
$$= (J + K) + (N + P)$$
$$= (J + N) + (K + P),$$

which means that $(I, J) + (M, N) \simeq (K, L) + (P, Q)$. For the multiplication,

$$(I, J) * (M, N) = ([I * M] + [J * N], [I * N] + [J * M]),$$

$$(K, L) * (P, Q) = ([K * P] + [L * Q], [K * Q] + [L * P]),$$

so that $(I, J) * (M, N) \simeq (K, L) * (P, Q)$ if and only if

$$([I * M] + [J * N]) + ([K * Q] + [L * P]) = ([I * N] + [J * M]) + ([K * P] + [L * Q]).$$

Additions and cancellations (theorem 4.55) give

$$I * M + J * N + K * Q + L * P = I * N + J * M + K * P + L * Q$$
$$\Updownarrow$$
$$I * M + I * Q + J * N + K * Q + L * P = I * N + I * Q + J * M + K * P + L * Q$$
$$\Updownarrow$$
$$I * (M + Q) + J * N + K * Q + L * P = I * N + J * M + K * P + (I + L) * Q$$
$$\Updownarrow$$
$$I * (N + P) + J * N + K * Q + L * P = I * N + J * M + K * P + (J + K) * Q$$
$$\Updownarrow$$
$$I * P + J * N + L * P = J * M + K * P + J * Q$$
$$\Updownarrow$$
$$I * P + J * N + J * P + L * P = J * M + K * P + J * P + J * Q$$
$$\Updownarrow$$
$$I * P + J * (N + P) + L * P = J * M + (K + J) * P + J * Q$$
$$\Updownarrow$$
$$I * P + J * (M + Q) + L * P = J * M + (I + L) * P + J * Q$$
$$\Updownarrow$$
$$0 = 0,$$

which is universally valid, whence $(I, J) * (M, N) \simeq (K, L) * (P, Q)$.   □

Hence the following definition specifies an arithmetic with equivalence classes based on the arithmetic of representative pairs.

4.77 Definition (Integer addition and multiplication).

$$\big([(M, N)]_\simeq\big) + \big([(P, Q)]_\simeq\big) := \big([(M + P, \ N + Q)]_\simeq\big),$$

$$\big([(M, N)]_\simeq\big) * \big([(P, Q)]_\simeq\big) := \big([(M * P + N * Q, \ M * Q + N * P)]_\simeq\big).$$

4.78 Theorem. *For* $\mathbb{Z}$, *and for all* $K, L, M, N, P, Q \in \mathbb{N}$,
addition commutes,
addition is associative,
$[(0, 0)]_\simeq$ *is the additive unit:* $[(M, N)]_\simeq + [(0, 0)]_\simeq = [(M, N)]_\simeq,$
$[(N, M)]_\simeq$ *is the additive inverse:* $[(M, N)]_\simeq + [(N, M)]_\simeq = [(0, 0)]_\simeq,$

multiplication commutes,
multiplication is associative,
multiplication distributes over addition, and
$[(1, 0)]_\simeq$ *is a multiplicative unit:* $[(M, N)]_\simeq * [(1, 0)]_\simeq = [(M, N)]_\simeq.$

Proof. The proof forms the object of exercises.   □

The following definition specifies a subtraction.

4.79 Definition (Integer subtraction). Define a subtraction of integers by

$$\big([(M, N)]_\simeq\big) - \big([(P, Q)]_\simeq\big) := \big([(M, N)]_\simeq\big) + \big([(Q, P)]_\simeq\big),$$

which is a binary operation $-$ from $\mathbb{Z} \times \mathbb{Z}$ to $\mathbb{Z}$. Define an "opposite" by

$$-\big([(P, Q)]_\simeq\big) := \big([(Q, P)]_\simeq\big),$$

which is a unary operation (unfortunately also denoted by $-$) from $\mathbb{Z}$ to $\mathbb{Z}$.

4.80 Example. Reducing every pair (I, J) to either form $(I - J, 0)$ or $(0, J - I)$ can facilitate calculations:

$$\begin{aligned}
[(2, 3)]_\simeq + [(5, 1)]_\simeq &= [(0, 1)]_\simeq + [(4, 0)]_\simeq \\
&= [(0 + 4, 1 + 0)]_\simeq \\
&= [(4, 1)]_\simeq \\
&= [(3, 0)]_\simeq;
\end{aligned}$$

$$\begin{aligned}
[(2, 1)]_\simeq - [(7, 3)]_\simeq &= [(2, 1)]_\simeq + [(3, 7)]_\simeq \\
&= [(1, 0)]_\simeq + [(0, 4)]_\simeq \\
&= [(1, 4)]_\simeq \\
&= [(0, 3)]_\simeq \\
&= -[(3, 0)]_\simeq;
\end{aligned}$$

$$\begin{aligned}
[(3, 1)]_\simeq * [(2, 5)]_\simeq &= [(2, 0)]_\simeq * [(0, 3)]_\simeq \\
&= [(2 * 0 + 0 * 3, 2 * 3 + 0 * 0)]_\simeq \\
&= [(0, 6)]_\simeq \\
&= -[(6, 0)]_\simeq.
\end{aligned}$$

4.5.3 Order on the Integers

The order $<$ on $\mathbb{N}$ extends to $\mathbb{Z}$ by a definition of "positive" and "negative" on $\mathbb{Z}$. By theorem 4.73, every integer $[(M, N)]_\simeq$ is the equivalence class of a pair $(I, 0)$ or $(0, I)$ for some $I \in \mathbb{N}$. In the first case $M - N = I - 0 \geq 0$, so that $M - N$

is a nonnegative difference. In the second case $N - M = I - 0 \geq 0$, so that the opposite, $N - M$, is a nonnegative difference, and then $M - N$ represents a non-positive difference.

4.81 Definition. An integer $[(M, N)]_\simeq$ is **positive** if and only if $M > N$; an integer $[(M, N)]_\simeq$ is **negative** if and only if $M < N$.

4.82 Definition. The sets of nonzero integers ($\mathbb{Z}^*$), negative integers ($\mathbb{Z}_-^*$), positive integers ($\mathbb{Z}_+^*$), non-positive integers ($\mathbb{Z}_-$), and nonnegative integers ($\mathbb{Z}_+$) are

$$\mathbb{Z}^* := \mathbb{Z} \setminus \{[(0, 0)]_\simeq\},$$

$$\begin{aligned}
\mathbb{Z}_- &:= \{[(M, N)]_\simeq \in \mathbb{Z} : M \leq N\}, \\
&:= \{[(M, N)]_\simeq \in \mathbb{Z} : \exists I\{(I \in \mathbb{N}) \wedge [(M, N) \simeq (0, I)]\}\},
\end{aligned}$$

$$\begin{aligned}
\mathbb{Z}_+ &:= \{[(M, N)]_\simeq \in \mathbb{Z} : M \geq N\}, \\
&:= \{[(M, N)]_\simeq \in \mathbb{Z} : \exists I\{(I \in \mathbb{N}) \wedge [(M, N) \simeq (I, 0)]\}\},
\end{aligned}$$

$$\begin{aligned}
\mathbb{Z}_-^* &:= \{[(M, N)]_\simeq \in \mathbb{Z} : M < N\}, \\
&:= \{[(M, N)]_\simeq \in \mathbb{Z} : \exists I\{(I \in \mathbb{N}^*) \wedge [(M, N) \simeq (0, I)]\}\},
\end{aligned}$$

$$\begin{aligned}
\mathbb{Z}_+^* &:= \{[(M, N)]_\simeq \in \mathbb{Z} : M > N\} \\
&:= \{[(M, N)]_\simeq \in \mathbb{Z} : \exists I\{(I \in \mathbb{N}^*) \wedge [(M, N) \simeq (I, 0)]\}\}.
\end{aligned}$$

The following relation on pairs will lead to an order on equivalence classes.

4.83 Definition. Define a relation $<$ on $\mathbb{N} \times \mathbb{N}$ as follows:

$$(M, N) < (0, 0)$$
$$\updownarrow$$
$$M < N$$

and

$$(M, N) < (P, Q)$$
$$\updownarrow$$
$$(M, N) + (Q, P) < (0, 0)$$
$$\updownarrow$$
$$M + Q < N + P.$$

The following theorem verifies that the ordering does not depend on the choice of the pairs representing the equivalence classes.

4.84 Theorem. *For all $I, J, K, L, M, N, P, Q \in \mathbb{N}$, if*

$$(I, J) \simeq (K, L),$$

$$(M, N) \simeq (P, Q),$$

then

$$(I, J) < (M, N)$$
$$\Updownarrow$$
$$(K, L) < (P, Q).$$

Proof. If $(I, J) \simeq (K, L)$, then $I + L = J + K$. If also $(M, N) \simeq (P, Q)$, then also $M + Q = N + P$. Consequently, adding $L + P$ to each side of the inequalities yields

$$(I, J) < (M, N)$$
$$\Updownarrow$$
$$I + N < J + M$$
$$\Updownarrow$$
$$(I + L) + (N + P) < (L + P) + (J + M)$$
$$\Updownarrow$$
$$(J + K) + (M + Q) < (L + P) + (J + M)$$
$$\Updownarrow$$
$$(J + M) + (K + Q) < (L + P) + (J + M)$$
$$\Updownarrow$$
$$K + Q < L + P$$
$$\Updownarrow$$
$$(K, L) < (P, Q).$$

$\square$

A definition of an order on $\mathbb{Z}$ can thus use any pair from an equivalence class.

4.85 Definition. Define an order $<$ on $\mathbb{Z}$ by

$$[(M, N)]_{\simeq} < [(P, Q)]_{\simeq}$$
$$\Updownarrow$$
$$(M, N) < (P, Q)$$
$$\Updownarrow$$
$$M + Q < N + P.$$

Theorem 4.86 shows that $<$ is a strict total order (definition 3.187).

4.86 Theorem.   *The relation $<$ is a strict total order on $\mathbb{Z} \times \mathbb{Z}$. In particular, if*

$$[(K, L)]_\backsimeq < [(M, N)]_\backsimeq$$

and

$$[(M, N)]_\backsimeq < [(P, Q)]_\backsimeq,$$

then

$$[(K, L)]_\backsimeq < [(P, Q)]_\backsimeq.$$

Proof. This proof verifies that $<$ is irreflexive, connected, and transitive.

Irreflexivity

(The relation does not relate any element to itself.) For each pair $(M, N) \in \mathbb{N} \times \mathbb{N}$, $M + N \not< N + M$, whence $([(M, N)]_\backsimeq) \not< ([(M, N)]_\backsimeq)$.

Connectedness

(Each element is related to every *different* element.) If $([(M, N)]_\backsimeq) \neq ([(P, Q)]_\backsimeq)$, then $M + Q \neq N + P$ whence either $M + Q < N + P$, and then $([(M, N)]_\backsimeq) < ([(P, Q)]_\backsimeq)$, or $M + Q > N + P$, and then $([(M, N)]_\backsimeq) > ([(P, Q)]_\backsimeq)$.

Transitivity

If $([(K, L)]_\backsimeq) < ([(M, N)]_\backsimeq)$, then

$$K + N < L + M;$$

if also $([(M, N)]_\backsimeq) < ([(P, Q)]_\backsimeq)$, then also

$$M + Q < N + P.$$

Consequently, adding the preceding two inequalities gives

$$(K + N) + (M + Q) < (L + M) + (N + P)$$

whence the commutativity and associativity of addition yields

$$K + [(M + N) + Q] < L + [(M + N) + P]$$

whence (by theorem 4.56) cancelling $M + N$ yields

$$K + Q < L + P$$

so that $[(K, L)]_\simeq < [(P, Q)]_\simeq.$ □

The next theorem shows that multiplication by a positive integer keeps the order.

4.87 Theorem. *If*

$$[(M, N)]_\simeq < [(P, Q)]_\simeq,$$

$$[(0, 0)]_\simeq < [(I, J)]_\simeq < [(K, L)]_\simeq,$$

then

$$([(I, J)]_\simeq * [(M, N)]_\simeq) < ([(I, J)]_\simeq * [(P, Q)]_\simeq),$$

whence, if also $(P, Q) > (0, 0)$, then

$$([(I, J)]_\simeq * [(M, N)]_\simeq) < ([(K, L)]_\simeq * [(P, Q)]_\simeq).$$

Proof. The proof uses the equivalence $(I, J) \simeq (I - J, 0)$. First, from $[(M, N)]_\simeq < [(P, Q)]_\simeq$ it follows that $M + Q < N + P$ by definition of $<$ on the pairs. Hence a multiplication throughout by $I - J > 0$ yields

$$(I - J) * (M + Q) < (I - J) * (N + P)$$

whence

$$\begin{aligned}
(I, J) * (M, N) &\simeq (I - J, 0) * (M, N) \\
&= ([I - J] * M, [I - J] * N) \\
&< ([I - J] * P, [I - J] * Q) \\
&= (I - J, 0) * (P, Q) \\
&\simeq (I, J) * (P, Q).
\end{aligned}$$

Swapping the roles of (I, J) with (K, L), and (M, N) with (P, Q), then yields

$$(I, J) * (P, Q) = (P, Q) * (I, J) < (P, Q) * (K, L) = (K, L) * (P, Q).$$

Consequently,

$$(I, J) * (M, N) < (I, J) * (P, Q) < (K, L) * (P, Q).$$

□

The following theorem shows that the square of every integer is nonnegative.

4.88 Theorem. *For every* $X \in \mathbb{Z}$, $X * X \geq 0$.

Proof. For every $X \in \mathbb{Z}$ there exists $K \in \mathbb{N}$ such that $X = [(K, 0)]_{\simeq}$ or $X = [(0, K)]_{\simeq}$, by theorem 4.73. However,

$$
\begin{aligned}
(K, 0) * (K, 0) &= (K * K + 0 * 0, K * 0 + 0 * K) \\
&= (K * K, 0) \\
&= (0 * 0 + K * K, 0 * K + K * 0) \\
&= (0, K) * (0, K)
\end{aligned}
$$

and in either case $X * X = [(K * K, 0)]_{\simeq} \geq [(0, 0)]_{\simeq}$ because $K * K + 0 \geq 0 + 0$. □

4.89 Example. $(-[(1, 0)]_{\simeq}) * (-[(1, 0)]_{\simeq}) = ([(0, 1)]_{\simeq}) * ([(0, 1)]_{\simeq}) = [(1, 0)]_{\simeq}$.

An alternative method to define all the integers specifies from $\mathbb{N}$ a similar but disjoint set $\mathbb{Z}^*_-$ for all the "negative" integers, and then defines $\mathbb{Z} := \mathbb{Z}^*_- \cup \mathbb{N}$. For instance, apply the axioms of the power set, pairing, and separation to define

$$
\begin{aligned}
\mathbb{Z}^*_- &:= \{K \in \mathscr{P}(\mathbb{N}) : \exists N \, [(N \in \mathbb{N}) \wedge (N \neq 0) \wedge (K = \{N\})]\} \\
&= \{\, \{N\} : N \in \mathbb{N}^* \,\} \\
&= \{\ldots, \{\{\varnothing, \{\varnothing\}\}, \{\{\varnothing\}\}\} \\
&= \{\ldots, \{3\}, \{2\}, \{1\}\} \,.
\end{aligned}
$$

Then change the notation to define $-N := \{N\}$ for each $N \in \mathbb{N}^*$. In particular, $\mathbb{Z}^*_- \cap \mathbb{N} = \varnothing$. Indeed, $\varnothing \in N$ by theorem 4.45 and $\varnothing \notin \{N\}$ for each positive natural number $N \in \mathbb{N} \setminus \{\varnothing\}$. Consequently, $\{N\} \notin \mathbb{N}$ for every $\{N\} \in \mathbb{Z}^*_-$.

4.90 Definition (Landau's definition of integer arithmetic). The set $\mathbb{Z} := \mathbb{Z}^*_- \cup \mathbb{N}$ is the set of **integers**. Arithmetic extends from $\mathbb{N}$ to $\mathbb{Z}$ by setting

$$
\begin{aligned}
(-M) + (-N) &:= -(M + N), \\
(-N) + M := M + (-N) &:= -(N - M) \quad \text{if } M < N, \\
(-N) + M := M + (-N) &:= M - N \qquad \text{if } N < M.
\end{aligned}
$$

Similarly,

$$
\begin{aligned}
(-N) * (-M) &:= M * N, \\
(-N) * M := M * (-N) &:= -(M * N).
\end{aligned}
$$

Moreover, the ordering $\in$ on $\mathbb{N}$ extends to an ordering $\mathbb{Z}$ by the definition

$$
\begin{aligned}
(-N) &< (-M) \text{ if and only } M < N, \\
(-N) &< K
\end{aligned}
$$

for all $K \in \mathbb{N}$ and $M, N \in \mathbb{N}^*$.

The addition and multiplication thus defined for integers remain associative and commutative, multiplication distributes over addition, $N + 0 = N$ and $1 * N = N$ for each integer $N \in \mathbb{Z}$. Proofs proceed by cases, depending on the sign of the operands, and are straightforward but lengthy. See [76, Ch. IV] or the exercises.

4.91 Remark. Common usage abbreviates each nonnegative integer $[(I, 0)]_{\simeq}$ by I. Also, any variable can denote an integer, for instance, $M = [(P, Q)]_{\simeq}$.

4.5.4 Nonnegative Integral Powers of Integers

The J-th power M^J of an integer $M \in \mathbb{Z}$ is the product $M * \cdots * M$ of J factors M. Specifically, from the convention $M^0 := 1$, induction produces higher powers.

4.92 Definition (Integral powers). For each integer M, define

$$M^0 := 1.$$

Then for each integer M and for each nonnegative integer J define

$$M^{J+1} := (M^J) * M.$$

In M^J, the number M is the **base** while J is the **exponent**.

4.93 Remark. According to definition 4.92, the Jth power of M amounts to J multiplications of M, beginning with 1. With an alternative but logically equivalent notation, for each $M \in \mathbb{N}$ definition 4.92 uses a recursive definition with G in theorem 4.21 replaced by the multiplication function $H^{(M)}$ from remark 4.29:

$$G := H^{(M)} : \mathbb{N} \to \mathbb{N},$$
$$L \mapsto M * L,$$

to specify by induction an **exponentiation** function $E^{(M)} : \mathbb{N} \to \mathbb{Z}, J \mapsto M^J$, with

$$G(L) := H^{(M)}(L) = M * L,$$
$$E^{(M)}(0) := 1,$$
$$E^{(M)}(J + 1) := G\big[E^{(M)}(J)\big].$$

4.94 Example. Here are the first four nonnegative powers of the integer 2:

$$2^0 = 1;$$
$$2^1 = (2^0) * 2 = 1 * 2 = 2;$$
$$2^2 = (2^1) * 2 = 2 * 2 = 4;$$
$$2^3 = (2^2) * 2 = 4 * 2 = 8.$$

The following theorem establishes relations between product of bases, sums of exponents, and integral powers.

4.95 Theorem. *For all* $M \in \mathbb{Z}$ *and* $N \in \mathbb{Z}$, *and for all* $I \in \mathbb{N}$ *and* $J \in \mathbb{N}$,

$$(M * N)^J = (M^J) * (N^J),$$
$$N^{I+J} = (N^I) * (N^J).$$

Proof. This proof proceeds by induction with the exponent J. The first equation, $(M * N)^J = (M^J) * (N^J)$, holds for all integers M and N, and for $J \in \{0, 1\}$:

$$(M * N)^0 = 1 = 1 * 1 = (M^0) * (N^0);$$

$$(M * N)^1 = M * N = (M^1) * (N^1).$$

Hence, to prove that $(M * N)^J = (M^J) * (N^J)$ for every $J \in \mathbb{N}$, let

$$S := \{J \in \mathbb{N} : \forall M \forall N\{[(M \in \mathbb{N}) \wedge (N \in \mathbb{N})] \Rightarrow [(M * N)^J = (M^J) * (N^J)]\}\} .$$

Thus, *assume* that there exists $K \in S$, or, equivalently, $K \in \mathbb{N}$ such that the theorem holds for $J := K$; then

$$
\begin{aligned}
(M * N)^{K+1} &= [(M * N)^K] * (M * N) &\quad \text{definition 4.92,} \\
&= [(M^K) * (N^K)] * (M * N) &\quad \text{induction hypothesis,} \\
&= [(M^K) * M] * [(N^K) * N] &\quad \text{associativity and commutativity,} \\
&= (M^{K+1}) * (N^{K+1}) &\quad \text{definition 4.92 twice,}
\end{aligned}
$$

whence $K + 1 \in S$, and hence, $S = \mathbb{N}$, which means that the first equation holds.

The second equation, $N^{I+J} = (N^I) * (N^J)$, holds for each integer N, for each nonnegative integer I, and for $J \in \{0, 1\}$:

$$N^{I+0} = N^I = (N^I) * 1 = (N^I) * (N^0),$$
$$N^{I+1} = (N^I) * N = (N^I) * (N^1).$$

Next, let

$$S := \{J \in \mathbb{N} : \forall I \forall N[N^{I+J} = (N^I) * (N^J)]\}$$

and *assume* that there exists $K \in S$, or, equivalently, $K \in \mathbb{N}$ such that the theorem holds for $J := K$, so that $N^{I+K} = (N^I) * (N^K)$ for all $I \in \mathbb{N}$ and $N \in \mathbb{N}$; then

$$
\begin{aligned}
N^{I+(K+1)} &= N^{(I+1)+K} & \text{associativity and commutativity of } +, \\
&= (N^{I+1}) * (N^K) & \text{induction hypothesis,} \\
&= [(N^I) * N] * (N^K) & \text{definition 4.92,} \\
&= (N^I) * [(N^K) * N] & \text{associativity and commutativity of } *, \\
&= (N^I) * (N^{K+1}) & \text{definition 4.92,}
\end{aligned}
$$

whence $K + 1 \in S$, and, consequently, $S = \mathbb{N}$, which proves the second equation.

$\square$

4.5.5 Exercises on Integers with Induction

4.71 . Calculate $[(2, 4)]_{\simeq} + [(6, 3)]_{\simeq}$.

4.72 . Calculate $[(5, 3)]_{\simeq} + [(1, 6)]_{\simeq}$.

4.73 . Calculate $[(3, 1)]_{\simeq} - [(5, 2)]_{\simeq}$.

4.74 . Calculate $[(7, 3)]_{\simeq} * [(2, 5)]_{\simeq}$.

4.75 . Prove that Kunen's addition commutes.

4.76 . Prove that Landau's addition commutes.

4.77 . Prove that Kunen's addition is associative.

4.78 . Prove that Landau's addition is associative.

4.79 . Prove that Kunen's multiplication commutes.

4.80 . Prove that Landau's multiplication commutes.

4.81 . Prove that Kunen's multiplication is associative.

4.82 . Prove that Landau's multiplication is associative.

4.83 . Prove that Kunen's multiplication distributes over addition.

4.84 . Prove that Landau's multiplication distributes over addition.

4.85 . Prove that subtraction does *not* commute.

4.86 . Prove that subtraction is *not* associative.

4.87 . Prove that multiplication distributes over subtraction.

4.88 . Prove that subtraction does *not* distribute over multiplication.

4.89 . Prove that subtraction does *not* distribute over addition.

4.90 . Prove that addition does *not* distribute over subtraction.

4.6 Rational Numbers

4.6.1 Definition of Rational Numbers

Some practical situations involve comparisons of proportions. Integer arithmetic does not allow for proportions, but a method similar to that for passing from natural numbers to differences also leads from integers to proportions. As a pair of natural numbers (I, J) can represent a difference, a pair of integers (P, Q) can represent a proportion.

For example, the density of the universe at one location involves the mass or number of particles P in a volume Q, which can be summarized by the ordered pair (P, Q). The density at another location involves the mass or number of particles M in a volume N, as summarized by the pair (M, N). A comparison of the densities (P, Q) and (M, N) can proceed through a multiplication by a common factor, for instance, Q or N.

Specifically, in a volume N times larger than Q, the first density of P particles in a volume Q becomes $P * N$ particles in a volume $Q * N$, or $(P * N, Q * N)$.

Likewise in a volume Q times larger than N, the second density of M particles in a volume N becomes $Q * M$ particles in a volume $Q * N$, or $(Q * M, Q * N)$.

Both densities $(P*N, Q*N)$ and $(Q*M, Q*N)$ refer to the same volume $Q*N$; thus, they are identical if and only if their masses equal each other: $P * N = Q * M$.

The foregoing reasoning leads to a relation between pairs of integers.

4.96 Definition. On the set $\mathbb{Z} \times \mathbb{Z}^*$ of all pairs of integers (P, Q) with $Q \neq 0$, define a relation $\equiv$ by

$$(P, Q) \equiv (M, N)$$
$$\Updownarrow$$
$$P * N = Q * M$$

The relation $\equiv$ on $\mathbb{Z} \times \mathbb{Z}^*$ in definition 4.96 *differs* from the relation $\backsimeq$ on $\mathbb{N} \times \mathbb{N}$ in definition 4.66. Nevertheless, it is also an equivalence relation (definition 3.150).

4.97 Theorem. *The relation $\equiv$ in definition 4.96 is an equivalence relation on $\mathbb{Z} \times \mathbb{Z}^*$.*

Proof. This proof checks algebraically that $\equiv$ is reflexive, symmetric, and transitive.

Reflexivity

For each pair, $(I, J) \equiv (I, J)$ because $I * J = J * I$.

Symmetry

If $(I, J) \equiv (K, L)$, then $I * L = J * K$ by definition of $\equiv$ whence $K * J = L * I$ by commutativity, so that $(K, L) \equiv (I, J)$.

Transitivity

If

$$(I, J) \equiv (K, L),$$
$$(K, L) \equiv (M, N),$$

then

$$I * L = J * K,$$
$$K * N = L * M,$$

whence multiplying the left-hand sides and the right-hand sides together gives

$$(I * L) * (K * N) = (J * K) * (L * M),$$
$$(I * N) * (L * K) = (J * M) * (K * L),$$

and hence cancelling the nonzero factor $L * K = K * L$ yields

$$I * N = J * M$$

which means that $(I, J) \equiv (M, N)$.   □

4.98 Definition.   The set of **rational** numbers, denoted by $\mathbb{Q}$ (for "quotients"), is the set of all equivalence classes for the relation $\equiv$. A **rational number** is an element of $\mathbb{Q}$.

The equivalence class of a pair (I, J) is called the **ratio** of I to J, and it is denoted by $[(I, J)]$, or also by I/J, or also by $\frac{I}{J}$. A **fraction** is a pair $(I, J) \in (\mathbb{Z} \times \mathbb{Z}^*)$, where I is the **numerator** and J is the **denominator**.

The following theorem provides a means to select a specific fraction from a rational number, or, equivalently, a specific pair (M, N) from an equivalence class P/Q.

4.99 Theorem.   *For each $P/Q \in \mathbb{Q}$, there exists $M/N \equiv P/Q$ such that*

$$M = \min\{I \in \mathbb{N} : \exists J[(J \in \mathbb{N}) \wedge (I/J \equiv P/Q)]\}.$$

Proof. This proof applies the Well-Ordering Principle to the set of all nonnegative numerators of a rational number. To this end, for each $P/Q \in \mathbb{Q}$ define the set

$$E := \{I \in \mathbb{N} : \exists J[(J \in \mathbb{N}) \wedge (I/J \equiv P/Q)]\}.$$

Then $E \neq \varnothing$, because if $P \geq 0$ then $P/Q \equiv P/Q$, whence $P \in E$, whereas if $P < 0$ then $(-P)/(-Q) \equiv P/Q$, whence $(-P) \in E$. The Well-Ordering Principle (theorem 4.47) then guarantees the existence of a first element in E.   □

4.6.2   Arithmetic with Rational Numbers

The comparison of two rational numbers P/Q and M/N can proceed through equivalent fractions with a common denominator, for instance, $(P*N)/(Q*N)$ and $(Q*M)/(Q*N)$. Common denominators also lead to an arithmetic with fractions (ordered pairs) and then with rational numbers (equivalence classes).

4.100 Definition. For all pairs (I, J) and (K, L) in $\mathbb{Z} \times \mathbb{Z}^*$, define functions $+$ and $*$ on $(\mathbb{Z} \times \mathbb{Z}^*) \times (\mathbb{Z} \times \mathbb{Z}^*)$ by their counterparts $+$ and $*$ already defined on $\mathbb{Z} \times \mathbb{Z}^*$:

$$(I, J) + (K, L) := ([I * L] + [J * K],\ J * L),$$

$$(I, J) * (K, L) := (I * K,\ J * L).$$

The symbols $+$ and $*$ on the left-hand sides are the functions being defined, whereas the symbols $+$ and $*$ on the right-hand sides are the addition and multiplication of integers. Yet common usage employs $+$ and $*$ for both.

The following theorem shows that equivalent fractions lead to equivalent results.

4.101 Theorem. *If*

$$(I, J) \equiv (K, L),$$

$$(M, N) \equiv (P, Q),$$

then

$$(I, J) + (M, N) \equiv (K, L) + (P, Q),$$

$$(I, J) * (M, N) \equiv (K, L) * (P, Q).$$

Proof. This proof proceeds through algebraic verifications. By hypotheses, $I * L = J * K$ and $M * Q = N * P$, whence

$$(I, J) + (M, N) = (I * N + J * M, J * N),$$

$$(K, L) + (P, Q) = (K * Q + L * P, L * Q),$$

where

$$(I * N + J * M) * (L * Q) - (J * N) * (K * Q + L * P)$$

$$= [I * N * L * Q + J * M * L * Q] - [J * N * K * Q + J * N * L * P]$$

$$= [(I * L) * (N * Q) + (J * L) * (M * Q)]$$
$$-[(J * K) * (N * Q) + (J * L) * (N * P)]$$

$$= [(I * L) * (N * Q) + (J * L) * (M * Q)]$$
$$-[(I * L) * (N * Q) + (J * L) * (M * Q)]$$

$$= 0.$$

Thus, $(I, J) + (M, N) \equiv (K, L) + (P, Q)$. Similarly,

$$(I, J) * (M, N) = (I * M, J * N),$$
$$(K, L) * (P, Q) = (K * P, L * Q),$$

where, by hypotheses,

$$(I * M) * (L * Q) - (J * N) * (K * P)$$
$$= (I * L) * (M * Q) - (J * K) * (N * P)$$
$$= (I * L) * (M * Q) - (I * L) * (M * Q)$$
$$= 0.$$

Thus, $(I, J) * (M, N) \equiv (K, L) * (P, Q)$.   □

4.102 Definition. Define the addition and the multiplication of rational numbers by

$$\frac{I}{J} + \frac{K}{L} := \frac{I * L + J * K}{J * L},$$

$$\frac{I}{J} * \frac{K}{L} := \frac{I * K}{J * L}.$$

The following theorems establish algebraic characteristics of rational arithmetic.

4.103 Theorem. *The addition of rational numbers commutes.*

Proof. For all I/J and K/L in $\mathbb{Q}$,

$$\frac{I}{J} + \frac{K}{L} = \frac{I * L + J * K}{J * L} = \frac{L * I + K * J}{L * J} = \frac{K}{L} + \frac{I}{J}.$$

□

4.104 Theorem. *The multiplication of rational numbers commutes.*

Proof. For all I/J and K/L in $\mathbb{Q}$,

$$\frac{I}{J} * \frac{K}{L} = \frac{I * K}{J * L} = \frac{K * I}{L * J} = \frac{K}{L} * \frac{I}{J}.$$

□

4.105 Theorem. *The addition of rational numbers is associative.*

Proof. For all I/J, K/L, and M/N in $\mathbb{Q}$,

$$\left(\frac{I}{J} + \frac{K}{L}\right) + \frac{M}{N} = \frac{I * L + J * K}{J * L} + \frac{M}{N}$$

$$= \frac{[(I * L + J * K) * N] + [(J * L) * M]}{(J * L) * N}$$

$$= \frac{[I * (L * N) + J * (K * N)] + [J * (L * M)]}{J * (L * N)}$$

$$= \frac{[I * (L * N)] + [J * (K * N + L * M)]}{J * (L * N)}$$

$$= \frac{I}{J} + \frac{K * N + L * M}{L * N}$$

$$= \frac{I}{J} + \left(\frac{K}{L} + \frac{M}{N}\right).$$

□

4.106 Theorem. *The multiplication of rational numbers is associative.*

Proof. For all I/J, K/L, and M/N in $\mathbb{Q}$,

$$\left(\frac{I}{J} * \frac{K}{L}\right) * \frac{M}{N} = \frac{I * K}{J * L} * \frac{M}{N}$$

$$= \frac{(I * K) * M}{(J * L) * N}$$

$$= \frac{I * (K * M)}{J * (L * N)}$$

$$= \frac{I}{J} * \frac{K * M}{L * N}$$

$$= \frac{I}{J} * \left(\frac{K}{L} * \frac{M}{N} \right).$$

□

4.107 Theorem. *For each $N \in \mathbb{N}^*$ and each $I/J \in \mathbb{Q}$, multiplications of the numerator and denominator by a nonzero common factor yields the same rational number: $I/J = (I * N)/(J * N)$.*

Proof. Verify the criterion for equivalent fractions: $I * (J * N) = J * (I * N)$.   □

4.108 Theorem. *The multiplication of rational numbers distributes over addition.*

Proof. For all I/J, K/L, and M/N in $\mathbb{Q}$,

$$\left(\frac{I}{J} + \frac{K}{L} \right) * \frac{M}{N} = \frac{I * L + J * K}{J * L} * \frac{M}{N}$$

$$= \frac{(I * L + J * K) * M}{(J * L) * N}$$

$$= \frac{(I * L) * M + (J * K) * M}{J * (L * N)}$$

$$= \frac{(I * M) * L + J * (K * M)}{J * (L * N)}$$

$$= \frac{(I * M) * (L * N) + (J * N) * (K * M)}{(J * N) * (L * N)}$$

$$= \frac{I * M}{J * N} + \frac{K * M}{L * N}$$

$$= \left(\frac{I}{J} * \frac{M}{N} \right) + \left(\frac{K}{L} * \frac{M}{N} \right).$$

□

The following theorem shows that adding $0/1$ does not produce any change.

4.109 Theorem. *For each $K/L \in \mathbb{Q}$, $(K/L) + (0/1) = (K/L)$.*

Proof. $(K/L) + (0/1) = ([K * 1 + L * 0]/L * 1) = (K/L)$.   □

Thus, the rational number $0/1$ plays the same role as the integer 0 in additions. The following theorem shows that each rational number has an additive inverse.

4.110 Theorem. *Each $I/J \in \mathbb{Q}$ has an additive inverse:* $(I/J) + ([-I]/J) = (0/1)$.

Proof.

$$\frac{I}{J} + \frac{-I}{J} = \frac{I * J + J * (-I)}{J * J}$$

$$= \frac{J * I + J * (-I)}{J * J}$$

$$= \frac{J * [I + (-I)]}{J * J}$$

$$= \frac{J * 0}{J * J}$$

$$= \frac{0}{J * J}$$

$$= \frac{0 * (J * J)}{1 * (J * J)}$$

$$= \frac{0}{1}.$$

□

The following theorem shows that multiplying by $1/1$ changes nothing.

4.111 Theorem. *For each $K/L \in \mathbb{Q}$, $(K/L) * (1/1) = (K/L)$.*

Proof. $(K/L) * (1/1) = ([K * 1]/[L * 1]) = (K/L)$.   □

Thus, the rational number $1/1$ plays the same role as the integer 1 in multiplications. Similarly, each nonzero rational number has a multiplicative inverse.

4.112 Theorem. *Each $I/J \in \mathbb{Q}$ such that $I \neq 0$ has a multiplicative inverse:* $(I/J) * (J/I) = (1/1)$.

Proof.

$$\frac{I}{J} * \frac{J}{I} = \frac{I * J}{J * I} = \frac{I * J}{I * J} = \frac{1 * (I * J)}{1 * (I * J)} = \frac{1}{1}.$$

□

Rational arithmetic thus satisfies the algebraic properties in table 4.115.

4.113 Definition (Field). A **field** (of numbers) is a set $\mathbb{F}$ with at least two different elements 0 and 1, so that $0 \neq 1$, and binary operations $+$ and $*$, satisfying the algebraic properties in table 4.115.

Table 4.115 These properties hold for all $I/J, K/L, P/Q \in \mathbb{Q}$.

(1)	Associativity of $+$	$[(I/J) + (K/L)] + (P/Q) = (I/J) + [(K/L) + (P/Q)]$
(2)	Commutativity of $+$	$(I/J) + (K/L) = (K/L) + (I/J)$
(3)	Additive identity	$(K/L) + (0/1) = (K/L) = (0/1) + (K/L)$
(4)	Additive inverse	$(K/L) + ([-K]/L) = (0/1)$
(5)	Associativity of $*$	$[(I/J)(K/L)](P/Q) = (I/J)[(K/L)(P/Q)]$
(6)	Commutativity of $*$	$(I/J)(K/L) = (K/L)(I/J)$
(7)	Multiplicative identity	$(K/L)(1/1) = (K/L) = (1/1)(K/L)$
(8)	Multiplicative inverse	If $(K/L) \neq 0,$ then $(K/L)(L/K) = (1/1)$
(9)	Distributivity	$(I/J)[(K/L) + (P/Q)] = [(I/J)(K/L)] + [(I/J)(P/Q)]$

Technically, the pair $(+, *)$ suffices to identify the set $\mathbb{F}$ and the elements 0 and 1, but for emphasis a field can be defined as the quintuple $(\mathbb{F}, +, 0, *, 1)$.

4.114 Example. The quintuple $(\mathbb{Q}, +, 0, *1)$ is a field (of numbers).

The next theorem forms the basis for a concept of division of rational numbers.

4.116 Theorem. *For each $K/L \in \mathbb{Q}$, and for each $I/J \in \mathbb{Q}$ such that $I \neq 0$,* $[(K/L) * (J/I)] * (I/J) = (K/L).$

Proof.

$$\left(\frac{K}{L} * \frac{J}{I} \right) * \frac{I}{J} = \frac{K*J}{L*I} * \frac{I}{J} = \frac{(K*J)*I}{(L*I)*J} = \frac{K*(J*I)}{L*(I*J)} = \frac{K*(I*J)}{L*(I*J)} = \frac{K}{L}.$$

$\square$

4.117 Definition. For each $K/L \in \mathbb{Q}$, and for each $I/J \in \mathbb{Q}$ such that $I \neq 0$, define

$$\frac{K}{L} \div \frac{I}{J} := \frac{K}{L} * \frac{J}{I}.$$

4.118 Definition. For each $K/L \in \mathbb{Q}$, and for each $J \in \mathbb{N}$, define

$$\left(\frac{K}{L} \right)^0 := \frac{1}{1},$$

$$\left(\frac{K}{L} \right)^{J+1} := \left(\frac{K}{L} \right)^J * \left(\frac{K}{L} \right).$$

Moreover, if $K/L \neq 0/1$, then for each $J \in \mathbb{Z}$, define

$$\left(\frac{K}{L} \right)^{J-1} := \left(\frac{K}{L} \right)^J * \left(\frac{L}{K} \right).$$

4.6.3 Notation for Sums and Products

The notation introduced here proves convenient to define and investigate sums and products of finite sequences of numbers.

4.119 Definition. A **finite sequence of rational numbers** is a function $S : N \to \mathbb{Q}$ defined on some $N \in \mathbb{N}$. The value $S(K)$ is also denoted by S_K; then the function S is also denoted by (S_K).

4.120 Example. The function $S : 9 \to \mathbb{Q}$ defined by $S_K := (2/3)^K$ is a finite sequence of numbers:

$$
\begin{aligned}
S_0 &= (2/3)^0 = 1, \\
S_1 &= (2/3)^1 = 2/3, \\
S_2 &= (2/3)^2 = 4/9, \\
S_3 &= (2/3)^3 = 8/27, \\
S_4 &= (2/3)^4 = 16/81, \\
S_5 &= (2/3)^5 = 32/243, \\
S_3 &= (2/3)^6 = 64/729, \\
S_4 &= (2/3)^7 = 128/2187, \\
S_5 &= (2/3)^8 = 256/6561.
\end{aligned}
$$

The next definition gives a notation for the product of a finite sequence of numbers.

4.121 Definition (Product notation). For each finite sequence of numbers $S : N \to \mathbb{Q}$, define the "empty product" to be 1:

$$
\prod_{K<0} S_K := 1.
$$

Then define the product of the first value to be the first value:

$$
\prod_{K=0}^{0} S_K := S_0.
$$

Hence for each $L \in \mathbb{N}$, such that $0 < L < N$, define the product of the first L values of the sequence S "inductively" [77, p. 5] or "recursively" [49, p. 133] by

$$
\prod_{K=0}^{L} S_K := \left(\prod_{K=0}^{L-1} S_K \right) * S_L.
$$

4.122 Example. Consider the finite sequence $S : 9 \to \mathbb{Q}$ defined by $S_K := (2/3)^K$:

$$\prod_{K<0} S_K = 1,$$

$$\prod_{K=0}^{0} S_K = S_0$$
$$= 1,$$

$$\prod_{K=0}^{1} S_K = \left(\prod_{K=0}^{0} S_K\right) * S_1$$
$$= (1) * 2/3,$$

$$\prod_{K=0}^{2} S_K = \left(\prod_{K=0}^{1} S_K\right) * S_2$$
$$= (1 * 2/3) * 4/9,$$

$$\prod_{K=0}^{3} S_K = \left(\prod_{K=0}^{2} S_K\right) * S_3$$
$$= (1 * 2/3 * 4/9) * 8/27,$$

$$\prod_{K=0}^{4} S_K = \left(\prod_{K=0}^{3} S_K\right) * S_4$$
$$= (1 * 2/3 * 4/9 * 8/27) * 16/81,$$
$$\vdots$$

$$\prod_{K=0}^{8} S_K = \left(\prod_{K=0}^{7} S_K\right) * S_8$$
$$= (1 * \tfrac{2}{3} * \tfrac{4}{9} * \tfrac{8}{27} * \tfrac{16}{81} * \tfrac{32}{243} * \tfrac{64}{729} * \tfrac{128}{2187}) * \tfrac{256}{6561}.$$

The next definition gives a notation for the sum of a finite sequence of numbers.

4.123 Definition (Sum notation). For each finite sequence of numbers $S : N \to \mathbb{Q}$, define the "empty sum" to be 0:

$$\sum_{K<0} S_K := 0.$$

Then define the sum of the first value to be the first value:

$$\sum_{K=0}^{0} S_K := S_0.$$

Hence for each $L \in \mathbb{N}$, such that $0 < L < N$, define the sum of the first L values of the sequence S inductively by

$$\sum_{K=0}^{L} S_K := \left(\sum_{K=0}^{L-1} S_K\right) + S_L.$$

4.124 Example. Consider the finite sequence $S : 9 \to \mathbb{Q}$ defined by $S_K := (2/3)^K$:

$$\sum_{K<0} S_K = 0,$$

$$\sum_{K=0}^{0} S_K = S_0$$
$$= 1,$$

$$\sum_{K=0}^{1} S_K = \left(\sum_{K=0}^{0} S_K \right) + S_1$$
$$= (1) + 2/3,$$

$$\sum_{K=0}^{2} S_K = \left(\sum_{K=0}^{1} S_K \right) + S_2$$
$$= (1 + 2/3) + 4/9,$$

$$\sum_{K=0}^{3} S_K = \left(\sum_{K=0}^{2} S_K \right) + S_3$$
$$= (1 + 2/3 + 4/9) + 8/27,$$

$$\sum_{K=0}^{4} S_K = \left(\sum_{K=0}^{3} S_K \right) + S_4$$
$$= (1 + 2/3 + 4/9 + 8/27) + 16/81,$$

$$\vdots$$

$$\sum_{K=0}^{8} S_K = \left(\sum_{K=0}^{7} S_K \right) + S_8$$
$$= \left(1 + \tfrac{2}{3} + \tfrac{4}{9} + \tfrac{8}{27} + \tfrac{16}{81} + \tfrac{32}{243} + \tfrac{64}{729} + \tfrac{128}{2187} \right) + \tfrac{256}{6561}.$$

The pattern in the foregoing example, called a **geometric series**, is amenable to an alternative formula, which expresses the entire sum as one ratio.

4.125 Theorem (geometric series). *For every $N \in \mathbb{N}^*$ and every $X \in \mathbb{Q} \setminus \{1\}$,*

$$\sum_{K=0}^{N-1} X^K = \frac{1 - X^N}{1 - X}.$$

Proof. This proof uses induction with N. For $N := 1$, and for every $X \in \mathbb{Q} \setminus \{1\}$,

$$\sum_{K=0}^{1-1} X^K = \sum_{K=0}^{0} X^K = X^0 = 1 = \frac{1 - X^1}{1 - X}.$$

Assume that there exists $I \in \mathbb{N}^*$ such that the theorem holds for $N := I$ and for every $X \in \mathbb{Q} \setminus \{1\}$, so that

$$\sum_{K=0}^{I-1} X^K = \frac{1 - X^I}{1 - X}.$$

Then

$$\sum_{K=0}^{(I+1)-1} X^K = \left(\sum_{K=0}^{I-1} X^K\right) + X^I$$

$$= \frac{1 - X^I}{1 - X} + X^I$$

$$= \frac{1 - X^I}{1 - X} + \frac{(1 - X) * X^I}{1 - X}$$

$$= \frac{(1 - X^I) + (X^I - X^{I+1})}{1 - X}$$

$$= \frac{1 - X^{I+1}}{1 - X}.$$

$\square$

An alternative proof of the same formula proceeds along the following outline:

$$\sum_{K=0}^{N-1} X^K \quad = 1 + X + \cdots + X^{N-2} + X^{N-1}$$

$$X * \sum_{K=0}^{N-1} X^K \quad = X + X^2 + \cdots + X^{N-1} + X^N$$

$$(1 - X) * \sum_{K=0}^{N-1} X^K = 1 + 0 + \cdots + 0 - X^N$$

whence dividing both sides by $(1 - X)$ yields

$$\sum_{K=0}^{N-1} X^K = \frac{1 - X^N}{1 - X}.$$

Yet such a proof also requires induction to rearrange the terms in the subtraction.

4.126 Example. Consider the finite sequence $S : 9 \to \mathbb{Q}$ defined by $S_K := (2/3)^K$:

$$\sum_{K=0}^{8} S_K = \left(1 + \tfrac{2}{3} + \tfrac{4}{9} + \tfrac{8}{27} + \tfrac{16}{81} + \tfrac{32}{243} + \tfrac{64}{729} + \tfrac{128}{2187}\right) + \tfrac{256}{6561}$$

$$= \frac{1 - (2/3)^9}{1 - 2/3}$$

$$= \frac{1 - {}^{512}/19{,}683}{1/3}$$

$$= \frac{{}^{19{,}171}/19683}{1/3}$$

$$= {}^{3}/_{1} * {}^{19171}/_{19683}$$

$$= {}^{19171}/_{6561}.$$

4.6.4 Order on the Rational Numbers

The determination of whether two rational numbers P/Q and M/N coincide can utilize equivalent fractions with a common denominator, for instance, $(P * N)/(Q * N)$ and $(Q * M)/(Q * N)$, and then with the comparison of the numerators $P * N$ and $Q * M$. The same comparison leads to a concept of order on the rational numbers.

4.127 Definition. Define a relation $<$ on $\mathbb{Q}$ as follows. First, $0 < (P/Q)$ if and only if $0 < P * Q$, so that either P and Q are both positive, or P and Q are both negative:

$$[0 < (P/Q)] \Leftrightarrow [0 < (P * Q)].$$

Second, $(I/J) < (P/Q)$ if and only if $0 < [(P/Q) - (I/J)]$:

$$[(I/J) < (P/Q)] \Leftrightarrow \{0 < [(P/Q) - (I/J)]\}.$$

The following theorem shows that the square of any nonzero rational number, and the sum and product of positive rational numbers, are positive rational numbers.

4.128 Theorem. *If $(M/N) > 0$ and $(I/J) > 0$, then $(M/N) + (I/J) > 0$ and $(M/N) * (I/J) > 0$. Moreover, $(K/L) * (K/L) > 0$ for every $(K/L) \neq (0/1)$.*

Proof. For the square, let $P/Q := (K/L)^2 = (K^2)/(L^2)$. Then $P * Q = (K^2) * (L^2) = (K * L)^2 > 0$ (theorem 4.88). Thus $(K/L)^2 = P/Q > 0$ (definition 4.127).

For the product, let $P/Q := (M/N) * (I/J) = (M * I)/(N * J)$. By the hypotheses, $M * N > 0$ and $I * J > 0$. Hence $P * Q = (M * I) * (N * J) = (M * N) * (I * J) > 0$, whence $(M/N) * (I/J) = P/Q > 0$ (definition 4.127). For the sum, let

$$\frac{P}{Q} := \frac{M}{N} + \frac{I}{J} = \frac{(M * J) + (N * I)}{N * J}.$$

Then

$$P * Q = [(M * J) + (N * I)] * (N * J) = (M * N) * (J * J) + (N * N) * (I * J) > 0.$$

Indeed, $J^2 > 0$ and $N^2 > 0$ (theorem 4.88), $(M*N)*(J^2) > 0$ and $(N^2)*(I*J) > 0$ by hypothesis, whence $P * Q = (M * N) * (J^2) + (N^2) * (I * J) > 0$.   □

The next theorem shows that $<$ is a strict total order (definition 3.187) on $\mathbb{Q}$.

4.129 Theorem. *The relation $<$ is a strict total order on the rational numbers.*

Proof. This proof verifies that $<$ is connected, irreflexive, and transitive.

Irreflexivity

$(P/Q) \not< (P/Q)$ because $0 \not< 0 = (P/Q) - (P/Q)$.

Connectedness

If $(M/N) \neq (P/Q)$, then $P * N \neq Q * M$ whence $(P * N - Q * M) \neq 0$, whence $(Q*N)*(P*N-Q*M) \neq 0$, and then either $(Q*N)*(P*N-Q*M) < 0$, in which case $(P/Q)-(M/N) < (0/1)$, so that $(P/Q) < (M/N)$, or $(Q*N)*(P*N-Q*M) > 0$, in which case $(P/Q) - (M/N) > (0/1)$, so that $(P/Q) > (M/N)$.

Transitivity

If $(I/J) < (K/L)$ and $(K/L) < (M/N)$, then $(K/L) - (I/J) > 0$ and $(M/N) - (K/L) > 0$, whence $(M/N) - (I/J) = [(M/N) - (K/L)] + [(K/L) - (I/J)] > 0$ (theorem 4.128) and then $(M/N) > (I/J)$ (definition 4.127).   □

4.130 Definition. The sets of nonzero rationals ($\mathbb{Q}^*$), negative rationals ($\mathbb{Q}_-^*$), positive rationals ($\mathbb{Q}_+^*$), non-positive rationals ($\mathbb{Q}_-$), and nonnegative rationals ($\mathbb{Q}_+$) are

$$\mathbb{Q}^* := \mathbb{Q} \setminus \{0/1\},$$
$$\mathbb{Q}_- := \{P/Q \in \mathbb{Q} : P/Q \leq 0/1\},$$
$$\mathbb{Q}_+ := \{P/Q \in \mathbb{Q} : P/Q \geq 0/1\},$$
$$\mathbb{Q}_-^* := \{P/Q \in \mathbb{Q} : P/Q < 0/1\},$$
$$\mathbb{Q}_+^* := \{P/Q \in \mathbb{Q} : P/Q > 0/1\}.$$

4.131 Definition. Define the **absolute value** $|P/Q|$ of a rational number P/Q by

$$|P/Q| := \begin{cases} P/Q & \text{if } P/Q \geq 0, \\ -(P/Q) & \text{if } P/Q < 0. \end{cases}$$

4.132 Theorem (Triangle Inequality). *For all $K/L \in \mathbb{Q}$ and $M/N \in \mathbb{Q}$,*

$$\left| \frac{K}{L} + \frac{M}{N} \right| \leq \left| \frac{K}{L} \right| + \left| \frac{M}{N} \right|$$

with equality if and only if $(K/L)(M/N) \geq 0$. Also,

$$\left| \left| \frac{K}{L} \right| - \left| \frac{M}{N} \right| \right| \leq \left| \frac{K}{L} - \frac{M}{N} \right|$$

with equality if and only if $(K/L)(M/N) \geq 0$.

Proof. Apply the definition of the absolute value to four cases. □

4.133 Theorem (Archimedean Property of the Rationals). *For each rational $P/Q \in \mathbb{Q}$ there exists a natural number $N \in \mathbb{N}$ such that $P/Q < N/1$.*

Proof. If $P/Q \leq 0$, let $N := 1$. If $P/Q > 0$, let $N := |P| + 1$; then

$$\frac{N}{1} - \frac{P}{Q} = \frac{N}{1} - \frac{|P|}{|Q|} = \frac{\{[(|P| + 1) * |Q|] - (1 * |P|)\}}{1 * |Q|} > 0,$$

because $[(|P| + 1) * |Q|] - (1 * |P|) = |P| * (|Q| - 1) + |Q| \geq |Q| > 0$. □

4.6.5 Exercises on Rational Numbers

4.91 . Calculate $(2/3) + (7/5)$.

4.92 . Calculate $(5/2) + (1/7)$.

4.93 . Calculate $(7/3) - (2/5)$.

4.94 . Calculate $(1/2) - (1/3)$.

4.95 . Calculate $(2/3) * (7/5)$.

4.96 . Calculate $(5/2) * (1/7)$.

4.97 . Calculate $(2/3) \div (7/5)$.

4.98 . Calculate $(5/2) \div (1/7)$.

4.99 . Prove that on $\mathbb{Q}$ the division does *not* commute.

4.100 . Prove that on $\mathbb{Q}$ division is *not* associative.

4.101 . Prove that on $\mathbb{Q}$ division does *not* distribute over addition.

4.102 . Prove that on $\mathbb{Q}$ division does *not* distribute over multiplication.

4.103 . Prove that on $\mathbb{Q}$ addition does *not* distribute over division.

4.104 . Prove that on $\mathbb{Q}$ multiplication does *not* distribute over division.

4.105 . Prove that if $0 < (I/J)$ and $0 < (P/Q)$, then $0 < [(I/J) + (P/Q)]$.

4.106 . Prove that if $0 < (I/J)$ and $0 < (P/Q)$, then $0 < [(I/J) * (P/Q)]$.

4.107 . Prove that if $Q > 0$ and $R > 0$, then $P/(Q/R) = (P * R)/Q$.

4.108 . Prove that for each $P/Q \in \mathbb{Q}$ there exists a *smallest* $N \in \mathbb{N}^*$ such that there exists $M \in \mathbb{Z}$ with $P/Q = M/N$.

4.109 . Prove that for each $K/L \in \mathbb{Q}_+$ there exists a *smallest* $N \in \mathbb{N}$ with $K/L \leq N/1$.

4.110 . Find rational numbers K/L and M/N with $(K/L)^2 + (M/N)^2 = 1$.

4.7   Finite Cardinality

4.7.1   Equal Cardinalities

The adjective "cardinal" means "principal" or "of greatest importance" . In the context of sets, the cardinal feature of sets is their size. One way to define the "size" of a set consists in establishing a correspondence with another set of known size, for instance, a natural number, as in figure 4.1. Such a natural number is then the "number" of elements in the set, and the correspondence amounts to an operation of counting. Thus the natural numbers constitute the "standard" sizes with which to "count" sets. More generally, two sets have the same cardinality if and only if there exists a bijection between those two sets.

4.134 Definition.   For all sets A and B, the sets A and B have **the same cardinality** if and only if there exists a bijection $F : A \rightarrow B$, a situation denoted by

$$A \approx B,$$
$$\bar{\bar{A}} = \bar{\bar{B}},$$
$$|A| = |B|,$$
$$\#(A) = \#(B),$$
$$\text{card}\,(A) = \text{card}\,(B).$$

Definition 4.134 does not yet define the concept of cardinality; it only defines the concept of *same cardinality*. Because such a definition leaves the notation #(A)

Fig. 4.1 Same cardinality: Earth and Moon, or Pluto and Charon, or $\{\varnothing, \{\varnothing\}\}$.

yet undefined, this exposition adopts the notation $A \approx B$, which merely means that there exists a bijection from A to B.

4.135 Example. All empty sets have the same cardinality. Indeed, by the axiom of extensionality there exists only one empty set, namely $\varnothing$, and the empty function $\varnothing : \varnothing \to \varnothing$ is a bijection, whence $\varnothing \approx \varnothing$.

4.136 Example. All singletons have the same cardinality. Indeed, for all sets X and Y and all singletons $\{X\}$ and $\{Y\}$, the function $F : \{X\} \to \{Y\}$ defined by $F := \{(X, Y)\}$ is a bijection. Thus $\{X\} \approx \{Y\}$.

4.137 Example. The sets $A := \{4, 9\}$ and $B := \{2, 3\}$ have the same cardinality, thanks to the bijection $F : A \to B$ with $F := \{(4, 2), (9, 3)\}$. The other bijection, $G := \{(4, 3), (9, 2)\}$, could also serve to prove that A and B have the same cardinality.

The following theorem forms the basis for the relation between the addition of natural numbers and the union of disjoint sets.

4.138 Theorem. *For all sets A, B, C, and D, if*

$$A \approx C,$$
$$B \approx D,$$
$$A \cap B = \varnothing,$$
$$C \cap D = \varnothing,$$

then

$$(A \dot\cup B) \approx (C \dot\cup D).$$

Proof. The hypotheses $A \approx C$ and $B \approx D$ mean that there exist bijections

$$F : A \to C,$$
$$G : B \to D.$$

Such bijections lead to a bijection

$$H : (A \dot\cup B) \to (C \dot\cup D)$$

defined by

$$H := (F \dot\cup G) \subseteq (A \dot\cup B) \times (C \dot\cup D)$$

so that

$$[(X, Y) \in H] \Leftrightarrow \begin{cases} Y = F(X) \text{ if } X \in A, \\ \\ Y = G(X) \text{ if } X \in B. \end{cases}$$

The relation H just defined is a function, because for each $X \in (A \dot\cup B)$ the relation H contains only one pair (X, Y). Indeed, thanks to $A \cap B = \varnothing$, either $X \in A$ and then $Y = F(X)$, or $X \in B$ and then $Y = G(X)$, but not both (definition 3.94).

To verify the injectivity of H, assume that $W \in (A \dot\cup B)$ and $X \in (A \dot\cup B)$ have the same image $H(W) = Y = H(X)$ in $C \cup D$. Because of $C \cap D = \varnothing$, either both images lie in C or both lie in D. In the first case, if both images lie in C, then $W \in A$ and $X \in A$, and then

$$F(W) = H(W) = H(X) = F(X),$$

whence $W = X$ by injectivity of F. In the second case, if both images lie in D, then $W \in B$ and $X \in B$, and then

$$G(W) = H(W) = H(X) = G(X),$$

whence $W = X$ by injectivity of G. To verify the surjectivity of H, assume that $Z \in (C \dot\cup D)$. Then either $Z \in C$ or $Z \in D$. In the first case, if $Z \in C$, then the surjectivity of F guarantees the existence of an element $W \in A$ such that

$$Z = F(W) = H(W).$$

In the second case, if $Z \in D$, then the surjectivity of G guarantees the existence of an element $X \in B$ such that

$$Z = G(X) = H(X).$$

Therefore, $H : (A \dot\cup B) \to (C \dot\cup D)$ is a bijection. $\square$

The following theorem forms the basis for the relation between multiplication of natural numbers and Cartesian products of sets.

4.139 Theorem. *For all sets A, B, C, and D, if*

$$A \approx C,$$
$$B \approx D,$$

then

$$(A \times B) \approx (C \times D).$$

Proof. The hypotheses $A \approx C$ and $B \approx D$ mean that there exist bijections

$$F : A \to C,$$
$$G : B \to D.$$

Such bijections lead to a bijection H defined by

$$H : (A \times B) \to (C \times D),$$
$$(W, X) \mapsto \big(F(W), G(X)\big).$$

The relation H is a function, because F and G are functions, so that for each $W \in A$ and each $X \in B$ there exists at most one $Y \in C$ and at most one $Z \in D$ with $(W, Y) \in F$ and $(X, Z) \in G$, so that there exists at most one $(Y, Z) \in C \times D$ with $\big((W, X), (Y, Z)\big) \in H$. To verify the injectivity of H, assume that $(W, X) \in (A \times B)$ and $(U, V) \in (A \times B)$ have the same image $H(W, X) = H(U, V)$ in $C \times D$:

$$H(W, X) = H(U, V) \text{ hypothesis,}$$
$$\updownarrow \text{ definition of } H,$$
$$(F(W), G(X)) = (F(U), G(V))$$
$$\updownarrow \text{ equality of pairs,}$$
$$[F(W) = F(U)] \wedge [G(X) = G(V)]$$
$$\updownarrow \text{ injectivity of } F \text{ and } G,$$
$$[W = U] \wedge [X = V]$$
$$\updownarrow \text{ equality of pairs.}$$
$$(W, X) = (U, V)$$

To verify the surjectivity of H, assume that $(R, S) \in (C \times D)$. Then the surjectivity of F guarantees the existence of an element $W \in A$ such that

$$R = F(W),$$

and the surjectivity of G guarantees the existence of an element $X \in B$ such that

$$S = G(X).$$

Consequently,

$$(R, S) = (F(W), G(X)) = H(W, X).$$

Therefore, $H : (A \times B) \to (C \times D)$ is a bijection.   □

4.7.2 Finite Sets

The following definition establishes the concept of cardinality for finite sets.

4.140 Definition. For each set S, the set S is **finite** if and only if there exists $N \in \mathbb{N}$ and a bijection $F : N \to S$. Such a natural number N is then called the **number of elements in S,** or the **cardinality** of S, which is denoted by $\#(S)$, $|S|$, or $\bar{\bar{S}}$.

4.141 Example. For each natural number $N \in \mathbb{N}$ the set N is finite and has N elements, because the identity function $I_N : N \to N$, $K \mapsto K$ is a bijection.

4.142 Remark. Because every bijection has an inverse function, a set S is finite if and only if there exist a natural number $N \in \mathbb{N}$ and a bijection $G : S \to N$, for instance, the inverse function $G := F^{\circ -1}$ for any bijection $F : N \to S$.

The following theorem shows that the insertion of a new element into a set corresponds to the arithmetic addition of 1 to its cardinality.

4.143 Theorem. *The equality $\#(A \dot\cup \{Z\}) = [\#(A)] + 1$ holds for each finite set A, and for each set $Z \notin A$.*

Proof. For each finite set A there is a natural number N and a bijection $F : N \to A$.

Moreover, for each set Z there exists a bijection of singletons $G : \{N\} \to \{Z\}$.

Consequently, because $A \cap \{Z\} = \varnothing$ by hypothesis, and because $N \cap \{N\} = \varnothing$ by theorem 4.42, it follows that theorem 4.138 gives a bijection $N + 1 \to A \dot\cup \{Z\}$:

$$H = (F \dot\cup G) : N + 1 = (N \dot\cup \{N\}) \to (A \dot\cup \{Z\}).$$

      □

The following theorem shows that the cardinality of the union of two disjoint finite sets equals the arithmetic sum of their two cardinalities.

4.144 Theorem. *The equality $\#(A \dot\cup B) = [\#(A)] + [\#(B)]$ holds for all disjoint finite sets A and B.*

Proof. This proof proceeds by induction with the cardinality of the second set.

If $\#(B) = 0$, then $B = \varnothing$ by definition, whence for each finite set A,

$$\#(A \dot\cup B) = \#(A \dot\cup \varnothing) = \#(A) = [\#(A)] + 0 = [\#(A)] + [\#(B)].$$

Hence, assume that there exists a natural number $N \in \mathbb{N}$ for which the theorem holds, so that the equality $\#(A \dot\cup B) = [\#(A)] + [\#(B)]$ holds for all disjoint finite sets A and B with $\#(B) = N$. For each set C with $N + 1$ elements, there exists a bijection $F : N + 1 \to C$. Consequently, the subset $B := F''(N) = F''(\{0, \ldots, N - 1\})$ has N elements, because the restriction $F|_N : N \to B$ is a bijection. Hence, with the element $Z := F(N)$, it follows that $C = B \dot\cup \{Z\}$ with $Z \notin B$, whence

$$
\begin{aligned}
&\#(A \dot\cup C) \\
&\quad = \text{because } C = B \dot\cup \{Z\}, \\
&\#(A \dot\cup [B \dot\cup \{Z\}]) \\
&\quad = \text{associativity of } \cup, \\
&\#[(A \dot\cup B) \dot\cup \{Z\}] \\
&\quad = \text{theorem 4.143}, \\
&\#(A \dot\cup B) + \#(\{Z\}) \\
&\quad = \text{induction hypothesis}, \\
&[\#(A) + \#(B)] + \#(\{Z\}) \\
&\quad = \text{associativity of } +, \\
&\#(A) + [\#(B) + \#(\{Z\})] \\
&\quad = \text{theorem 4.143}, \\
&\#(A) + \#(C).
\end{aligned}
$$

$\square$

The following two theorems confirm that every subset of a finite set is also finite.

4.145 Theorem. *For each $N \in \mathbb{N}$, every subset $S \subseteq N$ is also a finite set, with at most N elements. Moreover, each proper subset $S \subset N$ has fewer than N elements.*

Proof. This proof proceeds by induction with N.

If $N := 0$, then $N = \varnothing$, and the only subset $S \subseteq N$ is $S = \varnothing$, which is finite.

As an induction hypothesis, assume that there exists a natural number $K \in \mathbb{N}$ such that the theorem holds for $N := K$, so that each subset $S \subseteq K$ is finite with at most K elements, and that each proper subset $S \subset K$ has fewer than K elements. Hence, consider a subset $R \subseteq K + 1$. Two cases arise: either $K \notin R$ or $K \in R$.

If $K \notin R$, then $R \subseteq [(K + 1) \setminus \{K\}] = K$ whence R is finite with at most K elements by induction hypothesis.

If $K \in R$, then the set $C := R \setminus \{K\}$ is a subset of $[(K + 1) \setminus \{K\}] = K$, whence C is finite with at most K elements by induction hypothesis. Thus, there exists a natural number $L \leq K$ such that $C = R \setminus \{K\}$ has cardinality $L \leq K$, and $L = K$ if and only if $C = K$. Hence, theorem 4.143 shows that $R = C \dot{\cup} \{K\}$ is finite with cardinality $L + 1 \leq K + 1$, and $L + 1 = K + 1$ if and only if $R = K + 1$.   □

4.146 Theorem. *Every subset of a finite set is also finite, with at most as many elements.*

Proof. For each finite set A there exists a natural number $N \in \mathbb{N}$ and a bijection $F : A \to N$. Hence, for each subset $B \subseteq A$, the restriction $F|_B : B \to N$ is a bijection from B onto a subset $S := F''(B) \subseteq N$. Because every such subset $S \subseteq N$ is also finite with at most N elements, it follows that there exists a natural number $L \leq N$ and a bijection $G : S \to L$. Consequently, the composition $G \circ F|_B$ establishes a bijection from B onto L, which means that B is finite with $L \leq N$ elements.   □

4.147 Theorem. *The equality* $\#(A \setminus B) = [\#(A)] - [\#(B)]$ *holds for every subset* $B \subseteq A$ *of every finite set A.*

Proof. The result follows from the disjointness of B and $A \setminus B$, and from theorem 4.146, which ensures that both B and $A \setminus B$ are finite:

$$B \cap (A \setminus B) = \varnothing,$$
$$B \cup (A \setminus B) = A,$$
$$[\#(B)] + [\#(A \setminus B)] = \#(A),$$

whence $\#(A \setminus B) = [\#(A)] - [\#(B)]$ by definition (4.59) of subtraction.   □

Theorem 4.145 shows that for each subset $S \subseteq N$ there is a bijection $F : S \to L$ with $L \leq N$, but it does not yet prevent the existence of other bijections $G : S \to L$ with $L > N$. The following theorem confirms that there is no such bijection.

4.148 Theorem. *For all $K, N \in \mathbb{N}$ with $K < N$, there are no injections $F : N \to K$.*

Proof. This proof uses induction with N. For $N := 1$, the only smaller natural number is $K := 0$, and there exists no function $F : \{1\} \to \varnothing$, whence no injection either.

As an induction hypothesis, assume that there exists a positive integer $L \in \mathbb{N}^*$ such that the theorem holds for $N := L$, so that for every natural number $K < L$ there are no injections $F : L \to K$. The proof that for every $K < L + 1$ there are no injections $F : (L + 1) \to K$ proceeds by contraposition. Thus, assume that there is such an injection $F : (L + 1) \to K$. Let $Z := F(L)$ and $S := K \setminus \{Z\}$. By theorem 4.147, S is a finite set with $K - 1$ elements, and there is a bijection $G : S \to (K - 1)$. Then the restriction $F|_L : L \to S$ is an injection, and the composition $G \circ F|_L : L \to (K - 1)$ is an injection. Hence $K - 1 \geq L$ by induction hypotheses, whence $K \geq L + 1$.   □

The following theorem shows the equivalence of the concepts of injection, surjection, and bijection between sets with the *same finite* cardinality.

4.149 Theorem. *For all sets A and B with the* same finite *cardinality, and for each function F : A → B with domain $\mathscr{D}(F) = A$, the following conditions are mutually equivalent:*

(P)   *F is injective,*
(Q)   *F is surjective,*
(R)   *F is bijective.*

Proof. If F is bijective, then F is also injective and surjective, because $(R) \Leftrightarrow [(P) \wedge (Q)]$ by definition of bijectivity (definition 3.118); therefore both $(R) \Rightarrow (P)$ and $(R) \Rightarrow (Q)$ hold, by theorems 1.53 and 1.54.

For the reverse implications, because A and B have the same finite cardinality, there exist a natural number $N \in \mathbb{N}$ and a bijection $G : N \to A$.

If F is injective, then $F''(A) \subseteq B$ is a finite set with cardinality $L \leq N$, so that there exists a bijection $H : F''(A) \to L$. Then the composition

$$H \circ F \circ G : N \xrightarrow{G} A \xrightarrow{F} F''(A) \xrightarrow{H} L$$

is an injection, whence $L \geq N$ by theorem 4.148. From $L \leq N$ and $N \leq L$ it follows that $L = N$. Hence H, G, and $(H \circ F \circ G) : N \to N$ are bijections, whence F is also surjective, whence bijective, which proves the implication $(P) \Rightarrow (Q)$; therefore $(P) \Rightarrow [(P) \wedge (Q)]$ holds (by theorem 1.82), whence also $(P) \Rightarrow (R)$.

The proof of the converse, $(Q) \Rightarrow (P)$, uses the contraposition $[\neg(P)] \Rightarrow [\neg(Q)]$. If F is *not* injective, then A contains two distinct elements $X \neq Z$ such that $F(X) = F(Z)$. Let $S := A \setminus \{Z\}$, so that $F''(A) = F''(S)$ and there exists a bijection $J : (N - 1) \to S$, by theorem 4.147. Also, there is a bijection $I : B \to N$, because A and B have the same cardinality, N. Then the composition

$$I \circ F \circ J : (N - 1) \xrightarrow{J} S \xrightarrow{F} F''(S) \xrightarrow{I} N$$

cannot be a bijection, because by theorem 4.148 its inverse could not be an injection from N to $N - 1$. Hence F cannot be surjective, which proves $(Q) \Rightarrow (P)$.   □

The following theorem shows that the number of elements in a Cartesian product equals the product of the numbers of elements in its factors.

4.150 Theorem. *For all $K, L \in \mathbb{N}$, the Cartesian product $K \times L$ has $K * L$ elements.*

Proof. By induction with L, if $L = 0$, then $L = \varnothing$, whence for every $K \in \mathbb{N}$

$$K \times 0 = K \times \varnothing = \varnothing,$$
$$\#(K \times 0) = \#(\varnothing) = 0 = K * 0.$$

As induction hypothesis, assume that there exists a natural number $M \in \mathbb{N}$ such that the theorem holds for $L := M$, so that the equality

$$\#(K \times M) = K * M$$

holds for every $K \in \mathbb{N}$. Hence, from the disjoint union $M + 1 = M \dot\cup \{M\}$, and from the distributivity of Cartesian products over unions (theorem 3.63), it follows that

$$K \times (M + 1) = K \times (M \dot\cup \{M\})$$
$$= (K \times M) \dot\cup (K \times \{M\}),$$

$$\#[K \times (M + 1)] = \#[(K \times M) \dot\cup (K \times \{M\})]$$
$$= [\#(K \times M)] + [\#(K \times \{M\})]$$
$$= (K * M) + K$$
$$= K * (M + 1),$$

thanks to the disjoint union $(K \times M) \dot\cup (K \times \{M\})$.   □

4.7.3   Exercises on Finite Sets

4.111 . Determine $\#(\varnothing)$.

4.112 . Determine $\#[\mathscr{P}(\varnothing)]$.

4.113 . Determine $\#\big(\mathscr{P}[\mathscr{P}(\varnothing)]\big)$.

4.114 . Determine $\#\big[\mathscr{P}\big(\mathscr{P}[\mathscr{P}(\varnothing)]\big)\big]$.

4.115 . Determine $\#\Big(\mathscr{P}\big[\mathscr{P}\big(\mathscr{P}[\mathscr{P}(\varnothing)]\big)\big]\Big)$.

4.116 . Determine $\#\Big[\mathscr{P}\big(\big\{\varnothing, \{\varnothing\}, \{\varnothing, \{\varnothing\}\}\big\}\big)\Big]$.

4.117 . Prove or disprove that all ordered pairs have the same cardinality.

4.118 . For all finite sets A and B, prove that $(A \times B) \approx (B \times A)$.

4.119 . For all finite sets A, B, C, prove that $[(A \times B) \times C] \approx [A \times (B \times C)]$.

4.120 . For all finite sets A and B, prove $[\#(A \triangle B)] = [\#(A \cup B)] - [\#(A \cap B)]$.

4.8   Infinite Cardinality

4.8.1   Infinite Sets

Some sets are *not* finite, for they do *not* admit any bijection onto any natural number.

4.151 Definition (infinite sets). A set Z is **infinite** if and only if Z is **not finite**, which means that there are no bijections from Z onto any natural number.

For instance, the set $\mathbb{N}$ is infinite.

4.152 Theorem. *The set $\mathbb{N}$ of all natural numbers is* not *finite.*

Proof. For each natural number N and for each function $F : \mathbb{N} \rightarrow N$, the restriction $F|_{N+1}$ $(N + 1) \rightarrow N$ cannot be injective, by theorem 4.148, whence F cannot be injective. Therefore, there are no such bijections, which means that $\mathbb{N}$ is *not* finite.

□

As there exist finite sets with different cardinalities, there also exist infinite sets with different cardinalities. For instance, the following considerations lead to infinite sets with cardinalities different from the cardinality of the natural numbers.

4.153 Definition. For all sets X, Y, let Y^X denote the set of all functions from X to Y, with domain X.

4.154 Example. If $X := \varnothing$, then $2^X = \{\varnothing\}$, because $\varnothing : \varnothing \rightarrow \{0, 1\}$ is the only function from $X = \varnothing$ to $2 = \{0, 1\}$.

4.155 Example. If $X := \{S\}$ is a singleton, then 2^X consists of two elements, because there are exactly two functions from $X = \{S\}$ to $2 = \{0, 1\}$:

$$F : \{S\} \rightarrow \{0, 1\},$$
$$S \mapsto 0;$$

$$G : \{S\} \rightarrow \{0, 1\},$$
$$S \mapsto 1.$$

Thus, $2^{\{S\}} = \{F, G\}$ has exactly two elements, F and G.

The next theorem shows that each set X is "strictly smaller" than 2^X.

4.156 Theorem. *For each set X, there exists an injection from X to 2^X. Yet there does* not *exist any surjection from X to 2^X.*

Proof. To establish the existence of an injection $X \hookrightarrow 2^X$, consider the function $J : X \hookrightarrow 2^X$ defined as follows. The function J maps each element $N \in X$ to a function $J(N) : X \rightarrow 2$ specified by

$$[J(N)](K) := \begin{cases} 1 \text{ if } K = N, \\ 0 \text{ if } K \neq N. \end{cases}$$

In other words, the function $J(N)$ is the characteristic function $\chi_{\{N\}}$ of the singleton $\{N\}$ (from example 3.89). Consequently, J is injective; indeed if $M \neq N$, then $[J(M)](M) = 1$ but $[J(N)](M) = 0$, whence $J(M) \neq J(N)$.

To prove the absence of any surjection $X \twoheadrightarrow 2^X$, for each function $J : X \rightarrow 2^X$, this proof demonstrates a method known as **Cantor's diagonalization** to show that J is not surjective. Each such function $J : X \rightarrow 2^X$ maps each element $N \in X$ to a function $J(N) \in 2^X$, so that $J(N) : X \rightarrow 2$. In particular, $J(N)$ maps N to an element $[J(N)](N) \in 2 = \{0, 1\}$. Thus, define a function $F : X \rightarrow 2$ by

$$F(N) := \begin{cases} 0 \text{ if } [J(N)](N) = 1, \\ 1 \text{ if } [J(N)](N) = 0. \end{cases}$$

Thus, $F(N) \neq [J(N)](N)$ for every $N \in X$, whence $F \neq [J(N)]$ for every $N \in X$. Consequently $F \notin J''(X)$, whence J is not surjective.   □

4.157 Example. One among several methods to define the set $\mathbb{R}$ of all **real numbers** consists in defining $\mathbb{R}$ as a set of infinite sequences of digits [118, p. 565–566]. Thus the set of all real numbers between 0 and 1 can be defined as the set of all sequences $R \in 3^{\mathbb{N}}$ (subject to the constraint that there does *not* exist any $K \in \mathbb{N}$ such that $R(N) = 2$ for every $N \geq K$). Then every function $J : \mathbb{N} \to \mathbb{R}$ fails to be surjective. Indeed, as in the proof of theorem 4.156, for each function $J : \mathbb{N} \to \mathbb{R}$ define a function $G : \mathbb{N} \to \mathbb{R}$ by

$$G(N) := \begin{cases} 1 \text{ if } [J(N)](N) = 2, \\ 0 \text{ if } [J(N)](N) = 1. \\ 1 \text{ if } [J(N)](N) = 0. \end{cases}$$

Thus, $G \in \mathbb{R}$ but $G(N) \neq [J(N)](N)$ for every $N \in \mathbb{N}$, whence $G \neq [J(N)]$ for every $N \in \mathbb{N}$. Consequently $G \notin J''(\mathbb{N})$, and therefore J is not surjective. Hence there does not exist any bijection $J : \mathbb{N} \to \mathbb{R}$.

Example 4.157 reveals that the set $\mathbb{R}$ of all real numbers is "more infinite" than the set $\mathbb{N}$ of all natural numbers. Moreover, applying theorem 4.156 to $X := \mathbb{R}$ shows that $2^{\mathbb{R}}$ is also "more infinite" than $\mathbb{R}$. Then $2^{(2^{\mathbb{R}})}$ is also "more infinite" than $2^{\mathbb{R}}$. And so forth, thus there exists an "infinite" variety of "infinite" sets.

4.8.2 Denumerable Sets

There exist several infinite sets that have the same cardinality as $\mathbb{N}$ has. For instance, using only addition and multiplication from integer arithmetic, this subsection presents a proof that the set of all nonnegative integers $\mathbb{N}$ and the set of all integers $\mathbb{Z}$ have the same cardinality; similarly, $\mathbb{N}$ and the Cartesian product $\mathbb{N} \times \mathbb{N}$ have the same cardinality. The following terminology conforms to [8, p. 152], [30, p. 47], and [128, p. 151].

4.158 Definition. A set is **denumerable** — or has cardinality $\aleph_0$ (read "aleph zero") — if and only if it has the same cardinality as the set $\mathbb{N}$ of all natural numbers.
  A set is **countable** if and only if it is either finite or denumerable.

The following theorem shows that the set $\mathbb{Z}$ of *all* the integers has the same cardinality as the set $\mathbb{N}$ of all the nonnegative integers.

4.159 Theorem. *The sets $\mathbb{N}$ and $\mathbb{Z}$ have the same cardinality.*

Proof. Define $F : \mathbb{N} \to \mathbb{Z}$ by

$$F(N) := \begin{cases} N/2 & \text{if there exists } K \in \mathbb{N} \text{ with } N = 2 * K, \\ -(N+1)/2 & \text{if there exists } K \in \mathbb{N} \text{ with } N + 1 = 2 * K. \end{cases}$$

The function F is surjective. Indeed, if $L \in \mathbb{N}$, then $L = F(2 * L)$. Similarly, if $L \in \mathbb{Z} \setminus \mathbb{N}$, then $L = F([2 * L] + 1)$.

The function F is also injective. Indeed, if $F(M) = F(N)$, then either both or neither of $F(M)$ and $F(N)$ are elements of $\mathbb{N}$. If $F(M) \in \mathbb{N}$ and $F(N) \in \mathbb{N}$, then $M/2 = F(M) = F(N) = N/2$, whence $M = N$. If $F(M) \notin \mathbb{N}$ and $F(N) \notin \mathbb{N}$, then $-(M+1)/2 = F(M) = F(N) = -(N+1)/2$, whence $M = N$. $\square$

4.160 Theorem. *For all disjoint denumerable sets A and B, $A \dot\cup B$ is denumerable.*

Proof. By the hypotheses there exist bijections $I : A \to \mathbb{N}$ and $J : B \to \mathbb{N}$. The function $K : \mathbb{N} \to \mathbb{Z}$ with $K(N) := -(N+1)$ is injective, and hence the composition $K \circ I$ is also injective. Therefore, the function $G := (K \circ I) \dot\cup J$ is a bijection from $A \dot\cup B$ to $\mathbb{Z}$. Hence $F^{\circ -1} \circ G : A \dot\cup B \to \mathbb{N}$ is a bijection, with F as in theorem 4.159. $\square$

To facilitate the proof that $\mathbb{N}$ and the Cartesian product $\mathbb{N} \times \mathbb{N}$ have the same cardinality, the following definition specifies an inductive method to define and compute the sum $1 + 2 + \cdots + (N - 1) + N$, known as an **arithmetic series**.

4.161 Definition. Define a function $T : \mathbb{N} \to \mathbb{N}$ by

$$T(0) := 0,$$
$$T(N + 1) := T(N) + (N + 1).$$

Also, define the notation $0 + 1 + \cdots + N := T(N)$.

Thus,

$$T(0) := 0,$$
$$T(1) := T(0) + (0 + 1) = 0 + (0 + 1) = 1,$$
$$T(2) := T(1) + (1 + 1) = 1 + (1 + 1) = 3,$$
$$T(3) := T(2) + (2 + 1) = 3 + (2 + 1) = 6,$$
$$\vdots$$

The values of T are called **triangular numbers** because they correspond to the number of elements in the following patterns:

$$T(0) \quad T(1) \qquad T(2) \qquad\qquad T(3) \qquad \cdots$$

The following theorem provides a different formula to compute the same function T.

4.162 Theorem (arithmetic series). *For each natural number $N \in \mathbb{N}$,*

$$0 + 1 + 2 + \cdots + (N-1) + N = T(N) = \frac{N * (N+1)}{2}.$$

Proof. This proof uses induction with N. If $N := 0$, then $T(0) = 0 = 0 * (0+1)/2$. Assume that there exists $M \in \mathbb{N}$ such that the theorem holds for $N := M$, so that $T(M) \in \mathbb{N}$ and $2 * T(M) = M * (M+1)$. Then $T(M+1) = T(M) + (M+1) \in \mathbb{N}$, and

$$\begin{aligned}
2 * T(M+1) &= 2 * [T(M) + (M+1)] \\
&= [2 * T(M)] + [2 * (M+1)] \\
&= M * (M+1) + 2 * (M+1) \\
&= (M+2) * (M+1) \\
&= (M+1) * [(M+1) + 1].
\end{aligned}$$

Consequently, $2 * T(M+1) = (M+1) * [(M+1) + 1]$, but $T(M+1) \in \mathbb{N}$, whence $(M+1) * [(M+1) + 1]/2 \in \mathbb{N}$ and $T(M+1) = (M+1) * [(M+1) + 1]/2$. $\square$

An alternative proof of the same formula proceeds along the following outline:

$$\begin{aligned}
T(N) &= & 0 &+ & 1 &+ \cdots + & (N-1) + & N, \\
T(N) &= & N &+ & (N-1) &+ \cdots + & 1 &+ & 0, \\
T(N) + T(N) &= & (N+1) &+ & (N+1) &+ \cdots + & (N+1) + & (N+1),
\end{aligned}$$

whence $2 * T(N) = N * (N+1)$. However, a proof along this outline also requires induction to rearrange the terms of the sum with associativity and commutativity.

The following definition provides a formula for **Cantor's diagonal enumeration** of $\mathbb{N} \times \mathbb{N}$, and the subsequent theorems will verify that indeed it enumerates $\mathbb{N} \times \mathbb{N}$.

4.163 Definition. Define a function $\mathscr{T} : (\mathbb{N} \times \mathbb{N}) \to \mathbb{N}$ by

$$\mathscr{T}(M, N) := M + T(M+N)$$

$$= M + \frac{(M+N) * (M+N+1)}{2}.$$

The value $\mathscr{T}(M,N)$ corresponds to the sum of the number of elements in the triangular pattern counted by the "triangular number" $T(M+N)$ and a last partial row with M elements (instead of a complete last row of $M+N$ elements for the next triangular number). For example, with $M := 1$ and $N := 2$,

$$\mathscr{T}(1,2)$$

$$T(M+N)$$

$$M$$

The following theorem shows that the function $\mathscr{T} : (\mathbb{N} \times \mathbb{N}) \to \mathbb{N}$ is surjective.

4.164 Theorem. *For each $I \in \mathbb{N}$ there exist $M \in \mathbb{N}$ and $N \in \mathbb{N}$ such that*

$$I = M + \frac{(M+N) * (M+N+1)}{2}.$$

Proof. This proof proceeds by induction with I.

First, if $I := 0$, then $I = 0 + T(0+0)$ with $M := 0$ and $N := 0$.

Second, assume that there exists $K \in \mathbb{N}$ such that the theorem holds for $I := K$, so that there exist $M, N \in \mathbb{N}$ with $K = M + T(M+N)$.

If $N > 0$, then $K+1 = 1 + [M + T(M+N)] = (M+1) + T([M+1]+[N-1])$.

If $N = 0$, then $K + 1 = 1 + [M + T(M+0)] = (M+1) + T(M+0) = 0 + T(0 + [M+1])$. □

The following theorem shows that the function $\mathscr{T} : (\mathbb{N} \times \mathbb{N}) \to \mathbb{N}$ is injective.

4.165 Theorem. *For all $K, L, M, N \in \mathbb{N}$, if $K + T(K+L) = M + T(M+N)$, then both $K = M$ and $L = N$.*

Proof. If

$$K + T(K+L) = M + T(M+N),$$

then subtracting M and $T(K+L)$ from both sides gives

$$K - M = T(M+N) - T(K+L).$$

If $K > M$, then $K - M > 0$, so that the left-hand side is positive, but then the right-hand side must also be positive: $T(M+N) - T(K+L) > 0$. Hence $M+N > K+L$, but then

$$T(M+N) - T(K+L) \geq T(M+N) - T(M+N-1) = M+N$$

by definition of T. Thus

$$K - M = T(M + N) - T(K + L) \geq M + N$$

whence $K \geq (2 * M) + N$. With $M + N > K + L$, this gives

$$M + N > K + L \geq [(2 * M) + N] + L$$

whence subtracting $(2 * M) + N$ from all sides gives

$$-M \geq L,$$

which contradicts the hypothesis that $L \geq 0$. □

The following theorem confirms that $\mathbb{N} \times \mathbb{N}$ is denumerable.

4.166 Theorem. *The function $\mathscr{T} : (\mathbb{N} \times \mathbb{N}) \to \mathbb{N}$ is bijective.*

Proof. The bijectivity results from theorems 4.165 and 4.164. □

The computation of the inverse function $\mathscr{T}^{-1} : \mathbb{N} \to \mathbb{N} \times \mathbb{N}$ can proceed according to the straightforward algorithm provided by the proof of theorem 4.164: for each $I \in \mathbb{N}$, observe that $T(0) = 0 \leq I$, and compute $T(0), T(1), \ldots, T(L)$ until $T(L - 1) \leq I < T(L)$. Then let $M := I - T(L - 1)$ and $N := (L - 1) - M$.

4.8.3 The Bernstein–Cantor–Schröder Theorem

The following theorem guarantees the existence of a bijection between two sets, provided that there exist injections from one set to the other and vice versa. According to Suppes [128, p. 95], Cantor conjectured the theorem and then Bernstein and Schröder proved it independently of each other in the 1890s. Fraenkel [37, p. 102] credits the following proof to J. M. Whitaker.

4.167 Theorem (Bernstein–Cantor–Schröder). *For all sets A and B, if there are injections $F : A \to B$ and $G : B \to A$, then there is a bijection $H : A \to B$.*

Proof. The strategy of this proof consists in producing a subset $E \subseteq A$ such that

$$G''[B \setminus F''(E)] = (A \setminus E)$$

and then in setting

$$H := (F|_E) \,\dot{\cup}\, \left(G^{\circ -1}|_{A \setminus E} \right).$$

To this end, define

$$\mathscr{D} := \{C \subseteq A : G''[B \setminus F''(C)] \subseteq (A \setminus C)\}$$
$$= \{C \subseteq A : C \subseteq A \setminus G''[B \setminus F''(C)]\};$$

the proof will verify that $\bigcup \mathscr{D}$ satisfies the requirements.

First, for all subsets $V, W \subseteq A$, if $V \subseteq W$, then $F''(V) \subseteq F''(W)$, whence $[B \setminus F''(W)] \subseteq [B \setminus F''(V)]$, and hence $G''[B \setminus F''(W)] \subseteq G''[B \setminus F''(V)]$, so that

$$\{A \setminus G''[B \setminus F''(V)]\} \subseteq \{A \setminus G''[B \setminus F''(W)]\}.$$

In particular, if $V \in \mathscr{D}$, then $V \subseteq A \setminus G''[B \setminus F''(V)]$ by definition of $\mathscr{D}$, and $V \subseteq W := \bigcup \mathscr{D}$ by definition of $\bigcup \mathscr{D}$; consequently

$$V \subseteq \{A \setminus G''[B \setminus F''(V)]\} \subseteq \left\{A \setminus G''\left[B \setminus F''\left(\bigcup \mathscr{D}\right)\right]\right\}.$$

Because these inclusions hold for every element $V \in \mathscr{D}$, it follows that they also hold for their union:

$$\bigcup \mathscr{D} \subseteq \left\{A \setminus G''\left[B \setminus F''\left(\bigcup \mathscr{D}\right)\right]\right\}.$$

Let

$$E := A \setminus G''\left[B \setminus F''\left(\bigcup \mathscr{D}\right)\right].$$

From $\bigcup \mathscr{D} \subseteq E$ it follows that

$$A \setminus G''\left[B \setminus F''\left(\bigcup \mathscr{D}\right)\right] \subseteq \{A \setminus G''[B \setminus F''(E)]\},$$

so that

$$E \subseteq \{A \setminus G''[B \setminus F''(E)]\}$$

whence $E \in \mathscr{D}$. Consequently, $E \subseteq \bigcup \mathscr{D}$, whence $E = \bigcup \mathscr{D}$, but then the definition of E gives

$$E = A \setminus G''[B \setminus F''(E)].$$

$\square$

The next theorem shows a use of the Bernstein–Cantor–Schröder Theorem.

4.168 Theorem.   *There exists a bijection* $H : \mathbb{Z} \to \mathbb{Q}$.

Proof.   First, there exists an injection $F : \mathbb{Z} \to \mathbb{Q}$ with $F(N) := N/1$.

Second, there exists an injection $I : \mathbb{Q} \to (\mathbb{Z} \times \mathbb{Z})$ such that $I(P/Q) := (P, Q)$ with $P \in \mathbb{N}$ minimum (theorem 4.99).

Also, there exists an injection $J : (\mathbb{Z} \times \mathbb{Z}) \to \mathbb{Z}$, by theorem 4.166.

Consequently, the composition $G := J \circ I$ is an injection $\mathbb{Q} \to \mathbb{Z}$.

Therefore, the Bernstein–Cantor–Schröder Theorem guarantees the existence of a bijection $H : \mathbb{Z} \to \mathbb{Q}$.   □

Theorem 4.169 shows that every infinite subset of a denumerable set is denumerable. Hence every subset of a countable set is also countable.

4.169 Theorem. *For each* infinite *subset* $S \subseteq \mathbb{N}$, *there exists an injection* $H : \mathbb{N} \to S$. *Hence* S *is denumerable.*

Proof. Define $G : \mathscr{P}(S) \to \mathscr{P}(S)$ by $G(S) := S$ and $G(B) := B \cup \{\min(S \setminus B)\}$ for every $B \neq S$.

Define $F : \mathbb{N} \to \mathscr{P}(S)$ by $F(0) := \{\min(S)\}$, and $F(N + 1) := G[F(N)]$. Thus $F(N + 1)$ inserts the smallest element of S that has not yet been included by $F(0), \ldots, F(N)$.

Let $H(M) := \max[F(M)]$.

Also, the inclusion function $\iota : S \to \mathbb{N}$ defined in example 3.93 by $\iota(X) = X$ is injective by example 3.121.

Hence there exists a bijection between S and $\mathbb{N}$, by the Cantor–Bernstein–Schröder Theorem (4.167).   □

Theorem 4.170 shows that every subset of the range of a function defined on a countable domain is also countable.

4.170 Theorem. *For each set* E, *if there exists a surjection* $Q : \mathbb{N} \to E$, *then* E *is countable.*

Proof. If E is finite, then E is countable, by definition.

If E is not finite, then define an injection $G : E \to \mathbb{N}$ by $G(X) := \min[Q^{\circ -1"}(\{X\})]$, so that $G(X)$ is the smallest natural number mapped to X by Q. The function G is well-defined, because Q is surjective, so that the pre-image $Q^{\circ -1"}(\{X\}) \neq \varnothing$ contains a unique smallest element. The function G is also injective; indeed, if $X \neq Y$, then $Q^{\circ -1"}(\{X\}) \cap Q^{\circ -1"}(\{Y\}) = \varnothing$, because Q is a function. Consequently, the image $S := G''(E) \subseteq \mathbb{N}$ is an infinite subset of $\mathbb{N}$.

By theorem 4.169 and the Bernstein–Cantor–Schröder Theorem (4.167), it follows that S and hence E are denumerable.   □

4.8.4 Denumerability of all Finite Sequences of Natural Numbers

This subsection shows that the set of all finite sequences of natural numbers is denumerable.

Theorem 4.171 shows that allowing the number of variables to change from one function to another still produces a *set* of functions.

4.171 Theorem. *For all sets E and C, all the functions $F : E^K \to C$ of any finite number of variables from the domain E and with values in the co-domain C form a set*

$$\left\{ C^{(E^K)} : K \in \mathbb{N} \right\}.$$

Proof. Let $A := \mathbb{N}$, $B := E$, and $\mathscr{D} := \mathbb{N}$. Thus $\mathscr{D} = \mathbb{N} \subseteq \mathscr{P}(\mathbb{N}) = \mathscr{P}(A)$ because $K \subset \mathbb{N}$ for each $K \in \mathbb{N}$, by theorem 4.20. Then, with the notation of theorem 3.136,

$$\mathscr{F}_{\mathbb{N},E} := \left\{ E^K : K \in \mathbb{N} \right\}$$

is the set of all finite sequences of elements of E.

Next, again with the notation of theorem 3.136, let $A := \bigcup \mathscr{F}_{\mathbb{N},E}$. Then $E^K \in \mathscr{F}_{\mathbb{N},E}$ whence $E^K \subseteq A$ for each $K \in \mathbb{N}$, by theorem 3.40, and hence $\mathscr{F}_{\mathbb{N},E} \subseteq \mathscr{P}(A)$. Also let $\mathscr{D} := \mathscr{F}_{\mathbb{N},E}$ and $B := C$, the set C in the hypothesis of the theorem. Thus $C^{(E^K)}$ is the set of all functions $F : E^K \to C$ defined on all of E^K, with $E^K \in \mathscr{D}$ for each $K \in \mathbb{N}$, and

$$\mathscr{F}_{\mathscr{D},C} := \left\{ C^{(E^K)} : K \in \mathbb{N} \right\}$$

is the *set* of all functions defined on any E^K into C.

4.172 Theorem. *For each $K \in \mathbb{N}^*$, the set $\mathbb{N}^K$ of all finite natural sequences of length K is denumerable.*

Proof. Following [86, p. 799], this proof proceeds by induction with K.

For $K = 1$, the identity function $I_\mathbb{N} : \mathbb{N} \to \mathbb{N}$, $N \mapsto N$ defined in example 3.86 is a bijection, by example 3.120.

For $K = 2$, Cantor's diagonal enumeration $\mathscr{T} : \mathbb{N} \times \mathbb{N} \to \mathbb{N}$ from definition 4.163 is a bijection, by theorem 4.166.

Denote the identity function $I_\mathbb{N}$ by $\mathscr{T}_1$, and Cantor's diagonal enumeration $\mathscr{T}$ by $\mathscr{T}_2$.

As an induction hypothesis, assume that for some $K = M$ there exists a bijection $\mathscr{T}_M : \mathbb{N}^M \to \mathbb{N}$.

For the induction step from $K = M$ to $K = M + 1$, define $\mathscr{T}_{M+1} : \mathbb{N}^{M+1} \to \mathbb{N}$ by

$$\mathscr{T}_{M+1}(N_0, \ldots, N_{M-1}, N_M) := \mathscr{T}_2[\mathscr{T}_M(N_0, \ldots, N_{M-1}), N_M].$$

Equivalently, $\mathscr{T}_{M+1} = \mathscr{T}_2 \circ (\mathscr{T}_M \boxtimes I_\mathbb{N})$, with $\boxtimes$ from definition 3.116, and where $\mathscr{T}_M$ is bijective by induction hypothesis, whence $\mathscr{T}_M \boxtimes I_\mathbb{N}$ is also bijective, by theorem 3.117, and hence the composition of the bijection $\mathscr{T}_2$ preceded by the bijection $\mathscr{T}_M \boxtimes I_\mathbb{N}$ is again bijective, by theorem 3.123. □

4.173 Theorem. *The set* $\bigcup_{K \in \mathbb{N}^*} \mathbb{N}^K$ *of all finite sequences of natural numbers is denumerable.*

Proof. For each $K \in \mathbb{N}^*$, denote by $\mathscr{T}_K : \mathbb{N}^K \to \mathbb{N}$ any enumeration of all natural sequences of length K, as in theorem 4.172. Again following [86, p. 799], this proof pieces together all the bijections $\mathscr{T}_K$ of $\mathbb{N}^K$ by means of K and $\mathscr{T}_2$. Specifically, define $\mathscr{S} : \bigcup_{K \in \mathbb{N}^*} \mathbb{N}^K \to \mathbb{N}$ by

$$\mathscr{S}(N_0, \ldots, N_{K-1}) := \mathscr{T}_2[\mathscr{T}_K(N_0, \ldots, N_{K-1}), K].$$

The map $\mathscr{S}$ is injective: indeed, if $\mathscr{S}(N_0, \ldots, N_{K-1}) = \mathscr{S}(J_0, \ldots, J_{L-1})$, then $\mathscr{T}_2[\mathscr{T}_K(N_0, \ldots, N_{K-1}), K] = \mathscr{T}_2[\mathscr{T}_L(J_0, \ldots, J_{L-1}), L]$, whence $K = L$ and $\mathscr{T}_K(N_0, \ldots, N_{K-1}) = \mathscr{T}_L(J_0, \ldots, J_{L-1})$, because $\mathscr{T}_2$ is injective by theorem 4.165. Consequently, $\mathscr{T}_L(N_0, \ldots, N_{L-1}) = \mathscr{T}_L(J_0, \ldots, J_{L-1})$, whence $(N_0, \ldots, N_{L-1}) = (J_0, \ldots, J_{L-1})$, because $\mathscr{T}_L$ is injective, by theorem 4.172.

The map $\mathscr{S}$ is also surjective: indeed, for each $M \in \mathbb{N}$ there exists $(I, K) \in \mathbb{N}^2$ such that $\mathscr{T}_2(I, K) = M$, because $\mathscr{T}_2$ is surjective by theorem 4.164. Consequently, there exists $(N_0, \ldots, N_{K-1}) \in \mathbb{N}^K$ such that $\mathscr{T}_K(N_0, \ldots, N_{K-1}) = I$, because $\mathscr{T}_K$ is surjective, also by theorem 4.172. Thus $\mathscr{S}(N_0, \ldots, N_{K-1}) = M$.   □

4.8.5   Other Infinite Sets

The present text has defined a set to be finite if and only if there exists a bijection onto a natural number, and infinite if and only if there does not exist any such bijection. There exists a *different* definition of infinite sets, called "Dedekind-infinite" [128, p. 107], corresponding to Dedekind's definition [25, §V, #64, p. 63].

4.174 Definition. A set Z is **Dedekind-infinite** if and only if there exists a proper subset $Y \subsetneqq Z$ and a bijection $F : Z \to Y$, or, equivalently, $F^{\circ -1} : Y \to Z$.

Thus, a set is Dedekind-infinite if and only if it has the same cardinality as that of one of its proper subsets.

4.175 Example. The set $\mathbb{N}$ is Dedekind-infinite because there is a proper subset $\mathbb{N}^* \subset \mathbb{N}$ and a bijection defined by the successor function:

$$G : \mathbb{N}^* \to \mathbb{N},$$
$$X \mapsto X \cup \{X\}.$$

4.176 Theorem. *Every Dedekind-infinite set is also infinite.*

Proof. This proof proceeds by contradiction. If a set Z is *not* infinite, in other words, if Z is *finite*, then there exist a natural number $N \in \mathbb{N}$ and a bijection $G : N \to Z$.

If Z is also Dedekind-infinite , then there exists a proper subset $Y \subsetneqq Z$ and a bijection $F : Z \to Y$. Then the composition

$$H := G^{\circ-1} \circ F \circ G : N \overset{G}{\to} Z \overset{F}{\to} Y \overset{G^{\circ-1}}{\to} N$$

would be a bijection from N onto a proper subset $G^{\circ-1}"(Y) \subsetneqq N$, which would contradict theorem 4.148.   □

The proof of the converse requires the Axiom of Choice (exercises 4.139, 4.140).

4.177 Theorem. *Every infinite set contains a denumerable subset.*

Proof. Apply recursion (theorem 4.21). If Z is infinite, then $Z \neq \varnothing$, whence there exists some $X \in Z$. Define $F_0 : \{0\} \to Z$ by $0 \mapsto X$. Assume that there exists an injection $F_N : N \to Z$. Then $Z \neq F_N"(N)$ because Z is infinite, whence $Z \setminus F_N"(N) \neq \varnothing$ and there exists $X \in Z \setminus F_N"(N)$. Let $H_N : \{N\} \to \{X\}$ and let $F_{N+1} := F_N \dot{\cup} H_N$. Finally, let $F := \bigcup_{N \in \mathbb{N}} F_N$. Then $F : \mathbb{N} \to Z$ is an injection.   □

Theorem 4.177 explains the subscript 0 in the notation $\aleph_0$ for the cardinality of $\mathbb{N}$. Because in the Zermelo-Frænkel set theory with the Axiom of Choice every infinite set Z contains a denumerable subset, there exists an injection $F : \mathbb{N} \to Z$, so that the cardinality of $\mathbb{N}$ cannot exceed that of Z. If there is also an injection $G : Z \to \mathbb{N}$, then the Bernstein–Cantor–Schröder Theorem (theorem 4.167) guarantees that there is also a bijection $H : \mathbb{N} \to Z$; thus, the cardinality of Z cannot be strictly smaller than that of $\mathbb{N}$. Thus $\aleph_0$ represents the "smallest" infinite cardinality.

4.178 Theorem. *Every infinite set is also Dedekind-infinite.*

Proof. If W is infinite, then W contains a denumerable subset $Z \subseteq W$, by theorem 4.177. By example 4.175, there exists a bijection $F : Z \to Y$ onto a proper subset $Y \subsetneqq Z$. Extend F to all of W by setting $H := F \dot{\cup} \left(I_W|_{W \setminus Z} \right)$.   □

4.179 Theorem. *Every denumerable union of disjoint denumerable sets is denumerable.*

Proof. If $\mathscr{F}$ is denumerable, then there is a bijection $A : \mathbb{N} \to \mathscr{F}$ with $I \mapsto A(I)$. If each $A(I)$ is denumerable, then by the Axiom of Choice there is a bijection $F_I : \mathbb{N} \to A(I)$ with $J \mapsto F_I(J) \in A(I)$. Hence the function $G : \mathbb{N} \times \mathbb{N} \to \bigcup \mathscr{F}$ defined by $G(I, J) := F_I(J)$ is a bijection.   □

4.8.6 Further Issues in Cardinality

4.8.6.1 Other Axioms of Infinity

Bolzano [10] and Dedekind [25, §V, Theorem 66, p. 64] argued for the "existence" of an infinite set from practical considerations. Yet the "existence" of an infinite set does *not* follow from the other axioms, but requires an axiom of infinity [8, p. 21].

Zermelo [145] introduced such an axiom with a set such that if X is an element, then $\{X\}$ is also an element. The variant adopted here, with $\{X\}$ replaced by $X \cup \{X\}$, is attributed to John von Neumann [135] and has proved more convenient [8, p. 22]. There also exist "infinitely" many other nonequivalent axioms of infinity [18, §57, p. 342–346].

4.8.6.2   Peano's Axioms

There is an alternative method to introduce natural numbers, published by Giuseppe Peano, which does not involve set theory. Instead, Peano's system consists of the following axioms, stated here beginning with 0, as done in [46, §2, #II], [128, p. 121].

Axiom A0 0 is a natural number.

Axiom A1 $K = L$ means that K and L are the same natural numbers.

Axiom A2 For each natural number N there exists exactly one natural number denoted by N' and called the successor of N.

Axiom A3 $N' \neq 0$ for each natural number N.

Axiom A4 If $K' = L'$, then $K = L$.

Axiom A5 For each set S of natural numbers, if

- 0 is an element of S, and
- if N belongs to S, then K' belongs to S,

then every natural number is an element of S.

From Peano's axioms, and two additional axioms for recursive definitions of addition and multiplication [128, p. 136], the same proofs as in this chapter verify all the algebraic properties of arithmetic and ordering with natural numbers [76, Ch. 1, p. 1–18]. However, in situations that involve other topics, for examples, rational numbers or cardinality of sets, the use of Peano's axioms would require some theoretical link between Peano's natural numbers and other sets, in other words, some means of including Peano's arithmetic and applications within the same framework. This theoretical link can be the development of arithmetic from within set theory, as done here.

4.8.6.3   Alternative Sequence of Developments

The progression from $\mathbb{N}$ to $\mathbb{Z}$ and then to $\mathbb{Q}$ allows for subtractions with integers without requiring rational numbers, and then allows for divisions and subtractions with rational numbers without requiring the development of "real" numbers.

An alternative development proceeds from $\mathbb{N}$ to $\mathbb{Q}_+$, then to the nonnegative "real" numbers $\mathbb{R}_+$ and finally to the "real" numbers $\mathbb{R}$ and the "complex" numbers $\mathbb{C}$, as outlined by Edmund Landau [76]. Yet other constructions of the set of real numbers from the axioms are detailed by Michael Henle [58] and John Stillwell [120].

4.8.6.4 The Generalized Continuum Hypothesis

Because there is *no* surjection from $\mathbb{N}$ to $2^{\mathbb{N}}$ (theorem 4.156), the question arises whether there is any set S with a cardinality between the cardinalities of $\mathbb{N}$ and $2^{\mathbb{N}}$. In other words, the question pertains to the existence of a set S for which there exist injections

$$\mathbb{N} \overset{I}{\hookrightarrow} S \overset{J}{\hookrightarrow} 2^{\mathbb{N}}$$

but no injections from $2^{\mathbb{N}}$ back to S and no injections from S back to $\mathbb{N}$. The hypothesis that no such set as S exists is called the **continuum hypothesis**.

More generally, because for every infinite set X there does *not* exist any surjection from X to 2^{X}, the question arises whether there exists any set S with a cardinality strictly between the cardinalities of X and 2^{X}. In other words, the question pertains to the existence of a set S for which there exist injections

$$X \overset{I}{\hookrightarrow} S \overset{J}{\hookrightarrow} 2^{X}$$

but no injections from 2^{X} back to S and no injections from S back to X. The hypothesis that no such set as S exists is called the **generalized continuum hypothesis**.

The axioms of set theory (S1–S8) are consistent with both the generalized continuum hypothesis (as proved by K. Gödel [44, 45]) and with the negation of the generalized continuum hypothesis (as proved by P. J. Cohen [19, 20, 22]). Therefore, if the axioms of set theory are consistent, then the generalized continuum hypothesis can be neither proved nor disproved within set theory. The generalized continuum hypothesis is thus logically independent from set theory. Hence there exist two mutually exclusive extensions of set theory, one extension with the generalized continuum hypothesis, the other extension with the negation of the generalized continuum hypothesis.

4.8.7 *Exercises on Infinite Sets*

4.121 . Prove that if A is infinite and if $A \subseteq B$, then B is also infinite.

4.122 . Prove that if A is denumerable and $Z \notin A$, then $A \dot{\cup} \{Z\}$ is denumerable.

4.123 . With A denumerable, B finite, disjoint, prove that $A \dot\cup B$ is denumerable.

4.124 . Prove that if A is denumerable and B finite, then $A \cup B$ is denumerable.

4.125 . Prove that if A and B are denumerable, then $A \cup B$ is denumerable.

4.126 . Prove that there exists a bijection from $2^{\mathbb{N}}$ to $\mathscr{P}(\mathbb{N})$.

4.127 . Prove that $\mathscr{P}(\mathbb{N})$ is *not* countable.

4.128 . Prove: if A is denumerable and B is finite, then $A \times B$ is countable.

4.129 . For each nonempty set X prove that there is no injection $2^X \hookrightarrow X$.

4.130 . Prove that if $[\#(X)] = [\#(Y)]$, then $[\#(2^X)] = [\#(2^Y)]$.

4.131 . Prove: if $[\#(X)] < [\#(Y)]$ are both finite, then $[\#(2^X)] < [\#(2^Y)]$.

4.132 . Prove that if X is a finite set, then 2^X is also finite.

4.133 . Prove that if X is a finite set, then $[\#(X)] < [\#(2^X)]$.

4.134 . Prove that every infinite subset $S \subseteq \mathbb{N}$ is denumerable.

4.135 . For each denumerable set A prove that every subset $S \subseteq A$ is countable.

4.136 . Prove: if A is uncountable and $A \subseteq B$, then B is uncountable.

4.137 . Prove that if A, B, and C are denumerable, then so is $(A \times B) \times C$.

4.138 . Prove that if $A \times B$ is denumerable, then A and B are countable.

4.139 . Show where the proof of theorem 4.179 invokes the Axiom of Choice.

4.140 . Show where the proof of theorem 4.177 invokes the Axiom of Choice.

Chapter 5
Well-Formed Sets: Proofs by Transfinite Induction with Already Well-Ordered Sets

5.1 Introduction

This chapter focuses on "well-formed" sets, which are defined by means restricted to the axioms of Zermelo-Fraenkel set theory from chapters 1, 2, 3, and 4. The main result states that no two well-formed sets are members of each other, and consequently that every well-formed set is *not* an element of itself.

This chapter shows the dependence of one axiom on the others for sets — called well-formed sets — defined in specific ways solely through the axioms of set theory. Such well-formed sets suffice for most of logic, mathematics, computer science, and their applications to the sciences and engineering. Among other features, the result shows that no well-formed set is an element of itself, and that this result is provable from the other axioms of set theory [74]. This chapter also provides a way to revisit induction (chapter 4) on a different level.

5.2 Transfinite Methods

5.2.1 Transfinite Induction

Transfinite methods lead to an example of decidability in set theory. On the set $\mathbb{N}$, the Principle of Mathematical Induction (theorem 4.15) is logically equivalent to the Well-Ordering Principle (theorem 4.47), which states that every nonempty subset $S \subseteq \mathbb{N}$ has a smallest element. All well-ordered sets also lend themselves to a method of proof known as transfinite *induction*, which relies on "initial intervals" in well-ordered sets. *All sets in this chapter are already well-ordered.*

5.1 Definition. For each set W well-ordered (definition 3.194) by a relation $\prec$ and for each $C \in W$, the **initial interval** determined by C is the subset

© Springer Science+Business Media New York 2015
Y. Nievergelt, *Logic, Mathematics, and Computer Science*,
DOI 10.1007/978-1-4939-3223-8_5

$$W_C := \{B \in W : (B \prec C) \wedge (B \neq C)\},$$

which consists of all elements preceding C but different from C.

5.2 Example. If $W = \mathbb{N}$, with $<$ for $\prec$, then $\mathbb{N}_N = N$ for each $N \in \mathbb{N}$.

The Principle of Mathematical Induction extends to all well-ordered sets.

5.3 Theorem (Transfinite Induction). *For each set W well-ordered by $\prec$ and for each set V, if $V \subseteq W$, then $V = W$ if and only if*

$$\forall C\{[(C \in W) \wedge (W_C \subseteq V)] \Rightarrow (C \in V)\}.$$

Proof. If $V = W$, then the formula is a tautology: $[(P) \wedge (Q)] \Rightarrow (P)$. For the converse, let $U := W \setminus V$. If $U \neq \varnothing$, then U has a first element $A \in U$. Thus, if $B \prec A$ but $B \neq A$, then $B \notin U$, whence $B \in W \setminus U = V$. Hence the initial interval $W_A \subseteq V$, but then $A \in V$ by hypothesis on V, which contradicts $A \in U = W \setminus V$. $\square$

Well-ordered sets also lend themselves to a method of *definition* known as transfinite *construction,* which relies on the concept of "ideal" in a well-ordered set.

5.4 Definition (ideal). For each set W well-ordered by a relation $\prec$ a subset $V \subseteq W$ is an **ideal** of W if and only if V contains every element preceding any of its elements. Thus V is an ideal if and only if $W_C \subseteq V$ for every $C \in V$:

$$\forall B \forall C\{[(B \in W) \wedge (B \prec C) \wedge (C \in V)] \Rightarrow (B \in V)\}.$$

The set of all ideals of W relative to $\prec$ is denoted by $\mathscr{I}_{\prec}(W)$.

5.5 Example. If $W = \mathbb{N}$, with $<$ for $\prec$, then N is an ideal for each $N \in \mathbb{N}$.

The following two theorems yield relations between ideals and initial intervals.

5.6 Theorem. *For each set W well-ordered by a relation $\prec$ and for each ideal $V \subseteq W$, if $B \in W \setminus V$, then $V \subseteq W_B$.*

Proof. By definition of an ideal, if $C \in V$ and $B \in W$ with $B \prec C$, then $B \in V$. By contraposition, $C \notin V$ follows from $B \in W \setminus V$, and $C \in W$ with $B \prec C$. Because $\prec$ totally orders W (remark 3.195), it follows that if $B \in W \setminus V$, then $(W \setminus V) \subseteq [W \setminus (W_B \cup \{B\})]$, whence $V \subseteq W_B \cup \{B\}$. Yet $B \notin V$, whence $V \subseteq W_B$. $\square$

5.7 Theorem. *For each set W well-ordered by a relation $\prec$ and for each ideal $V \subsetneqq W$, there exists a smallest element $A \in W \setminus V$, and for this element $W_A = V$.*

Proof. If $V \subsetneqq W$, then $W \setminus V \neq \varnothing$. Because W is well-ordered by $\prec$ it follows that $W \setminus V$ contains a smallest element $A \in W \setminus V$. Thus for every $B \in W$ such that $B \prec A$ and $B \neq A$, it follows that $B \in V$ by minimality of A in $W \setminus V$. Therefore $W_A \subseteq V$. Moreover, $V \subseteq W_A$ by theorem 5.6. $\square$

The following theorems show that the set of all ideals is well-ordered by inclusion.

5.8 Theorem. *For each set W well-ordered by a relation $\prec$ and for each nonempty set $\mathscr{F}$ of ideals of W, the intersection $\bigcap \mathscr{F}$ is also an ideal of W.*

Proof. If $C \in \bigcap \mathscr{F}$, then $C \in V$ for each ideal $V \in \mathscr{F}$. Hence, if $B \in W$ and $B \prec C$, then $B \in V$ because V is an ideal. This conclusion holds for each $V \in \mathscr{F}$, whence $B \in \bigcap \mathscr{F}$. Thus $\bigcap \mathscr{F}$ is an ideal of W. ☐

5.9 Theorem. *For each set W well-ordered by a relation $\prec$ and for each nonempty set $\mathscr{F}$ of ideals of W, the smallest element of $\mathscr{F}$ is $\bigcap \mathscr{F}$. Therefore the set $\mathscr{I}_{\prec}(W)$ is well-ordered by inclusion ($\subseteq$).*

Proof. For each nonempty set $\mathscr{F}$ of ideals of W, there exists at least one ideal $U \in \mathscr{F}$, and the intersection $\bigcap \mathscr{F}$ is also an ideal, by theorem 5.8. Let

$$Z := \left\{ C \in W : C \in \left(\bigcup \mathscr{F} \right) \setminus \left(\bigcap \mathscr{F} \right) \right\}.$$

If $Z = \varnothing$, then $\bigcup \mathscr{F} = \bigcap \mathscr{F}$, whence $\mathscr{F}$ contains only one ideal, in effect $U = \bigcap \mathscr{F}$.

If $Z \neq \varnothing$ then it has a smallest element $A \in Z \subseteq W$. Because $\bigcap \mathscr{F}$ is an ideal by theorem 5.8, and because $A \notin \bigcap \mathscr{F}$, it follows that $W_A = \bigcap \mathscr{F}$ by theorem 5.7.

Also because $A \notin \bigcap \mathscr{F}$, there exists an ideal $V \in \mathscr{F}$ with $A \notin V$. Hence $V \subseteq W_A$ by theorem 5.6.

From $V \subseteq W_A$ and $W_A = \bigcap \mathscr{F}$ it follows that $V \subseteq \bigcap \mathscr{F}$, but $\bigcap \mathscr{F} \subseteq V$. Consequently $\bigcap \mathscr{F} = V \in \mathscr{F}$.

Therefore every nonempty set $\mathscr{F} \subseteq \mathscr{I}_{\prec}(W)$ has a smallest element, in effect $\bigcap \mathscr{F}$, so that $\mathscr{I}_{\prec}(W)$ is well-ordered by set inclusions. ☐

5.2.2 Transfinite Construction

As the Principle of Mathematical Induction leads to a method of definition by induction (theorem 4.21) — also called a recursive definition or recursion — similarly transfinite induction also yields a method of *definition* by transfinite induction.

5.10 Theorem (Transfinite Construction). *For each nonempty set W well-ordered by a relation $\prec$ with first element $A \in W$, and for each nonempty set E, let*

$$Y := \bigcup_{B \in W} E^{W_B}$$

denote the set of all functions with domain equal to an initial interval W_B and with range in E (definition 4.153). For each $Z \in E$, and for each function $P : Y \to E$, there exists exactly one function $F : W \to E$ such that $F(A) = Z$ and $F(B) = P(F|_{W_B})$.

Proof. This proof establishes the uniqueness and existence separately.

Uniqueness

There is at most one such function. Indeed if $F : W \to E$ and $G : W \to E$ are two such functions, with $F(A) = Z = G(A)$ and $F(B) = P(F|_{W_B})$, $G(B) = P(G|_{W_B})$, then let

$$S := \{C \in W : F(C) \neq G(C)\}.$$

If $S \neq \emptyset$, then S has a smallest element, $D \in S$. Hence $F(B) = G(B)$ for every $B \prec D$ in W, which means that $F|_{W_D} = G|_{W_D}$, but then

$$F(D) = P(F|_{W_D}) = P(G|_{W_D}) = G(D)$$

would contradict $D \in S$. Consequently, $S = \emptyset$, so that $F(B) = G(B)$ for every $B \in W$, which means that $F = G$.

Existence

Let $\mathscr{F}$ denote the set of all ideals $V \subseteq W$ for which there exists a function $F_V : V \to E$ such that $F_V(A) = Z$ and $F_V(B) = P(F_V|_{W_B})$.

For each ideal $U \in \mathscr{F}$, applying the uniqueness just proved to the well-ordered set $U \cap V$ instead of W shows that the functions F_U and F_V coincide on the well-ordered subset $U \cap V$ in W.

Hence, define a function $F_{\mathscr{F}} : \bigcup \mathscr{F} \to E$ by setting $F_{\mathscr{F}}(B) := F_U(B)$ for any ideal $U \in \mathscr{F}$ with $B \in U$. In other terms, $F_{\mathscr{F}} = \bigcup_{U \in \mathscr{F}} F_U$. The preceding argument confirms that this definition does not depend on which ideal U contains B, because if $B \in U$ and $B \in V$, then $F_U(B) = F_V(B)$.

Next, if an ideal U is an initial interval, $U = W_B$ for some $B \in W$, and if $W_B \in \mathscr{F}$, then $W_B \cup \{B\} \in \mathscr{F}$. Indeed, a function $F_U : U \to E$ extends to $W_B \cup \{B\}$ by the definition $F_{W_B \cup \{B\}}(B) := P(F_U)$.

Suppose that $W \notin \mathscr{F}$. Then let $\mathscr{G}$ denote the set of all the ideals of W that are not elements of $\mathscr{F}$. In particular, $W \in \mathscr{G}$. Define $V := \bigcap \mathscr{G}$, which is then the smallest ideal of W in $\mathscr{G}$. If V had a *last* element D, then $V = W_D \cup \{D\}$ by definition of W_D and of an ideal; however, $W_D \in \mathscr{F}$, otherwise $V \neq \bigcap \mathscr{G}$, but from $W_D \in \mathscr{F}$ it follows that $W_D \cup \{D\} \in \mathscr{F}$. Thus V cannot have a last element.

If V does not have a last element, then $V = \bigcup_{B \in V} W_B$ by definition of an ideal. Again, it follows that $W_B \in \mathscr{F}$ for each $B \in V$, whence $V = \bigcup_{B \in V} W_B \subseteq \bigcup \mathscr{F}$ and then $V \in \mathscr{F}$ because of the existence of $F_{\mathscr{F}}$, contradicting the definition of V. Therefore, $W \in \mathscr{F}$, which means that F extends to all of W.   □

5.2.3 Exercises on Transfinite Methods

5.1 . Prove that in each well-ordered set each initial interval is an ideal.

5.2 . Provide an example of an ideal that is *not* an initial interval in a well-ordered set.

5.3 . Prove that if $\mathbb{Z}$ is well-ordered by $\preccurlyeq$, then $\preccurlyeq$ differs from $\leq$.

5.4 . Prove that if $\mathbb{Q}$ is well-ordered by $\preccurlyeq$, then $\preccurlyeq$ differs from $\leq$.

5.5 . Provide an example of a well-order $\prec$ on a set of modular integers $\mathbb{Z}_M = \{[0]_M, \ldots, [M-1]_M\}$ and modular integers $[I]_M, [K]_M, [L]_M$, such that $[K]_M \prec [L]_M$ but $[I]_M + [K]_M \not\prec [I]_M + [L]_M$.

5.6 . Provide an example of a well-order $\prec$ on a set of modular integers $\mathbb{Z}_M = \{[0]_M, \ldots, [M-1]_M\}$ and modular integers $[I]_M, [K]_M, [L]_M$, such that $[0]_M \prec [I]_M$ and $[K]_M \prec [L]_M$ but $[I]_M * [K]_M \not\prec [I]_M * [L]_M$.

5.7 . Prove that every subset of a well-ordered set is also well-ordered.

5.8 . Determine whether for each set $\mathscr{F}$ of *ideals* in W the union $\bigcup \mathscr{F}$ is also an ideal in W.

5.9 . Determine whether for each set $\mathscr{G}$ of *initial intervals* in W the union $\bigcup \mathscr{G}$ is also an initial interval in W.

5.10 . Determine whether for each set $\mathscr{G}$ of *initial intervals* in W the intersection $\bigcap \mathscr{G}$ is also an initial interval in W.

5.3 Transfinite Sets and Ordinals

5.3.1 Transitive Sets

Sets defined exclusively through the axioms of set theory adopted here are called well-formed sets. They have the advantage of avoiding certain contradictions that would arise from defining sets by means not so strict. The definition of well-formed sets involves the concept of sets that are "transitive" relative to the relation $\in$.

5.11 Definition (Transitive Sets). A set A is **transitive** if and only if every element of A is also a subset of A, so that $\forall X[(X \in A) \Rightarrow (X \subseteq A)]$, or, equivalently,

$$\forall Y \forall X\{[(Y \in X) \wedge (X \in A)] \Rightarrow (Y \in A)\}.$$

5.12 Example. The following sets are transitive:

$$\varnothing,$$
$$\{\varnothing\},$$
$$\{\varnothing, \{\varnothing\}\},$$
$$\Big\{ \varnothing, \{\varnothing\}, \{\varnothing, \{\varnothing\}\} \Big\},$$
$$\Big\{ \varnothing, \{\varnothing\}, \{\{\varnothing\}\}, \{\varnothing, \{\varnothing\}\} \Big\}.$$

5.13 Counterexample. The set $A := \big\{\{\varnothing\}\big\}$ is *not* transitive, because it contains an element $X := \{\varnothing\}$ that is *not* a subset of A: $X \nsubseteq A$, because $\varnothing \in X$ but $\varnothing \notin A$.

Power sets, unions, and intersections of transitive sets are also transitive.

5.14 Theorem. *If a set A is transitive, then $\mathscr{P}(A)$ is also transitive.*

Proof. If $S \in \mathscr{P}(A)$, then $S \subseteq A$. Thus if $X \in S$, then $X \in A$, and $X \subseteq A$ by transitivity of A. Hence $X \in \mathscr{P}(A)$ for each $X \in S$, whence $S \subseteq \mathscr{P}(A)$. □

5.15 Theorem. *If a set $\mathscr{F}$ is transitive, then $\bigcup \mathscr{F}$ is also transitive.*

Proof. If $S \in \bigcup \mathscr{F}$, then there exists $A \in \mathscr{F}$ with $S \in A$. Yet $A \subseteq \mathscr{F}$ by transitivity of $\mathscr{F}$. From $S \in A$ and $A \subseteq \mathscr{F}$ follows $S \in \mathscr{F}$, whence $S \subseteq \bigcup \mathscr{F}$. □

5.16 Theorem. *If $\mathscr{F}$ is a nonempty set of transitive sets, then $\bigcup \mathscr{F}$ and $\bigcap \mathscr{F}$ are also transitive.*

Proof. If $S \in \bigcup \mathscr{F}$, then $S \in A$ for some $A \in \mathscr{F}$, whence $S \subseteq A$ by transitivity of A, and hence $S \subseteq A \subseteq \bigcup \mathscr{F}$, so that $\bigcup \mathscr{F}$ is transitive.

If $S \in \bigcap \mathscr{F}$, then $S \in A$ for every $A \in \mathscr{F}$, whence $S \subseteq A$ by transitivity of A, and hence $S \subseteq \bigcap \mathscr{F}$, so that $\bigcap \mathscr{F}$ is transitive. □

5.3.2 Ordinals

Well-formed sets will rely on the concept of ordinals (also called "ordinal numbers" [30, p. 42]). The following definition conforms to Kunen's [74, p. 16].

5.17 Definition (ordinals). A set A is an **ordinal** if and only if it is a transitive set, and the relation $\in$ is an *irreflexive* well-ordering on the set A.

5.18 Example. The following sets are ordinals:

$$\varnothing,$$
$$\{\varnothing\},$$
$$\{\varnothing, \{\varnothing\}\},$$
$$\Big\{\varnothing, \{\varnothing\}, \{\varnothing, \{\varnothing\}\}\Big\}.$$

5.19 Counterexample. The set

$$A := \Big\{\varnothing, \{\varnothing\}, \{\{\varnothing\}\}, \{\varnothing, \{\varnothing\}\}\Big\} = \mathscr{P}(\{\varnothing, \{\varnothing\}\})$$

is transitive but *not* an ordinal. Indeed, if

$$X := \varnothing,$$
$$Y := \{\{\varnothing\}\},$$

then $X \in A$ and $Y \in A$, so that $\{X, Y\} \subseteq A$, but $\{X, Y\}$ does not have any smallest element relative to the relation $\in$, because $X \notin Y$ and $Y \notin X$.

In particular the subset $\{X, Y\}$ is *not* an ordinal.

5.20 Theorem. *The empty set is an element of every nonempty ordinal.*

Proof. By definition, every ordinal A is transitive, so that if $X \in A$ then $X \subseteq A$. Consequently, if $Y \in X$ and $X \in A$, then $Y \in A$. Therefore, if $X \in A$ and $X \neq \varnothing$, then X is not the smallest element of A. Yet every nonempty ordinal A has a smallest element. Hence contraposition shows that the smallest element must be $\varnothing$. □

5.21 Theorem. *If A is an ordinal, then $A \notin A$. Moreover, $A \notin X$ for each $X \in A$. In particular, if A and X are ordinals, then $A \notin X$ or $X \notin A$.*

Proof. If A is an ordinal, then $\in$ is connected (definitions 3.185, 3.194) and irreflexive (definition 5.17): exactly one of $X \in Y$, $X = Y$, or $Y \in X$ holds for all $X, Y \in A$. Because $A = A$, it follows that $A \notin A$. Moreover, if A is an ordinal and $X \in A$, then $X \subseteq A$, whence if $Y \in X$ then $Y \in A$. With $Y := A$, it follows by contraposition that $A \notin X$. □

5.22 Theorem. *Every element of an ordinal is an ordinal.*

Proof. If A is an ordinal and $X \in A$, then $X \subseteq A$ because A is a transitive set. Hence $\in$ well-orders X, because $\in$ well-orders A. If moreover $Z \in Y$ and $Y \in X$, then $Z \in X$ because A is a transitive set, whence $Y \subseteq X$. Thus X is also a transitive set. Furthermore, the relation $\in$ remains irreflexive on the subset $X \subseteq A$. □

5.23 Theorem. *If A is an ordinal, then either A contains a last element D and $A = D \cup \{D\}$, or $A = \bigcup A$ is the union of all its elements.*

Proof. If A is an ordinal and $B \in A$, then $B \subseteq A$; consequently, $\bigcup A \subseteq A$.

Let $D := \bigcup A$. If $D \neq A$, then there exists $X \in A \setminus D$. Then $X \subseteq D$ because $X \in A$; hence X is an ordinal by theorem 5.22, whence $X \neq A$ by theorem 5.21, and $X \subseteq \bigcup A = D$.

Conversely, still with $X \in A \setminus D$, for each $B \in A$ it follows that $X \notin B$, whence $B \in X$, and hence $B \subseteq X$, so that $\bigcup_{B \in A} B \subseteq X$. Thus, $D = \bigcup A = \bigcup_{B \in A} B \subseteq X$.

Therefore, if $D \neq A$, then $D \subseteq X \subseteq D$, whence $X = D$ is the only element of $A \setminus D$, and hence $A = D \cup \{D\}$. $\square$

5.24 Theorem. *If B is an ordinal, then $B \cup \{B\}$ is also an ordinal.*

Proof. First, $B \cup \{B\}$ is transitive. Indeed, if $X \in B \cup \{B\}$, then either $X \in B$, whence $X \subseteq B \subseteq B \cup \{B\}$, or $X \in \{B\}$, whence $X = B \subseteq B \cup \{B\}$.

Second, $\in$ well-orders $B \cup \{B\}$. Indeed, for each nonempty subset $S \subseteq B \cup \{B\}$, two situations can occur: $S \subseteq B$ or $S \cap \{B\} \neq \varnothing$. If $S \subseteq B$ then S has a smallest element, because B is an ordinal. If $S \cap \{B\} \neq \varnothing$, then either $S = \{B\}$ has the smallest element B, or $S \cap B \neq \varnothing$ and then $S \cap B$ is a subset of B and hence has a smallest element, which is then also a smallest element of $S = (S \cap B) \cup \{B\}$.

Moreover, the relation $\in$ remains irreflexive and transitive on $B \cup \{B\}$. Indeed, $B \notin B$ by theorem 5.21. Furthermore, if $X \in Y$ and $Y \in Z$ in $B \cup \{B\}$, then $Y \in B$ from either $Z \in B$ or $Z \in \{B\}$. Also, $Y \neq B$ by theorem 5.21, which forbids $B \in Z$ and $Z \in B \cup \{B\}$, and hence $X \neq B$ also by theorem 5.21, which forbids $B \in Y$ $Y \in Z$, and $Z \in B \cup \{B\}$. Consequently only two cases can occur: $Z \in B$ or $Z = B$.

If $Z = B$, then $X \neq B$ and $Y \neq B$, whence $X \in Z$.

If $Z \in B$, then X, Y, and Z all three lie in B, whence $X \in Z$ because $\in$ is transitive on the well-ordered set B.

Finally, $\in$ is strict on $B \cup \{B\}$. Indeed, because $\in$ is strict on B, it follows that if $X \in B \cup \{B\}$ and $Y \in B \cup \{B\}$, then two different cases can arise.

If $X \in B$ and $Y \in B$, then $X \in Y$ and $Y \in X$ cannot both hold, for transitivity would yield $X \in X$ which cannot hold by strictness of $\in$ on B.

If $X \in B$ and $Y \in \{B\}$, then $X \in Y$ and $Y \in X$ cannot both hold. Otherwise $Y = B$ and then $X \in B$ and $B \in X$. However, X is also an ordinal by theorem 5.22, whence $\in$ is also transitive on X, so that $B \in X$ and $X \in B$ yield $B \in B$, which cannot hold by strictness of $\in$ on X. $\square$

5.3.3 Well-Ordered Sets of Ordinals

The following theorems show that every set of ordinals is well-ordered. First $\subseteq$ is strongly connected (definition 3.184) on every set of ordinals.

5.25 Theorem. *For all ordinals A and B, either $A \subset B$, or $A = B$, or $B \subset A$.*

Proof. If $B \nsubseteq A$, then there exists $X \in B \setminus A$, and hence there exists a smallest such element: $X \in B \setminus A$, so that $X \in Y$ for every $Y \in B \setminus A$ with $Y \neq X$. Also, $X \neq \varnothing$, because $X \notin A$ but $\varnothing \in A$. From $X \in B$ it follows that $X \subseteq B$, whence if $Y \in X$, then $Y \in B$, but then $Y \in A$ because of the minimality of X. Thus $X \subseteq A \cap B$.

Conversely, if $Y \in A \cap B$, then $Y \in B$, whence either $Y \in X$ or $X \in Y$, because $X \in B$ also. However, $X \in Y$ cannot occur, because $X \in Y$ and $Y \in A \cap B$ would yield $X \in A$. Thus, if $Y \in A \cap B$, then $Y \in X$, which means that $A \cap B \subseteq X$.

Consequently, $X = A \cap B$.

The foregoing argument with A and B switched shows that if $A \nsubseteq B$, then $Z := B \cap A$ is the smallest element in $A \setminus B$. In particular, $Z = B \cap A = A \cap B = X$.

Consequently, if $A \nsubseteq B$ and $B \nsubseteq A$ both held, then $X := A \cap B =: Z$ would be the smallest element in both $A \setminus B$ and $B \setminus A$.

However, because $(A \setminus B) \cap (B \setminus A) = \varnothing$, it follows that $X \notin (A \setminus B) \cap (B \setminus A)$. Thus at least one of $B \nsubseteq A$ or $A \nsubseteq B$ must fail to hold, which means that $B \subseteq A$ or $A \subseteq B$ or both, so that either $A \subset B$, or $A = B$, or $B \subset A$. □

The second theorem shows that $\in$ is strongly connected on every set of ordinals.

5.26 Theorem. *For all ordinals A and B, either $A = B$, or $A \in B$, or $B \in A$.*

Proof. Consider the sets $C := A \cup \{A\}$ and $D := B \cup \{B\}$, which are ordinals by theorem 5.24. Applying theorem 5.25 to C and D instead of A and B shows that $D \subseteq C$ or $C \subseteq D$. If $C \subseteq D$, then $A \in C \subseteq D = B \cup \{B\}$, so that either $A \in B$ or $A = B$. If $D \subseteq C$, then $B \in D \subseteq C = A \cup \{A\}$, so that either $B \in A$ or $B = A$. □

5.27 Theorem. *Every set $\mathscr{F}$ of ordinals is well-ordered by $\in$.*

Proof. The relation $\in$ is irreflexive on $\mathscr{F}$ by theorem 5.21, and it is connected by theorem 5.26. The relation $\in$ is also transitive on $\mathscr{F}$. Indeed, for all $A, B, C \in \mathscr{F}$, if $A \in B$ and $B \in C$, then $B \subseteq C$ whence $A \in C$.

Moreover, for each nonempty subset $\mathscr{G} \subseteq \mathscr{F}$, there exists some $C \in \mathscr{G}$. For each $B \in \mathscr{G}$, either $B \in C$, or $B = C$, or $C \in B$, by theorem 5.26. Define

$$E := \{B \in \mathscr{G} : B \in C\} = C \cap \mathscr{G}.$$

If $E = \varnothing$, then C is the smallest element of $\mathscr{G}$, because then $C \in B$ for each $B \in \mathscr{G}$ with $B \neq C$. If $E \neq \varnothing$, then E has a smallest element $A \in E$, because E is a subset of the ordinal C. If $B \in \mathscr{G}$, then $B \notin A$, because $B \in A$ would yield $B \in C$, contradicting the minimality of A. Hence $A \in B$ for every $B \in \mathscr{G}$ with $B \neq A$. □

5.3.4 Unions and Intersections of Sets of Ordinals

The union and the intersection of every nonempty set of ordinals is an ordinal.

5.28 Theorem. *For each set $\mathscr{F}$ of ordinals, $\bigcup \mathscr{F}$ is also an ordinal.*

Proof. The union $\bigcup \mathscr{F}$ is transitive, by theorem 5.15.

The union $\bigcup \mathcal{F}$ is well-ordered. Indeed, for each nonempty subset $S \subseteq \bigcup \mathcal{F}$, let

$$E := \{A \in \mathcal{F} : A \cap S \neq \varnothing\}.$$

From $S \neq \varnothing$ follows $E \neq \varnothing$. Hence $E \subseteq \mathcal{F}$ has a smallest element A, because $\mathcal{F}$ is an ordinal, by theorem 5.27. Thus $S \cap A \neq \varnothing$. Therefore, $S \cap A \subseteq A$ also has a smallest element $B \in S \cap A$. Also, for each $C \in S$ there exists $D \in \mathcal{F}$ with $C \in D$. Then either $A = D$, or $A \in D$, or $D \in A$. Yet $D \in A$ cannot occur, by minimality of A. From $B \in A$, with $A \in D$ or $A = D$, follows $B \in D$; hence $B \in D$ and $C \in D$. Consequently, either $B \in C$ or $C \in B$ or $B = C$, but $C \in B$ cannot occur by minimality of B and because $B \in A$. Thus, $B \in C$ or $B = C$, which shows that B is the smallest element of S. By theorem 5.21 $\in$ is strict on $\bigcup \mathcal{F}$, for every element is an ordinal by theorem 5.22. □

5.29 Theorem. *For each nonempty set $\mathcal{F}$ of ordinals, $\bigcap \mathcal{F}$ is also an ordinal.*

Proof. The intersection $\bigcap \mathcal{F}$ is a transitive set by theorem 5.16. The intersection $\bigcap \mathcal{F}$ is also well-ordered. Indeed, each nonempty subset $S \subseteq \bigcap \mathcal{F}$ is also a subset $S \subseteq A$ of some $A \in \mathcal{F}$ and hence S has a smallest element, because A is an ordinal.

The relation $\in$ is a strict total order on $\bigcap \mathcal{F}$, as it is on every subset of $A \in \mathcal{F}$, where it is strict. Indeed, if $X \in \bigcap \mathcal{F}$ and $Y \in \bigcap \mathcal{F}$, then $X \in A$ and $Y \in A$ whence $X \in Y$ and $Y \in X$ cannot both hold. □

5.3.5   Exercises on Ordinals

5.11 . Prove that $\mathbb{N}$ is a transitive set.

5.12 . Prove that every natural number $N \in \mathbb{N}$ is a transitive set.

5.13 . Prove that $\mathbb{N}$ is an ordinal.

5.14 . Prove that every natural number $N \in \mathbb{N}$ is an ordinal.

5.15 . Investigate whether $\in$ is a transitive relation on every transitive set.

5.16 . Prove that every ordinal is an element of some ordinal.

5.17 . Prove that every ordinal is a subset of some ordinal.

5.18 . Determine whether every ordinal is a subset of some transitive set.

5.19 . Determine whether every transitive set is a subset of some ordinal.

5.20 . Verify that it is not necessary to assume that $V \neq \varnothing$ for Transfinite Induction. In other words, prove that for each set W well-ordered by a relation $\prec$ and for each subset $V \subseteq W$, if the following formula is True,

$$\forall C\{[(C \in W) \wedge (W_C \subseteq V)] \Rightarrow (C \in V)\},$$

then either $W = \varnothing$ or $V \neq \varnothing$.

5.21 . Determine whether every singleton $\{A\}$ with an ordinal A is an ordinal.

5.22 . Determine whether $\{A, B\}$ is an ordinal for all ordinals A and B.

5.23 . Determine whether every set of ordinals is an ordinal.

5.24 . Determine whether every subset of every ordinal is an ordinal.

5.25 . Prove that there is an ordinal whose power set is not an ordinal.

5.26 . Prove that every countable set admits a well-ordering.

5.4 Regularity of Well-Formed Sets

5.4.1 Well-Formed Sets

The following definition establishes sets that contain all the well-formed sets.

5.30 Definition. For each ordinal A, define a set $R(A)$ by transfinite construction:

- $R(\varnothing) := \varnothing$;
- $R(A \cup \{A\}) := \mathscr{P}[R(A)]$;
- $R(A) := \bigcup_{B \in A} R(B)$ if there does not exist any ordinal B such that $A = B \cup \{B\}$, but if $R(B)$ has been defined for every $B \in A$.

5.31 Definition. A set X is **well-formed** if and only if there exists an ordinal A such that $X \in R(A)$.

5.32 Remark. The transfinite construction proceeds as follows. For each ordinal W, let

$$E := \mathscr{P}[\mathscr{P}(W)],$$
$$Y := \bigcup_{A \in W} E^{W_A},$$
$$P : Y \to E,$$
$$P(R|_{W_A}) := \bigcup_{B \in A} R(B).$$

The values of P remain in E, because if $R|_{W_A} : W_A \to E$, then $R(B) \in E = \mathscr{P}[\mathscr{P}(W)]$ for each $B \in W_A = A$, whence $R(B) \subseteq \mathscr{P}(W)$, and hence $\bigcup_{B \in A} R(B) \subseteq \mathscr{P}(W)$, so that $\bigcup_{B \in A} R(B) \in \mathscr{P}[\mathscr{P}(W)]$. Yet for all ordinals U and V, $U \subseteq V$ or $V \subseteq U$ by theorem 5.25, so that $\mathscr{P}[\mathscr{P}(U)] \subseteq \mathscr{P}[\mathscr{P}(V)]$ or $\mathscr{P}[\mathscr{P}(V)] \subseteq \mathscr{P}[\mathscr{P}(U)]$. Therefore, the transfinite construction defined with $W := U$ or $W := V$ gives the same definition on $U \cap V$. In other words, for each ordinal A, the definition of $R(A)$ can proceed from any ordinal W containing A, for example, $W := A \cup \{A\}$.

5.33 Example. The first few sets of the form $R(A)$ are also ordinals:

$$R(\varnothing) = \varnothing,$$

$$R(\{\varnothing\}) = R(\varnothing \cup \{\varnothing\}) = \mathscr{P}[R(\varnothing)] = \mathscr{P}[\varnothing] = \{\varnothing\},$$

$$R\left(\{\varnothing, \{\varnothing\}\}\right) = R\left(\{\varnothing\} \cup \{\{\varnothing\}\}\right) = \mathscr{P}[R(\{\varnothing\})] = \mathscr{P}[\{\varnothing\}] = \{\varnothing, \{\varnothing\}\}.$$

The next ordinal,

$$A := \Big\{ \varnothing, \{\varnothing\}, \{\varnothing, \{\varnothing\}\} \Big\},$$

is not of the form $R(B)$ for any set B, but $A = B \cup \{B\}$ for the ordinal $B := \{\varnothing, \{\varnothing\}\}$. Hence, the list continues with

$$R\left(\Big\{ \varnothing, \{\varnothing\}, \{\varnothing, \{\varnothing\}\} \Big\}\right) = R\left(\{\varnothing, \{\varnothing\}\} \cup \Big\{\{\varnothing, \{\varnothing\}\}\Big\}\right)$$

$$= \mathscr{P}\left[R\left(\{\varnothing, \{\varnothing\}\}\right)\right]$$

$$= \mathscr{P}\left[\{\varnothing, \{\varnothing\}\}\right]$$

$$= \Big\{ \varnothing, \{\varnothing\}, \{\{\varnothing\}\}, \{\varnothing, \{\varnothing\}\} \Big\},$$

which is of the form $R(A)$, but it is not well-ordered by $\in$ and hence not an ordinal, by counterexample 5.19. The set $R(A)$ is also different from the ordinal

$$\Big\{ \varnothing, \{\varnothing\}, \{\varnothing, \{\varnothing\}\}, \big\{\varnothing, \{\varnothing\}, \{\varnothing, \{\varnothing\}\}\big\} \Big\}.$$

5.34 Theorem. *For each ordinal Q, the set $R(Q)$ is transitive.*

Proof. This proof proceeds by transfinite induction (theorem 5.3).

Choose an ordinal W with $Q \in W$, for instance, $W := Q \cup \{Q\}$. Then consider the subset $V \subseteq W$ of all the elements $C \in W$ for which $R(C)$ is transitive.

Consider any element $C \in W$ such that $W_C \subseteq V$. By transfinite induction, it suffices to verify that $C \in V$ to establish that $V = W$. Because

$$W_C = \{B \in W : (B \in C) \wedge (B \neq C)\}$$

by definition 5.1 for the relation $\in$ instead of $\prec$, it follows that $W_C = C$. From $C = W_C \subseteq V$ it then follows that $R(B)$ is transitive for each $B \in C$ by definition of V. Consequently, $\bigcup_{B \in C} R(B)$ is also transitive, by theorem 5.15.

Two cases can arise, either C does not have a last element, or C has a last element. If C does not have a last element, then $R(C) = \bigcup_{B \in C} R(B)$ is transitive.

If C has a last element $Z \in C$, then $C = Z \cup \{Z\}$. From $Z \in C = W_C \subseteq V$, it follows that $R(Z)$ is transitive by definition of V. Hence $\mathscr{P}[R(Z)]$ is also transitive, by theorem 5.14. Yet $R(C) = R(Z \cup \{Z\}) = \mathscr{P}[R(Z)]$, whence $R(C)$ is transitive. Thus in either case $R(C)$ is transitive, whence $C \in V$, and thence $V = W$.

Finally, from $Q \in W = V$ it follows that $R(Q)$ is transitive by definition of V. $\square$

5.4.2 Regularity

The following theorems confirm that every element, subset, pairing, power set, union, intersection, and Cartesian product of well-formed sets is again a well-formed set.

5.35 Theorem. *For each well-formed set X there is a* smallest *ordinal A with* $X \in R(A)$.

Proof. If X is a well-formed set, then there exists an ordinal C such that $X \in R(C) \in \mathscr{P}[R(C)] = R(C \cup \{C\})$. Hence $X \in R(C \cup \{C\})$ by transitivity (theorem 5.34). By theorem 5.27, there exists a smallest ordinal $A \in C \cup \{C\}$ such that $X \in R(A)$.

For every ordinal D, either $D = A$, or $D \in A \in C \cup \{C\}$ and then $X \notin R(D)$, or $A \in D$ and then D is not the smallest such ordinal. $\square$

5.36 Theorem. *If X is well-formed, then every $Y \in X$ is well-formed.*

Proof. If X is well-formed, then there exists an ordinal A such that $X \in R(A)$.

If $A = \varnothing$, then $X \in R(A) = R(\varnothing) = \varnothing$, whence $Y \in X$ is vacuously well-formed.

If there exists an ordinal B such that $A = B \cup \{B\}$, then $X \in R(A) = \mathscr{P}[R(B)]$, whence $X \subseteq \mathscr{P}[R(B)]$ by transitivity of $\mathscr{P}[R(B)]$, Consequently, $Y \in R(A) = \mathscr{P}[R(B)]$ is well-formed for each $Y \in X$. If $R(A) = \bigcup_{B \in A} R(B)$ and the theorem holds for each $Z \in R(B)$ for each $B \in A$, then for each $X \in R(A)$ there exists $B \in A$ such that $X \in R(B)$ whence every $Y \in X$ is also well-formed. $\square$

5.37 Theorem. *If X and Y are well-formed sets, then so are $\{X, Y\}$, $\mathscr{P}(X)$, $\bigcup X$, $X \times Y$, every subset of X, and $\bigcap X$ provided $X \neq \varnothing$.*

Proof. If X and Y are well-formed sets, then there exist ordinals A and B such that $X \in R(A)$ and $Y \in R(B)$. Either $A = B$ (whence $R(A) = R(B)$), or $A \in B$ (whence $R(A) \subseteq R(B)$), or $B \in A$ (whence $R(B) \subseteq R(A)$). For instance, assume that $R(A) \subseteq R(B)$. Thus $X \in R(B)$ and $Y \in R(B)$, so that $\{X, Y\} \in \mathscr{P}[R(B)] = R(B \cup \{B\})$, whence $\{X, Y\}$ is well-formed, because $B \cup \{B\}$ is an ordinal.

Similarly, if $X \in R(B)$ is a well-formed set, then $X \subseteq R(B)$ by transitivity, whence $\mathscr{P}(X) \subseteq \mathscr{P}[R(B)] = R(B \cup \{B\})$ and hence $\mathscr{P}(X) \in \mathscr{P}[R(B \cup \{B\})] = R(A \cup \{A\})$ with $A := B \cup \{B\}$. Thus, $\mathscr{P}(X)$ is well-formed.

Consequently, from $\mathscr{P}(X) \in R(A \cup \{A\})$ it follows that $\mathscr{P}(X) \subseteq \mathscr{P}[R(A \cup \{A\})]$, whence if $S \subseteq X$, then $S \in \mathscr{P}[R(A \cup \{A\})]$ is well-formed.

In particular, $\bigcap X$ is well-formed because $\bigcap X \subseteq \bigcup X$ and $\bigcup X$ is well-formed.

Also, if $X \in R(B)$ is a well-formed set, then $X \subseteq R(B)$. Hence, if $Z \in Y$ and $Y \in X$, then $Y \subseteq X$ whence $Z \in X$ and $Z \in R(B)$. This shows that $\bigcup X \subseteq R(B)$. Consequently, $\bigcup X \in \mathscr{P}[R(B)] = R(B \cup \{B\})$ is well-formed.

In particular, because $\{X, Y\}$ is well-formed, it follows that $X \cup Y = \bigcup \{X, Y\}$ is well-formed, whence $\mathscr{P}(X \cup Y)$, $\mathscr{P}[\mathscr{P}(X \cup Y)]$, and $\mathscr{P}\{\mathscr{P}[\mathscr{P}(X \cup Y)]\}$ are also well-formed. Therefore, $X \times Y \in \mathscr{P}\{\mathscr{P}[\mathscr{P}(X \cup Y)]\}$ is also well-formed. $\square$

5.38 Theorem. *For all well-formed sets X and Y, $\neg[(X \in Y) \wedge (Y \in X)]$. In particular, $X \notin X$ for each well-formed set X.*

Proof. If $X = \varnothing$, then $Y \notin X$ and $X \notin X$.

If $X \neq \varnothing$ and $Y \in X$, then it suffices to verify that $X \notin Y$. To this end, let A be the first ordinal such that $X \in R(A)$; in particular, $X \subseteq R(A)$ by transitivity of $R(A)$.

If $A = Z \cup \{Z\}$, then $X \in R(A) = R(Z \cup \{Z\}) = \mathscr{P}[R(Z)]$, so $X \subseteq R(Z)$, whence $Y \in X \subseteq R(Z)$, and hence $Y \subseteq R(Z)$. However, $X \notin R(Z)$ because A is the first ordinal with $X \in R(A)$. If A does not contain a last element, then $R(A) = \bigcup_{B \in A} R(B)$, whence $X \in R(A)$ means that there exists $B \in A$ with $X \in R(B)$, and then $X \subseteq R(B)$ by transitivity of $R(B)$. Yet $X \notin R(B)$ because $B \in A$ and A is the first ordinal with $X \in R(A)$;

Thus, in either case there exists $C \in A$ such that $X \subseteq R(C)$ and $X \notin R(C)$ and $Y \in R(C)$. (In the first case $C := Z$, while $C := B$ in the second case.)

Therefore $X \notin Y$, because $X \in Y \in R(C)$ and the transitivity of $R(C)$ would yield $X \in R(C)$. In particular, $X \notin X$, because the foregoing argument applied to $Y := X$ shows that if $X \in X$, then $X \notin X$, contrary to the axioms governing $\in$, which state that for all sets X and Y either $X \in Y$ or $X \notin Y$ but not both.   □

5.4.2.1 Independence of the axiom of regularity

As an extension of theorem 5.38, every nonempty well-formed set X contains an element Y that does not contain any element of X, so that $Y \cap X = \varnothing$, or, equivalently, so that there does not exist any $Z \in Y$ such that $Z \in X$ (exercise 5.31). Outside of well-formed sets, however, there exist systems of set theory in which the relation $A \in A$ may hold for some set. One way of preventing the relation $A \in A$ from holding for any set consists in the axiom of regularity,

$$\forall X\{(X \neq \varnothing) \Rightarrow \big(\exists Y[(Y \in X) \wedge \{\forall Z[(Z \in Y) \Rightarrow (Z \notin X)]\}]\big)\},$$

attributed independently to Zermelo and von Neumann [128, p. 53]. The axiom of regularity has the disadvantage of asserting a condition about sets already defined by previous axioms. Yet within the theory of well-formed sets, exercise 5.31 confirms that the axiom of regularity is *not* independent but is a theorem derivable from the other axioms. Because well-formed sets suffice for most of logic, mathematics, computer science, and their applications, the foundations of these fields can restrict themselves to well-formed sets [74]. In contrast to the derivability of the axiom of regularity from the other axioms of the theory of well-formed sets, neither the generalized continuum hypothesis nor its negation are derivable from the other axioms of the theory of well-formed sets [19, 20, 44, 45]. Thus, the "axiom of regularity" is an example of an axiom that is "dependent" on the other axioms, whereas the generalized continuum hypothesis is an example of an axiom that is "independent" from the other axioms.

5.4.3   *Exercises on Well-Formed Sets*

5.27 . Prove that the set $\mathbb{N}$ of all natural numbers is a well-formed set.

5.28 . Prove that every natural number $N \in \mathbb{N}$ is a well-formed set.

5.29 . Prove that the set $\mathbb{Z}$ of all integers is a well-formed set.

5.30 . Prove that the set $\mathbb{Q}$ of all rational numbers is a well-formed set.

5.31 . Prove that for each well-formed set X there exists $Y \in X$ such that $Y \cap X = \varnothing$.

5.32 . Prove that if X is a well-formed set, then so is $\{X\}$.

5.33 . For each well-formed set X, prove that if A is the smallest ordinal such that $X \in R(A)$, then there exists an ordinal B such that $A = B \cup \{B\}$.

5.34 . Prove that every finite set of well-formed sets is a well-formed set.

5.35 . Determine whether every countable set of well-formed sets is a well-formed set.

5.36 . Determine whether every set of well-formed sets is a well-formed set.

Chapter 6
The Axiom of Choice: Proofs by Transfinite Induction

6.1 Introduction

This chapter presents several statements, which are called "principles" because they are well-formed formulae but not propositions, in the sense that neither of them nor their negations are theorems, in the Zermelo-Frænkel set theory. The first sections show how Zorn's Maximal-Element Principle implies Zermelo's Well-Ordering Principle, which in turn implies the Choice Principle. Thus any extension of the Zermelo-Frænkel set theory that includes Zorn's Maximal-Element Principle as an axiom also includes the other two principles as theorems. From the Choice Principle, subsequent sections demonstrate the converse implications, known as Zorn's Lemma and Zermelo's Theorem, so that all three principles are logically equivalent within the Zermelo-Frænkel set theory. Hence all three principles are theorems in the Zermelo-Frænkel-Choice set theory, which includes the Choice Principle as the Axiom of Choice. The material also introduces yet other principles that are logically equivalent to the Axiom of Choice, for example, the principle of the distributivity of intersections over unions of families of sets. Any theory that requires any such equivalent principle thus also requires the Axiom of Choice. Other consequences of the Axiom of Choice include the existence of extrema for continuous functions on closed and bounded sets in Euclidean spaces.

6.2 The Choice Principle

This section presents several mutually equivalent forms of the Choice Principle.

© Springer Science+Business Media New York 2015
Y. Nievergelt, *Logic, Mathematics, and Computer Science*,
DOI 10.1007/978-1-4939-3223-8_6

6.2.1   The Choice-Function Principle

One version of the Choice Principle relies on the concept of a "choice function":

6.1 Definition (choice function). For each set $\mathscr{F}$ of nonempty sets, a **choice function** is a function $C : \mathscr{F} \to \bigcup \mathscr{F}$ such that $C(A) \in A$ for every set $A \in \mathscr{F}$.

6.2 Example. If $\mathscr{F} = \varnothing$, then $\bigcup \mathscr{F} = \bigcup \varnothing = \varnothing$: there exists exactly one function $C : \mathscr{F} \to \bigcup \mathscr{F}$, namely $\varnothing : \varnothing \to \varnothing$, which is "vacuously" a choice function.

6.3 Example. With $1 = \{\varnothing\}$, if $\mathscr{F} = \{1\}$, then $\bigcup \mathscr{F} = \bigcup\{1\} = 1 = \{\varnothing\}$: there exists exactly one function $C : \mathscr{F} \to \bigcup \mathscr{F}$, namely $C : \{1\} \to \{\varnothing\}$ with $C(1) = \varnothing$, which is a choice function because $C(1) \in 1$.

6.4 Example. For each nonempty set A, if $\mathscr{F} = \{A\}$, then there exists $X \in A$, whence $(A, X) \in \mathscr{F} \times \bigcup \mathscr{F} = \{A\} \times A$, and $C := \{(A, X)\}$ is a choice function.

6.5 Example. With $1 = \{\varnothing\}$ and $2 = \{\varnothing, 1\}$, if $\mathscr{F} = \{1, 2\}$, then $\bigcup \mathscr{F} = \bigcup\{1, 2\} = 1 \cup 2 = \{\varnothing, 1\} = 2$. There are two choice functions $C : \{1, 2\} \to \{\varnothing, 1\}$:
    The requirement that $C(1) \in 1 = \{\varnothing\}$ imposes that $C(1) = \varnothing$.
    The requirement that $C(2) \in 2 = \{\varnothing, 1\}$ allows for the two possibilities:
    either $C(2) = \varnothing \in \{\varnothing, 1\}$, or $C(2) = 1 \in \{\varnothing, 1\}$.
    Thus the two choice functions are $F : \mathscr{F} \to \bigcup \mathscr{F}$ with $F(1) = 0$ and $F(2) = 0$, or $G : \mathscr{F} \to \bigcup \mathscr{F}$ with $G(1) = 0$ and $G(2) = 1$.

More generally, each finite set of nonempty sets has a choice function.

6.6 Theorem. *Within the Zermelo-Frænkel set theory, each finite set of nonempty sets has a choice function.*

Proof. This proof proceeds by induction on the number $N \in \mathbb{N}$ of sets.
    For $N = 0$, example 6.2 shows that the empty set has a choice function.
    Example 6.4 proves the theorem for $N = 1$.
    As an induction hypothesis, assume that the theorem holds for some $N = M \in \mathbb{N}$ and every finite set $\mathscr{F}$ with exactly M elements, all nonempty. For each set $\mathscr{G}$ with exactly $M + 1$ elements, all nonempty, there exists $B \in \mathscr{G}$, so that $\mathscr{F} := \mathscr{G} \setminus \{B\}$ has exactly M elements, all nonempty. By the induction hypothesis, there exists a choice function $F : \mathscr{F} \to \bigcup \mathscr{F}$. Example 6.4 shows that there exists a choice function $G : \{B\} \to \bigcup\{B\} = B$. Because $\mathscr{F} \cap \{B\} = \varnothing$, the union $C := F \cup G : \mathscr{F} \cup \{B\} = \mathscr{G} \to (\bigcup \mathscr{F}) \cup B = \bigcup \mathscr{G}$ is a function. Also, $C(A) = F(A) \in A$ for each $A \in \mathscr{F}$ and $C(B) = G(B) \in B$ for each $B \in \{B\}$, whence C is a choice function for $\mathscr{G}$. □

Still, theorem 6.6 applies only to finite sets.

6.7 Example. No choice functions are known for $\mathscr{G} = \mathscr{P}[\mathscr{P}(\mathbb{N}) \setminus \{\varnothing\}]$, which is the set of all nonempty sets of subsets of the set $\mathbb{N}$ of natural numbers, which corresponds to the set of all nonempty sets of real numbers between 0 and 1.

The statement of the existence of choice functions is one version of the Choice Principle, called the Choice-Function Principle to distinguish it from other versions.

6.8 Definition (Choice-Function Principle). Each set of nonempty sets has a choice function. Specifically, the Choice-Function Principle is formula (6.1):

$$\forall \mathscr{F} \left(\{\forall A \, [(A \in \mathscr{F}) \Rightarrow (A \neq \varnothing)]\} \right. \tag{6.1}$$

$$\left. \Rightarrow \left[\exists C \left\{ \left(C : \mathscr{F} \rightarrow \bigcup \mathscr{F} \right) \wedge \left[(\forall A \, \{(A \in \mathscr{F}) \Rightarrow [C(A) \in A]\}) \right] \right\} \right] \right).$$

Neither the Choice-Function Principle 6.8 nor its negation are propositions in the Zermelo-Frænkel set theory: there it is merely formula (6.1).

The concept of a "family" of sets provides a convenient way to specify more than one operation on a set, for instance, to choose more than one element from a set.

6.9 Definition (family of sets). A **family** of sets $\mathscr{F} = \{A_i : i \in \mathscr{I}\}$ is a set (of sets) with a function $I : \mathscr{I} \rightarrow \mathscr{F}, i \mapsto A_i$, from a set $\mathscr{I}$, called the **indexing set** or **set of indices**, to $\mathscr{F}$.

6.10 Example (Self-Indexed Family of Sets). Each set (of sets) $\mathscr{F}$ is a family of sets: the same set $\mathscr{I} := \mathscr{F}$ serves as an indexing set, and the identity function $\iota : \mathscr{F} \rightarrow \mathscr{F}, E \mapsto A_E := E$ shows that $\mathscr{F} = \{E : E \in \mathscr{F}\} = \{A_E : E \in \mathscr{I}\}$.

Still other versions of the Choice Principle rely on families of sets, as in definitions 6.11 and 6.12.

6.11 Definition (family choice-function). For each family $\{A_i : i \in \mathscr{I}\}$ of sets, a **family choice-function** is a function $C : \mathscr{I} \rightarrow \bigcup_{i \in \mathscr{I}} A_i$ such that $C(i) \in A_i$ for each index $i \in \mathscr{I}$.

6.12 Definition (Family Choice-Function Principle). Each family of nonempty sets has a family choice-function.

The Family Choice-Function Principle 6.12 and its negation are formulae but not propositions in the Zermelo-Frænkel set theory. However, theorem 6.13 shows that choice functions are equivalent to family choice-functions.

6.13 Theorem. *Within the Zermelo-Frænkel set theory, the Choice-Function Principle 6.8 is logically equivalent to the Family Choice-Function Principle 6.12*

Proof. If the Family Choice-Function Principle 6.12 holds and $\mathscr{F}$ is a set of nonempty sets, then the self-indexed family $\mathscr{F} = \{E : E \in \mathscr{F}\}$ from example 6.10 has a family choice-function $C : \mathscr{F} \rightarrow \bigcup_{E \in \mathscr{F}} E = \bigcup \mathscr{F}$ such that $C(E) \in E$ for each $E \in \mathscr{F}$. Thus C is a choice function for $\mathscr{F}$.

Conversely, if the Choice-Function Principle 6.8 holds and $\mathscr{F} := \{A_i : i \in \mathscr{I}\}$ is a nonempty family of nonempty sets, indexed by a function $I : \mathscr{I} \rightarrow \mathscr{F}$, then there exists a choice function $C : \mathscr{F} \rightarrow \bigcup \mathscr{F}$ such that $C(E) \in E$ for each $E \in \mathscr{F}$. The composite function $C \circ I : \mathscr{I} \rightarrow \bigcup_{i \in \mathscr{I}} A_i, i \mapsto C[I(i)] = C[A_i] \in A_i$, is a family choice-function.   □

Another version of the Choice Principle relies on choice functions only for pairwise disjoint sets, as specified by definition 6.14.

6.14 Definition (Choice-Function Principle for Pairwise Disjoint Sets). Each set of *pairwise disjoint* nonempty sets has a choice function.

Yet another version of the Choice Principle relies on functions and relations, as specified by definition 6.15.

6.15 Definition (Choice-Relation Principle). For each relation R there exists a function $F \subseteq R$ such that F and R have the same domain [128, p. 243, AC$_3$].

In the Zermelo-Frænkel set theory, a relation is a subset of a Cartesian product: $R \subseteq A \times B$. There is a *different* principle, called the Relational Axiom of Choice, where R may be a relation defined for all sets [88, p. 22].

6.2.2 The Choice-Set Principle

This subsection provides different statements of the Choice Principle that are logically mutually equivalent within the Zermelo-Frænkel set theory. One version of the Choice Principle relies on "choice sets" rather than choice functions.

6.16 Definition (choice set). For each set $\mathscr{F}$ of nonempty sets, a **choice set** is a set $S \subseteq \bigcup \mathscr{F}$ such that for each $A \in \mathscr{F}$, $S \cap A$ contains exactly one element.

As with the Choice-Function Principle 6.8, neither the Choice-Set Principle nor its negation are propositions in the Zermelo-Frænkel set theory: it is merely a formula, stated in words in definition 6.17.

6.17 Definition (Choice-Set Principle). Each set of nonempty sets has a choice set.

6.18 Example. If $\mathscr{F} = \varnothing$, then $\bigcup \mathscr{F} = \bigcup \varnothing = \varnothing$: there exists exactly one subset $S \subseteq \bigcup \mathscr{F} = \varnothing$, namely $S = \varnothing$, which is "vacuously" a choice set, because $S \cap A$ "vacuously" contains exactly one element for each $A \in \mathscr{F} = \varnothing$.

6.19 Example. With $1 = \{\varnothing\}$, if $\mathscr{F} = \{1\}$, then $\bigcup \mathscr{F} = \bigcup \{1\} = 1 = \{\varnothing\}$, whence there exists exactly one choice set $S \subseteq \bigcup \mathscr{F} = 1$, namely $S = 1$. Indeed, the only element in $\mathscr{F} = \{1\}$ is 1, and $S \cap 1 = 1 \cap 1 = \{\varnothing\}$, which contains exactly one element, namely $\varnothing \in 1$. In contrast $\varnothing \subseteq \bigcup \mathscr{F} = 1$ is not a choice set.

Theorem 6.20 shows that choice sets are equivalent to choice functions.

6.20 Theorem. *Within the Zermelo-Frænkel set theory, the Choice-Function Principle 6.8 is logically equivalent to the Choice-Set Principle 6.17.*

Proof. Each choice set $S \subseteq \bigcup \mathscr{F}$ corresponds to the choice function F_S defined by

$$F_S := \left\{ (A, X) \in \mathscr{F} \times \bigcup \mathscr{F} : X \in A \cap S \right\}.$$

Conversely, each choice function is a subset of a Cartesian product: $F \subseteq \mathscr{F} \times \bigcup \mathscr{F}$. Thus F consists of pairs (A, X) with exactly one $X \in A$ for each $A \in \mathscr{F}$. Hence

$$S_F := \left\{ X \in \bigcup \mathscr{F} : \exists A \{ (A \in \mathscr{F}) \wedge [(A, X) \in F] \} \right\}$$

is a choice set, obtained from the second projection (example 3.87) of F. $\square$

Still another version of the Choice Principle relies on choice sets only for pairwise disjoint sets, as specified by definition 6.21.

6.21 Definition ("Pairwise Disjoint" Choice-Set Principle). Each set of *pairwise disjoint* nonempty sets has a choice set.

Theorem 6.22 shows that choice sets for any sets, and choice sets for pairwise disjoints sets, are mutually equivalent in the Zermelo-Frænkel set theory.

6.22 Theorem. *Within the Zermelo-Frænkel set theory, the Choice-Set Principle 6.17 is logically equivalent to the "Pairwise Disjoint" Choice-Set Principle 6.21.*

Proof. The "Pairwise Disjoint" Choice-Set Principle 6.21 is a particular case of the Choice-Set Principle 6.17. Thus the latter implies the former.

To establish the converse implication, to each set $\mathscr{E}$ of nonempty sets corresponds the set $\mathscr{F}$ of *pairwise disjoint* nonempty sets defined by

$$A' := \{ (A, X) : (X \in A) \} \subseteq \mathscr{E} \times \left(\bigcup \mathscr{E} \right),$$

$$\mathscr{F} := \{ A' : A \in \mathscr{E} \} \subseteq \mathscr{P} \left[\mathscr{E} \times \left(\bigcup \mathscr{E} \right) \right].$$

The elements of $\mathscr{F}$ are pairwise disjoint, because if $A, B \in \mathscr{E}$ and $A \neq B$, then $(A, X) \neq (B, Y)$ regardless of $X \in A$ and $Y \in B$, whence $A' \cap B' = \varnothing$. Also, $A' \neq \varnothing$ for each $A' \in \mathscr{F}$, because A contains at least one element X by hypothesis, so that A' contains at least one element (A, X). If the "Pairwise Disjoint" Choice-Set Principle 6.21. holds, then there exists a choice set $S \subseteq \bigcup \mathscr{F} \subseteq \mathscr{E} \times (\bigcup \mathscr{E})$ such that $S \cap A'$ is a singleton for each $A' \in \mathscr{F}$: there exists a unique $X \in A$ for which $(A, X) \in A' \cap S$. The second projection of S yields a choice set $T \subseteq \bigcup \mathscr{E}$ such that $T \cap A = \{X\}$ is a singleton for each $A \in \mathscr{F}$. $\square$

The statements of the Choice-Set Principle 6.17 and the "Pairwise Disjoint" Choice-Set Principle 6.21 with a proof of their mutual equivalence in the Zermelo-Frænkel set theory similar to theorem 6.22 are in Zermelo's [145].

6.23 Definition (Axiom of Choice). The **Axiom of Choice** is any of the principles 6.8, 6.14, 6.15, 6.17, 6.21, included as an axiom in a set theory.

6.2.3   Exercises on Choice Principles

6.1 . Find the number of choice functions for each finite set $\mathscr{F} = \{A_0, \ldots, A_{N-1}\}$ of nonempty finite sets where each element A_j has exactly N_j elements.

6.2 . Find the number of choice sets for each finite set $\mathscr{F} = \{A_0, \ldots, A_{N-1}\}$ of nonempty finite sets where each element A_j has exactly N_j elements.

6.3 . Prove that each finite set of nonempty finite sets has a choice set.

6.4 . Prove that the Choice-Function Principle 6.8 is logically equivalent to the "Pairwise Disjoint" Choice-Function Principle 6.14. within the Zermelo-Frænkel set theory.

6.5 . Translate the Choice-Set Principle 6.17 into a logical formula, similar to formula (6.1) for the Choice-Function Principle 6.8.

6.6 . Translate the "Pairwise Disjoint" Choice-Set Principle 6.21. into a logical formula, similar to formula (6.1) for the Choice-Function Principle 6.8.

6.7 . Prove that the Choice-Relation Principle 6.15 is logically equivalent to the Choice-Function Principle 6.8 in the Zermelo-Frænkel set theory.

6.8 . Translate the "Pairwise Disjoint" Choice-Function Principle 6.14 into a logical formula, similar to formula (6.1) for the Choice-Function Principle 6.8.

6.3   Maximality and Well-Ordering Principles

This section introduces two principles and shows that each of them implies the Choice Principle.

6.3.1   Zermelo's Well-Ordering Principle

As the Choice-Function Principle 6.8, neither Zermelo's Well-Ordering Principle nor its negation are propositions in the Zermelo-Frænkel set theory: it is merely a formula, stated in words in definition 6.24.

6.24 Definition (Zermelo's Well-Ordering Principle). Each set has a well-order. [88, p. 117].

6.25 Example. The order $\leq$ is a well-order on the set $\mathbb{N}$ of all natural numbers.

6.26 Example. Every finite set has a well-order. Indeed, by definition of "finite" for each finite set E there exists a natural number $N \in \mathbb{N}$ and a bijection $F : E \to N$. The subset $N \subset \mathbb{N}$ is well-ordered by example 6.25 and theorem 3.197. Hence the relation $\preceq$ defined on E by $X \preceq Y$ if and only if $F(X) \leq F(Y)$ well-orders E.

Theorem 6.27 shows that the existence of a choice function on a set follows from the existence of a well-ordering on its union.

6.27 Theorem. *For each set $\mathscr{F}$ of nonempty sets with a well-ordered union $\bigcup \mathscr{F}$, there exists a function $C : \mathscr{F} \to \bigcup \mathscr{F}$ such that $C(A) \in A$ for every set $A \in \mathscr{F}$.*

Proof. Let $C(A)$ be the first element of A relative to the well-order on $\bigcup \mathscr{F}$. ☐

6.28 Example. If $\mathscr{F} = \mathscr{P}(\mathbb{N}) \setminus \{\varnothing\}$, which is the set of all nonempty subsets of the set $\mathbb{N}$ of natural numbers, then $\bigcup \mathscr{F} = \mathbb{N}$, where $\leq$ is a well-order. By the proof of theorem 6.27, the function $C : \mathscr{F} \to \mathbb{N}$ with $C(A) = \min(A)$ chooses the smallest element of A and thus is a choice function for $\mathscr{F}$.

6.29 Example. No well-orders are known for $\mathscr{R} = \mathscr{P}(\mathbb{N})$, which is the set of all subsets of the set $\mathbb{N}$ of natural numbers, which is also isomorphic to the set of all real numbers from 0 through 1. Because $\mathscr{R} = \bigcup \mathscr{G}$ with $\mathscr{G} = \mathscr{P}[\mathscr{P}(\mathbb{N}) \setminus \{\varnothing\}]$ from example 6.7, any specific well-order on $\mathscr{R}$ would yield a specific choice function on $\mathscr{G}$, by theorem 6.27.

Theorem 6.30 shows a logical relation between the foregoing principles.

6.30 Theorem. *Zermelo's Well-Ordering Principle 6.24 implies the Choice-Function Principle 6.8.*

Proof. Theorem 6.27 shows that *if every set* admits a well-ordering relation, in particular, $\bigcup \mathscr{F}$, *then every set $\mathscr{F}$* of nonempty sets has a choice function. ☐

As sets (of pairs), relations are partially ordered by inclusion. Theorem 6.31 shows that every total order is maximal relative to inclusion among partial orders.

6.31 Theorem. *Each reflexive total order on a set is maximal among all partial orders on that set.*

Proof. If R is a reflexive total order on a set A, and if T is a relation on A such that $R \subsetneq T$, then there are $X, Y \in A$ such that $(X, Y) \in T$ but $(X, Y) \notin R$, whence $(Y, X) \in R$ by totality of R. Hence $X \neq Y$ by reflexivity of R. Consequently, $(Y, X) \in R \subset T$ and $(X, Y) \in T$ with $X \neq Y$. Thus T is not anti-symmetric and therefore not a partial order. Therefore, R is not properly contained in any partial order on A. ☐

The concept of maximality in a partially ordered set leads to Zorn's Maximal-Element Principle.

6.3.2 Zorn's Maximal-Element Principle

As the Choice-Function Principle 6.8 and Zermelo's Well-Ordering Principle 6.24, neither Zorn's Maximal-Element Principle nor its negation are propositions in the Zermelo-Frænkel set theory: it is merely a formula, stated in words in definition 6.32.

6.32 Definition (Zorn's Maximal-Element Principle). In each partially ordered set where each chain has an upper bound, there is a maximal element [88, p. 117], [128, p. 248, Z_2].

Theorem 6.33 shows another logical relation between the foregoing principles.

6.33 Theorem. *Zorn's Maximal-Element Principle 6.32 implies Zermelo's Well-Ordering Principle 6.24.*

Proof. This proof follows Dugundji's [30, p. 34, (2) $\Rightarrow$ (3)]. For each set X, let $\mathscr{X}$ be the set of all pairs (A, R) where R weakly well-orders a subset $A \subseteq X$. Thus $\mathscr{X} \subseteq [\mathscr{P}(X)] \times [\mathscr{P}(X \times X)]$ is well-formed. Also, $\mathscr{X} \neq \varnothing$, because $(\varnothing, \varnothing) \in \mathscr{X}$.

Define a relation $\preceq$ on $\mathscr{X}$ by $(A, R) \preceq (B, S)$ if and only if $A \subseteq B$ and $R \subseteq S$ with $(X, Y) \in S$ for all $X \in A$ and $Y \in B \setminus A$. Then $\preceq$ partially orders $\mathscr{X}$:

Reflexivity. $(A, R) \preceq (A, R)$ because $A \subseteq A$ and $R \subseteq R$ with $A \setminus A = \varnothing$.

Antisymmetry.   If $(A, R) \preceq (B, S)$ with $(B, S) \preceq (A, R)$, then $A \subseteq B$ and $R \subseteq S$ with $B \subseteq A$ and $S \subseteq R$, whence $A = B$ and $R = S$, so that $(A, R) = (B, S)$.

Transitivity.   If $(A, R) \preceq (B, S)$ and $(B, S) \preceq (C, T)$, then $A \subseteq B \subseteq C$ and $R \subseteq S \subseteq T$. If also $X \in A$ and $Z \in C \setminus A$, then two cases can arise: either $Z \in B \setminus A$, or $Z \in C \setminus B$.

In the first case, if $Z \in B \setminus A$, then $(X, Z) \in S \subseteq T$ because $X \in A$.

In the second case, if $Z \in C \setminus B$, then $(X, Z) \in T$ because $X \in B$.

Thus $(A, R) \preceq (C, T)$.

For each chain $\mathscr{Y} \subseteq \mathscr{X}$ linearly ordered by $\preceq$ in $\mathscr{X}$ define

$$C := \bigcup_{(A,R) \in \mathscr{Y}} A,$$

$$T := \bigcup_{(A,R) \in \mathscr{Y}} R.$$

Then T weakly well-orders C:

Reflexivity.   For each $X \in C$ there exists $(A, R) \in \mathscr{Y}$ such that $X \in A$, whence $(X, X) \in R \subseteq T$ by reflexivity of R.

Antisymmetry.   For all $X, Y \in C$ there exist $(A, R), (B, S) \in \mathscr{Y}$ such that $X \in A$ and $Y \in B$. Because $\mathscr{Y}$ is a chain, two cases can arise: either $(A, R) \preceq (B, S)$, or $(B, S) \preceq (A, R)$. In the first case, if $(A, R) \preceq (B, S)$, then $X, Y \in B$; if also

$(X, Y) \in T$ and $(Y, X) \in T$, then $(X, Y) \in S$ and $(Y, X) \in S$ by maximality of S on B from theorem 6.31, whence $X = Y$ by anti-symmetry of S. The second case is similar.

Transitivity. For all $X, Y, Z \in C$ there exist $(A, R), (B, S), (D, U) \in \mathscr{Y}$ such that $X \in A, Y \in B, Z \in D$. Because $\mathscr{Y}$ is a chain, assume $A \subseteq B \subseteq D$ and $R \subseteq S \subseteq U$. (The other five orders are similar.) If also $(X, Y), (Y, Z) \in T$, then there exist $(E, V), (F, W) \in \mathscr{Y}$ such that $(X, Y) \in V$ and $(Y, Z) \in W$. Again because $\mathscr{Y}$ is a chain, assume $V \subseteq W$, whence $(X, Y), (Y, Z) \in W$, and hence $(X, Y), (Y, Z) \in U$ by maximality of U on D from theorem 6.31, whence $(X, Z) \in U \subseteq T$ by transitivity of U.

Well-ordering. Each nonempty subset $E \subseteq C$ contains at least one element $Z \in E \subseteq C = \bigcup_{(A,R) \in \mathscr{Y}} A$. Hence there exists $(A, R) \in \mathscr{Y}$ such that $Z \in A$. Thus $A \cap E \neq \varnothing$, whence $A \cap E$ has a least element $X \in A \cap E$, because R well-orders A.

Hence X is also the least element of E. Indeed, if $Y \in E$, then there exists $(B, S) \in \mathscr{Y}$ such that $Y \in B$. Either $Y \in A$, or $Y \in B \setminus A$.

In the first case, if $Y \in A$, then $(X, Y) \in R \subseteq T$, because X is the least element of $A \cap E$.

In the second case if $Y \in B \setminus A$, then $(X, Y) \in S$ by definition of $\preceq$ and hence $(X, Y) \in S \subseteq T$.

In either case $(X, Y) \in T$. Therefore X is the least element of E.

Thus $(C, T) \in \mathscr{X}$, and (C, T) is an upper bound relative to $\preceq$ for $\mathscr{Y}$.

Hence, if Zorn's Maximal-Element Principle holds, then $\mathscr{X}$ contains a maximal element (D, U). In particular, $D = X$: by contraposition, if $D \neq X$, then there would exist $K \in X \setminus D$. Setting $E := D \cup \{K\}$ and $V := U \cup \{(H, K) : H \in D\}$ would define a strict upper bound such that $(D, U) \prec (E, V)$, contradicting the maximality of (D, U). Therefore, U well-orders $D = X$. $\qquad \square$

Definition 6.34 states another version of Zorn's Maximal Principle.

6.34 Definition (Zorn's Maximal-Set Principle). In each nonempty set (of sets) that contains the union of each chain of elements (which are sets) relative to inclusion, there is a maximal element relative to inclusion: an element that is not a proper subset of any other element. [128, p. 245, Z_1].

Max Zorn *stated* the Maximal-Set Principle 6.34 for sets of sets partially ordered by inclusion in reference [147, p. 667, (MP)], where the axiom of the power-set is also in question [147, p. 669, (MP)], but did not prove it there from any other axioms. Indeed, Max Zorn also stated that he would show its relation and equivalence with the Axiom of Choice in "another paper" [147, p. 669, (MP)]. In other words, "Zorn's Lemma" is *not* in [147].

The Axiom of Choice is logically independent from the Zermelo-Frænkel axioms of set theory [21, 66, Ch. 5]. Consequently, appending Zorn's Maximal-Element Principle to the Zermelo-Frænkel axioms of set theory yields a larger "Zermelo-Frænkel-Choice" set theory that also includes Zermelo's Well-Ordering Principle and the Axiom of Choice.

6.3.3   Exercises on Maximality and Well-Orderings

6.9 . Prove that the hypothesis of Zorn's Maximal-Element Principle 6.32 implies that the partially ordered set is not empty.

6.10 . Prove that the conclusion of Zorn's Maximal-Set Principle 6.34 requires the hypothesis that the set be not empty.

6.11 . Translate Zorn's Maximal-Set Principle 6.34 into a logical formula, similar to formula (6.1) for the Choice-Function Principle 6.8.

6.12 . Translate Zermelo's Well-Ordering Principle 6.24 into a logical formula, similar to formula (6.1) for the Choice-Function Principle 6.8.

6.13 . For each partial order $\preceq$ on a set E, let $\mathscr{C} \subseteq \mathscr{P}(E)$ be the set of all chains relative to $\preceq$ in E. Thus each element of $\mathscr{C}$ is a subset of E on which $\preceq$ is a linear order. Partially order $\mathscr{C}$ by inclusion. Prove that the union of each chain in $\mathscr{C}$ relative to inclusion is an element of $\mathscr{C}$.

6.14 . Prove that Zorn's Maximal-Set Principle 6.34 is logically equivalent to Zorn's Maximal-Element Principle 6.32 within the Zermelo-Frænkel set theory.

6.4   Unions, Intersections, and Products of Families of Sets

This section shows that the Choice Principle is equivalent to the distributivity of intersections over unions of sets.

6.4.1   The Multiplicative Principle

Yet other versions of the Choice Principle rely on Cartesian products, as in definitions 6.35 and 6.36.

6.35 Definition (Cartesian Product). The **Cartesian product** $\prod_{i \in \mathscr{I}} A_i$ of a family of sets $\{A_i : i \in \mathscr{I}\}$ is the set of all functions $C : \mathscr{I} \rightarrow \bigcup_{i \in \mathscr{I}} A_i$, such that $C(i) \in A_i$ for each $i \in \mathscr{I}$

6.36 Definition (Multiplicative Principle). For each nonempty family of nonempty sets $\{A_i : i \in \mathscr{I}\}$, the Cartesian product $\prod_{i \in \mathscr{I}} A_i$ is not empty [88, p. 117].

The Multiplicative Principle 6.36 and its negation are formulae but not propositions in the Zermelo-Frænkel set theory. However, theorem 6.40 shows that the Multiplicative Principle 6.36 is equivalent to the foregoing choice principles.

6.37 Theorem. *Within the Zermelo-Frænkel set theory, the Family Choice-Function Principle 6.12 is logically equivalent to the Multiplicative Principle 6.36.*

Proof. Definitions 6.11 and 6.35 show that the Cartesian product consists of all the family choice-functions as its elements. Thus the Cartesian product is nonempty if and only if there exists a family choice-function.　□

6.4.2 The Distributive Principle

This subsection shows that the Choice Principle is equivalent to the distributivity of intersections over unions of families of sets. The following development follows Dugundji's [30, Ch. I, § 9.7–9.8, p. 25].

6.38 Definition (Distributive Principle). For each family of sets $\{A_i : i \in \mathscr{I}\}$ and each partition $\{\mathscr{I}_\ell : \ell \in \mathscr{L}\}$ of the indexing set $\mathscr{I}$, which sets up equivalence classes in $\mathscr{I}$, let $\mathscr{K} := \prod_{\ell \in \mathscr{L}} \mathscr{I}_\ell$; then

$$\bigcup_{k \in \mathscr{K}} \left(\bigcap_{\ell \in \mathscr{L}} A_{k(\ell)} \right) = \bigcap_{\ell \in \mathscr{L}} \left(\bigcup_{i \in \mathscr{I}_\ell} A_i \right). \tag{6.2}$$

6.39 Example. For the self-indexed family $\mathscr{F} := \{A, B, C, D\}$ and the self-indexed partition $\mathscr{G} := \{\{A, B\}, \{C, D\}\}$, the right-hand side of equation (6.2) becomes

$$\bigcup_{E \in \{A,B\}} E = A \cup B,$$

$$\bigcup_{E \in \{C,D\}} E = C \cup D,$$

$$\bigcap_{\ell \in \mathscr{L}} \left(\bigcup_{i \in \mathscr{I}_\ell} A_i \right) = (A \cup B) \cap (C \cup D).$$

The Cartesian product of the partitioning equivalence classes is $\mathscr{K} := \{A, B\} \times \{C, D\} = \{(A, C), (A, D), (B, C), (B, D)\}$, and the left-hand side of equation (6.2) becomes

$$\bigcup_{k \in \mathscr{K}} \left(\bigcap_{\ell \in \mathscr{L}} A_{k(\ell)} \right) = (A \cap C) \cup (A \cap D) \cup (B \cap C) \cup (B \cap D).$$

The Distributive Principle 6.38 and its negation are not propositions in the Zermelo-Frænkel set theory; there they are mere formulae. Nevertheless, theorem 6.37 shows that one of the inclusions, but only one, implied by equation (6.2) is a theorem in the Zermelo-Frænkel set theory.

6.40 Theorem. *Within the Zermelo-Frænkel set theory, for each family of sets*
$\{A_i : i \in \mathscr{I}\}$ *and each partition* $\{\mathscr{I}_\ell : \ell \in \mathscr{L}\}$ *of the indexing set* $\mathscr{I}$*, which
sets up equivalence classes in* $\mathscr{I}$*, let* $\mathscr{K} := \prod_{\ell \in \mathscr{L}} \mathscr{I}_\ell$*; then*

$$\bigcup_{k \in \mathscr{K}} \left(\bigcap_{\ell \in \mathscr{L}} A_{k(\ell)} \right) \subseteq \bigcap_{\ell \in \mathscr{L}} \left(\bigcup_{i \in \mathscr{I}_\ell} A_i \right). \tag{6.3}$$

Proof. This proof unravels the definitions of unions, intersections, and Cartesian
products:

$X \in \bigcup_{k \in \mathscr{K}} \left(\bigcap_{\ell \in \mathscr{L}} A_{k(\ell)} \right)$

$\Updownarrow$ definition of $\bigcup$

$\exists k \left[(k \in \mathscr{K}) \wedge \left(X \in \bigcap_{\ell \in \mathscr{L}} A_{k(\ell)} \right) \right]$

$\Updownarrow$ definition of $\bigcap$

$\exists k \left[(k \in \mathscr{K}) \wedge \left\{ \forall \ell \left[(\ell \in \mathscr{L}) \Rightarrow \left(X \in A_{k(\ell)} \right) \right] \right\} \right]$

$\Downarrow$ $(k \in \mathscr{K}) \Rightarrow$
                                                $(\forall \ell \{ (\ell \in \mathscr{L}) \Rightarrow [k(\ell) \in \mathscr{I}_\ell] \})$

$\forall \ell \left[(\ell \in \mathscr{L}) \Rightarrow \{ \exists i \left[(i \in \mathscr{I}_\ell) \wedge (X \in A_i) \right] \} \right]$

$\Updownarrow$ definition of $\bigcup$

$\forall \ell \left[(\ell \in \mathscr{L}) \Rightarrow \left(X \in \bigcup_{i \in \mathscr{I}_\ell} A_i \right) \right]$

$\Updownarrow$ definition of $\bigcap$

$X \in \bigcap_{\ell \in \mathscr{L}} \left(\bigcup_{i \in \mathscr{I}_\ell} A_i \right)$

$\square$

Theorem 6.41 shows that in the Zermelo-Frænkel set theory, the Distributive
Principle 6.38 is equivalent to the foregoing choice principles.

6.41 Theorem. *Within the Zermelo-Frænkel set theory, the Multiplicative
Principle 6.36 is logically equivalent to the Distributive Principle 6.38.*

Proof. Reversing the last two equivalences in the proof of theorem 6.40 gives

$$\left[X \in \bigcap_{\ell \in \mathscr{L}} \left(\bigcup_{i \in \mathscr{I}_\ell} A_i \right) \right] \Leftrightarrow \{ \forall \ell \left[(\ell \in \mathscr{L}) \Rightarrow \{ \exists i \left[(i \in \mathscr{I}_\ell) \wedge (X \in A_i) \right] \} \right] \},$$

which implies that for every $\ell \in \mathscr{L}$,

$$\{ i \in \mathscr{I}_\ell : X \in A_i \} \neq \varnothing.$$

If the Multiplicative Principle 6.36 holds, then the Cartesian product contains a fam-
ily choice function $k \in \prod_{\ell \in \mathscr{L}} \{ i \in \mathscr{I}_\ell : X \in A_i \}$, so that $k(\ell) \in \{ i \in \mathscr{I}_\ell : X \in A_i \}$,

whence $X \in A_{k(\ell)}$, for each $\ell \in \mathscr{L}$, which is precisely the converse of the implication in the proof of theorem 6.40. Thus equation (6.2) holds, and so does the Distributivity of Intersection over Union Principle 6.38.

Conversely, for each nonempty family of nonempty sets $\{\mathscr{I}_\ell : \ell \in \mathscr{L}\}$, let $\mathscr{K} := \prod_{\ell \in \mathscr{L}} \mathscr{I}_\ell$, and for each $i \in \bigcup_{\ell \in \mathscr{L}} \mathscr{I}_\ell$ define $A_i := \{\varnothing\}$ or any other singleton. Thus for every $\ell \in \mathscr{L}$,

$$\bigcup_{i \in \mathscr{I}_\ell} A_i = \{\varnothing\},$$

$$\bigcap_{\ell \in \mathscr{L}} \left(\bigcup_{i \in \mathscr{I}_\ell} A_i \right) = \{\varnothing\}.$$

If the Distributivity of Intersection over Union Principle 6.38 holds, then equation (6.2) gives

$$\bigcup_{k \in \mathscr{K}} \left(\bigcap_{\ell \in \mathscr{L}} A_{k(\ell)} \right) = \{\varnothing\},$$

which is not empty, whence neither is $\mathscr{K} = \prod_{\ell \in \mathscr{L}} \mathscr{I}_\ell$, so that the Multiplicative Principle 6.36 holds.   □

6.4.3   Exercises on the Distributive and Multiplicative Principles

6.15 . Translate the Multiplicative Principle 6.36 into a logical formula, similar to formula (6.1) for the Choice-Function Principle 6.8.

6.16 . Translate the Distributive Principle 6.38 into a logical formula, similar to formula (6.1) for the Choice-Function Principle 6.8.

6.5   Equivalence of the Choice, Zorn's, and Zermelo's Principles

Based on Dugundji's proof [30, Ch. II, § 2], this section proves that the Choice-Function Principle 6.8 implies Zorn's Maximal-Element Principle 6.32 by means of the concept of a "tower" of sets relative to a function.

6.5.1 Towers of Sets

The subsection introduces the concept of "towers" of sets relative to a function.

6.42 Definition (Tower). [30, Ch. II, p. 32, def. 2.2] A nonempty set $\mathcal{T} \subseteq \mathcal{P}(X)$ of subsets of a set X is a **tower** relative to a function $F : \mathcal{T} \to X$ if and only if

(T.A) $\varnothing \in \mathcal{T}$,
(T.B) if $\mathcal{A} \subseteq \mathcal{T}$ is linearly ordered by inclusion, then $\bigcup \mathcal{A} \in \mathcal{T}$, and
(T.C) if $A \in \mathcal{T}$, then $A \cup \{F(A)\} \in \mathcal{T}$.

A **sub-tower** is a nonempty subset $\mathcal{S} \subseteq \mathcal{T}$ that is also a tower relative to the restriction $F|_{\mathcal{S}} : \mathcal{S} \to X$.

An element $M \in \mathcal{S}$ is **medial** in a sub-tower $\mathcal{S}$ if and only if $A \subseteq M$ or $M \subseteq A$ for every element $A \in \mathcal{S}$.

A sub-tower is **minimal** if and only if it does not properly contain any sub-tower.

6.43 Example. The empty set $\varnothing$ is medial in every tower $\mathcal{T}$: indeed, $\varnothing \in \mathcal{T}$ by the definition 6.42 of towers, and $\varnothing \subseteq A$ for every $A \in \mathcal{T}$. $\qquad\square$

6.44 Theorem. *For each nonempty set $\mathcal{F}$ of towers relative to a common function $F : \bigcup \mathcal{F} \to X$, the intersection $\mathcal{M} := \bigcap \mathcal{F}$ is also a tower relative to $F|_{\mathcal{M}}$.*

In particular, the set $\mathcal{F}$ of all sub-towers of a tower $\mathcal{T}$ contains a unique smallest sub-tower $\mathcal{M} := \bigcap \mathcal{F}$.

Proof. This proof verifies that $\mathcal{M} = \bigcap \mathcal{F}$ satisfies the definition 6.42 of towers.

(T.A) From $\varnothing \in \mathcal{T}$ for every $\mathcal{T} \in \mathcal{F}$ follows $\varnothing \in \bigcap \mathcal{F} = \mathcal{M}$.
(T.B) If $\mathcal{A} \subseteq \mathcal{M}$, then $\mathcal{A} \subseteq \mathcal{T}$ for every $\mathcal{T} \in \mathcal{F}$; if $\mathcal{A}$ is also linearly ordered by inclusion, then $\bigcup \mathcal{A} \in \mathcal{T}$ for every $\mathcal{T} \in \mathcal{F}$, whence $\bigcup \mathcal{A} \in \bigcap \mathcal{F} = \mathcal{M}$.
(T.C) If $A \in \mathcal{M}$, then $A \in \mathcal{T}$ for every $\mathcal{T} \in \mathcal{F}$, whence $A \cup \{F(A)\} \in \mathcal{T}$ for every $\mathcal{T} \in \mathcal{F}$, and hence $A \cup \{F(A)\} \in \bigcap \mathcal{F} = \mathcal{M}$.

In particular, the set $\mathcal{F}$ of all sub-towers of a tower $\mathcal{T}$ contains the sub-tower $\mathcal{M} = \bigcap \mathcal{F} \subseteq \bigcup \mathcal{F} = \mathcal{T}$, because $\mathcal{T} \in \mathcal{F}$ and. $\mathcal{S} \subseteq \mathcal{T}$ for every $\mathcal{S} \in \mathcal{F}$. Also, $\mathcal{M} = \bigcap \mathcal{F} \subseteq \mathcal{S}$ for every $\mathcal{S} \in \mathcal{F}$, so that $\mathcal{M}$ is the smallest sub-tower. $\quad\square$

6.45 Theorem. *If $M \in \mathcal{M}$ is medial in the minimal sub-tower $\mathcal{M}$ of a tower $\mathcal{T}$, then either $A \subseteq M$ or $M \cup \{F(M)\} \subseteq A$ for every element $A \in \mathcal{M}$.*

Proof. This proof verifies that the set $\mathcal{S} := \{A \in \mathcal{M} : (A \subseteq M) \vee (M \cup \{F(M)\} \subseteq A)\}$ is also a sub-tower of the smallest sub-tower $\mathcal{M}$.

(T.A) $\varnothing \in \mathcal{S}$ because $\varnothing \subseteq M$ regardless of M. Thus $\mathcal{S} \neq \varnothing$.
(T.B) If $\mathcal{R}$ is a chain in $\mathcal{S}$, then for each $A \in \mathcal{R} \subseteq \mathcal{S}$, either $M \cup \{F(M)\} \subseteq A$ or $A \subseteq M$. Hence two cases can arise:
In the first case, if there exists any $A \in \mathcal{R}$ such that $M \cup \{F(M)\} \subseteq A$, then $M \cup \{F(M)\} \subseteq A \subseteq \bigcup \mathcal{R}$.
In the second case, if there does not exist any $A \in \mathcal{R}$ such that $M \cup \{F(M)\} \subseteq A$, then $A \subseteq M$ for every $A \in \mathcal{R} \subseteq \mathcal{S}$ by definition of $\mathcal{S}$, whence $\bigcup \mathcal{R} \subseteq M$.

In either case $\bigcup \mathscr{R} \in \mathscr{S}$.

(T.C) For each $A \in \mathscr{S}$, either $M \cup \{F(M)\} \subseteq A$ or $A \subseteq M$, by definition of $\mathscr{S}$. In the first case, if $M \cup \{F(M)\} \subseteq A$, then $M \cup \{F(M)\} \subseteq A \subseteq A \cup \{F(A)\}$. In the second case, if $A \subseteq M$ and $A \in \mathscr{S} \subseteq \mathscr{M}$, then $A \cup \{F(A)\} \in \mathscr{M}$ because $\mathscr{M}$ is a tower, whence either $M \cup \{F(M)\} \subseteq A \cup \{F(A)\}$ or $A \cup \{F(A)\} \subseteq M$ because M is medial in $\mathscr{M}$. Thus $A \cup \{F(A)\} \in \mathscr{S}$.

Thus $\mathscr{S}$ is a sub-tower of the smallest sub-tower $\mathscr{M}$. Hence $\mathscr{S} = \mathscr{M}$, so that $A \subseteq M$ or $M \cup \{F(M)\} \subseteq A$ for each medial element $M \in \mathscr{M}$ and each $A \in \mathscr{M}$. □

6.46 Theorem. *Every element of the smallest sub-tower of any tower is medial in the smallest sub-tower. The smallest sub-tower is linearly ordered by inclusion.*

Proof. This proof consists of verifying that the set $\mathscr{V}$ of all medial element of $\mathscr{M}$ is a sub-tower of $\mathscr{M}$.

(T.A) $\varnothing \in \mathscr{V}$ because $\varnothing$ is medial in $\mathscr{M}$ by example 6.43.

(T.B) If $\mathscr{U} \subseteq \mathscr{V}$ is a chain in $\mathscr{V}$ and $B \in \mathscr{U} \subseteq \mathscr{V}$, then B is medial in $\mathscr{M}$, so that $B \cup \{F(B)\} \subseteq A$ or $A \subseteq B$ for each $A \in \mathscr{M}$, by theorem 6.45.
In the first case, if there exists any $A \in \mathscr{U}$ such that $B \cup \{F(B)\} \subseteq A$, then $B \cup \{F(B)\} \subseteq A \subseteq \bigcup \mathscr{U}$.
In the second case, if there does not exist any $A \in \mathscr{U}$ such that $B \cup \{F(B)\} \subseteq A$, then $A \subseteq B$ for every $A \in \mathscr{U} \subseteq \mathscr{V}$, by definition of $\mathscr{V}$ whence $\bigcup \mathscr{U} \subseteq B$.
In either case $\bigcup \mathscr{U} \in \mathscr{V}$.

(T.C) Every $A \in \mathscr{V}$ is medial in $\mathscr{M}$ by definition of $\mathscr{V}$. By theorem 6.45 for every $B \in \mathscr{M}$ either $B \subseteq A$, in which case $B \subseteq A \subset A \cup \{F(A)\}$, or $A \cup \{F(A)\} \subseteq B$. In either case $A \cup \{F(A)\}$ is medial in $\mathscr{M}$, so that $A \cup \{F(A)\} \in \mathscr{V}$.

Thus $\mathscr{V}$ is a sub-tower of the minimal sub-tower $\mathscr{M}$. Hence $\mathscr{V} = \mathscr{M}$, so that every element of $\mathscr{M}$ is medial in $\mathscr{M}$. Hence $\mathscr{M}$ is linearly ordered by inclusion. □

6.47 Theorem. *For every tower $\mathscr{T} \subseteq \mathscr{P}(X)$ relative to a function $F : \mathscr{T} \to X$ there exists $A \in \mathscr{T}$ such that $F(A) \in A$.*

Proof. Let $\mathscr{M}$ be the smallest sub-tower, and let $A := \bigcup \mathscr{M}$. Then $A \in \mathscr{M}$ by definition 6.42 of a tower and because $\mathscr{M} \subseteq \mathscr{M}$ and $\mathscr{M}$ is linearly ordered, by theorem 6.46. Hence also $A \cup \{F(A)\} \in \mathscr{M}$ by definition 6.42 of a tower, whence $A \cup \{F(A)\} \subseteq \bigcup \mathscr{M} = A$, and hence $F(A) \in A$. □

6.5.2 Zorn's Maximality from the Choice Principle

This subsection shows that Zorn's Maximality Principle follows from the Choice Principle.

6.48 Theorem. *The Choice-Function Principle 6.8 implies Zorn's Maximal-Element Principle 6.32.*

Proof. For each set X with a reflexive partial order $\preceq$ let $\mathscr{C}$ be the set of all chains relative to $\preceq$ in X.

For each nonempty chain $C \in \mathscr{C}$, let U_C be the set of all upper bounds for C relative to $\preceq$ in X. The hypotheses of Zorn's Maximal-Element Principle 6.32 imply that $U_C \neq \varnothing$.

If the Choice-Function Principle 6.8 holds, then there exists a function $F : \{U_C : C \in \mathscr{C}\} \to X$ such that $F(U_C) \in U_C$, which chooses one upper bound $F(U_C)$ for each nonempty chain $C \in \mathscr{C}$. Hence define $D_C := \{K \in X : [F(U_C) \preceq K] \wedge \{\neg[K \preceq F(U_C)]\}\}$, which is the set of all the elements of X that strictly follow the upper bound $F(U_C)$. If there exists $C \in \mathscr{C}$ such that $D_C = \varnothing$, then $F(U_C)$ is a maximal element of X.

Alternatively, if $D_C \neq \varnothing$ for every $C \in \mathscr{C}$, then, again by the Axiom of Choice 6.8, there is a function $G : \{D_C : C \in \mathscr{C}\} \to X$ such that $G(D_C) \in D_C$ for each $C \in \mathscr{C}$. The hypotheses of Zorn's Maximal-Element Principle 6.32 also imply $X \neq \varnothing$, so that there exists $Z \in X$. Define $H(\varnothing) := Z$ and $H(C) := G(D_C)$. The remainder of the proof verifies that $\mathscr{T} := \mathscr{C} \cup \{\varnothing\}$ is a tower relative to H.

(T.A) $\varnothing \in \mathscr{T}$ by the definition $\mathscr{T} := \mathscr{C} \cup \{\varnothing\}$.

(T.B) If $\mathscr{S}$ is a chain in $\mathscr{T}$ relative to inclusion, then $\bigcup \mathscr{S}$ is a chain relative to $\preceq$ in X. Indeed, for all $K, L \in \bigcup \mathscr{S}$ there exist $A, B \in \mathscr{S}$ such that $K \in A$ and $L \in B$. Also, $\mathscr{S}$ is linearly ordered by inclusion, so that $A \subseteq B$ or $B \subseteq A$. Hence $K, L \in B$ or $K, L \in A$. Moreover, $A, B \in \mathscr{S} \subseteq \mathscr{T}$ are chains relative to $\preceq$ in X, whence A and B are linearly ordered, and hence $K \preceq L$ or $L \preceq K$, in B or A. Thus $\bigcup \mathscr{S} \in \mathscr{T}$.

(T.C) If $C \in \mathscr{T}$, then $H(C)$ is an upper bound for C relative to $\preceq$ in X, whence $C \cup \{H(C)\}$ is also a chain relative to $\preceq$ in X. Hence $C \cup \{H(C)\} \in \mathscr{T}$.

By theorem 6.47 there is a chain $C \in \mathscr{T}$ such that $H(C) \in C$. Hence $H(C) \preceq F(U_C)$ because $F(U_C)$ is an upper bound for C. Yet $H(C) \in D_C$ is a strict upper bound with $\neg[H(C) \preceq F(U_C)]$. This contradiction completes the proof: there is a chain $C \in \mathscr{C}$ for which $D_C = \varnothing$, so that $F(D_C)$ is maximal relative to $\preceq$ in X. □

Zermelo's Theorem [144] consists of the logical implication that the Well-Ordering Principle follows from the Axiom of Choice within the Zermelo-Frænkel-Choice set theory.

Zorn's Lemma consists of the logical implication that Zorn's Maximal-Element Principle follows from the Axiom of Choice within the Zermelo-Frænkel-Choice set theory.

6.49 Definition (Interval). In a set E pre-ordered by a relation $\preceq$ a subset $S \subseteq E$ is an **interval** if and only if for all $U, W \in S$ and every $V \in E$, if $U \preceq V \preceq W$, then $V \in S$.

6.50 Example. The empty subset $\varnothing$ and the whole set E are intervals.

6.5.3 Exercises on Towers of Sets

6.17. Define $F : \mathscr{P}(\mathbb{N}) \to \mathbb{N}$ by $F(\varnothing) := 0$ and $F(A) := \min(A)$ for $A \neq \varnothing$. Prove that the set $\mathscr{T}$ of all intervals of $\mathbb{N}$ is a tower relative to F.

6.18. For each set E well-ordered by a reflexive relation $\preceq$ define $F : \mathscr{P}(E) \to E$ by $F(\varnothing) := \min(E)$ and $F(A) := \min(A)$ for $A \neq \varnothing$. Prove that the set $\mathscr{T}$ of all intervals of E is a tower relative to F.

6.19. Prove that $\mathscr{P}(E)$ is a tower relative to any function $\mathscr{P}(E) \to E$.

6.20. Prove or disprove that every union of towers is a tower.

6.6 Yet Other Principles Related to the Axiom of Choice

This section states principles that are related to the Axiom of Choice, in particular, in the sense that they are not propositions within the Zermelo-Frænkel set theory.

6.6.1 Yet Other Principles Equivalent to the Axiom of Choice

This subsection states principles that are equivalent to the Choice Principle in the Zermelo-Frænkel set theory.

6.51 Definition (Hausdorff's Particular Maximal-Chain Principle). In a set of sets partially ordered by inclusion, each chain is a subset of a maximal chain [128, p. 248, H_1].

6.52 Definition (Hausdorff's Maximal-Chain Principle). Each set of sets has a maximal chain relative to inclusion [88, p. 118], [128, p. 248, H_2].

6.53 Definition (Hausdorff's Maximal-Subset Principle). In each partially ordered set there is a maximal linearly ordered subset [56, 57, p. 339], [91, p. 69].

6.54 Definition (Zermelo's Principle). For each partition $\mathscr{F}$ of a set (of sets) A, there exists a subset $C \subseteq A$ such that $C \cap B$ is a singleton for each $B \in \mathscr{F}$ [88, p. 117].

6.55 Definition (Counting Principle). For each set E there is an ordinal O and a bijective function $F : O \to E$ with domain O [88, p. 117], [128, p. 241].

6.56 Definition (Kuratowski's Maximal-Order Principle). For each partial order R and for each linear order $L \subseteq R$ there exists a maximal (relative to inclusion) linear order M such that $L \subseteq M \subseteq R$ [88, p. 118].

6.57 Definition (Trichonomy Principle). For all sets A and B, there exists an injection F with $F : A \hookrightarrow B$ with domain A, or $F : B \hookrightarrow A$ with domain B [88, p. 118], [128, p. 241].

6.58 Definition (Mapping Principle). For all *nonempty* sets A and B, there exists a surjection F with $F : A \twoheadrightarrow B$, or $F : B \twoheadrightarrow A$ [88, p. 118].

6.59 Definition. A set (of sets) A is of **finite character** if and only if

    (FC.A) $A \neq \varnothing$,
    (FC.B) if $X \in A$ and $B \subseteq X$ is finite, then $B \in A$, and
    (FC.C) for each set E, if every finite subset of E is a member of A, then $E \in A$.

6.60 Definition (Teichmüller-Tukey Maximal-Element Principle). Every set of finite character has a maximal element [128, p. 249].

6.6.2   Consequences of the Axiom of Choice

This subsection states principles that follow from the Choice Principle in the Zermelo-Frænkel set theory.

6.61 Definition (Finite Sets and Infinite Sets). A set E is **finite** if and only if there is $N \in \mathbb{N}$ and a bijection $F : N \to E$; it is **infinite** if and only if it is not finite.

6.62 Theorem. *In the Zermelo-Frænkel-Choice set theory, every infinite set contains a denumerable subset.*

Proof. For each set E, let $\mathscr{E}$ be the set of all injections from any subset of $\mathbb{N}$ into E. Thus each element $F \in \mathscr{E}$ is a subset of $\mathbb{N} \times E$, and $\mathscr{E} \subseteq \mathscr{P}(\mathbb{N} \times E)$ is partially ordered by inclusion. If E is infinite, then for each $N \in \mathbb{N}$ there exists an injection $F : N \hookrightarrow E$, which is also a subset of $\mathbb{N} \times E$. In particular $\mathscr{E} \neq \varnothing$.

    Each chain $\mathscr{F} \subseteq \mathscr{E}$ defines an injection $\bigcup \mathscr{F} \in \mathscr{E}$. Indeed, for each $(X, Y) \in \bigcup \mathscr{F}$, there exists $F \in \mathscr{F}$ such that $(X, Y) \in F$. If also $(X, Z) \in G \in \mathscr{F}$, then $F \subseteq G$ or $G \subseteq F$, because $\mathscr{F}$ is a chain, whence $Y = Z$. Moreover, if $(U, Y), (X, Y) \in \bigcup \mathscr{F}$, then there exist $F, G \in \mathscr{F}$ such that $(X, Y) \in F$ and $(U, Y) \in G$, with $F \subseteq G$ or $G \subseteq F$, whence $U = X$ by injectivity of F or G. Thus $\bigcup \mathscr{F}$ is an upper bound for $\mathscr{F}$ and $\bigcup \mathscr{F} \in \mathscr{E}$. By Zorn's Maximal-Element Principle 6.32 the set $\mathscr{E}$ contains a maximal element F.

    If the domain of F was finite, then a reparametrization of its domain would give an injection $G : N \hookrightarrow \mathbb{N}$, which would not be injective, by definition of infinite. Hence the domain of F is an infinite subset of $\mathbb{N}$.

Therefore, the image of F is a denumerable subset of E.   □

Without the Axiom of Choice, theorem 6.62 fails [66, p. 141].

6.63 Definition (Dedekind Infinite Sets). A set D is **Dedekind infinite** if and only if there exist a *proper* subset $B \subsetneq D$ and an *injection* $F : D \hookrightarrow B$. A set is **Dedekind finite** if there are no injections from it into any of its proper subsets.

6.64 Theorem. *In the Zermelo-Frænkel set theory, every Dedekind infinite set is infinite.*

Proof. By contraposition, if a set is finite, then there are no injections from it into any of its proper subsets, whence it is Dedekind finite.   □

6.65 Theorem. *In the Zermelo-Frænkel-Choice set theory, every infinite set is Dedekind infinite.*

Proof. If a set E is infinite, then by theorem 6.62 it contains a denumerable subset $D \subseteq E$. In particular, there is a bijection $C : \mathbb{N} \to D$. Define $B := E \setminus \{C(0)\}$, and $F : E \to B$ with $F(X) := X$ for every $X \in E \setminus D$ and $F[C(N)] := F[C(N + 1)]$ if $X = C(N) \in D$. Then F is an injection from E to B.   □

6.66 Definition (Countable Axiom of Choice). Each *countable* set of nonempty sets has a choice function.

Proofs of the equivalence of compactness and sequential compactness, or continuity and sequential continuity — and hence of the existence of extrema of continuous functions on closed and bounded intervals — invoke the Countable Axiom of Choice [66, p. 21, § 2.4.3] but do not require (the results do not logically imply) the Axiom of Choice [111, p. 178].

6.6.3   *Exercises on Related Principles*

6.21 . Prove that in the Zermelo-Frænkel-Choice set theory every denumerable union of pairwise disjoint denumerable sets is denumerable.

6.22 . Following the R. L. Moore method [15, 146], prove all logical implications between all "principles" stated in the present chapter.

Chapter 7
Applications: Nobel-Prize Winning Applications of Sets, Functions, and Relations

7.1 Introduction

This chapter shows concrete applications of sets, functions, and relations:

1. Arrow's Impossibility Theorem. Kenneth J. Arrow received the Nobel Prize in Economic Science in 1972, mainly for his Impossibility Theorem, from work at the RAND Corporation in 1948 [4, p. 328, footnote 1].
2. Gale and Shapley's Matching Algorithm. Gale and Shapley's Ph.D. advisor was Princeton's Albert William Tucker; Lloyd S. Shapley received the Nobel Prize in Economic Science in 2012, for work that can be traced back to a lecture by John von Neumann in 1948 at the RAND Corporation [68, p. 384].
3. Nash's Equilibrium. Nash's Ph.D. advisor was Princeton's Albert William Tucker; John Forbes Nash, Jr., received the Nobel Prize in Economic Sciences in 1994, for work that can be traced to Melvin Dresher and Merrill Meeks Flood in 1950 at the RAND Corporation [12, 124, 125]. He received the Abel Prize in 2015.

David Gale and Lloyd S. Shapley pointed out that readers without a technical background may miss the logic in their Nobel-Prize winning work even though it is written in plain English:

> "Yet any mathematician will recognize the argument as mathematical, while people without mathematical training will probably find difficulty in following the argument" — [41, p. 391].

This chapter shows what mathematicians recognize in the argument: links between applications and mathematical sets, functions, and relations. Expositions considered concise and elegant by mathematicians are listed in the references.

© Springer Science+Business Media New York 2015
Y. Nievergelt, *Logic, Mathematics, and Computer Science*,
DOI 10.1007/978-1-4939-3223-8_7

7.2   Game Theory

The mathematical analysis of games, called **game theory**, explains why and predicts how people will rationally make decisions against their interest. Examples include decisions whether to arm one's self. For their contributions to game theory, John Harsanyi, Reinhard Selten, and John Forbes Nash, Jr., received the Nobel Prize in Economics in 1994 [93–95]. This introduction to game theory relies on mathematical functions and ordering relations.

7.2.1   Introduction

Publicized by Albert W. Tucker as **A Two-Person Dilemma** [131], but now known as **The Prisoner's Dilemma** [132], a prototype of the subject of game theory originated through the work of Melvin Dresher and Merrill Meeks Flood in 1950 at the RAND Corporation [12, 124, 125]. Similar games explain the nuclear arm race (see Figure 7.1) [13, 126].

Fig. 7.1 Early days of the nuclear arm race. Left: a Soviet modified SS-6 (Sapwood) with Sputnik 2 on 3 November 1957, photograph courtesy NASA. Right: an Atlas rocket lifts off, photograph courtesy U.S. Air Force. (http://www.history.nasa.gov/SP-4408pt1.pdf) (http://www.nationalmuseum.af.mil/shared/media/photodb/photos/050406-F-1234P-014.jpg)

7.1 Example (The Prisoner's Dilemma [131, 132]). The police charge Al and Bo with a crime and hold them in separate cells, so that Al and Bo cannot communicate with each other. Al and Bo believe that they have at least a first strategy:

(PD.1) If Al and Bo do not confess to the crime, then they both go free.

However, the police gives each of them another strategy, to confess:

(PD.2) If both Al and Bo confess to the crime, then each gets sentenced to prison.
(PD.3) If one confesses but the other does not confess, then whoever confesses gets a reward and goes frees, while whoever does not confess gets a death sentence.

Al and Bo's dilemma is whether to confess or not to confess.

The Prisoner's Dilemma also applies to two countries, or their rulers, or any two individuals, Al and Bo, deciding whether to *cooperate* not to arm themselves, or to *defect* the agreement of cooperation and arm themselves. The Prisoner's Dilemma predicts that they will both arm themselves, as shown in figure 7.1, and explains why [13, 126], as demonstrated in subsection 7.2.2. Game theory also explains animal behavior, for example musth in male African elephants [87]. To get a sense of Nash's work, the reader may at this stage try to design a method to analyze mathematical games such as The Prisoner's Dilemma but with any numbers of players and strategies.

7.2.2 Mathematical Models for The Prisoner's Dilemma

This subsection describes ways to design a mathematical model and analyze games such as The Prisoner's Dilemma.

7.2 Example (The Prisoner's Dilemma, continued [131, 132]). What Al and Bo eventually get (reward, freedom, prison, or death) is called their **payoff**. The first step in designing a mathematical model of the game consists in modeling the players' preferences with ordering relations on each player's set of payoffs. In this example, it seems reasonable to assume that Al and Bo rank all four payoffs in the same order of preference, with $>$ meaning "better than":

$$\text{reward \& freedom} > \text{freedom} > \text{prison} > \text{death}.$$

The second step in designing a mathematical model of the game consists in modeling every way the players may play by organizing their actions, called **strategies**, and corresponding payoffs in a table. Table 7.3 summarizes Al's and Bo's strategies and their resulting payoffs in The Prisoner's Dilemma. Al has only two strategies available: either to confess or not to confess. Similarly, Bo may either confess or not confess. Moreover, The Prisoner's Dilemma is a **noncooperative game** in the sense that neither player knows the other player's action in advance.

The analysis of the game then examines every combination of strategies in the table. For instance, Al may examine each of Al's own strategies and seek to avoid the worst payoff.

Al confesses.   If Al confesses, then meanwhile two cases can occur: Bo can either confess or not confess.

Bo confesses.   If Bo also confesses, then Al faces a prison term,
Bo does not confess   If Bo does not confess, then Al goes free with a reward.

Thus, if Al confesses, then the worst that can happen to Al is a prison term.
Al does not confess.   Similarly, if Al does not confess, then two cases can occur: Bo can either confess or not confess.

Bo confesses.   If Bo confesses, then Al faces death.
Bo does not confess   If Bo does not confess either, then Al goes free.

Thus, if Al does not confess, then the worst that can happen to Al is death.

Consequently, to avoid death with certainty but without knowing Bo's action, Al must confess.

Table 7.3 is **symmetric** in the sense that swapping Al's and Bo's rôles does not change their payoffs. Thus Al's reasoning also applies to Bo, who must also confess to avoid death with certainty.

Therefore, to avoid death with certainty, Al and Bo both confess, even though they would both be better off by not confessing.

7.4 Example (Arm race [13, 126]). If two individuals or countries cooperate not to arm themselves, then they remain free and safe. If both defect and arm themselves, then they remain safe but incur a penalty with the cost of armament. If either one defects and arms itself while the other one does not arm itself, then whoever arms itself stays free and gets a reward by conquering and looting the other, who suffers from death, destruction, and loosing the property to the conqueror. The ordering of the outcomes just described are the same as those in table 7.3 for The Prisoner's Dilemma. Therefore, the analysis of The Prisoner's Dilemma in example 7.2 predicts and explains why they will both arm themselves.

Example 7.5 shows another way to analyze The Prisoner's Dilemma.

Table 7.3 Payoffs for the Prisoner's Dilemma.

	BO'S STRATEGIES			
	CONFESS		NOT CONFESS	
	PAYOFFS TO		PAYOFFS TO	
	AL	BO	AL	BO
CONFESS	prison	prison	reward	death
AL'S STRATEGIES				
NOT CONFESS	death	reward	freedom	freedom

7.5 Example (The Prisoner's Dilemma, continued [131, 132]). In contrast to example 7.2, Al may focus first on Bo's strategies and seek to get the best payoff. Again, Bo may either confess or not confess.

Bo confesses.   If Bo confesses, then Al has two choices.

Al confesses.   If Al confesses, too, then Al gets a prison term.
Al does not confess.   If Al does not confess, then Al gets death.

Thus, if Bo confesses, then Al gets the better of the two payoffs available (prison or death) by confessing.
Bo does not confess.   If Bo does not confess, then Al has two choices.

Al confesses.   If Al confesses, then Al goes free with a reward.
Al does not confess.   If Al does not confess, then Al goes free.

Thus, if Bo does not confess, then Al *also* gets the better payoff by confessing.

Thus, regardless of Bo's action, Al gets the better payoff available by confessing. Therefore, Al confesses. The same analysis applies to Bo, who also confesses.

7.2.3   Dominant Strategies

Some of the modeling and analysis of The Prisoner's Dilemma carry over to other games. To this end, this subsection shows that a special type of strategy contains an *equilibrium*, from which no players has any incentive to switch to another strategy. For a game with exactly two players, Al and Bo, assume that Al has a nonempty set of strategies S_{Al} and a nonempty set of payoffs P_{Al}, while Bo has a nonempty set of strategies S_{Bo} and a nonempty set of payoffs P_{Bo}. Assume also that Al and Bo have a linear order "$\geq$" on their sets of payoffs. If Al plays a strategy A $\in S_{Al}$ and Bo plays a strategy B $\in S_{Bo}$, then Al gets a payoff $p_{A,B}^{Al}$ while Bo gets a payoff $p_{A,B}^{Bo}$. Thus for each Ci $\in \{Al, Bo\}$, p^{Ci} is a function $p^{Ci} : S_{Al} \times S_{Bo} \to P_{Ci}$. Table 7.6 shows the payoff table of a game for two players with two strategies.

7.7 Definition (dominant strategy). A strategy is **weakly dominant** for a player (Al, for instance) if and only if, for each fixed combination of the other players' strategies, that player (Al) cannot get a higher payoff by switching to

Table 7.6 Payoffs for noncooperative two-person games with two strategies.

	BO'S STRATEGIES			
	L ("LEFT")		R ("RIGHT")	
	PAYOFFS TO		PAYOFFS TO	
	AL	BO	AL	BO
T ("TOP")	$p_{T,L}^{Al}$	$p_{T,L}^{Bo}$	$p_{T,R}^{Al}$	$p_{T,R}^{Bo}$
AL'S STRATEGIES				
B ("BOTTOM")	$p_{B,L}^{Al}$	$p_{B,L}^{Bo}$	$p_{B,R}^{Al}$	$p_{B,R}^{Bo}$

another strategy. A strategy is **strongly dominant** for a player (Al) if and only if, for each fixed combination of the other players' strategies, that player (Al) always gets a lower payoff by switching to another strategy.

7.8 Example. In The Prisoner's Dilemma (examples 7.1 and 7.5), Al's strongly dominant strategy is to confess: If Bo confesses, then Al gets a higher payoff by confessing (prison) than by not confessing (death). Similarly, if Bo does not confess, then Al also gets a higher payoff by confessing (freedom and a reward) than by not confessing (freedom). By symmetry of The Prisoner's Dilemma, Bo's strongly dominant strategy is also to confess.

7.9 Definition (dominant strategy equilibrium). A **position** is an array of strategies, with one strategy for each player. A **weakly** (respectively **strongly**) **dominant strategy equilibrium** is a position where each player plays a weakly (respectively strongly) dominant strategy.

7.10 Example. In The Prisoner's Dilemma, the position where Al and Bo both confess is a dominant strategy equilibrium, where their dominant strategies meet. Neither Al nor Bo has any incentive to switch to another strategy (not to confess), because whoever decides not to confess while the other still confesses faces death.

7.11 Definition (Nash equilibrium). A **Nash equilibrium** is a position from which no players can get a higher payoff by changing only their own strategy, while all the other players' strategies remain fixed [93, p. 287].

7.12 Example. In The Prisoner's Dilemma, the position where Al and Bo both confess is a Nash equilibrium: If Al decides not to confess but Bo still confesses, then Al's payoff drops. Similarly, if Bo decides not to confess but Al still confesses, then Bo's payoff drops. In contrast, the position where neither Al nor Bo confess is *not* a Nash equilibrium, because whoever decides to confess gets a higher payoff if the other one still does not confess.

7.13 Theorem. *Every weakly dominant equilibrium is a Nash equilibrium.*

Proof. Table 7.14 shows the conditions for either of Bo's strategic to be weakly dominant in a game with only one other player (Al).

Table 7.14 Bo's potentially weakly dominant strategies.

		BO'S STRATEGIES					
		L IS WEAKLY DOMINANT: PAYOFFS TO		OR	R IS WEAKLY DOMINANT: PAYOFFS TO		
		BO	BO		BO		BO
	T	$p_{T,L}^{Bo}$	$\geq$	$p_{T,R}^{Bo}$	$p_{T,L}^{Bo}$	$\leq$	$p_{T,R}^{Bo}$
AL'S STRATEGIES			AND	OR		AND	
	B	$p_{B,L}^{Bo}$	$\geq$	$p_{B,R}^{Bo}$	$p_{B,L}^{Bo}$	$\leq$	$p_{B,R}^{Bo}$

If R is Bo's single weakly dominant strategy (Bo's R column in table 7.14), then Bo plays R. Indeed, if Al plays T , then R gets Bo a larger payoff, because $p_{T,L}^{Bo} \leq p_{T,R}^{Bo}$; if Al plays B , then R also gets Bo a larger payoff, because $p_{B,L}^{Bo} \leq p_{B,R}^{Bo}$.

Since Al knows the game, Al knows that R is Bo's single weakly dominant strategy. Thus, Al knows that Bo will play R. Consequently, Al has only two payoffs available: $p_{T,R}^{Al}$ or $p_{B,R}^{Al}$. Therefore, Al chooses a strategy that yields the larger available payoff. For instance, if $p_{T,R}^{Al} \leq p_{B,R}^{Al}$, then Al plays B.

Hence the game ends at (B,R) with payoffs $p_{B,R}^{Al}$ to Al and $p_{B,R}^{Bo}$ to Bo. From (B,R) Bo cannot get a higher payoff by switching to L because $p_{B,L}^{Bo} \leq p_{B,R}^{Bo}$. From (B,R) Al cannot get a higher payoff by switching to T because $p_{T,R}^{Al} \leq p_{B,R}^{Al}$. Thus (B,R) is a Nash equilibrium.

If R and L are both weakly dominant strategies for Bo, then Bo's payoffs are identical in R and L. If $p_{B,R}^{Al}$ is the largest of Al's four payoffs in Bo's weakly dominant strategies, then (B,R) is still a Nash equilibrium. Yet Al no longer knows which of R or L Bo chooses: the game might not end at a Nash equilibrium.   □

7.2.4   Mixed Strategies

Some games or situations might not occur more than once in the players' life time, for instance, The Prisoner's Dilemma. Yet other games or situations may occur more than once in the players' life time. Each player may then adopt a **mixed strategy**, defined by choosing one strategy some of the time and another strategy at other times. Each player's payoff may then be a weighted average of the payoffs from the different strategies in the mixed strategy. For emphasis, strategies that are *not* mixed, for instance, as in The Prisoner's Dilemma, are called **pure strategies**.

7.15 Example (The Battle of the Sexes [92]). Al and Bo would like to go to a show together. Al prefers a Jazz concert while Bo prefers a play at the theater. Nevertheless, they would prefer to go to the same show together, rather than to different shows separately. After some negotiation they agree on a show. On the day of the show, however, they each forget which show they had agreed on, but cannot communicate with each other. So they each must decide to which show to go, without knowing in advance where the other will go. Table 7.16 shows their payoffs. In hope of finding Bo, Al goes to the Jazz concert one half of the time, and

Table 7.16 Payoffs for the Battle of the Sexes.

		BO GOES TO			
		JAZZ (J)		PLAY (P)	
		PAYOFFS TO		PAYOFFS TO	
		AL	BO	AL	BO
	JAZZ (J)	3	2	1	1
AL GOES TO					
	PLAY (P)	0	0	2	3

Table 7.17 Payoffs for the Battle of the Sexes over four days.

		BO GOES TO					
		JAZZ (J)			PLAY (P)		
		PAYOFFS TO			PAYOFFS TO		
		AL		BO	AL		BO
	JAZZ (J)	3	(DAY 1)	2	1	(DAY 3)	1
AL GOES TO							
	PLAY (P)	0	(DAY 2)	0	2	(DAY 4)	3

to the play at the theater the other half of the time. In hope of finding Al, Bo does the same. Table 7.17 shows the four combinations of strategies; for this example, assume that they occur equally frequently. Thus Al's and Bo's average payoffs over four days are

$$p^{Al}(1/2, 1/2) = \frac{3 + 0 + 1 + 2}{4} = \frac{3}{2};$$
$$p^{Bo}(1/2, 1/2) = \frac{2 + 0 + 1 + 3}{4} = \frac{3}{2}.$$

Al now decides to go to the Jazz concert every day, while Bo still goes to Jazz every other day and to the play every other day. There are then only two kinds of days, and their average payoffs become

$$p^{Al}(1, 0) = \frac{3 + 1}{2} = 2;$$
$$p^{Bo}(1/2, 1/2) = \frac{2 + 1}{2} = \frac{3}{2}.$$

Hence arises the question whether other mixed strategies can increase both Al's and Bo's payoffs.

7.2.5 Existence of Nash Equilibria for Two Players with Two Mixed Strategies

This section shows that if neither player has any weakly dominant pure strategy, then the game still has a Nash equilibrium for some combination of mixed strategies.

7.18 Definition (mixed strategy). The same players might play multiple rounds of the same game and independently choose any available strategy for each round. Over M rounds, Al may play T exactly T times and B exactly B times, with $T + B = M$. Thus Al plays T with a frequency $t = T/M$ and B with a frequency $b = B/M = 1 - t$. The ordered pair of frequencies $(t, 1 - t)$ is called a **mixed**

strategy for Al. Meanwhile, Bo may play L exactly L times and R exactly R times, with $L+R = M$. Thus Bo plays L with a frequency $\ell = L/M$ and R with a frequency $r = R/M = 1 - \ell$. Thus $(\ell, 1 - \ell)$ is a mixed strategy for Bo. Mixed strategies are also subject to the condition that Al and Bo play (T,L) with frequency $t \cdot \ell$ and hence also receive payoffs $p_{\text{T,L}}^{\text{Al}}$ and $p_{\text{T,L}}^{\text{Bo}}$ with frequency $t \cdot \ell$. Similar products of frequencies apply to the other plays (T,R), (B,L), and (B,R). The sum of such payoffs to Al and Bo are

$$p^{\text{Al}}(t, \ell) = t \cdot \ell \cdot p_{\text{T,L}}^{\text{Al}} + t \cdot (1 - \ell) \cdot p_{\text{T,R}}^{\text{Al}} + (1 - t) \cdot \ell \cdot p_{\text{B,L}}^{\text{Al}} + (1 - t) \cdot (1 - \ell) \cdot p_{\text{B,R}}^{\text{Al}},$$

$$p^{\text{Bo}}(t, \ell) = t \cdot \ell \cdot p_{\text{T,L}}^{\text{Bo}} + t \cdot (1 - \ell) \cdot p_{\text{T,R}}^{\text{Bo}} + (1 - t) \cdot \ell \cdot p_{\text{B,L}}^{\text{Bo}} + (1 - t) \cdot (1 - \ell) \cdot p_{\text{B,R}}^{\text{Bo}}.$$

7.19 Theorem. *Every game for two players with two pure strategies has a Nash equilibrium.*

Proof. For Bo's payoff, collecting similar powers of t and ℓ leads after some algebra to the equivalent formula

$$p^{\text{Bo}}(t, \ell) = \ell \cdot \{t \cdot [(p_{\text{T,L}}^{\text{Bo}} - p_{\text{T,R}}^{\text{Bo}}) + (p_{\text{B,R}}^{\text{Bo}} - p_{\text{B,L}}^{\text{Bo}})] - (p_{\text{B,R}}^{\text{Bo}} - p_{\text{B,L}}^{\text{Bo}})\} + t \cdot (p_{\text{T,R}}^{\text{Bo}} - p_{\text{B,R}}^{\text{Bo}}) + p_{\text{B,R}}^{\text{Bo}}.$$

Al has a problem: if Al can choose a frequency t^* such that

$$t^* \cdot [(p_{\text{T,L}}^{\text{Bo}} - p_{\text{T,R}}^{\text{Bo}}) + (p_{\text{B,R}}^{\text{Bo}} - p_{\text{B,L}}^{\text{Bo}})] - (p_{\text{B,R}}^{\text{Bo}} - p_{\text{B,L}}^{\text{Bo}}) = 0,$$

then Bo's payoff becomes

$$p^{\text{Bo}}(t^*, \ell) = t^* \cdot (p_{\text{T,R}}^{\text{Bo}} - p_{\text{B,R}}^{\text{Bo}}) + p_{\text{B,R}}^{\text{Bo}},$$

which does not depend on ℓ and thus strips away from Bo all controls over Bo's own payoff. In particular, Bo cannot get a higher payoff by switching to a different frequency ℓ. Yet if Bo can choose a frequency ℓ^* so that Al cannot control Al's own payoff, then they are at a Nash equilibrium (t^*, ℓ^*). The following considerations show that such a Nash equilibrium exists provided that neither Al nor Bo has any weakly dominant strategy. Table 7.20 shows the conditions for Bo not to have any weakly dominant strategy, obtained from the logical negation of Table 7.14. Two cases emerge: either $p_{\text{B,L}}^{\text{Bo}} < p_{\text{B,R}}^{\text{Bo}}$ and $p_{\text{T,L}}^{\text{Bo}} > p_{\text{T,R}}^{\text{Bo}}$, or $p_{\text{T,L}}^{\text{Bo}} < p_{\text{T,R}}^{\text{Bo}}$ and $p_{\text{B,L}}^{\text{Bo}} > p_{\text{B,R}}^{\text{Bo}}$.

In the first case, $p_{\text{B,L}}^{\text{Bo}} < p_{\text{B,R}}^{\text{Bo}}$ and $p_{\text{T,L}}^{\text{Bo}} > p_{\text{T,R}}^{\text{Bo}}$. Hence $p_{\text{B,R}}^{\text{Bo}} - p_{\text{B,L}}^{\text{Bo}} > 0$ and $p_{\text{T,L}}^{\text{Bo}} - p_{\text{T,R}}^{\text{Bo}} > 0$. Consequently,

$$(p_{\text{T,L}}^{\text{Bo}} - p_{\text{T,R}}^{\text{Bo}}) + (p_{\text{B,R}}^{\text{Bo}} - p_{\text{B,L}}^{\text{Bo}}) > (p_{\text{T,L}}^{\text{Bo}} - p_{\text{T,R}}^{\text{Bo}}) > 0.$$

Thus, Al can choose a frequency t^* between 0 and 1 where Bo has no controls over Bo's payoff:

Table 7.20 Conditions for Bo to have no weakly dominant pure strategies.

		BO'S STRATEGIES					
		L IS NOT WEAKLY DOMINANT: PAYOFFS TO		AND	R IS NOT WEAKLY DOMINANT: PAYOFFS TO		
		BO	BO		BO		BO
	T	$p_{T,L}^{Bo}$	$<$	$p_{T,R}^{Bo}$		$p_{T,L}^{Bo}$	$>$	$p_{T,R}^{Bo}$
AL'S STRATEGIES			OR		AND		OR
	B	$p_{B,L}^{Bo}$	$<$	$p_{B,R}^{Bo}$		$p_{B,L}^{Bo}$	$>$	$p_{B,R}^{Bo}$

$$1 > t^* = \frac{(p_{B,R}^{Bo} - p_{B,L}^{Bo})}{(p_{T,L}^{Bo} - p_{T,R}^{Bo}) + (p_{B,R}^{Bo} - p_{B,L}^{Bo})} > 0.$$

Similarly, if Al has no weakly dominant strategies, then Bo can choose a frequency ℓ^* between 0 and 1 where Al has no control over Al's payoff, because

$$1 > \ell^* = \frac{(p_{B,R}^{Al} - p_{T,R}^{Al})}{(p_{B,L}^{Al} - p_{T,L}^{Al}) + (p_{B,R}^{Al} - p_{T,R}^{Al})} > 0.$$

With such frequencies,

$$p^{Al}(t^*, \ell^*) = \ell^* \cdot p_{B,L}^{Al} + (1 - \ell^*) \cdot p_{B,R}^{Al},$$
$$p^{Bo}(t^*, \ell^*) = t^* \cdot p_{T,R}^{Bo} + (1 - t^*) \cdot p_{B,R}^{Bo}.$$

Neither Al nor Bo can change their own payoff by switching to another frequency while the other player's frequency is fixed. Thus (t^*, ℓ^*) is a Nash equilibrium. The second case, where $p_{T,L}^{Bo} < p_{T,R}^{Bo}$ and $p_{B,L}^{Bo} > p_{B,R}^{Bo}$, is entirely similar.

If a player has a dominant strategy, then such a strategy contains a Nash equilibrium, by theorem 7.19.   □

For Nash's equilibria with two players and two strategies, see also [106, p. 138–139], [107], [119, p. 155–168].

7.21 Example (The Professors' Problem). Al and Bo are on the tenured faculty at King Game's College. Each may either teach students or sit on committees. Teaching does not hamper any one's work, but committees hamper other faculty members' work, to the extent summarized by Al's and Bo's end-of-the-year bonus payoffs in table 7.22. The sum of Al's and Bo's payoffs reflects the value of their work to the College. Al and Bo know table 7.22 but work in different offices and thus do not cooperate with each other; hence they must choose a strategy without knowing in advance what the other is doing.

Bo has exactly one weakly dominant strategy, which is to sit on committees; consequently, Bo declines to teach and decides to sit. Al knows table 7.22 and thus

Table 7.22 Payoffs for The Professors' Problem.

		BO'S STRATEGIES			
		TEACH		SIT	
		PAYOFFS TO		PAYOFFS TO	
		AL	BO	AL	BO
AL'S STRATEGIES	TEACH	6	9	2	9
	SIT	5	1	3	2

Table 7.23 Payoffs for The Administration's Response.

		BO'S STRATEGIES			
		TEACH		SIT	
		PAYOFFS TO		PAYOFFS TO	
		AL	BO	AL	BO
AL'S STRATEGIES	TEACH	4	9	4	5
	SIT	5	1	3	2

knows that Bo will sit on committees; therefore, to get the higher payoff available to Al under Bo's decision to sit, Al also declines to teach and decides to sit on committees.

Not only do Al and Bo choose the Nash equilibrium where they are both worse off than in the other Nash equilibrium, but the sum of the values of their work to the College is the worst of all possibilities.

The administration's challenge is to entice the faculty to teach by modifying the payoff table.

Table 7.23 shows the new game on campus after the administration capped payoffs from committees to 5 units. If Bo sits on committees, then Al may also sit for a payoff of 4 rather than teach for 6: Al now values teaching only up to 4.

To get a deeper sense of one of Nash's many contributions, the reader may attempt proving the existence of a Nash equilibrium in games for any number of players with any number of strategies.

7.24 Example (Blue against Red [26]). In Melvin Dresher's initial context, the Blue and Red commanders led opposing armed forces [26, p. 4], but they might also lead sports teams [12]. Table 7.25 shows only the payoffs to the Blue commander.

7.2.6 *Exercises on Mathematical Games*

7.1 . Identify all the weakly dominant pure strategies in The Battle of the Sexes (table 7.16 in example 7.15).

Table 7.25 Payoffs *to Blue* for a noncooperative two-commander game with three strategies (adapted from [26, p. 4]).

		RED'S STRATEGIES		
		L ("LEFT")	C ("CENTER")	R ("RIGHT")
	T ("TOP")	Failure	Success	Success
BLUE'S				
	M ("MIDDLE")	Draw	Success	Draw
STRATEGIES				
	B ("BOTTOM")	Success	Draw	Failure

7.2 . Identify all weakly dominant pure strategies in the Administration's Solution to the Professors' Problem (table 7.23 in example 7.21).

7.3 . Identify all Nash equilibria with pure strategies in The Battle of the Sexes (table 7.16 in example 7.15).

7.4 . Identify all Nash equilibria with mixed strategies in the Administration's Solution to the Professors' Problem (table 7.23 in example 7.21).

7.5 . Prove or disprove that every Nash equilibrium is a dominant strategy equilibrium.

7.6 . Prove or disprove that every two-player game restricted to pure strategies has a Nash equilibrium.

7.7 . For each function $f : A \times B \to C$ such that C is linearly ordered and for each nonempty subset $E \subseteq A \times B$ the image $f''(E)$ has a first element $\min[f''(E)]$ and a last element $\max[f''(E)]$, prove that

$$\max\{\min\{f(x,y) : y \in B\} : x \in A\} \leq \min\{\max\{f(x,y) : x \in A\} : y \in B\}.$$

7.8 . Design a function $f : A \times B \to C$ such that C is linearly ordered and for each nonempty subset $E \subseteq A \times B$ the image $f''(E)$ has a first element $\min[f''(E)]$ and a last element $\max[f''(E)]$, with a strict inequality

$$\max\{\min\{f(x,y) : y \in B\} : x \in A\} < \min\{\max\{f(x,y) : x \in A\} : y \in B\}.$$

Denote by "Al's strategies" and "Bo's strategies" the set of all strategies available to Al and Bo respectively.

7.9 . Prove that if each player plays so as to avoid the worst payoff available, then they get the payoffs

$$\text{payoff to Al} = \max\{\min\{p^{Al}_{S_{Al},S_{Bo}} : S_{Bo} \in \text{Bo's strategies}\} : S_{Al} \in \text{Al's strategies}\},$$

$$\text{payoff to Bo} = \max\{\min\{p^{Bo}_{S_{Al},S_{Bo}} : S_{Al} \in \text{Al's strategies}\} : S_{Bo} \in \text{Bo's strategies}\}.$$

7.10 . Assume that in a game for two players with two strategies $p^{Al}_{S_{Al},S_{Bo}} \geq p^{Al}_{S'_{Al},S'_{Bo}}$ if and only if $p^{Bo}_{S_{Al},S_{Bo}} \leq p^{Bo}_{S'_{Al},S'_{Bo}}$ for all positions (S_{Al}, S_{Bo}) and (S'_{Al}, S'_{Bo}). Prove that if each player plays so as to avoid the worst payoff available, then they get the payoffs

$$\text{payoff to Al} = \min\{\max\{p^{Al}_{S_{Al},S_{Bo}} : S_{Bo} \in \text{Bo's strategies}\} : S_{Al} \in \text{Al's strategies}\},$$

$$\text{payoff to Bo} = \min\{\max\{p^{Bo}_{S_{Al},S_{Bo}} : S_{Al} \in \text{Al's strategies}\} : S_{Bo} \in \text{Bo's strategies}\}.$$

7.11 . For noncooperative games with any number of players and any number of pure strategies, prove or disprove that if at least one player has at least one weakly dominant strategy, then such a strategy contains a Nash equilibrium.

7.12 . For noncooperative games with two players but any number of pure strategies, prove or disprove that if at least one player has at least one weakly dominant strategy, then such a strategy contains a Nash equilibrium.

7.13 . Analyze the game of Blue against Red defined by table 7.25 in example 7.24.

7.14 . Prove or disprove that the formulae from exercise 7.10 still hold for more than two strategies.

7.3   Match Making

Match making pairs up items or individuals from two groups. For example, the National Resident Matching Program (NRMP) uses an algorithm developed by David Gale, Alvin E. Roth, and Lloyd S. Shapley to match medical doctors with internships in hospitals [68, 96]. For their work on such problems, Alvin E. Roth and Lloyd S. Shapley received the Nobel Prize in Economics in 2012. The precise statements and proofs of their algorithms involve the mathematical concepts of sets, functions, relations, and induction.

7.3.1   Introduction

The problems considered here have been documented for millennia since Plato's time (Figure 7.2): how to arrange for proper marriages [97, p. 27], and how to admit students to schools [127, p. 44, footnote 11]. The problems consist in matching in some "optimal" way individuals from two groups, for example, boys and girls for marriage, students and schools for education, doctors and hospitals for internships, or, more generally, beggars and choosers (table 7.26). "Optimality" may mean, for instance, that there are no beggar from one couple and chooser from another couple that prefer each other to their current mate. A similar optimality applies to schools

Fig. 7.2 Heracles and
Athena, the goddess of
wisdom, 480–470 B.C. (about
the time of the Battles of
Thermopylae and Salamis
between Greek and Persian
forces), by an olive tree
presumably at the then future
site of Plato's Academy;
photograph courtesy
"User:Bibi Saint-Pol" via
Wikipedia. (http://en.
wikipedia.org/wiki/File:
Athena_Herakles_Staatliche_
Antikensammlungen_2648.
jpg)

Table 7.26 Applications of
Gale and Shapley's
algorithms [68].

BEGGARS B	CHOOSERS C	RELATION M
Boy	Girl	Marriage
Student	School	Admission
Doctor	Hospital	Residency
Recipient	Donor	Transplant

admitting several students. To get a sense of Gale and Shapley's work, the reader
may at this stage try to design such optimal match-making procedures.

7.3.2   A Mathematical Model for Optimal Match Making

The first step in producing a successful match-making method consists in designing
a mathematical model of the situation. With marriages, for instance, the group of
boys may be modeled by a set B and the girls by a set C. To allow for applications
more general than marriages and to shorten the prose, call the elements of B beggars
and those of C choosers. Moreover, in the contexts considered here, the two sets B
and C must be disjoint: $B \cap C = \varnothing$. The goal consists in marrying each boy exactly
one girl and each girl to exactly one boy. In general, the goal consists in establishing
a bijection between the two sets B and C, or, yet more generally, a relation $M \subseteq B \times$
C. Yet not every relation leads to successful marriages, because each chooser prefers
some beggars over others while each beggar prefers some choosers over others. Thus
a mathematical model of the situation must also include such preferences.

To specify preferences for certain choosers over others, each beggar $b \in B$ ranks
all choosers by a strict well-order $\overset{\succ}{_b}$ on C. Thus relative to $\overset{\succ}{_b}$ each nonempty
subset $W \subseteq C$ has a *unique* first-ranked element, denoted by $\mathrm{first}_b(W)$.

To specify preferences for certain beggars over others, each chooser $c \in C$ ranks all beggars by a strict well-order $\overset{\succ}{_c}$ on B. Thus relative to $\overset{\succ}{_c}$ each nonempty subset $V \subseteq B$ has a *unique* first-ranked element, denoted by $\text{first}_c(V)$.

The two disjoint sets B and C, with B well-ordered by each chooser and C well-ordered by each beggar, complete the mathematical model of the situation. The second step consists in designing a model of a successful relation.

For each relation $M \subseteq B \times C$, denote its domain by $\text{Domain}(M)$ and its range by $\text{Range}(M)$. Also, call "single" each beggar $b' \in B'_M := B \setminus \text{Domain}(M)$ and each chooser $c' \in C'_M := C \setminus \text{Range}(M)$: neither b' nor c' are related to anyone by M. Definition 7.27 specifies the notion of a successful relation by a concept of stability.

7.27 Definition. A relation $M \subseteq B \times C$ is **unstable** if and only if at least one of the following conditions holds:

(US.1) A beggar and a chooser prefer each other to their current mates: there are different couples $(b_1, c_1), (b_2, c_2) \in M$ for whom $b_2 \overset{\succ}{_{c_1}} b_1$ and $c_1 \overset{\succ}{_{b_2}} c_2$, a condition denoted by $(b_1, c_1) \bowtie (b_2, c_2)$.

(US.2) There is a couple $(b, c) \in M$ and a "single" beggar $b' \in B'_M = B \setminus \text{Domain}(M)$ for whom $b' \overset{\succ}{_c} b$ and $c \overset{\succ}{_{b'}} c'$ for every "single" chooser $c' \in C'_M = C \setminus \text{Range}(M)$, a condition denoted by $(b, c) \bowtie b'$.

(US.3) There is a couple $(b, c) \in M$ and a "single" chooser $c' \in C'_M = C \setminus \text{Range}(M)$ for whom $c' \overset{\succ}{_b} c$ and $b \overset{\succ}{_{c'}} b'$ for every "single" beggar $b' \in B'_M = B \setminus \text{Domain}(M)$, a condition denoted by $c' \bowtie (b, c)$.

A relation $M \subseteq B \times C$ is **stable** if and only if it is not unstable, so that none of the preceding three conditions holds.

7.28 Definition. A stable relation $M \subseteq B \times C$ is **total** if and only if $\text{Domain}(M) = B$ and $\text{Range}(M) = C$.

7.29 Example. The empty relation $\varnothing \subseteq B \times C$ is vacuously stable. It is total if and only if $B = \varnothing = C$.

Specifications of the situation and goal do not yet suffice to reach the goal. To this end, the next step in mathematical modeling consists in developing an algorithm, method, or procedure to reach the goal from the current situation. One algorithm uses a match maker, another algorithm is carried out by the participants without a match maker.

7.3.3 An Algorithm for Optimal Match Making with a Match Maker

This subsection defines and demonstrates an algorithm for a match maker (person or machine) to find a stable relation. First, a match maker with access to all well-orders from all beggars and choosers can extend any stable relation.

7.30 Theorem. *For each stable relation $M \subset B \times C$ with* $\mathrm{Domain}(M) \subset B$ *and* $\mathrm{Range}(M) \subset C$ *there exists a stable proper extension $\hat{M}$ with $M \subset \hat{M} \subseteq B \times C$.*

Proof. By the hypothesis that $\mathrm{Domain}(M) \subset B$, there exists a "single" beggar $b' \in B'_M = B \setminus \mathrm{Domain}(M) \neq \varnothing$. Since the well-order $\overset{\succ}{b'}$ restricts to a well-order on $C'_M = C \setminus \mathrm{Range}(M) \neq \varnothing$, the "single" beggar b' has a first-ranked single chooser $c' = \mathrm{first}_{b'}(C'_M)$. In particular, the set C_1 of those single choosers in C'_M that are ranked first among C'_M by some single beggar in B'_M is not empty:

$$C_1 := \{c_1 \in C'_M : \exists b'(b' \in B'_M) \wedge [c_1 = \mathrm{first}_{b'}(C'_M)]\} \neq \varnothing.$$

Also, for each first-ranked single chooser $c_1 \in C_1$, the set $B_{c_1} \subseteq B'_M$ of single beggars who rank c_1 first in C'_M,

$$B_{c_1} := \{b' \in B'_M : c_1 = \mathrm{first}_{b'}(C'_M)\},$$

is not empty by definition and hence has a *unique* first element $b_{c_1} := \mathrm{first}_{c_1}(B_{c_1})$. Moreover, if $c_1 \neq c_2 \in C_1$, then $b_{c_1} \neq b_{c_2}$, because b_{c_1} ranks c_1 first, ahead of c_2. Thus the function $C \hookrightarrow B'_M$, $c_1 \to b_{c_1}$ is injective. Define

$$M_1 := \{(b_{c_1}, c_1) \in B'_M \times C'_M : (c_1 \in C_1) \wedge [b_{c_1} = \mathrm{first}_{c_1}(B_{c_1})]\};$$

$$\hat{M} := M \cup M_1 \supset M.$$

First, the relation M_1 is stable in $B'_M \times C'_M$:

If $(b_{c_1}, c_1), (b'_{c'_1}, c'_1) \in M_1$, then $c_1 = \mathrm{first}_{b_{c_1}}(C'_M)$, whence $c_1 \overset{\succ}{b_{c_1}} c'$ for every $c' \in C'_M$, so that $(b'_{c'_1}, c'_1) \bowtie (b_{c_1}, c_1)$ and $c' \bowtie (b_{c_1}, c_1)$ do not occur. Moreover, $b_{c_1} = \mathrm{first}_{c_1}(B_{c_1})$, so that if $b' \in B'_M$ and $b' \overset{\succ}{c_1} b_{c_1}$, then $b' \notin B_{c_1}$, which means that there exists $c' \in C'_M$ with $c' \overset{\succ}{b'} c_1$, so that $(b_{c_1}, c_1) \bowtie b'$ does not occur either.

Second, $\hat{M}$ is stable in $B \times C$: if $(b, c) \in M$ and $(b_{c_1}, c_1) \in M_1$, then either $b \overset{\succ}{c} b_{c_1}$, in which case $(b, c) \bowtie (b_{c_1}, c_1)$ fails, or $b_{c_1} \overset{\succ}{c} b$ but then by the assumed stability of M in $B \times C$ there exists a "single" chooser $c' \in C'_M$ for whom $c' \overset{\succ}{b_{c_1}} c$, yet $c_1 = \mathrm{first}_{b_{c_1}}(C'_M)$, so that either $c_1 = c'$ or $c_1 \overset{\succ}{b_{c_1}} c' \overset{\succ}{b_{c_1}} c$. In either case $(b, c) \bowtie (b_{c_1}, c_1)$ fails.

Similarly, if $c \overset{\succ}{b} c_1$, then $(b_{c_1}, c_1) \bowtie (b, c)$ fails; also, if $c_1 \overset{\succ}{b} c$, then by the assumed stability of M there exists some "single" beggar $b' \in C'_M$ such that $b' \overset{\succ}{c} b$, whence either $b' = b_{c_1}$ or $b_{c_1} \overset{\succ}{c} b' \overset{\succ}{c} b$, and then $(b_{c_1}, c_1) \bowtie b$ fails. □

Second, if the sets B of beggars and C of choosers have the same finite number of elements, then the algorithm that starts from the stable empty relation and iterates theorem 7.30 yields a total stable relation after finitely many iterations.

7.31 Algorithm (Match Maker's Algorithm).
  Initially, for all disjoint sets B and C, set $M := \varnothing \subseteq B \times C$.
  While $\mathrm{Domain}(M) \neq B$ and $\mathrm{Range}(M) \neq C$, find a proper extension $\hat{M}$ by Theorem 7.30, and re-set $M := \hat{M}$.

Table 7.33 Gale and
Shapley's unlabeled example,
adapted from [41, p. 389].

BEGGARS	CHOOSERS C			
B	c_1	c_2	c_3	c_4
b_1	1, 3	2, 2	3, 1	4, 3
b_2	1, 4	2, 3	3, 3	4, 4
b_3	3, 1	1, 4	2, 3	4, 2
b_4	2, 2	3, 1	1, 4	4, 1

7.32 Example. Consider Gale and Shapley's unlabeled example [40, 41, p. 389], adapted in table 7.33. At the intersection of the row for b_i and the column for c_j, the ordered pair (m_i, n_j) states that b_i ranks c_j in position m_i, whereas c_j ranks b_i in position n_j. For instance, at the intersection of the row for b_3 and the column for c_2, the ordered pair $(1, 4)$ states that b_3 ranks c_2 in position 1, whereas c_2 ranks b_3 in position 4.

Start from the stable empty relation $\varnothing \subseteq B \times C$. Thus $B'_\varnothing = B$ and $C'_\varnothing = C$. The set of first-ranked choosers is $C_1 = \{c_1, c_2, c_3\}$. They are ranked first by

$$B_{c_1} = \{b_1, b_2\}, \qquad B_{c_2} = \{b_3\}, \qquad B_{c_3} = \{b_4\}.$$

Among those beggars who ranked them first, their first-ranked beggars are

$$b_{c_1} = \text{first}_{c_1}(\{b_1, b_2\}) = b_1, \qquad b_{c_2} = b_3, \qquad b_{c_3} = b_4.$$

Theorem 7.30 extends $M := \varnothing$ to

$$\hat{M} = \{(b_1, c_1), (b_3, c_2), (b_4, c_3)\},$$

where all beggars have their first choice. Theorem 7.30 then extends $\hat{M}$ to

$$\hat{\hat{M}} = \{(b_1, c_1), (b_3, c_2), (b_4, c_3)\} \cup \{(b_2, c_4)\},$$

where b_2 and c_4 have their worst choice, but then c_4 was ranked last by every beggar.

7.3.4   An Algorithm for Optimal Match Making Without a Match Maker

This subsection describes Gale & Shapley's algorithm to find a stable relation without any match maker [40, 41]. In the context of marriages, boys and girls carry out the algorithm themselves through rounds of proposals and rejections.

7.34 Algorithm (Gale & Shapley's Deferred Acceptance Procedure).
Initially, for each girl $c \in C$, the set B_c of boys who proposed to her is empty. Similarly, for each boy $b \in B$, the set C_b of girls who rejected him is empty. Then each round of proposals and rejections proceeds as follows:

(BG.1) Each girl $c \in C$ has a set B_c, which is either empty, or is a singleton containing only (the name of) her top-ranked boy among those boys who have already proposed to her.

(BG.2) Each boy $b \in B$ has a set C_b, which is either empty, or contains all the girls who have already rejected him.

(BG.3) Each boy $b \in B$ proposes to his top-ranked girl $c_b := \text{first}(C \setminus C_b)$ among those girls who have not yet rejected him.

(BG.4) From the boys' proposals, each girl $c \in C$ receives a set B'_c of proposals, which may be empty, she forms the union $B''_c := B'_c \cup B_c$, and rejects all but her top-ranked boy in that set, in effect re-setting it to $B_c := \{\text{first}_c(B''_c)\}$.

(BG.5) Each boy $b \in B$ who receives a rejection from a girl $c \in C$ appends her to his set of rejections, in effect replacing C_b by $C_b \cup \{c\}$.

If the sets B of beggars and C of choosers have the same finite number of elements, then Gale & Shapley's algorithm yields a stable relation after finitely many iterations, as proved in exercises 7.15 and 7.16.

7.3.5   Exercises on Gale & Shapley's Algorithms

7.15. For disjoints sets B and C with the same finite number N of elements, prove that if in some round any boy $b \in B$ receives his $(N-1)$-th rejection, then Gale & Shapley's algorithm terminates at the next round.

7.16. For disjoints sets B and C with the same finite number N of elements, prove that Gale & Shapley's algorithm terminates after at most $N^2 + 2 - 2N$ rounds.

7.17. Prove or disprove that for all disjoint sets B and C with the same *infinite* cardinality there must exist a total stable relation.

7.18. Carry out Gale & Shapley's algorithm with Gale & Shapley's example 7.32.

7.19. Extend algorithm 7.31 (with a match maker) to stable relations that may relate each chooser to more than one beggar, with a quota q_c of beggars for chooser c (polygamy, polyandry, schools admitting more than one student, etc.).

7.20. Extend algorithm 7.34 (without a match maker) to stable relations that may relate each chooser to more than one beggar, with a quota q_c of beggars for chooser c (polygamy, polyandry, schools admitting more than one student, etc.).

7.21. Denote by $\mathscr{S}$ the subset of the power set $\mathscr{P}(B \times C)$ consisting of all stable relations, partially ordered by inclusion. Prove that for each chain $\mathscr{T} \subseteq \mathscr{S}$, its union $U_{\mathscr{T}} := \bigcup \mathscr{T} \in \mathscr{S}$ is an upper bound for $\mathscr{T}$ in $\mathscr{S}$.

7.22 . Determine the maximum number of iterations of theorem 7.30 (match making with a match maker) necessary to complete a total stable relation between disjoints sets B and C with the same finite number N of elements.

7.23 . Design and test a program to iterate theorem 7.30 (match making with a match maker) for disjoints sets B and C with the same finite number N of elements.

7.24 . Design and test a program to iterate theorem 7.30 (match making with a match maker) for disjoints finite sets B and C with quota.

7.3.6 Projects

7.35 Project. Develop concepts, theorems, and algorithms for match making where beggars might order any subset, not necessarily the whole set, of choosers, and choosers might order any subset, not necessarily the whole set, of beggars. For example, the National Resident Matching Program (NRMP) evidently uses an algorithm to this effect [96].

7.36 Project. Develop concepts, theorems, and algorithms for tri-partite match making. (Tri-partite reproduction occurs in Isaac Asimov's novel *The Gods Themselves* [5]. See also the "three-parent" therapy [2].)

7.4   Arrow's Impossibility Theorem

Arrow's Impossibility Theorem exposes some of the limitations inherent to voting: several desired features of voting procedures are mutually incompatible. Jointly with John R. Hicks, Kenneth J. Arrow received the Nobel Prize in Economics in 1972, in part for his Impossibility Theorem. Its precise statement and proof involve the mathematical concepts of sets, functions, and ranking relations.

7.4.1 Introduction

The problem considered here consists in designing a procedure to rank several mutually exclusive alternatives, or merely to choose exactly one such alternative. By law or otherwise, the voting procedure may have to take into account many decision criteria from the voters, and may have to conform to rules imposed in advance by the voters.

   In the political arena, the decision criteria may be ballots submitted by voters, while the alternatives may be persons who are candidates for public office. Different voting procedures may lead to the election of different candidates.

Fig. 7.3 The Mars Climate Orbiter is conjectured to have followed too low a trajectory and burned up in Mars atmosphere; art work courtesy Jet Propulsion Laboratory (JPL) and NASA. (http://www.jpl.nasa.gov/ jplhistory/the90/images/ climate-orbiter-browse.jpg)

7.37 Example. In the United States presidential election of 1824, the popular and electoral votes had ranked Andrew Jackson (Democrat-Republican Party) ahead of John Quincy Adams (Coalition Party), William H. Crawford, and Henry Clay (Whig Party); nevertheless the House of Representatives elected John Quincy Adams ahead of Andrew Jackson (http://www.archives.gov/federal-register/electoral-college/ scores.html).

In the scientific arena, the decision criteria may be measurements from sensors, which may act as voters, while the alternatives, which play the rôles of candidates, may be decisions on how to proceed with a mission. In some missions, a simple majority of votes may fail to select the best alternative (Figure 7.3).

7.38 Example. The Mars Climate Orbiter spacecraft was lost on 23 September 1999, due to small thrusters controlled by faulty unit conversions that compounded during the year-long flight, which had been tracked by three telemetric systems: Doppler only, range only, and Doppler and range combined. Two out of the three systems submitted the same vote and won. Range only, and Doppler and range combined, both showed a flight path clearing the planet, allowing the mission to proceed without corrections. Both systems were wrong and the probe crashed. The minority vote was right: "The Doppler-only solutions consistently indicated a flight path insertion closer to the planet. These discrepancies were not resolved" [78, p. 13]. In some situations, discrepancies must be investigated, not voted away.

Among other procedures, the voting procedures considered here are based on voters ranking all the candidates.

7.39 Example. Table 7.40 shows Judges' (voters') ranking of three skaters (candidates) at the 1994 Winter Olympic Games in Lillehammer, Norway [112, p. 22]. The judges' **plurality voting** procedure selects the skater rated first by the largest number of judges, here Baiul ahead of Kerrigan.

Table 7.40 Judges' rankings of skaters [112, p. 22].

SKATERS	VOTERS								
	D	PL	CZ	UA	PRC	USA	J	CDN	UK
Baiul	1	1	1	1	1	2	2	3	3
Kerrigan	2	2	2	2	2	1	1	1	1
Chen Lu	3	3	3	3	3	3	3	2	2

Table 7.42 Voters' rankings of three candidates [113, p. 448].

RANKS	VOTERS' RANKINGS OF CANDIDATES										
	0	1	2	3	4	5	6	7	8	9	X
Top	Al	Al	Al	Al	Al	Bo	Bo	Ci	Ci	Ci	Ci
Second	Bo	Bo	Bo	Ci	Ci	Ci	Ci	Bo	Bo	Bo	Bo
Last	Ci	Ci	Ci	Bo	Bo	Al	Al	Al	Al	Al	Al

Table 7.43 Voters' rankings of the two remaining candidates.

RANKS	VOTERS' RANKINGS OF CANDIDATES										
	0	1	2	3	4	5	6	7	8	9	X
Top	Al	Al	Al	Al	Al	Bo	Bo	Bo	Bo	Bo	Bo
Second	Bo	Bo	Bo	Bo	Bo	Al	Al	Al	Al	Al	Al

Example 7.41 exposes some peculiarities of plurality voting.

7.41 Example. Table 7.42 shows hypothetical rankings of three candidates by eleven voters, adapted from [113, p. 448].

The method of election called **Borda's count** attempts to take into account voters' rankings by allocating a candidate two points for each top choice and one point for each second choice from the voters. Table 7.42 shows that Ci leads with with $(4 \times 2) + (4 \times 1) = 12$ points, followed by Bo $(2 \times 2) + (7 \times 1) = 11$ points, and Al with $5 \times 2 = 10$ points. The press might list the outcome as Ci > Bo > Al. Thus Ci is elected.

The method of election called **plurality voting** ignores voters' second and subsequent choices, takes into account only each voter's top choice, and elects the candidate who is the top choice of most voters. Table 7.42 shows Al leading with 5 top choices, followed by Ci with 4, and Bo with 2. Thus Al > Ci > Bo, and Al is elected.

Suppose now that Ci drops out (as did Ross Perrot in 1992). Table 7.43 shows the voters' rankings of Al and Bo from table 7.42. With plurality voting, Bo wins with 6 votes and Al loses with 5 votes: Bo > Al, and Bo is elected.

Such a reversal of the election result from Al to Bo, caused by a change in the ranking of a seemingly irrelevant third candidate, Ci, is one of the features of plurality voting that is deemed undesirable by some voters.

7.4.2   A Mathematical Model for Arrow's Impossibility Theorem

Many features have been deemed desirable from a voting procedure, for instance, the following features.

(AIT.1)   **Unrestricted Domain**. Each voter may submit *any* ranking of the candidates: no rankings are forbidden.

(AIT.2)   **Unanimity**. If every voter prefers candidate Al to Bo, then the voting procedure must rank Al ahead of Bo.

(AIT.3)   **Independence of Irrelevant Alternatives**. If every voter prefers candidate Al to Bo in two different elections, then the voting procedure must rank Al ahead of Bo in both elections, independently (regardless) of changes in the ranking of any candidate Ci other than Al and Bo between the two elections.

(AIT.4)   **Absence of Dictators**. The ranking from the voting procedure does not coincide with the ranking of any fixed voter: no voters can dictate the outcome of all elections.

(AIT.5)   **Anonymity**. The ranking from the voting procedure does not depend on who cast what vote. (Anonymity may be desirable for public votes, but undesirable for the votes of elected representatives or scientific sensors.)

To get a sense of Arrow's work, the reader may at this stage try to design a voting procedure with all the features just listed to elect one among three candidates: either find such a voting procedure, or prove that there are none.

Arrow's Impossibility Theorem states that the first four features, (AIT.1), (AIT.2), (AIT.3), and (AIT.4) are already mutually incompatible: no such voting procedures are possible. As just stated, the four features are somewhat vague. For instance, a voting procedure might need only to accommodate finitely many voters. A precision sufficient for rigorous reasoning may *have* to be inserted and culminate in a mathematical model of the descriptions of the features. To get a deeper sense of Arrow's work, the reader may at this stage try to formulate those four features mathematically.

In general, a set $\mathscr{V}$ of voters must rank a set $\mathscr{C}$ of mutually exclusive alternatives, for instance, candidates, decisions, laws, policies, etc. To this end, each voter submits one ranking of the set $\mathscr{C}$ of candidates. From the rankings submitted by the voters a "social welfare function" $\mathscr{F}$ produces a final "aggregate" ranking of the set $\mathscr{C}$ of candidates (Table 7.44).

7.45 Definition.   A **weak ranking** allowing for ties on a set $\mathscr{C}$ is a *transitive* and *strongly connected* relation $R \subseteq \mathscr{C} \times \mathscr{C}$.

To allow for ties, weak rankings need not be anti-symmetric; thus weak rankings are not partial orders.

The notation $(X, Y) \in R$, also abbreviated by XRY, means that R ranks X before or tied with Y.

The *transitivity* of R is defined by $\forall X \forall Y \forall Z \{[(XRY) \wedge (YRZ)] \Rightarrow (XRZ)\}$.

Table 7.44 Symbols from logic.

SYMBOL	DEFINITION
$\neg(P)$	"not P": True if and only if P is False.
$(P) \wedge (Q)$	"P and Q": True if and only if P and Q are both True.
$(P) \vee (Q)$	"P or Q": False if and only if P and Q are both False.
$(P) \Rightarrow (Q)$	"P implies Q": False if and only if P is True and Q is False.
$\forall$	"for each";
$\exists$	"there exists".

The *strong connectedness* of R is defined by

$$\forall X \forall Y \{[(X \in \mathscr{C}) \wedge (Y \in \mathscr{C})] \Rightarrow [(XRY) \vee (YRX)]\}.$$

In particular, for all elements X and Y in $\mathscr{C}$, if $X = Y$, then XRX, so that R is reflexive.

The inverse ranking is the inverse relation $R^{\circ -1} := \{(Y, X) \in \mathscr{C} \times \mathscr{C} : (X, Y) \in R\}$.

The **preference** P_R associated with the ranking R is the set-theoretic difference $P_R := R \setminus R^{\circ -1}$.

A weak ranking R may also be denoted by such a symbol as $\succeq$ so that $X \succeq Y$ means XRY.

The associated preference P_R may then be denoted by $\succ$ so that $X \succ Y$ means $(X \succeq Y) \wedge [\neg(Y \succeq X)]$.

The set of all weak rankings of a set $\mathscr{C}$ is denoted by $\mathscr{R}(\mathscr{C})$.

A **voters' profile** is a function $r : \mathscr{V} \to \mathscr{R}(\mathscr{C})$ with domain $\mathscr{V}$.

To each voter $V \in \mathscr{V}$, a voters' profile r associates that voter's ranking, $r(V) \in \mathscr{R}(\mathscr{C})$, also denoted by $\underset{V}{\succeq}$.

The voter's preference associated with that voter's ranking $r(V)$ is then also denoted by $P_{r(V)}$ or by $\underset{V}{\succ}$.

The set of all voters' profiles, which are all functions from $\mathscr{V}$ to $\mathscr{R}(\mathscr{C})$ with domain $\mathscr{V}$, is denoted by $\mathscr{R}(\mathscr{C})^{\mathscr{V}}$.

A **social welfare function** is a function $\mathscr{F} : \mathscr{R}(\mathscr{C})^{\mathscr{V}} \to \mathscr{R}(\mathscr{C})$ with domain $\mathscr{R}(\mathscr{C})^{\mathscr{V}}$.

To each voters' profile r, a social welfare function $\mathscr{F}$ associates an "aggregate" ranking $\mathscr{F}(r) \in \mathscr{R}(\mathscr{C})$.

7.46 Definition. A ranking $R \subseteq \mathscr{C} \times \mathscr{C}$ induces a ranking $R_{X,Y} := R \cap (\{X, Y\} \times \{X, Y\})$ of two candidates $X, Y \in \mathscr{C}$ **relative to each other**. Two rankings $R, S \subseteq \mathscr{C} \times \mathscr{C}$ rank two candidates $X, Y \in \mathscr{C}$ **in the same order** if and only if $R_{X,Y} = S_{X,Y}$. Two voters rank two candidates in the same order if and only if so do their rankings.

The problem considered here consists in investigating the feasibility and design of a social welfare function.

Table 7.47   Symbols for Arrow's Impossibility Theorem.

SYMBOL	DEFINITION
$\mathscr{C}$	Set of all candidates.
$\mathscr{R}(\mathscr{C})$	Set of all rankings on $\mathscr{C}$.
$\mathscr{V}$	Set of all voters.
r	Voters' profile: a function $r : \mathscr{V} \to \mathscr{R}(\mathscr{C})$ with domain $\mathscr{V}$.
$\mathscr{R}(\mathscr{C})^{\mathscr{V}}$	Set of all voters' profiles: all functions $\mathscr{V} \to \mathscr{R}(\mathscr{C})$ with domain $\mathscr{V}$.
$\mathscr{F}$	Social welfare function: a function $\mathscr{F} : \mathscr{R}(\mathscr{C})^{\mathscr{V}} \to \mathscr{R}(\mathscr{C})$.
$\mathscr{R}(\mathscr{C})^{[\mathscr{R}(\mathscr{C})^{\mathscr{V}}]}$	Set of all social welfare functions: all functions $\mathscr{R}(\mathscr{C})^{\mathscr{V}} \to \mathscr{R}(\mathscr{C})$.
$X \overset{\succ}{_V} Y$	Alternative notation for $(X, Y) \in P_{r(V)}$: voter V prefers X to Y.

7.4.3   Statement and Proof of Arrow's Impossibility Theorem

Not all social welfare functions are considered here: only those that satisfy the following four conditions.

(AIT.1) **Unrestricted Domain**. The domain of each social welfare function $\mathscr{F}$ is *all* of $\mathscr{R}(\mathscr{C})^{\mathscr{V}}$. This means that each voter may submit *any* ranking of the candidates for the social welfare function to produce an aggregate ranking (Table 7.47).

(AIT.2) **Unanimity**. For all candidates X and Y in $\mathscr{C}$, if in a voters' profile r each voter V in $\mathscr{V}$ prefers X to Y, then so does the aggregate ranking from the social welfare function $\mathscr{F}$:

$$\forall X \forall Y \forall r \left(\{ [r \in \mathscr{R}(\mathscr{C})^{\mathscr{V}}] \wedge \forall V [(X, Y) \in P_{r(V)}] \} \Rightarrow [(X, Y) \in P_{\mathscr{F}(r)}] \right).$$

(AIT.3) **Independence of Irrelevant Alternatives**. If every voter $V \in \mathscr{V}$ ranks candidates X and Y in the same order in two voting profiles r and s, then the aggregate rankings $\mathscr{F}(r)$ and $\mathscr{F}(s)$ also rank X and Y in the same order:

$$\forall X \forall Y \forall r \forall s \left\{ \left(\forall V \left\{ [r, s \in \mathscr{R}(\mathscr{C})^{\mathscr{V}}] \wedge [(X, Y) \in r(V) \cap s(V)] \right\} \right) \right.$$
$$\left. \Rightarrow [(X, Y) \in \mathscr{F}(r) \cap \mathscr{F}(s)] \right\}.$$

(AIT.4) **Absence of Dictators**. The social welfare function does not coincide with the ranking of any voter:

$$\neg \left\{ \exists V \left[\forall r \left(\{ [r \in \mathscr{R}(\mathscr{C})^{\mathscr{V}}] \Rightarrow [\mathscr{F}(r) = r(V)] \} \right) \right] \right\}.$$

7.48 Definition.   A **permutation** of a set $\mathscr{V}$ is a bijection $\sigma : \mathscr{V} \to \mathscr{V}$ with domain $\mathscr{V}$ and range $\mathscr{V}$.

The set of all permutations of a set $\mathscr{V}$ is denoted by $\mathscr{S}_{\mathscr{V}}$ and called the **symmetric group** of $\mathscr{V}$.

For all elements Bo, Ci $\in \mathscr{C}$, the **transposition** $\tau_{\text{Bo,Ci}}$ is the permutation $\tau_{\text{Bo,Ci}} \in \mathscr{S}_{\mathscr{V}}$ that swaps Bo and Ci but fixes every other element Al $\in \mathscr{C} \setminus \{\text{Bo, Ci}\}$:

$$\tau_{\text{Bo,Ci}}(\text{Bo}) := \text{Ci},$$

$$\tau_{\text{Bo,Ci}}(\text{Ci}) := \text{Bo};$$

$$\forall \text{Al}\{(\text{Al} \in \mathscr{C} \setminus \{\text{Bo, Ci}\}) \Rightarrow [\tau_{\text{Bo,Ci}}(\text{Al}) := \text{Al}]\}.$$

A voter $V \in \mathscr{V}$ **swaps** two candidates Bo and Ci by changing from a ranking $\overset{\succeq}{V}$ to the ranking defined by

$$\{(X, Y) \in \mathscr{C} \times \mathscr{C} : \tau_{\text{Bo,Ci}}(X) \,\overset{\succeq}{V}\, \tau_{\text{Bo,Ci}}(Y)\}.$$

Arrow's Impossibility Theorem applies to every set $\mathscr{V}$ of voters that is, or can be, strictly well-ordered by a relation $\prec$ so that each nonempty subset $\mathscr{E} \subseteq \mathscr{V}$ has a first element first($\mathscr{E}$) and $\mathscr{V}$ has a last element last($\mathscr{V}$). The strict well-order $\prec$ remains fixed for the entire proof. For instance, some sets of voters may be in alphabetical order.

7.49 Theorem (Arrow's Impossibility Theorem). *No social welfare functions satisfy all four conditions (AIT.1)–(AIT.4).*

Proof. This proof expands on [4, 6, 42, 143], showing that every function satisfying (AIT.1), (AIT.2), (AIT.3) violates (AIT.4). For all distinct candidates Al, Bo $\in \mathscr{C}$, at one extreme focus on any profile r where every voter $V \in \mathscr{V}$ prefers Al to Bo, so that $(\text{Al}, \text{Bo}) \in P_{r(V)}$ for every $V \in \mathscr{V}$:

$$\forall V\{(V \in \mathscr{V}) \Rightarrow [(\text{Al}, \text{Bo}) \in P_{r(V)}]\}.$$

By the rule of *unanimity* so does the aggregate ranking: $(\text{Al}, \text{Bo}) \in P_{\mathscr{F}(r)}$.

At the other extreme, focus on any profile s where every voter $V \in \mathscr{V}$ prefers Bo to Al, so that $(\text{Al}, \text{Bo}) \in P_{s(V)}$ for every $V \in \mathscr{V}$:

$$\forall V\{(V \in \mathscr{V}) \Rightarrow [(\text{Bo}, \text{Al}) \in P_{s(V)}]\}.$$

By the rule of *unanimity* so does the aggregate ranking: $(\text{Bo}, \text{Al}) \in P_{\mathscr{F}(s)}$.

Between extremes, for each voter $V \in \mathscr{V}$ define a profile $r_V \in \mathscr{R}(\mathscr{C})$ such that each voter $U \preceq V$ prefers Bo to Al, so that Bo $\overset{\succeq}{U}$ Al for each $U \preceq V$, while each voter $W \succ V$ prefers Al to Bo, so that Al $\overset{\succeq}{W}$ Bo for each $W \succ V$:

$$U \preceq V \Rightarrow (\text{Bo}, \text{Al}) \in P_{r_V(U)};$$

$$V \prec W \Rightarrow (\text{Al}, \text{Bo}) \in P_{r_V(W)}.$$

If $V = \text{last}(\mathscr{V})$, then $r_V = s$ and $\mathscr{F}(r_V) = \mathscr{F}(s)$ ranks Bo ahead of Al.

Let $\mathscr{E}_{\mathrm{Bo/Al}}$ be the subset of voters for whom $\mathscr{F}(r_V)$ ranks Bo ahead of Al:

$$\mathscr{E}_{\mathrm{Bo/Al}} := \{V \in \mathscr{V} : (\mathrm{Bo}, \mathrm{Al}) \in P_{\mathscr{F}(r_V)}\}.$$

Thus last($\mathscr{V}$) $\in \mathscr{E}_{\mathrm{Bo/Al}} \neq \varnothing$ and $\mathscr{E}_{\mathrm{Bo/Al}}$ has a first elements $V_{\mathrm{Bo/Al}} := \mathrm{first}(\mathscr{E}_{\mathrm{Bo/Al}})$. This first element $V_{\mathrm{Bo/Al}}$ is called a **pivotal voter for** Bo **over** Al.

By the rule of *Independence of Irrelevant Alternatives* $V_{\mathrm{Bo/Al}}$ is a pivotal voter for Bo over Al starting from every profile r' where every voter ranks Al before Bo. Indeed, for each voter $V \in \mathscr{V}$, in r' and in r, also in s' and in s, as well as in r'_V and in r_V, each voter $U \preceq V$, V, and $W \succ V$ ranks Al and Bo relative to each other in the same way.

The following considerations show that $V_{\mathrm{Bo/Al}}$ can also dictate the ranking of Bo relative to any third candidate Ci $\in \mathscr{C} \setminus \{\mathrm{Al}, \mathrm{Bo}\}$. To this end, consider the voters' profile q_V defined for $V := V_{\mathrm{Bo/Al}}$ by

$$q_V \begin{cases} U \prec V : & \mathrm{Bo} \underset{U}{\succ} \mathrm{Ci} \underset{U}{\succ} \mathrm{Al}, \\ V : & \mathrm{Al} \underset{V}{\succ} \mathrm{Bo} \underset{V}{\succ} \mathrm{Ci}, \\ V \prec W : & \mathrm{Al} \underset{W}{\succ} \mathrm{Bo} \underset{W}{\succ} \mathrm{Ci}. \end{cases}$$

Since $V = V_{\mathrm{Bo/Al}}$ is pivotal for Bo over Al, the aggregate ranking $\mathscr{F}(q_V)$ still ranks Al $\succ$ Bo. Also, by *unanimity* $\mathscr{F}(q_V)$ ranks Bo $\succ$ Ci. By transitivity the aggregate ranking $\mathscr{F}(q_V)$ ranks Al $\succ$ Bo $\succ$ Ci. If $V = V_{\mathrm{Bo/Al}}$ swaps Al and Bo, then the voting profile changes from q_V to r_V, defined by

$$r_V \begin{cases} U \prec V : & \mathrm{Bo} \underset{U}{\succ} \mathrm{Ci} \underset{U}{\succ} \mathrm{Al}, \\ V : & \mathrm{Bo} \underset{V}{\succ} \mathrm{Al} \underset{V}{\succ} \mathrm{Ci}, \\ V \prec W : & \mathrm{Al} \underset{W}{\succ} \mathrm{Bo} \underset{W}{\succ} \mathrm{Ci}. \end{cases}$$

Since $V = V_{\mathrm{Bo/Al}}$ is pivotal for Bo over Al, the aggregate ranking $\mathscr{F}(r_V)$ now ranks Bo $\succ$ Al. By *Independence of Irrelevant Alternatives* $\mathscr{F}(r_V)$ still ranks Al $\succ$ Ci, because each voter ranks Al and Ci in the same order in both profiles q_V and r_V. By transitivity, the aggregate ranking $\mathscr{F}(r_V)$ ranks Bo $\succ$ Al $\succ$ Ci. If any, some, or all voters other than $V = V_{\mathrm{Bo/Al}}$ swap Bo and Ci, then the voting profile changes from r_V to s_V, defined by

$$s_V \begin{cases} U \prec V : & \mathrm{Ci} \underset{U}{\succ} \mathrm{Bo} \underset{U}{\succ} \mathrm{Al} \text{ or } \mathrm{Bo} \underset{U}{\succ} \mathrm{Ci} \underset{U}{\succ} \mathrm{Al}, \\ V : & \mathrm{Bo} \underset{V}{\succ} \mathrm{Al} \underset{V}{\succ} \mathrm{Ci}, \\ V \prec W : & \mathrm{Al} \underset{W}{\succ} \mathrm{Ci} \underset{W}{\succ} \mathrm{Bo} \text{ or } \mathrm{Al} \underset{W}{\succ} \mathrm{Bo} \underset{W}{\succ} \mathrm{Ci}. \end{cases}$$

Then Al and Bo are ranked in the same order by every voter in r_V and s_V, whence the aggregate ranking $\mathscr{F}(s_V)$ still ranks Bo $\succ$ Al. Similarly, Al and Ci are ranked in the same order by every voter in r_V and s_V, whence the aggregate ranking $\mathscr{F}(s_V)$ still ranks Al $\succ$ Ci. By transitivity the aggregate ranking $\mathscr{F}(s_V)$ ranks Bo $\succ$ Al $\succ$ Ci, even though every voter other than V ranks Ci $\succ$ Bo.

Candidate Al plays only an auxiliary rôle in the proof that $V_{\mathrm{Bo/Ci}}$ is a dictator for Bo over Ci. If any, some, or all voters swap Al and any candidate Ig other than Bo and Ci, then the voting profile changes from s_V to p_V, defined by

$$p_V \begin{cases} U \prec V: & \mathrm{Ci} \underset{U}{\overset{>}{}} \mathrm{Bo} \underset{U}{\overset{>}{}} \mathrm{Al\ or} \cdots \overset{>}{\underset{U}{}} \mathrm{Ci} \underset{U}{\overset{>}{}} \mathrm{Bo} \underset{U}{\overset{>}{}} \cdots, \\ V: & \mathrm{Bo} \underset{V}{\overset{>}{}} \mathrm{Al} \underset{V}{\overset{>}{}} \mathrm{Ci\ or} \cdots \overset{>}{\underset{V}{}} \mathrm{Bo} \underset{V}{\overset{>}{}} \mathrm{Ci} \underset{V}{\overset{>}{}} \cdots, \\ V \prec W: & \mathrm{Al} \underset{W}{\overset{>}{}} \mathrm{Ci} \underset{W}{\overset{>}{}} \mathrm{Bo\ or} \cdots \overset{>}{\underset{W}{}} \mathrm{Ci} \underset{W}{\overset{>}{}} \mathrm{Bo} \underset{W}{\overset{>}{}} \cdots. \end{cases}$$

Then Bo and Ci are ranked in the same order by every voter in s_V and p_V, whence the aggregate ranking $\mathscr{F}(p_V)$ still ranks Bo $\succ$ Ci, even though every voter other than V ranks Ci $\succ$ Bo.

For the uniqueness of the dictator, in the strict well-ordering $\prec$ of the voters $\mathscr{V}$, a pivotal voter $V_{\mathrm{Bo/Ci}}$ for Bo over Ci cannot appear later than such a dictator for Bo over Ci as $V_{\mathrm{Bo/Al}}$, because such an earlier dictator $V_{\mathrm{Bo/Al}}$ would determine the ranking of Bo over Ci before a pivotal voter $V_{\mathrm{Bo/Ci}}$ does; consequently, $V_{\mathrm{Bo/Ci}} \preceq V_{\mathrm{Bo/Al}}$.

Similarly, a pivotal voter $V_{\mathrm{Ci/Bo}}$ for Ci over Bo cannot come earlier than such a dictator for Bo over Ci as $V_{\mathrm{Bo/Al}}$, for otherwise such a later dictator could reverse a pivotal vote from the pivotal voter; therefore $V_{\mathrm{Bo/Al}} \preceq V_{\mathrm{Ci/Bo}}$.

Combining $V_{\mathrm{Bo/Ci}} \preceq V_{\mathrm{Bo/Al}}$ with $V_{\mathrm{Bo/Al}} \preceq V_{\mathrm{Ci/Bo}}$ gives $V_{\mathrm{Bo/Ci}} \preceq V_{\mathrm{Bo/Al}} \preceq V_{\mathrm{Ci/Bo}}$ by transitivity.

Reversing the rôles of Bo and Ci gives the reverse ranks $V_{\mathrm{Ci/Bo}} \preceq V_{\mathrm{Ci/Al}} \preceq V_{\mathrm{Bo/Ci}}$.

Consequently, $V_{\mathrm{Bo/Ci}} = V_{\mathrm{Bo/Al}} = V_{\mathrm{Ci/Bo}}$.

Therefore there is exactly one pivotal voter, who is the pivotal voter and the dictator for every pair of candidates.

Thus every social welfare function satisfying (AIT.1), (AIT.2), (AIT.3) violates (AIT.4).   □

Another rule that might be imposed on a social welfare function pertains to the *anonymity* of voters. One way to model the concept of anonymity of voters consists in requiring that a social welfare function remains invariant under all permutations of the voters.

7.50 Definition. A social welfare function $\mathscr{F}: \mathscr{R}(\mathscr{C})^{\mathscr{V}} \to \mathscr{R}(\mathscr{C})$ is **invariant under permutations** if and only if $\mathscr{F}(r \circ \sigma) = \mathscr{F}(r)$ for every profile $r: \mathscr{V} \to \mathscr{R}(\mathscr{C})$ and every permutation $\sigma \in \mathscr{S}_{\mathscr{V}}$.

(AIT.5) **Anonymity.** A social welfare function respecting voters' anonymity is invariant under permutations.

Yet another condition attempts to model the condition that no voters have a right of veto.

(AIT.6) **No Veto.** Except perhaps for voter Val, if none of the other voters prefer any other candidate to Al, in other words, if all the other voters prefer Al to all the other candidates, or are indifferent between Al and all the other candidates, then the social welfare function does not rank any other candidate ahead of Al, regardless of Val's rankings [115, p. 386].

There is an alternative variation of the concept of unanimity.

(AIT.7) **Weak Unanimity**. Except perhaps for voter Val, if all the other voters do not prefer Bo to Al, so that they might prefer Al to Bo or be indifferent between Al and Bo, but Val prefers Al to Bo, then the social welfare function prefers Al to Bo:

$$\forall X \forall Y \forall r \left(\left\{ [r \in \mathscr{R}(\mathscr{C})^{\mathscr{V}}] \wedge [(X, Y) \in P_{r(\mathrm{Val})}] \wedge \forall V[(Y, X) \notin P_{r(V)}] \right\} \right.$$
$$\left. \Rightarrow [(X, Y) \in P_{\mathscr{F}(r)}] \right).$$

7.4.4   Exercises on Arrow's Impossibility Theorem

7.25 . Prove that for each relation $R \subseteq \mathscr{C} \times \mathscr{C}$ the relation $P_R := R \setminus R^{\circ-1}$ is irreflexive.

7.26 . Prove that for each relation $R \subseteq \mathscr{C} \times \mathscr{C}$ the relation $P_R := R \setminus R^{\circ-1}$ is asymmetric.

7.27 . Prove that if a relation $R \subseteq \mathscr{C} \times \mathscr{C}$ is transitive, then $P_R := R \setminus R^{\circ-1}$ is also transitive.

7.28 . Prove that for each relation $R \subseteq \mathscr{C} \times \mathscr{C}$ the relation $P_R := R \setminus R^{\circ-1}$ is anti-symmetric.

7.29 . Prove that if a relation $R \subseteq \mathscr{C} \times \mathscr{C}$ is transitive, then $P_R := R \setminus R^{\circ-1}$ is a partial order.

7.30 . Prove that if a relation $R \subseteq \mathscr{C} \times \mathscr{C}$ is transitive, then the inverse relation $R^{\circ-1}$ is also transitive.

7.31 . With at least two voters and at least two alternatives, prove that every social welfare function conforming to the rules of *Unrestricted Domain* and *Anonymity* has no dictators.

7.32 . With at least two voters and at least three alternatives, prove that every social welfare function conforming to the rules of *Unrestricted Domain*, *Unanimity*, and *Independence of Irrelevant Alternatives* is *not* invariant under permutations of the voters.

7.33 . Determine whether either of *Unanimity* and *Weak Unanimity* implies the other.

7.34 . Determine whether *Weak Unanimity* is compatible with *No Veto*.

Solutions to Some Odd-Numbered Exercises

Exercises from Chapter 1

1.1 Use axioms P1 and P2, theorem 1.12, and *Modus Ponens*:

$$\vdash (L) \Rightarrow [(K) \Rightarrow (L)] \qquad \text{axiom P1,}$$
$$\vdash (H) \Rightarrow \{(L) \Rightarrow [(K) \Rightarrow (L)]\} \qquad \text{theorem 1.12,}$$
$$\vdash \big[(H) \Rightarrow \{(L) \Rightarrow [(K) \Rightarrow (L)]\}\big] \Rightarrow \big([(H) \Rightarrow (L)] \Rightarrow \{(H) \Rightarrow [(K) \Rightarrow (L)]\}\big) \qquad \text{axiom P2,}$$
$$\vdash [(H) \Rightarrow (L)] \Rightarrow \{(H) \Rightarrow [(K) \Rightarrow (L)]\} \qquad \textit{Modus Ponens.}$$

1.3 Use the reflexivity of implications and the law of commutation:

$$\vdash \{[(A) \Rightarrow (B)] \Rightarrow (B)\} \Rightarrow \{[(A) \Rightarrow (B)] \Rightarrow (B)\} \qquad P := \{[(A) \Rightarrow (B)]$$
$$\Rightarrow (B)\} \text{ in 1.14,}$$
$$\vdash \underbrace{[(A) \Rightarrow (B)]}_{Q} \Rightarrow \Big[\underbrace{\{[(A) \Rightarrow (B)] \Rightarrow (B)\}}_{P} \Rightarrow \underbrace{(B)}_{R}\Big] \qquad \begin{array}{l}\text{substitution in the law}\\\text{of commutation.}\end{array}$$

1.5 Substitute P for A and also P for B in the preceding tautology:

$$\vdash [(A) \Rightarrow (B)] \Rightarrow \big[\{[(A) \Rightarrow (B)] \Rightarrow (B)\} \Rightarrow (B)\big] \qquad \text{preceding tautology,}$$
$$\vdash [(P) \Rightarrow (P)] \Rightarrow \big[\{[(P) \Rightarrow (P)] \Rightarrow (P)\} \Rightarrow (P)\big] \qquad \text{substitutions,}$$
$$\vdash (P) \Rightarrow (P) \qquad \text{theorem 1.14,}$$
$$\vdash \{[(P) \Rightarrow (P)] \Rightarrow (P)\} \Rightarrow (P) \qquad \textit{Modus Ponens.}$$

1.7 The formula $\{[(P) \Rightarrow (Q)] \Rightarrow (P)\} \Rightarrow (P)$ cannot be proved using only implications.

1.9 Use the tautology $[(H) \Rightarrow (L)] \Rightarrow \{(H) \Rightarrow [(K) \Rightarrow (L)]\}$:

$$\vdash \{\underbrace{[(R) \Rightarrow (Q)]}_{H} \Rightarrow \underbrace{(P)}_{L}\} \Rightarrow \{\underbrace{[(R) \Rightarrow (Q)]}_{H} \Rightarrow [\underbrace{(S)}_{K} \Rightarrow \underbrace{(P)}_{L}]\}$$

© Springer Science+Business Media New York 2015
Y. Nievergelt, *Logic, Mathematics, and Computer Science*,
DOI 10.1007/978-1-4939-3223-8

1.11 See also [72, p. 34]:

$\vdash (P) \Rightarrow \{[(P) \Rightarrow (P)] \Rightarrow (P)\}$ substitution in axiom 1a,

$\vdash \{(P) \Rightarrow [(P) \Rightarrow (P)]\}$
$\Rightarrow \big([(P) \Rightarrow \{[(P) \Rightarrow (P)] \Rightarrow (P)\}]$
$\Rightarrow [(P) \Rightarrow (P)]\big)$ substitution in axiom 1b,

$\vdash (P) \Rightarrow [(P) \Rightarrow (P)]$ substitution in axiom 1a,
$\vdash [(P) \Rightarrow \{[(P) \Rightarrow (P)] \Rightarrow (P)\}] \Rightarrow [(P) \Rightarrow (P)]$ *Detachment.*
$\vdash [(P) \Rightarrow (P)]$ *Detachment.*

1.13 Apply exercise 1.11 and *Detachment*:

$\vdash [(P) \Rightarrow (P)] \Rightarrow \big[\{[(P) \Rightarrow [(P) \Rightarrow (Q)]\}$ substitution in 1b,
$\Rightarrow [(P) \Rightarrow (Q)]\big]$
$\vdash \{[(P) \Rightarrow [(P) \Rightarrow (Q)]\} \Rightarrow [(P) \Rightarrow (Q)]$ *Detachment.*

1.15 Apply exercise 1.12 and *Detachment*:

$\vdash T$ hypothesis,
$\vdash (A) \Rightarrow (T)$ exercise 1.12,
$\vdash [(A) \Rightarrow (T)] \Rightarrow \big(\{(A) \Rightarrow [(T) \Rightarrow (C)]\}$ substitution in 1b,
$\Rightarrow [(A) \Rightarrow (C)]\big)$
$\vdash \{(A) \Rightarrow [(T) \Rightarrow (C)]\} \Rightarrow [(A) \Rightarrow (C)]$ *Detachment.*

1.17

$\vdash [(B) \Rightarrow (C)] \Rightarrow \{(A) \Rightarrow [(B) \Rightarrow (C)]\}$ 1a,
$\vdash \big([(B) \Rightarrow (C)] \Rightarrow \{(A) \Rightarrow [(B) \Rightarrow (C)]\}\big) \Rightarrow$
$\Big(\{[(B) \Rightarrow (C)] \Rightarrow \big(\{(A) \Rightarrow [(B) \Rightarrow (C)]\} \Rightarrow [(A) \Rightarrow (C)]\big)\}$
$\Rightarrow \{[(B) \Rightarrow (C)] \Rightarrow [(A) \Rightarrow (C)]\}\Big)$ 1b,
$\vdash \{[(B) \Rightarrow (C)] \Rightarrow \big(\{(A) \Rightarrow [(B) \Rightarrow (C)]\} \Rightarrow [(A) \Rightarrow (C)]\big)\}$
$\Rightarrow \{[(B) \Rightarrow (C)] \Rightarrow [(A) \Rightarrow (C)]\}$ *Detachment.*

1.19 Apply exercises 1.15, 1.18, and *Detachment*:

$\vdash [(A) \Rightarrow (B)] \Rightarrow \big(\{(A) \Rightarrow [(B) \Rightarrow (C)]\} \Rightarrow [(A) \Rightarrow (C)]\big)$ 1b,
$\vdash [(B) \Rightarrow (C)]$
$\Rightarrow \{[(A) \Rightarrow (B)] \Rightarrow \big(\{(A) \Rightarrow [(B) \Rightarrow (C)]\} \Rightarrow [(A) \Rightarrow (C)]\big)\}$ 1.12,
$\vdash \{[(B) \Rightarrow (C)] \Rightarrow [(A) \Rightarrow (B)]\}$
$\Rightarrow \Big\{[(B) \Rightarrow (C)] \Rightarrow \big(\{(A) \Rightarrow [(B) \Rightarrow (C)]\} \Rightarrow [(A) \Rightarrow (C)]\big)\Big\}$

1.21 $\vdash [(A) \Rightarrow (B)] \Rightarrow \{[(B) \Rightarrow (C)] \Rightarrow [(A) \Rightarrow (B)]\}$ by substitution in 1a.

1.23 Apply Tarski's axiom III and *Detachment*:

$\vdash (H) \Rightarrow (K)$	hypothesis,
$\vdash [(H) \Rightarrow (K)] \Rightarrow \{[(K) \Rightarrow (L)] \Rightarrow [(H) \Rightarrow (L)]\}$	substitution in axiom III,
$\vdash [(K) \Rightarrow (L)] \Rightarrow [(H) \Rightarrow (L)]$	*Detachment*,
$\vdash (K) \Rightarrow (L)$	hypothesis,
$\vdash (H) \Rightarrow (L)$	*Detachment*.

1.25 $\vdash (P) \Rightarrow \{[(P) \Rightarrow (Q)] \Rightarrow (P)\}$ by substitution in axiom I.

1.27 Apply exercises 1.23, 1.25, and 1.26:

$$(\underbrace{\;P\;}_{H}) \Rightarrow \underbrace{\big([(P) \Rightarrow (Q)] \Rightarrow \{[(P) \Rightarrow (Q)] \Rightarrow (Q)\}\big)}_{L}.$$

1.29 Apply exercises 1.27, 1.28, and 1.23:

$$(P) \Rightarrow \{[(P) \Rightarrow (Q)] \Rightarrow (Q)\}.$$

1.31 Apply exercise 1.29 and axiom III:

$\vdash \{(Q) \Rightarrow [(Q) \Rightarrow (R)] \Rightarrow (R)\} \Rightarrow$
$\Big[\big(\{[(Q) \Rightarrow (R)] \Rightarrow (R)\} \Rightarrow [(P) \Rightarrow (R)]\big)$
$\Rightarrow \{(Q) \Rightarrow [(P) \Rightarrow (R)]\}\Big]$ III,

$\vdash (Q) \Rightarrow [(Q) \Rightarrow (R)] \Rightarrow (R)$ 1.29,

$\vdash \big(\{[(Q) \Rightarrow (R)] \Rightarrow (R)\} \Rightarrow [(P) \Rightarrow (R)]\big)$
$\Rightarrow \{(Q) \Rightarrow [(P) \Rightarrow (R)]\}$ *Detachment*.

1.33 Apply axiom III, exercise 1.32, and *Detachment*:

$\vdash [(P) \Rightarrow (Q)] \Rightarrow \{[(Q) \Rightarrow (R)] \Rightarrow [(P) \Rightarrow (R)]\}$ axiom III,

$\vdash [(Q) \Rightarrow (R)] \Rightarrow \{[(P) \Rightarrow (Q)] \Rightarrow [(P) \Rightarrow (R)]\}$ 1.32, *Detachment*.

1.35 Apply axiom II with exercises 1.24, 1.33, and 1.23:

$\vdash \big([(P) \Rightarrow (Q)] \Rightarrow \{(P) \Rightarrow [(P) \Rightarrow (R)]\}\big)$
$\Rightarrow \big(\{(P) \Rightarrow [(P) \Rightarrow (R)]\} \Rightarrow [(P) \Rightarrow (R)]\big) \Rightarrow \{[(P) \Rightarrow (Q)] \Rightarrow [(P) \Rightarrow (R)]\}$ III,

$\vdash \big(\{(P) \Rightarrow [(P) \Rightarrow (R)]\} \Rightarrow [(P) \Rightarrow (R)]\big)$
$\Rightarrow \big([(P) \Rightarrow (Q)] \Rightarrow \{(P) \Rightarrow [(P) \Rightarrow (R)]\}\big) \Rightarrow \{[(P) \Rightarrow (Q)] \Rightarrow [(P) \Rightarrow (R)]\}$ 1.32,

$\vdash \{(P) \Rightarrow [(P) \Rightarrow (R)]\} \Rightarrow [(P) \Rightarrow (R)]$ II,

$\vdash \big([(P) \Rightarrow (Q)] \Rightarrow \{(P) \Rightarrow [(P) \Rightarrow (R)]\}\big) \Rightarrow \{[(P) \Rightarrow (Q)] \Rightarrow [(P) \Rightarrow (R)]\}$ *Detachment*,

$\vdash \{(Q) \Rightarrow [(P) \Rightarrow (R)]\} \Rightarrow \{[(P) \Rightarrow (Q)] \Rightarrow [(P) \Rightarrow (R)]\}$ 1.34.

$$[(P) \Rightarrow (Q)] \Rightarrow \big(\{(P) \Rightarrow [(Q) \Rightarrow (R)]\} \Rightarrow [(P) \Rightarrow (R)]\big).$$

1.37 Proceed as in theorem 1.40:

$\vdash (P) \Rightarrow [(Q) \Rightarrow (P)]$ axiom F1,

$\vdash [(Q) \Rightarrow (P)] \Rightarrow \{[\neg(P)] \Rightarrow [\neg(Q)]\}$ axiom F4,

$\vdash (P) \Rightarrow \{[\neg(P)] \Rightarrow [\neg(Q)]\}$ transitivity of implications (theorem 1.16).

1.39 A proof of $\{[\neg(Q)] \Rightarrow [\neg(P)]\} \Rightarrow [(P) \Rightarrow (Q)]$ can proceed as follows.

$\vdash \{[\neg(Q)] \Rightarrow [\neg(P)]\} \Rightarrow \left(\{\neg[\neg(P)]\} \Rightarrow \{\neg[\neg(Q)]\}\right)$ axiom F4,

$\vdash (P) \Rightarrow \{\neg[\neg(P)]\}$ axiom F6,

$\vdash \{[\neg(Q)] \Rightarrow [\neg(P)]\} \Rightarrow \left((P) \Rightarrow \{\neg[\neg(Q)]\}\right)$ transitivity (theorem 1.31),

$\vdash \{\neg[\neg(Q)]\} \Rightarrow (Q)$ axiom F5,

$\vdash \{[\neg(Q)] \Rightarrow [\neg(P)]\} \Rightarrow [(P) \Rightarrow (Q)]$ transitivity (theorem 1.32).

1.41

$\{[(B) \Rightarrow (F)] \Rightarrow [(A) \Rightarrow (F)]\} \Rightarrow \left(\{[(B) \Rightarrow (F)] \Rightarrow (A)\} \Rightarrow \{[(B) \Rightarrow (F)] \Rightarrow (F)\}\right)$ axiom C2,

$\{[(B) \Rightarrow (F)] \Rightarrow (F)\} \Rightarrow (B)$ axiom C3,

$\{[(B) \Rightarrow (F)] \Rightarrow [(A) \Rightarrow (F)]\} \Rightarrow \left[\{[(B) \Rightarrow (F)] \Rightarrow (A)\} \Rightarrow (B)\right]$ transitivity,

$(A) \Rightarrow \{[(B) \Rightarrow (F)] \Rightarrow (A)\}$ C1,

$\{[(B) \Rightarrow (F)] \Rightarrow [(A) \Rightarrow (F)]\} \Rightarrow [(A) \Rightarrow (B)]$ transitivity.

1.43 Substitutions of $\neg(B)$ for $(B) \Rightarrow (F)$ and $\neg(A)$ for $(A) \Rightarrow (F)$ into the solution of exercise 1.41 transform Church's third axiom into $\{[\neg(Q)] \Rightarrow [\neg(P)]\} \Rightarrow [(P) \Rightarrow (Q)]$. Consequently, all three axioms of classical logic remain valid in Church's logic, and, therefore, Church's logic allows for proofs of all the theorems of classical logic.

1.45 For the first theorem,

$\vdash [\neg(P)] \Rightarrow \{[(S) \Rightarrow (S)] \Rightarrow [\neg(P)]\}$ axiom P1,

$\vdash \{[(S) \Rightarrow (S)] \Rightarrow [\neg(P)]\} \Rightarrow \left(\{\neg[\neg(P)]\} \Rightarrow \{\neg[(S) \Rightarrow (S)]\}\right)$ law of contraposition,

$\vdash [\neg(P)] \Rightarrow \left(\{\neg[\neg(P)]\} \Rightarrow \{\neg[(S) \Rightarrow (S)]\}\right)$ transitivity,

$\vdash (P) \Rightarrow \{\neg[\neg(P)]\}$ converse double negation,

$\vdash [\neg(P)] \Rightarrow \left[(P) \Rightarrow \{\neg[(S) \Rightarrow (S)]\}\right]$ transitivity.

For the second theorem,

$\vdash (S) \Rightarrow (S)$ theorem 1.14,

$\vdash (P) \Rightarrow [(S) \Rightarrow (S)]$ theorem 1.12,

$\vdash [(P) \Rightarrow (Q)] \Rightarrow \left(\{(P) \Rightarrow [\neg(Q)]\} \Rightarrow [\neg(P)]\right)$ reduction ad absurdum,

$\vdash \{(P) \Rightarrow [(S) \Rightarrow (S)]\} \Rightarrow \left\{\left[(P) \Rightarrow \{\neg[(S) \Rightarrow (S)]\}\right] \Rightarrow [\neg(P)]\right\}$ substitution,

$\vdash \left[(P) \Rightarrow \{\neg[(S) \Rightarrow (S)]\}\right] \Rightarrow [\neg(P)]$ *Modus Ponens.*

1.47 Pierce's law:

$$\{[(P) \Rightarrow (Q)] \Rightarrow (P)\} \Rightarrow (P)$$

$\Updownarrow$ definition of $\vee$,

$$\left[\neg\big(\neg\{[\neg(P)] \vee (Q)\} \vee (P)\big) \vee (P)\right]$$

$\Updownarrow$ de Morgan's second law,

$$\{[\neg(\neg\{[\neg(P)] \vee (Q)\})] \wedge [\neg(P)]\} \vee (P)$$

$\Updownarrow$ double negations,

$$\big(\{[\neg(P)] \vee (Q)\} \wedge [\neg(P)]\big) \vee (P)$$

$\Updownarrow$ distributivity,

$$\left[\{[\neg(P)] \vee (Q)\} \vee (P)\right] \wedge \{[\neg(P)] \vee (P)\}$$

$\Updownarrow$ commutativity, associativity,
excluded middle,

$$[(Q) \vee (T)] \wedge (T)$$

$\Updownarrow$ identity,

$$(T) \wedge (T)$$

$\Updownarrow$ idempotence,

$$(T)$$

1.49

$$\{[(P) \Rightarrow (Q)] \Rightarrow (R)\} \Rightarrow \{[(R) \Rightarrow (P)] \Rightarrow (P)\}$$

$\Updownarrow$ definition of $\vee$,

$$\left[\neg\big(\neg\{[\neg(P)] \vee (Q)\} \vee (R)\}\big)\right] \vee \left[\neg\{[\neg(R)] \vee (P)\} \vee (P)\right]$$

$\Updownarrow$ de Morgan's second law,

$$\{[\neg(\neg\{[\neg(P)] \vee (Q)\})] \wedge [\neg(R)]\} \vee \left[(\{\neg[\neg(R)]\} \wedge [\neg(P)]) \vee (P)\right]$$

$\Updownarrow$ double negations,

$$\big(\{[\neg(P)] \vee (Q)\} \wedge [\neg(R)]\big) \vee \left[\{(R) \wedge [\neg(P)]\} \vee (P)\right]$$

$\Updownarrow$ distributivity,

$$\big(\{[\neg(P)] \wedge [\neg(R)]\} \vee \{(Q) \wedge [\neg(R)]\}\big) \vee \big([(R) \vee (P)] \wedge \{[\neg(P)] \vee (P)\}\big)$$

$\Updownarrow$ excluded middle,

$$\big(\{[\neg(P)] \wedge [\neg(R)]\} \vee \{(Q) \wedge [\neg(R)]\}\big) \vee \{[(R) \vee (P)] \wedge (T)\}$$

$\Updownarrow$ identity,

$$\big(\{[\neg(P)] \wedge [\neg(R)]\} \vee \{(Q) \wedge [\neg(R)]\}\big) \vee [(R) \vee (P)]$$

$\Updownarrow$ de Morgan's second law,

$$\big(\{\neg[(R) \vee (P)]\} \vee \{(Q) \wedge [\neg(R)]\}\big) \vee [(R) \vee (P)]$$

$\Updownarrow$ commutativity,
associativity,

$$\big(\{\neg[(R) \vee (P)]\} \vee [(R) \vee (P)] \vee \{(Q) \wedge [\neg(R)]\}\big)$$

$\Updownarrow$ excluded middle,

$$(T) \vee \{(Q) \wedge [\neg(R)]\})$$

$\Updownarrow$ identity,

$$(T)$$

1.51

$$\vdash (P) \Rightarrow \{[(P) \Rightarrow (Q)] \Rightarrow (Q)\} \qquad\qquad \text{theorem 1.26,}$$
$$\vdash \{[(P) \Rightarrow (Q)] \Rightarrow (Q)\} \Rightarrow \big([\neg(Q)] \Rightarrow \{\neg[(P) \Rightarrow (Q)]\}\big) \text{ theorem 1.44,}$$
$$\vdash (P) \Rightarrow \big([\neg(Q)] \Rightarrow \{\neg[(P) \Rightarrow (Q)]\}\big) \qquad\qquad \text{theorem 1.16.}$$

1.53 One implication is axiom P1 and the other implication is Pierce's law.

1.55 $[(P) \Rightarrow (Q)] \Leftrightarrow \big(\neg\{(P) \wedge [\neg(Q)]\}\big)$:

$$(P) \Rightarrow (Q)$$
$$\Updownarrow \qquad\qquad\qquad\qquad \text{double negations,}$$
$$(P) \Rightarrow \{\neg[\neg(Q)]\}$$
$$\Updownarrow \qquad\qquad\qquad\qquad \text{double negations,}$$
$$\neg\big(\neg[(P) \Rightarrow \{\neg[\neg(Q)]\}]\big)$$
$$\Updownarrow \qquad\qquad\qquad\qquad \text{definition of } \wedge,$$
$$\big(\neg\{(P) \wedge [\neg(Q)]\}\big)$$

1.57

$$\vdash (V) \Rightarrow (W) \qquad\qquad\qquad \text{hypothesis,}$$
$$\vdash (\neg(W)) \Rightarrow (\neg(V)) \qquad\qquad \text{contraposition,}$$
$$\vdash (R) \Rightarrow (S) \qquad\qquad\qquad \text{hypothesis,}$$
$$\vdash [\neg(V)] \Rightarrow [(R) \Rightarrow (S)] \qquad \text{theorem 1.12,}$$

$$\vdash \{[\neg(V)] \Rightarrow [(R) \Rightarrow (S)]\}$$
$$\qquad \Rightarrow \{[(\neg(V)) \Rightarrow (R)] \Rightarrow [(\neg(V)) \Rightarrow (S)]\} \text{ axiom P2,}$$

$$\vdash [(\neg(V)) \Rightarrow (R)] \Rightarrow [(\neg(V)) \Rightarrow (S)] \qquad \textit{Modus Ponens,}$$
$$\vdash [(\neg(V)) \Rightarrow (R)] \Rightarrow [(\neg(W)) \Rightarrow (S)] \qquad \text{second line and theorem 1.16,}$$
$$\vdash [(V) \vee (R)] \Rightarrow [(W) \vee (S)] \qquad\qquad \text{definition of } \vee.$$

1.59 No, the suggested rule fails if U and W are False but V is True.

1.61

$$\vdash \big([(P) \Rightarrow (S)] \wedge \{[\neg(Q)] \Rightarrow [\neg(S)]\}\big)$$
$$\qquad \Rightarrow \big(\{(P) \wedge [\neg(Q)]\} \Rightarrow \{(S) \wedge [\neg(S)]\}\big) \qquad \text{theorem,}$$

$$\vdash \big(\{(P) \wedge [\neg(Q)]\} \Rightarrow \{(S) \wedge [\neg(S)]\}\big) \Rightarrow$$
$$\qquad \big[(\neg\{(S) \wedge [\neg(S)]\}) \Rightarrow (\neg\{(P) \wedge [\neg(Q)]\})\big] \qquad \text{contraposition,}$$

$$\vdash \big[(\neg\{(S) \wedge [\neg(S)]\}) \Rightarrow (\neg\{(P) \wedge [\neg(Q)]\})\big]$$
$$\Rightarrow \big(\{[\neg(S)] \vee (S)\} \Rightarrow [(P) \Rightarrow (Q)]\big) \qquad \text{de Morgan, definition of } \wedge,$$

$$\vdash [\neg(S)] \vee (S) \qquad\qquad\qquad\qquad \text{excluded middle,}$$
$$\vdash (P) \Rightarrow (Q) \qquad\qquad\qquad\qquad\qquad \textit{Modus Ponens.}$$

1.63

$$\vdash (R) \Rightarrow [(T) \Rightarrow (R)] \qquad \text{axiom P1;}$$

$$\vdash (A) \Rightarrow \{[(A) \Rightarrow (B)] \Rightarrow (B)\} \qquad \text{theorem,}$$
$$\vdash (T) \Rightarrow \{[(T) \Rightarrow (R)] \Rightarrow (R)\} \qquad \text{substitutions in theorem,}$$
$$\vdash T \qquad \text{hypothesis,}$$
$$[(T) \Rightarrow (R)] \Rightarrow (R) \qquad \textit{Modus Ponens.}$$

1.65 Use $\{[\neg(R)] \Rightarrow (F)\} \Rightarrow \{[\neg(F)] \Rightarrow (R)\}$ and $[(T) \Rightarrow (R)] \Rightarrow (R)$.

1.67 Definition 1.51 of $(A) \Leftrightarrow (B)$ as $[(A) \Rightarrow (B)] \wedge [(B) \Rightarrow (A)]$ with theorems 1.57 $[(P) \wedge (Q)] \Rightarrow [(Q) \wedge (P)]$ and 1.52 $[(P) \wedge (Q)] \Rightarrow (Q)$ yield Tarski's axiom IV.

1.69 Theorem 1.61 gives a derivation of $(I) \Leftrightarrow (J)$. from $(I) \Rightarrow (J)$ and $(J) \Rightarrow (I)$. The Deduction Theorem (1.22) then yields a proof of $[(I) \Rightarrow (J)] \Rightarrow \{[(J) \Rightarrow (I)] \Rightarrow [(I) \Leftrightarrow (J)]\}$.

1.71 Subsections 1.3.10, 1.3.11, and 1.3.12 show that axioms P1 and P2 are derivable from Tarski's axioms I–III. Moreover, axiom P3 coincides with Tarski's axioms VII. Consequently, axioms P1, P2, and P3 are derivable from Tarski's axioms I–VII. Therefore, every theorem of the Classical Propositional Calculus is also derivable from Tarski's axioms I–VII.

1.73 Substitute R for S in theorem 1.82, which gives $\{[(P) \Rightarrow (Q)] \wedge [(R) \Rightarrow (R)]\} \Rightarrow \{[(P) \wedge (R)] \Rightarrow [(Q) \wedge (R)]\}$. Then apply the reflexivity of the logical implication (theorem 1.14), the law of contraposition (theorem 1.44), and transitivity, to derive Rosser's axiom R3 from axioms P1, P2, and P3.

1.75 Kleene's axiom 7 is the law of reductio ad absurdum (theorem 1.48).

1.77 $\{[(P) \Rightarrow (Q)] \Rightarrow (P)\} \Rightarrow (P)$ *is* a triadic tautology in Łukasiewicz's triadic logic.

1.79 $\{[(P) \Rightarrow (Q)] \Rightarrow (P)\} \Rightarrow (P)$ is *not* a triadic tautology in Church's system or in Łukasiewicz's.

1.81 $\{[(P) \Rightarrow (Q)] \Rightarrow (Q)\} \Rightarrow [(Q) \Rightarrow (P)] \Rightarrow (P)\}$ is a triadic tautology in Łukasiewicz's system, but not in Church's.

1.83 $\{(P) \Rightarrow [(Q) \Rightarrow (R)]\} \Rightarrow \{[(P) \Rightarrow (Q)] \Rightarrow [(P) \Rightarrow (R)]\}$ is a triadic tautology in Church's system, but not in Łukasiewicz's.

1.85 $\{(P) \Rightarrow [(Q) \Rightarrow (R)]\} \Rightarrow \{(Q) \Rightarrow [(P) \Rightarrow (R)]\}$ is a triadic tautology in both Church's system, and in Łukasiewicz's.

1.87

$$\vdash [\neg(P)] \Rightarrow (P) \text{ hypothesis,}$$
$$\vdash (P) \Rightarrow (P) \text{ theorem 1.14,}$$
$$\vdash P \text{ deduction rule.}$$

1.89 In the proof of $\{[\neg(P)] \Rightarrow (P)\} \vdash (P)$, the first step lists the hypothesis H, here $\vdash \{[\neg(P)] \Rightarrow (P)\}$. The Deduction Theorem replaces this step, $\vdash H$, by a complete proof of $(H) \Rightarrow (H)$, from theorem 1.14, with $[\neg(P)] \Rightarrow (P)$ substituted for H everywhere. The second step invokes theorem 1.14, which the Deduction Theorem replaces by a complete proof of theorem 1.14. The third step uses the deduction rule

$$\vdash [\neg(R)] \Rightarrow (S) \text{ hypothesis,}$$
$$\vdash (R) \Rightarrow (S) \text{ theorem 1.15,}$$
$$\vdash S \text{ deduction rule,}$$

which the Deduction Theorem replaces by a complete proof of this deduction rule.

1.95 The commutation law establishes the logical equivalence

$$\underbrace{\{[(P) \Rightarrow (Q)] \Rightarrow [(Q) \Rightarrow (R)]\}}_{H} \Rightarrow [\ \underbrace{(Q)}_{K}\ \Rightarrow\ \underbrace{(R)}_{L}\ ]$$

$$\Updownarrow$$

$$\underbrace{(Q)}_{K}\ \Rightarrow \Big[\underbrace{\{[(P) \Rightarrow (Q)] \Rightarrow [(Q) \Rightarrow (R)]\}}_{H} \Rightarrow\ \underbrace{(R)}_{L}\ \Big].$$

To prove either formula with the Deduction Theorem, this proof starts by assuming that the hypotheses H and K, here Q and $[(P) \Rightarrow (Q)] \Rightarrow [(Q) \Rightarrow (R)]$, are True.

$\vdash [(P) \Rightarrow (Q)] \Rightarrow [(Q) \Rightarrow (R)]$
$\vdash [(Q) \Rightarrow \{[(P) \Rightarrow (Q)] \Rightarrow [(Q) \Rightarrow (R)]\}] \Rightarrow (\{(Q) \Rightarrow [(P) \Rightarrow (Q)]\} \Rightarrow \{(Q) \Rightarrow [(Q) \Rightarrow (R)]\})$
$\vdash \{(Q) \Rightarrow [(P) \Rightarrow (Q)]\} \Rightarrow \{(Q) \Rightarrow [(Q) \Rightarrow (R)]\}$
$\vdash (Q) \Rightarrow [(P) \Rightarrow (Q)]$
$\vdash (Q) \Rightarrow [(Q) \Rightarrow (R)]$
$\vdash Q$
$\vdash (Q) \Rightarrow (R)$
$\vdash Q$
$\vdash R$

The result then follows from the Deduction Theorem.

1.97 The formula S defined by $\{[\neg(P)] \Rightarrow (P)\} \Rightarrow (P)$ has only one propositional variable, P. Moreover, S has the form $(V) \Rightarrow (W)$, with $[\neg(P)] \Rightarrow (P)$ for V, and P for W. Thus the first step consists in applying the Provability Theorem (theorem 1.125) to prove $P' \vdash S'$.

P True   If P is True, then so is W, and the single line $\vdash P$ constitutes a proof of $\vdash W$. Hence a copy of the proof of theorem 1.12 forms a complete proof of $P \vdash (V) \Rightarrow (W)$, which is $P' \vdash S'$.

P False   If P is False, then W is also False, but so is V. Thus, V' is $\neg(V)$, which is $\neg\{[\neg(P)] \Rightarrow (P)\}$. In this case the Provability Theorem calls for a proof of $P' \vdash V'$. However, V is False and has the form $(H) \Rightarrow (K)$, with $\neg(P)$ for H, and P for K. Hence the Provability Theorem calls for proofs of $P' \vdash H'$ and $P' \vdash K'$, which are here $[\neg(P)] \vdash [\neg(P)]$ and $[\neg(P)] \vdash [\neg(P)]$. In both cases the proof of $[\neg(P)] \vdash [\neg(P)]$ is a substitution in the proof of theorem 1.14. Hence follows a proof of $P' \vdash (H') \wedge (K')$, which is here $[\neg(P)] \vdash [\neg(P)] \wedge [\neg(P)]$, and, by definition of $\wedge$ the same proof shows that $[\neg(P)] \vdash \neg\{[\neg(P)] \Rightarrow (P)\}$, which is $[\neg(P)] \vdash [\neg(V)]$, or, equivalently, $P' \vdash V'$.

Thence follows a proof of $P' \vdash \{[\neg(W)] \Rightarrow [\neg(V)]\}$ and hence by contraposition a proof of $P' \vdash [(V) \Rightarrow (W)]$, which is again $P' \vdash S'$.

1.99 The law of reductio ad absurdum, S,

$$[(P) \Rightarrow (Q)] \Rightarrow \big(\{(P) \Rightarrow [\neg(Q)]\} \Rightarrow [\neg(P)]\big),$$

has the form $(V) \Rightarrow (W)$, with $(P) \Rightarrow (Q)$ for V, and $\{(P) \Rightarrow [\neg(Q)]\} \Rightarrow [\neg(P)]$ for W.

P True, Q True   If P is True and Q is True, then W is True. However, W has the form $(H) \Rightarrow (K)$, with $(P) \Rightarrow [\neg(Q)]$ for H, which is False, and $\neg(P)$ for K, which is also False. Hence the Provability Theorem calls for a proof of $P', Q' \vdash H'$, here $P, Q \vdash \neg(H)$. Because H has the form $(P) \Rightarrow [\neg(Q)]$, and hence $\neg(H)$ has the form $(P) \wedge \{\neg[\neg(Q)]\}$, the Provability Theorem calls for proofs of $P, Q \vdash P$ and $P, Q \vdash Q$, which follow from substitutions in the proof of theorem 1.14:

$\vdash (P) \Rightarrow (P)$	theorem 1.14,
$\vdash (Q) \Rightarrow (Q)$	theorem 1.14,
$\vdash (P) \Rightarrow \{(Q) \Rightarrow [(P) \wedge (Q)]\}$	theorem 1.82,
$\vdash (P) \Rightarrow \big[(Q) \Rightarrow \{\neg[(P) \Rightarrow [\neg(Q)]]\}\big]$	definition of $\wedge$,
$\vdash (P) \Rightarrow \{(Q) \Rightarrow [\neg(H)]\}$	substitution,
$\vdash (P) \Rightarrow \big[(Q) \Rightarrow \{[\neg(K)] \Rightarrow [\neg(H)]\}\big]$	axiom P1 and theorem 1.16,
$\vdash (P) \Rightarrow \{(Q) \Rightarrow [(H) \Rightarrow (K)]\}$	axiom P3 and theorem 1.16;
$\vdash (P) \Rightarrow [(Q) \Rightarrow (W)]$	substitution,
$\vdash (W) \Rightarrow [(V) \Rightarrow (W)]$	axiom P1,
$\vdash (P) \Rightarrow \{(Q) \Rightarrow [(V) \Rightarrow (W)]\}$	theorem 1.32,
$\vdash (P) \Rightarrow [(Q) \Rightarrow (S)]$	substitution.

P False   If *P* is False, then *W* is True, regardless of whether *Q* is True or False, and in either case the same proof of theorem 1.14 gives a proof of $[\neg(P)] \Rightarrow [\neg(P)]$, whence a proof of $[\neg(P)] \Rightarrow (W)$, and hence a proof of $[\neg(P)] \Rightarrow [(V) \Rightarrow (W)]$, which is $[\neg(P)] \Rightarrow (S)$.

P True, *Q* False   If *P* is True but *Q* is False, then *W* is False. Because *S* is a tautology, *V* is also False. Hence the Provability Theorem calls for a proof of $P', Q' \vdash V'$, here $P, [\neg(Q)] \vdash \neg(V)$. Because *V* has the form $(H) \Rightarrow (K)$, the Provability Theorem calls for proofs of $P', Q' \vdash H$ and $P', Q' \vdash [\neg(K)]$, here $P, [\neg(Q)] \vdash P$ and $P, [\neg(Q)] \vdash [\neg(Q)]$, both of which follow from substitutions in the proof of theorem 1.14. Thence the proof of theorem 1.82 gives a proof of $(H) \wedge [\neg(K)]$, which is $\neg(V)$:

$$\vdash (P) \Rightarrow (P) \qquad\qquad\qquad\qquad\qquad \text{theorem 1.14,}$$
$$\vdash [\neg(Q)] \Rightarrow [\neg(Q)] \qquad\qquad\qquad\qquad \text{theorem 1.14,}$$
$$\vdash (P) \Rightarrow \big([\neg(Q)] \Rightarrow \{(P) \wedge [\neg(Q)]\}\big) \qquad \text{theorem 1.82,}$$
$$\vdash (P) \Rightarrow \big([\neg(Q)] \Rightarrow \{\neg[(P) \Rightarrow (Q)]\}\big) \qquad \text{definition of } \wedge,$$
$$\vdash (P) \Rightarrow \{[\neg(Q)] \Rightarrow [\neg(V)]\} \qquad\qquad\quad \text{substitution,}$$
$$\vdash (P) \Rightarrow \big([\neg(Q)] \Rightarrow \{[\neg(W)] \Rightarrow [\neg(V)]\}\big) \text{ axiom P1 and theorem 1.16,}$$
$$\vdash (P) \Rightarrow \{[\neg(Q)] \Rightarrow [(V) \Rightarrow (W)]\} \qquad \text{axiom P3 and theorem 1.16.}$$

From the proofs of $(P) \Rightarrow [(Q) \Rightarrow (S)]$ and $(P) \Rightarrow \{[\neg(Q)] \Rightarrow (S)\}$ follows a proof of $(P) \Rightarrow (S)$, and thence from the proof of $[\neg(P)] \Rightarrow (S)$ follows a proof of *S*.

1.101 The propositional form $\{[(P) \Rightarrow (Q)] \Rightarrow (R)\} \Rightarrow \{[(R) \Rightarrow (P)] \Rightarrow (P)\}$ has the form $(V) \Rightarrow (W)$.

P True   If *P* is True, then axiom P1 gives a proof of $(P) \Rightarrow (W)$, and hence a proof of $(P) \Rightarrow (S)$.

P False, *R* False   If *P* is False, then $(P) \Rightarrow (Q)$ holds, by the law of denial of the antecedent (theorem 1.40):

$$\vdash [\neg(P)] \Rightarrow [(P) \Rightarrow (Q)].$$

If *R* is also False, then $[\neg(R)] \Rightarrow [\neg(R)]$, whence $[(P) \Rightarrow (Q)] \wedge [\neg(R)]$ holds, whence also $\neg\{[(P) \Rightarrow (Q)] \Rightarrow (R)\}$, which is $\neg(V)$.

P False, *R* True   If *P* is False and *R* is True, then $[(R) \Rightarrow (P)]$ is False, but so is *P*, whence $[(R) \Rightarrow (P)] \Rightarrow (P)$, which is *W*, is True.

1.103 The propositional form $\big[\{[(P) \Rightarrow (R)] \Rightarrow (Q)\} \Rightarrow (Q)\big] \Rightarrow \{[(Q) \Rightarrow (R)] \Rightarrow [(P) \Rightarrow (R)]\}$ has the form $(V) \Rightarrow (W)$.

R True   If *R* is True, then axiom P1 gives a proof of *W*, whence a proof of *S*.

R False, *P* False   If *P* is False, then the proof of $[\neg(P)] \Rightarrow [(P) \Rightarrow (R)]$ gives a proof of *W*, whence a proof of *S*.

R False, P True, Q True   If Q is True and R is False, then the definition of $\wedge$ gives a proof of $(Q) \Rightarrow \big([\neg(R)] \Rightarrow \{\neg[(Q) \Rightarrow (R)]\}\big)$ and hence a proof of W, whence a proof of S.

R False, P True, Q False   With R False, P True, Q False, the definition of $\wedge$ gives a proof of $(P) \Rightarrow \big([\neg(R)] \Rightarrow \{\neg[(P) \Rightarrow (R)]\}\big)$, and hence also a proof of $\{[(P) \Rightarrow (R)]\} \Rightarrow (Q)$, whence a proof of $\neg(V)$, because Q is False and V is $\big[\{[(P) \Rightarrow (R)]\} \Rightarrow (Q)\big] \Rightarrow (Q)$.

1.105 The propositional form U defined by

$$\{[(R) \Rightarrow (Q)] \Rightarrow [(S) \Rightarrow (P)]\} \Rightarrow \{[(R) \Rightarrow (P)] \Rightarrow [(S) \Rightarrow (P)]\}$$

has the form $(V) \Rightarrow (W)$.

P True   If P is True, then axiom P1 gives a proof of $(P) \Rightarrow [(S) \Rightarrow (P)]$, whence a proof of $(P) \Rightarrow (W)$, and hence a proof of $(P) \Rightarrow (U)$.

P False, S False   If S is False, then the law of denial of the antecedent, $[\neg(S)] \Rightarrow [(S) \Rightarrow (P)]$ gives a proof of $[\neg(S)] \Rightarrow (W)$ and hence of $[\neg(S)] \Rightarrow (U)$.

P False, S True, R False   With P False, S True, R False, the law of denial of the antecedent gives $[\neg(R)] \Rightarrow [(R) \Rightarrow (Q)]$, while $(S) \Rightarrow (P)$ is False, whence V is False, and then the law of denial of the antecedent gives $(V) \Rightarrow (W)$, which is U.

P False, S True, R True   With P False, S True, R True, $(R) \Rightarrow (P)$ is False, whence the law of denial of the antecedent gives $[(R) \Rightarrow (P)] \Rightarrow [(S) \Rightarrow (P)]$, which is W; hence the law of denial of the antecedent gives $(V) \Rightarrow (W)$, which is U.

P False, S True   The foregoing two cases give a proof of $[\neg(P)] \Rightarrow [(S) \Rightarrow (U)]$.

P False   The proofs of $[\neg(P)] \Rightarrow [(S) \Rightarrow (U)]$ and $[\neg(P)] \Rightarrow \{[\neg(S)] \Rightarrow (U)\}$ then combine into a proof of $[\neg(P)] \Rightarrow (U)$.

Finally, the proofs of $(P) \Rightarrow (U)$ and $[\neg(P)] \Rightarrow (U)$ combine into a proof of U.

Exercises from Chapter 2

2.1 $\exists X\{\forall Y[\neg(X \in Y)]\}$.

2.3 $\exists X\{(X \in A) \wedge [\neg(X \in B)]\}$.

2.5 $\exists X\big(\{(X \in C) \wedge [\neg(X \in A)] \wedge [\neg(X \in B)]\} \vee \{[\neg(X \in C)] \wedge [(X \in A) \vee (X \in B)]\}\big)$.

2.7 $\forall X[(X \in A) \Rightarrow (X \in B)]$.

2.9 $\forall X\{(X \in C) \Leftrightarrow [(X \in A) \wedge (X \in B)]\}$.

2.11 Theorem 2.46 establishes the equivalence $[\exists X(P)] \Leftrightarrow \big(\neg\{\forall X[\neg(P)]\}\big)$.

2.13 Axiom Q2 is a theorem derivable from Margaris's and Rosser's axioms:

$\vdash \{\forall X[(P) \Rightarrow (Q)]\} \Rightarrow \{[\forall X(P)] \Rightarrow [\forall X(Q)]\}$ axiom A4,

$\vdash (P) \Rightarrow [\forall X(P)]$ axiom A6, no free X in P,

$\vdash \{\forall X[(P) \Rightarrow (Q)]\} \Rightarrow \{(P) \Rightarrow [\forall X(Q)]\}$ derived rule.

2.15 Axiom Q4 follows from the abbreviation $\neg\{\forall X[\neg(P)]\}$, double negation and theorem 2.45:

$\vdash [\forall X(P)] \Leftrightarrow \{\forall X[\neg\neg(P)]\}$ theorem 2.45,

$\vdash \{\neg[\forall X(P)]\} \Leftrightarrow \left(\neg\{\forall X[\neg\neg(P)]\}\right)$ contraposition and its converse,

$\vdash \{\exists X[\neg(P)]\} \Leftrightarrow \left[\neg\left(\forall X\{\neg[\neg(P)]\}\right)\right]$ abbreviation, $(R) \Leftrightarrow (R)$,

$\vdash \{\exists X[\neg(P)]\} \Leftrightarrow \{\neg[\forall X(P)]\}$ transitivity.

2.17 Kleene's $\exists$-rule is derivable from the rules of inference with axioms Q1– Q4 and the propositional calculus.

$\vdash (P) \Rightarrow (Q)$ hypothesis,

$\vdash [(P) \Rightarrow (Q)] \Rightarrow \{[\neg(Q) \Rightarrow [\neg(P)]\}$ contraposition,

$\vdash [\neg(Q) \Rightarrow [\neg(P)]$ *Detachment*,

$\vdash \forall X\{[\neg(Q) \Rightarrow [\neg(P)]\}$ Generalization,

$\vdash [\neg(Q)] \Rightarrow \{\forall X[\neg(P)]\}$ theorem 2.29,

$\vdash \{\forall X[\neg(P)]\} \Rightarrow \{\neg[\exists X(P)]\}$ axiom Q3,

$\vdash [\neg(Q)] \Rightarrow \{\neg[\exists X(P)]\}$ transitivity,

$\vdash [\exists X(P)] \Rightarrow (Q)$ converse contraposition
and *Detachment*.

2.19 Kleene's $\exists$-schema is derivable from the rules of inference with axioms Q1– Q4 and the propositional calculus.

$\vdash \{\forall X[\neg(P)]\} \Rightarrow \{\mathrm{Subf}_Y^X[\neg(P)]\}$ axiom Q1,

$\vdash \{\mathrm{Subf}_Y^X[\neg(P)]\} \Rightarrow \{\neg[\mathrm{Subf}_Y^X(P)]\}$ remark 2.20,

$\vdash \{\forall X[\neg(P)]\} \Rightarrow \{\neg[\mathrm{Subf}_Y^X(P)]\}$ *Detachment*,

$\vdash [\mathrm{Subf}_Y^X(P)] \Rightarrow \left(\neg\{\forall X[\neg(P)]\}\right)$ contraposition and double negation,

$\vdash \left(\neg\{[\forall X[\neg(P)]\}\right) \Rightarrow [\exists X(P)]$ theorem 2.46,

$\vdash [\mathrm{Subf}_Y^X(P)] \Rightarrow [\exists X(P)]$ transitivity.

2.21

$\vdash (V) \Leftrightarrow (U)$ hypothesis,

$\vdash [(V) \Leftrightarrow (U)] \Rightarrow [(V) \Rightarrow (U)]$ theorem 1.62,

$\vdash (V) \Rightarrow (U)$ *Detachment*,

$\vdash [(V) \Rightarrow (U)] \Rightarrow \{[(U) \Rightarrow (W)] \Rightarrow [(V) \Rightarrow (W)]\}$ transitivity
(theorem 1.27),

$\vdash [(U) \Rightarrow (W)] \Rightarrow [(V) \Rightarrow (W)]$ *Detachment*.

2.23

$\vdash (V) \Leftrightarrow (U)$ hypothesis,

$\vdash [(V) \Leftrightarrow (U)] \Rightarrow [(U) \Rightarrow (V)]$ theorem 1.62,

$\vdash (U) \Rightarrow (V)$ *Detachment*,

$\vdash [(U) \Rightarrow (V)] \Rightarrow \{[(V) \Rightarrow (W)] \Rightarrow [(U) \Rightarrow (W)]\}$ transitivity (theorem 1.28),

$\vdash [(V) \Rightarrow (W)] \Rightarrow [(U) \Rightarrow (W)]$ *Detachment*.

2.25 Theorem 2.45 shows that if $\vdash (V) \Leftrightarrow (U)$, then $\vdash (P) \Rightarrow (Q)$.

2.27 If $\vdash (V) \Leftrightarrow (U)$, then $\vdash [\neg(V)] \Leftrightarrow [\neg(U)]$, by contraposition and transposition.

2.31

$$\{\exists X\,[(P) \vee (Q)]\} \Leftrightarrow \{[\exists X(P)] \vee [\exists X\,(Q)]\} \Leftrightarrow \{[\exists X(Q)] \vee [\exists X\,(P)]\}$$
$$\Leftrightarrow \left(\{\exists X\,[(Q) \vee (P)]\}\right)$$

2.33

$$\{\exists X[(P) \vee (P)]\} \Leftrightarrow \{[\exists X(P)] \vee [\exists X(P)]\} \Leftrightarrow [\exists X(P)]$$

2.35

$$\left(\forall X\,\{[(P) \wedge (Q)] \vee (R)\}\right) \qquad \Leftrightarrow$$
$$\left(\forall X\,\{[(P) \vee (R)] \wedge [(Q) \vee (R)]\}\right) \qquad \Leftrightarrow$$
$$\left(\{\forall X[(P) \vee (R)]\} \wedge \{\forall X[(Q) \vee (R)]\}\right)$$

2.37

$$[\exists X(Q)] \Leftrightarrow \left(\exists X\{\neg[\neg(Q)]\}\right)$$
$$\Leftrightarrow \left(\neg\,\{\forall X[\neg(Q)]\}\right)$$
$$\Leftrightarrow \left[\neg\left(\forall X\,\{\forall X[\neg(Q)]\}\right)\right]$$
$$\Leftrightarrow \left[\neg\left(\forall X\,\{\neg\,[\exists X(Q)]\}\right)\right]$$
$$\Leftrightarrow \left[\neg\left(\neg\,\{\exists X\,[\exists X(Q)]\}\right)\right]$$
$$\Leftrightarrow \{\exists X[\exists X(Q)]\}$$

2.39 The implication

$$\{\forall X[(P) \vee (Q)]\} \Rightarrow \{(P) \vee [\forall X(Q)]\}$$

is theorem 2.79. For the converse,

$$\vdash (P) \Rightarrow [\forall X(P)] \qquad\qquad\qquad \text{axiom Q1,}$$
$$\vdash [\forall X(Q)] \Rightarrow [\forall X(Q)] \qquad\qquad\quad \text{theorem 1.14,}$$
$$\vdash \{(P) \vee [\forall X(Q)]\} \Rightarrow \{[\forall X(P)] \vee [\forall X(Q)]\} \quad \text{theorem 2.39,}$$
$$\vdash \{[\forall X(P)] \vee [\forall X(Q)]\} \Rightarrow \{\forall X[(P) \vee (Q)]\} \quad \text{theorem 2.77,}$$
$$\vdash \{(P) \vee [\forall X(Q)]\} \Rightarrow \{\forall X[(P) \vee (Q)]\} \qquad \text{theorem 1.16.}$$

2.41 Invoke the reflexivity of the logical implication (theorem 1.63): by definition of $\mathscr{R}$,

$$[\mathscr{R}(A,B)] \Leftrightarrow \left(\forall X\{[\mathscr{E}(X,A)] \Rightarrow [\mathscr{E}(X,A)]\} \right),$$

which is in prenex form, and its matrix is an instance of theorem 1.63: $(P) \Rightarrow (P)$.

2.43 Invoke the transitivity of the logical implication (theorem 1.65):

$$\vdash [\mathscr{E}(X,A)] \Rightarrow [\mathscr{E}(X,B)] \qquad \text{specialization of the hypothesis} \mathscr{R}(A,B),$$
$$\vdash [\mathscr{E}(X,B)] \Rightarrow [\mathscr{E}(X,C)] \qquad \text{specialization of the hypothesis} \mathscr{R}(B,C),$$
$$\vdash [\mathscr{E}(X,A)] \Rightarrow [\mathscr{E}(X,C)] \qquad \text{transitivity of the implication (theorem 1.65),}$$

whence the conclusion $\mathscr{R}(A,C)$ follows by Generalization.

2.45 Invoke the reflexivity of the logical implication (theorem 1.63): by definition of $\mathscr{A}$,

$$[\mathscr{A}(A,A)] \Leftrightarrow \left(\forall X\{[\mathscr{E}(A,Y)] \Rightarrow [\mathscr{E}(A,Y)]\} \right),$$

which is in prenex form, and its matrix is an instance of theorem 1.63: $(P) \Rightarrow (P)$.

2.47 The equality predicate $\mathscr{I}$ defined as in example 2.85 for set theory is reflexive, as proved by the solutions to exercises 2.41 and 2.45, so that $\mathscr{I}$ satisfies axioms $\mathscr{J}1$. Also, the set theory described in example 2.85 has only one predicate, $\mathscr{E}$, and formula (2.2) shows that $\mathscr{I}$ satisfies axiom $\mathscr{J}2$.

2.49 Axiom $\mathscr{J}1$ from subsection 2.5.3 coincides with axiom $\mathscr{I}1$ from subsection 2.5.1. Axiom $\mathscr{J}2$ from subsection 2.5.3 is a special case of axiom $\mathscr{I}2$ from subsection 2.5.1, which allows for atomic formulae as particular cases of P and Q.

Exercises from Chapter 3

3.1 Negate the definition of the empty set: $\exists X(X \in S)$.

3.3 Negate the definition of supersets: $\exists X[(X \in B) \wedge (X \notin A)]$.

3.5 By definition of the empty set (axiom S2), the formula $\neg(X \in \varnothing)$ is universally valid. By specialization, with $\varnothing$ substituted for X, it follows that $\neg(\varnothing \in \varnothing)$ is True, whence $\varnothing \in \varnothing$ is False, by definition of False.

3.7 Use substitutions in the axiom of extensionality and the definition of equality.

3.9 For each set S, $\varnothing \subseteq S$, by theorem 3.11. Consequently, $\varnothing$ is a subset of $\varnothing$. Conversely, if S is a subset of $\varnothing$, so that $S \subseteq \varnothing$, then $S = \varnothing$ by theorem 3.10:

$\vdash S \subseteq \varnothing$ hypothesis on S,
$\vdash \varnothing \subseteq S$ theorem 3.11,
$\vdash S = \varnothing$ theorem 3.10.

3.11 This proof proceeds by contraposition, showing that if $S \neq \varnothing$, then S has a subset different from S. Because $\varnothing \subseteq S$ by theorem 3.11, it follows that if $S \neq \varnothing$, then S has a subset, $\varnothing$, different from S.

3.13 If S is a subset of every set, then S is a subset of the empty set: $S \subseteq \varnothing$. Moreover, $\varnothing \subseteq S$ by theorem 3.11. Consequently, $S = \varnothing$, by theorem 3.10.

3.15 If $A \subsetneqq B$ and $B \subsetneqq C$, then $A \subseteq B$ and $B \subseteq C$, whence $A \subseteq C$, by theorem 3.9. However, because $A \subsetneqq B$, there exists some $Z \in B$ such that $Z \notin A$. Consequently, $Z \in C$ but $Z \notin A$, so that $A \neq C$, whence $A \subsetneqq C$.

3.17 This proof establishes each implication ($\Rightarrow$ and $\Leftarrow$) independently.
    For one implication, assume $\vdash \forall Y[(A \subseteq Y) \Leftrightarrow (B \subseteq Y)]$.

$\vdash \forall Y[(A \subseteq Y) \Leftrightarrow (B \subseteq Y)]$ hypothesis,
$\Downarrow$ Subf_A^Y,
$\vdash (A \subseteq A) \Leftrightarrow (B \subseteq A)$
$\Downarrow$ $\Rightarrow$,
$\vdash (A \subseteq A) \Rightarrow (B \subseteq A)$
$\vdash A \subseteq A$ theorem 3.8,
$\vdash B \subseteq A$ *Modus Ponens*;

$\vdash \forall Y[(A \subseteq Y) \Leftrightarrow (B \subseteq Y)]$ hypothesis,
$\Downarrow$ Subf_B^Y,
$\vdash (A \subseteq B) \Leftrightarrow (B \subseteq B)$
$\Downarrow$ $\Rightarrow$,
$\vdash (A \subseteq B) \Leftarrow (B \subseteq B)$
$\vdash B \subseteq B$ theorem 3.8,
$\vdash A \subseteq B$ *Modus Ponens*;
$\vdash A = B$ theorem 3.10.

    For the converse, assume $\vdash A = B$, and begin with any superset Y.

$\vdash A \subseteq Y$ hypothesis,
$\vdash \forall X[(X \in A) \Rightarrow (X \in Y)]$ definition of subsets,
$\vdash B \subseteq A$ hypothesis,
$\vdash \forall X[(X \in B) \Rightarrow (X \in Y)]$ transitivity of $\Rightarrow$,
$\vdash B \subseteq Y$ definition of subsets.

Swapping A and B yields the converse. Hence $\vdash \forall Y[(A \subseteq Y) \Leftrightarrow (B \subseteq Y)]$.

3.19 If $C \supseteq D$ and $D \supseteq W$, then $D \subseteq C$ and $W \subseteq D$, whence $W \subseteq C$. For the converse, let $W := D$, whence $D \supseteq D$, the hypothesis $(D \supseteq D) \Rightarrow (C \supseteq D)$, and *Modus Ponens* yield $C \supseteq D$.

3.21 Let $X := \varnothing, Y := \{\varnothing\}, Z := \{\{\varnothing\}\}$.

3.23 Let $X := \varnothing$ and $Y := \{\{\varnothing\}\}$.

3.25 Let $X := \{\varnothing\}$ and $A := \{\{\varnothing\}\}$.

3.27 $(X \in \{\varnothing\}) \Leftrightarrow (X = \varnothing)$ whereas $(X \in \{\{\varnothing\}\}) \Leftrightarrow (X = \{\varnothing\})$. Yet $\varnothing \neq \{\varnothing\}$. Consequently, $\{\varnothing\}$ and $\{\{\varnothing\}\}$ have different elements. Therefore $\{\varnothing\} \neq \{\{\varnothing\}\}$.

3.29 $\{\varnothing\} \in \{\varnothing, \{\varnothing\}\}$. Yet $(X \in \{\varnothing\}) \Leftrightarrow (X = \varnothing)$ and $\varnothing \neq \{\varnothing\}$, whence $\{\varnothing\} \notin \{\varnothing\}$. Hence $\{\varnothing\} \neq \{\varnothing, \{\varnothing\}\}$.

3.31 From $\forall S(S \subseteq S)$ specialization with $S := \{\{\varnothing\}\}$ gives $\{\{\varnothing\}\} \subseteq \{\{\varnothing\}\}$.

3.33 The set $\{\{\varnothing\}\}$ has only one element $\{\varnothing\}$, which is also an element of the set $\{\varnothing, \{\varnothing\}\}$. Hence $\{\{\varnothing\}\} \subseteq \{\varnothing, \{\varnothing\}\}$.

3.35 For theorem 3.13, the word "and" in the informal proof corresponds to the logical connective $\vee$ in the formal proof through the universal quantifier, specializations with H first *and* then K, and the axioms of extensionality and pairing, along the following outline:

$\vdash (X = H) \Rightarrow [(X = H) \vee (X = K)]$ $(P) \Rightarrow [(P) \vee (Q)]$,
$\vdash [(X = H) \vee (X = K)] \Rightarrow (X \in L)$ pairing,
$\vdash (X = H) \Rightarrow (X \in L)$ transitivity,
$\vdash (X = H) \Rightarrow [(X \in L) \Rightarrow (H \in L)]$ extensionality,
$\vdash (X = H) \Rightarrow (H \in L)$ transitivity,
$\vdash (H = H) \Rightarrow (H \in L)$ specialization Subf_H^X,
$\vdash H = H$ extensionality,
$\vdash H \in L$ *Modus Ponens*;
$\vdash K \in L$ as for H;
$\vdash (H \in L) \wedge (K \in L)$ $(P) \Rightarrow \{(Q) \Rightarrow [(P) \wedge (Q)]\}$.

3.37 If $A = B$, and if $S \subseteq A$, then each element of S is also an element of B, whence $S \subseteq B$. Thus $\mathscr{P}(A) \subseteq \mathscr{P}(B)$, and conversely with the rôles of A and B switched. If $\mathscr{P}(A) = \mathscr{P}(B)$, and if $X \in A$, then $\{X\} \in \mathscr{P}(A)$, whence $\{X\} \in \mathscr{P}(B)$, so that $\{X\} \subseteq B$, and hence $X \in B$. Thus $A \subseteq B$, and conversely $B \subseteq A$ with the rôles of A and B switched. Therefore $A = B$.

3.39 $\{\{\varnothing\}\} \cap \{\varnothing, \{\varnothing\}\} = \{\{\varnothing\}\}$.

3.41 $\{\{\varnothing\}\} \cup \{\varnothing, \{\varnothing\}\} = \{\varnothing, \{\varnothing\}\}$ in the superset $\{\varnothing, \{\varnothing\}\}$

3.43 $\forall S[(S \cap \varnothing) = \varnothing]$.

3.45 By definition, $A \setminus \varnothing = \{X \in A : X \notin \varnothing\}$ where $X \notin \varnothing$ holds for every set X. Hence $\forall A\big(\forall X\{(X \in A) \Leftrightarrow [(X \in A) \wedge (X \notin \varnothing)]\}\big)$ whence $A = A \setminus \varnothing$ by extensionality.

3.47 By definition, $A \setminus B = \{X \in A : X \notin B\}$. Thus $(A \setminus B) = \varnothing$ if and only if $X \notin B$ fails, and hence $X \in B$ holds, for every $X \in A$, which is the definition of $A \subseteq B$.

3.49 The formula $\forall A \forall B[(A \setminus B) = (B \setminus A)]$ is False. Indeed, with $A := \varnothing$ and $B := \{\varnothing\}$, it follows that

$$
\begin{aligned}
A &:= \varnothing, \\
B &:= \{\varnothing\}, \\
A \setminus B &= \varnothing \setminus \{\varnothing\} \\
&= \varnothing \\
&\neq \{\varnothing\} \\
&= \{\varnothing\} \setminus \varnothing \\
&= B \setminus A.
\end{aligned}
$$

3.51 $\{2, 3, 7\} \cup \{3, 5, 7\} = \{2, 3, 5, 7\}$.

3.53 $\{2, 3, 7\} \cap \{3, 5, 7\} = \{3, 7\}$.

3.55 $\{2, 3, 7\} \triangle \{3, 5, 7\} = \{2, 5\}$.

3.57 $\bigcup \varnothing = \varnothing$.

3.59 $\bigcup\{\varnothing, \{\varnothing\}\} = \{\varnothing\} \neq \{\varnothing, \{\varnothing\}\}$.

3.61 $\{\varnothing\} \in \{\varnothing, \{\varnothing\}\}$ but $\{\varnothing\} \notin \{\varnothing\} = \bigcup\{\varnothing, \{\varnothing\}\}$.

3.63

$$
\begin{aligned}
X &:= \{\varnothing\}, \\
A &:= \{\{\varnothing\}\}; \\
Y &:= \{\{\varnothing\}\}, \\
B &:= \{\{\{\varnothing\}\}\}; \\
X \cup Y &= \{\varnothing, \{\varnothing\}\}, \\
A \cup B &= \{\{\varnothing\}, \{\{\varnothing\}\}\}.
\end{aligned}
$$

3.65

$$\left(X \in \bigcup\{A\}\right) \Leftrightarrow \{\exists Y[(Y \in \{A\}) \wedge (X \in Y)]\}$$
$$\Leftrightarrow \{\exists Y[(Y = A) \wedge (X \in Y)]\}$$
$$\Leftrightarrow [\exists Y(X \in A)]$$
$$\Leftrightarrow X \in A.$$

3.67 Use the tautology $\{[(P) \vee (Q)] \wedge (R)\} \Leftrightarrow \{[(P) \wedge (R)] \vee [(Q) \wedge (R)]\}$.

3.69 Use the tautology $[(P) \vee (F)] \Leftrightarrow (P)$:

$$[X \in (A \cup \varnothing)] \Leftrightarrow [(X \in A) \vee (X \in \varnothing)]$$
$$\Leftrightarrow X \in A.$$

3.71 Use the tautology $[(P) \vee (P)] \Leftrightarrow (P)$.

3.73 Use de Morgan's second law:

$$\{X \in [U \setminus (A \cap B)]\} \Leftrightarrow \Big[(X \in U) \wedge \{\neg[X \in (A \cap B)]\}\Big]$$
$$\Leftrightarrow \Big[(X \in U) \wedge \{\neg[(X \in A) \wedge (X \in B)]\}\Big]$$
$$\Leftrightarrow \{(X \in U) \wedge [(X \notin A) \vee (X \notin B)]\}$$
$$\Leftrightarrow \{[(X \in U) \wedge (X \notin A)] \vee [(X \in U) \wedge (X \notin B)]\}$$
$$\Leftrightarrow \{[(X \in (U \setminus A)] \vee [X \in (U \setminus B)]\}$$
$$\Leftrightarrow \{X \in [(U \setminus A)cup(U \setminus B)]\}$$

3.75 Use the tautology $(B) \vee [\neg(B)]$ and the contradiction $(B) \wedge [\neg(B)]$, with $X \in U$ for B:

$\{X \in [(A \setminus B) \setminus U]\}$
$\Leftrightarrow \{[X \in (A \setminus B)] \wedge [\neg(X \in U)]\}$
$\Leftrightarrow \{(X \in A) \wedge \{\neg[(X \in B)]\} \wedge [\neg(X \in U)]\}$
$\Leftrightarrow (\{(X \in A) \wedge [\neg(X \in U)]\} \wedge [\neg(X \in B)])$
$\Leftrightarrow \{[X \in (A \setminus U)] \wedge ([\neg(X \in B)] \vee \{(X \in U) \wedge [\neg(X \in U)]\})\}$
$\Leftrightarrow \big[\{(X \in A) \wedge [\neg(X \in U)]\} \wedge (\{[\neg(X \in B)] \vee (X \in U)\} \wedge \{[\neg(X \in B)] \vee [\neg(X \in U)]\})\big]$
$\Leftrightarrow \big[\{(X \in A) \wedge [\neg(X \in U)]\} \wedge (\{[\neg(X \in B)]\} \wedge \{[\neg(X \in B)] \vee [\neg(X \in U)]\})\big]$
$\Leftrightarrow \big[\{(X \in A) \wedge [\neg(X \in U)]\} \wedge (\{[\neg(X \in B)] \vee [\neg(X \in U)]\})\big]$
$\Leftrightarrow \big[\{(X \in A) \wedge [\neg(X \in U)]\} \wedge (\{\neg\{(X \in B) \wedge [\neg(X \in U)]\}\})\big]$
$\Leftrightarrow ([X \in (A \setminus U)] \wedge \{\neg[X \in (B \setminus U)]\})$
$\Leftrightarrow \{X \in [(A \setminus U) \setminus (B \setminus U)]\}.$

3.77

$$\{X \in [(A \cup B) \setminus U]\} \Leftrightarrow \{X \in (A \cup B) \wedge [\neg(X \in U)]\}$$
$$\Leftrightarrow \{[(X \in A) \vee (X \in B)] \wedge [\neg(X \in U)]\}$$
$$\Leftrightarrow (\{[(X \in A) \wedge [\neg(X \in U)]\} \vee \{[(X \in B) \wedge [\neg(X \in U)]\}\})$$
$$\Leftrightarrow \{X \in [(A \setminus U) \cup (B \setminus U)]\}.$$

3.79

$$\{X \in [U \cap (A \setminus B)]\}$$
$$\Leftrightarrow \{(X \in U) \wedge [X \in (A \setminus B)]\}$$
$$\Leftrightarrow ([(X \in U) \wedge \{(X \in A) \wedge [\neg(X \in B)]\}])$$
$$\Leftrightarrow [\{[(X \in U) \wedge [\neg(X \in B)]\} \wedge (X \in A)]$$
$$\Leftrightarrow \{[\{[(X \in U) \wedge [\neg(X \in B)]\} \wedge (X \in A)] \vee [\{[(X \in U) \wedge [\neg(X \in B)]\} \wedge [\neg(X \in U)]]\}$$
$$\Leftrightarrow ([\{[(X \in U) \wedge [\neg(X \in B)]\}] \wedge \{(X \in A) \wedge [\neg(X \in U)]\}])$$
$$\Leftrightarrow ([\{[(X \in U) \wedge [\neg(X \in B)]\}] \wedge [\neg\{[\neg(X \in A)] \vee (X \in U)\}])$$
$$\Leftrightarrow \{X \in [(U \setminus B) \setminus (U \setminus A)]\}.$$

3.81 If $C \subseteq (A \cap B)$, then $C \subseteq A$ and $C \subseteq B$.

3.83

$$A := \{\varnothing\},$$
$$B := \{\{\varnothing\}\},$$
$$A \cup B := \{\varnothing, \{\varnothing\}\},$$
$$C := A \cup B,$$
$$C \in \mathscr{P}(A \cup B),$$
$$C \notin \mathscr{P}(A),$$
$$C \notin \mathscr{P}(B),$$
$$C \notin \mathscr{P}(A) \cup \mathscr{P}(B).$$

3.85 $\varnothing \in \mathscr{P}(A \setminus B)$ but $\varnothing \in \mathscr{P}(B)$ whence $\varnothing \notin [\mathscr{P}(A) \setminus \mathscr{P}(B)]$.

3.87

$$\left(\bigcup \mathscr{F}\right) \cap B = \bigcup_{A \in \mathscr{F}} (A \cap B)$$
$$\Updownarrow$$
$$\forall X \left\{ \left[X \in \left(\bigcup \mathscr{F}\right) \cap B\right] \Leftrightarrow \left[X \in \bigcup_{A \in \mathscr{F}} (A \cap B)\right] \right\}$$
$$\Updownarrow$$
$$\forall X \left(\left\{ \left[X \in \left(\bigcup \mathscr{F}\right)\right] \wedge (X \in B) \right\} \Leftrightarrow \left[\exists A \left\{ (A \in \mathscr{F}) \wedge [X \in (A \cap B)] \right\} \right] \right)$$
$$\Updownarrow$$
$$\forall X \left(\left\{ [\exists A (A \in \mathscr{F}) \wedge (X \in A)] \wedge (X \in B) \right\} \right.$$
$$\left. \Leftrightarrow \left[\exists A \left\{ (A \in \mathscr{F}) \wedge [(X \in A) \wedge (X \in B)] \right\} \right] \right)$$

which is a tautology by associativity of $\wedge$.

3.89 $A \Delta A = (A \cup A) \setminus (A \cap A) = A \setminus A = \varnothing$.

3.91 $A \Delta B = (A \cup B) \setminus (A \cap B) = (B \cup A) \setminus (B \cap A) = B \Delta A$.

3.93 Yes, $[(A \Delta C) \cup (B \Delta C)] \supseteq [(A \cup B) \Delta C]$.

$$
\begin{aligned}
(A \cup B) \Delta C &= [(A \cup B) \cup C] \setminus [(A \cup B) \cap C] \\
&= [(A \cup B) \cup C] \setminus [(A \cap C) \cup (B \cap C)] \\
&= [(A \cup B \cup C) \setminus (A \cap C)] \cap [(A \cup B \cup C) \setminus (B \cap C)] \\
&= \{[(A \cup B \cup C) \setminus A] \cup [(A \cup B \cup C) \setminus C]\} \cap \{[(A \cup B \cup C) \setminus B] \\
&\quad \cup [(A \cup B \cup C) \setminus C]\} \\
&\subseteq \{[(A \cup C) \setminus A] \cup [(A \cup C) \setminus C]\} \cap \{[(B \cup C) \setminus B] \cup [(B \cup C) \setminus C]\} \\
&\subseteq \{[(A \cup C) \setminus A] \cup [(A \cup C) \setminus C]\} \cup \{[(B \cup C) \setminus B] \cup [(B \cup C) \setminus C]\} \\
&= [(A \cup C) \setminus (A \cap C)] \cup [(B \cup C) \setminus (B \cap C)] \\
&= [(A \Delta C) \cup (B \Delta C)].
\end{aligned}
$$

3.95 Yes, $[(A \Delta C) \cap (B \Delta C)] \supseteq [(A \cap B) \Delta C]$.

$$
\begin{aligned}
(A \cap B) \Delta C &= [(A \cap B) \cup C] \setminus [(A \cap B) \cap C] \\
&= \{[(A \cap B) \cup C] \setminus (A \cap B)\} \cup \{[(A \cap B) \cup C] \setminus C\} \\
&= \{[(A \cap B) \cup C] \setminus A\} \cup \{[(A \cap B) \cup C] \setminus B\} \cup \{[(A \cap B) \cup C] \setminus C\} \\
&\subseteq \{[(A \cup C) \setminus A] \cup [(A \cup C) \setminus C]\} \cap \{[(B \cup C) \setminus B] \cup [(B \cup C) \setminus C]\} \\
&= [(A \cup C) \setminus (A \cap C)] \cap [(B \cup C) \setminus (B \cap C)] \\
&= [(A \Delta C) \cap (B \Delta C)].
\end{aligned}
$$

3.97 Yes, $[(A \Delta C) \setminus (B \Delta C)] \supseteq [(A \setminus B) \Delta C]$.

3.99 No, $[(A \cup C)\Delta(B \cup C)] \not\supseteq [(A\Delta B) \cup C]$, because the left-hand side does not contain $(A \cup C) \cap (B \cup C)$ while the right-hand side contains all of C.

3.101 Yes, $[(A \cap C)\Delta(B \cap C)] \supseteq [(A\Delta B) \cap C]$.

3.103 Yes, $[(A \setminus C)\Delta(B \setminus C)] \supseteq [(A\Delta B) \setminus C]$.

3.105 No, $[(C \setminus A)\Delta(C \setminus B)] \not\supseteq [C \setminus (A\Delta B)]$, because the left-hand side does not contain $C \setminus (A \cup B)$ while the right-hand side contains all of $C \setminus (A \cup B)$.

3.107 No, $[\mathscr{P}(A)\Delta\mathscr{P}(B)] \not\supseteq [\mathscr{P}(A\Delta B)]$, because the left-hand side does not contain $\varnothing \in \mathscr{P}(A) \cap \mathscr{P}(B)$ while the right-hand side contains $\varnothing \in \mathscr{P}(A\Delta B)$.

3.109

$$(A\Delta B) \cap (A \cap B) = [(A \cup B) \setminus (A \cap B)] \cap (A \cap B)$$
$$= \varnothing;$$
$$(A\Delta B) \cup (A \cap B) = [(A \cup B) \setminus (A \cap B)] \cup (A \cap B)$$
$$= A \cup B.$$

3.111 As defined here the Cartesian product of sets is *not* associative (but a slightly different version of the Cartesian product is associative). For example, $(A\times B)\times C \neq A \times (B \times C)$ for the sets $A = \{0\}$, $B = \{1\}$, and $C = \{2\}$, because

$$(A \times B) \times C = \{((0, 1), 2)\} = \{\{(0, 1)\}, \{(0, 1), 2\}\}$$
$$A \times (B \times C) = \{(0, (1, 2))\} = \{\{\{0\}, \{0, (1, 2)\}\}\}$$

which reveals that $\{0\} \in (0, (1, 2)) \in A \times (B \times C)$ whereas $\{0\} \notin ((0, 1), 2) \in (A \times B) \times C$. Consequently, $(0, (1, 2)) \neq ((0, 1), 2)$ by extensionality, and then $(A\times B)\times C \neq A \times (B\times C)$ again by extensionality. (A slightly different definition of the Cartesian product through functions of integers will later provide an associative Cartesian product.)

3.113 (The last step still requires further symbolic substeps.)
$$(A = \varnothing) \vee (B = \varnothing)$$
$$\Updownarrow \quad \text{extensionality,}$$
$$\{\neg[\exists X(X \in A)]\} \vee \{\neg[\exists Y(Y \in B)]\}$$
$$\Updownarrow \quad \text{de Morgan's Law,}$$
$$\neg\{[\exists X(X \in A)] \wedge [\exists Y(Y \in B)]\}$$
$$\Updownarrow \quad \text{definition of } A \times B,$$
$$\neg\{\exists X \exists Y[(X, Y) \in (A \times B)]\}$$
$$\Updownarrow \quad \text{uniqueness of } \varnothing.$$
$$(A \times B) = \varnothing$$

3.115 The Cartesian product distributes over unions: for all sets A, B, and C,

$$[(A \times B) \cup (C \times B)] = [(A \cup C) \times B].$$

An informal proof can establish that $[(A \times B) \cup (C \times B)]$ and $[(A \cup C) \times (B)]$ have exactly the same elements. These two sets are Cartesian products, and, consequently their elements are ordered pairs.

- An ordered pair (X, Y) is an element of $(A \times B) \cup (C \times B)$ if and only if $(X, Y) \in (A \times B)$ or $(X, Y) \in (C \times B)$;
- hence $(X, Y) \in [(A \times B) \cup (C \times B)]$ if and only if $X \in A$ and $Y \in B$, or $X \in C$ and $Y \in B$,
- which is equivalent to $X \in A$ or $X \in C$, and $Y \in B$;
- thus $(X, Y) \in [(A \times B) \cup (C \times B)]$ if and only if $X \in (A \cup C)$ and $Y \in B$, which is equivalent to $(X, Y) \in [(A \cup C) \times B]$.

Just as the preceding informal proof concatenated the two occurrences of $Y \in B$ into one such occurrence, a formal proof can rely on the distributivity of $\wedge$ over $\vee$ by the tautology

$$\{[(P) \wedge (R)] \vee [(Q) \wedge (R)]\} \Leftrightarrow \{[(P) \vee (Q)] \wedge (R)\}.$$

$$(X, Y) \in [(A \times B) \cup (C \times B)]$$
$$\updownarrow \quad \text{definition of union,}$$
$$[(X, Y) \in (A \times B)] \vee [(X, Y) \in (C \times B)]$$
$$\updownarrow \quad \text{definition of Cartesian products,}$$
$$[(X \in A) \wedge (Y \in B)] \vee [(X \in C) \wedge (Y \in B)]$$
$$\updownarrow \quad \text{distributivity of } \wedge \text{ over } \vee,$$
$$[(X \in A) \vee (X \in C)] \wedge (Y \in B)$$
$$\updownarrow \quad \text{definition of union,}$$
$$[X \in (A \cup C)] \wedge [(Y \in B)]$$
$$\updownarrow \quad \text{definition of Cartesian products.}$$
$$(X, Y) \in [(A \cup C) \times B]$$

3.117

$$[A \times (B \setminus D)] = [(A \times B) \setminus (A \times D)]$$

$$\updownarrow$$

$$\forall X \forall Y \{(X, Y) \in [A \times (B \setminus D)] \Leftrightarrow (X, Y) \in [(A \times B) \setminus (A \times D)]\}$$

$$\updownarrow$$

$$\forall X \forall Y \{[(X \in A) \wedge (Y \in B) \wedge \{\neg(Y \in D)\}] \Leftrightarrow [(X \in A) \wedge (Y \in B) \wedge \{\neg[(X \in A) \wedge (Y \in D)]\}]\}$$

$$\updownarrow$$

$$\forall X \forall Y \{[(X \in A) \wedge (Y \in B) \wedge \{\neg(Y \in D)\}] \Leftrightarrow [(X \in A) \wedge (Y \in B) \wedge \{[\neg(X \in A)] \vee [\neg(Y \in D)]\}]\}$$

$$\updownarrow$$

$$\forall X \forall Y \{[(X \in A) \wedge (Y \in B) \wedge \{\neg(Y \in D)\}] \Leftrightarrow [(X \in A) \wedge (Y \in B) \wedge \{\neg(Y \in D)\}] \vee [(X \in A) \wedge (Y \in B) \wedge \{\neg(X \in A)\}]\}$$

$$(P) \Leftrightarrow [(P) \vee (\text{False})] \updownarrow$$

$$\forall X \forall Y \{[(X \in A) \wedge (Y \in B) \wedge \{\neg(Y \in D)\}] \Leftrightarrow [(X \in A) \wedge (Y \in B) \wedge \{\neg(Y \in D)\}]\}$$

3.119 No, $[(A \setminus C) \times (B \setminus D)] \not\supseteq [(A \times B) \setminus (C \times D)]$, because the right-hand side contains all of $(A \setminus C) \times B$.

3.121 No, $[(A \Delta C) \times (B \Delta D)] \not\supseteq [(A \times B) \Delta (C \times D)]$. For instance, if $B = D$, then $B \Delta D = \varnothing$, whence $(A \Delta C) \times (B \Delta D) = \varnothing$ on the left-hand side. Yet on the right-hand side, still with $B = D$, if $C = \varnothing$, then $C \times D = \varnothing$, whence $(A \times B) \Delta (C \times D) = (A \times B)$.

3.123 No, $([\mathscr{P}(A)] \times [\mathscr{P}(B)]) \not\supseteq \mathscr{P}(A \times B)$, because the left-hand side consists of pairs of subsets of A and B, whereas the right-hand side consists of subsets of $A \times B$. For instance, if $A = \varnothing = B$, then $A \times B = \varnothing$, whence $\mathscr{P}(A \times B) = \mathscr{P}(\varnothing) = \{\varnothing\}$, whereas $[\mathscr{P}(A)] \times [\mathscr{P}(B)] = [\mathscr{P}(\varnothing)] \times [\mathscr{P}(\varnothing)] = \{\varnothing\} \times \{\varnothing\} = \{(\varnothing, \varnothing)\}$.

3.125 If S denotes the relation of strict inclusion on $A := \mathscr{P}(H)$, then

$$S^{\circ-1} = \{(V, W) \in \mathscr{P}(H) \times \mathscr{P}(H) : W \subsetneq V\}.$$

3.127 From $\varnothing \subseteq A$ and $\varnothing \subseteq B$ it follows that $\varnothing \times \varnothing \in \mathscr{P}(A) \times \mathscr{P}(B)$, and also $\varnothing = \varnothing \times \varnothing \subseteq \varnothing \times \varnothing$. Thus $\varnothing \in \mathscr{D}$.

3.129 If $A = \varnothing = B$, then $\mathscr{P}(A) = \mathscr{P}(\varnothing) = \{\varnothing\} = \mathscr{D}$, because $\mathscr{P}(A) \times \mathscr{P}(B) = \mathscr{P}(\varnothing) \times \mathscr{P}(\varnothing) = \{\varnothing\} \times \{\varnothing\} = \{(\varnothing, \varnothing)\}$. From the solution to exercise 3.127, it follows that $\mathscr{P}(A) = \{\varnothing\}$ is all of $\mathscr{D}$.

3.131 No, F is *not* a function, because it contains two pairs, $(0, 1)$ and $(0, 4)$, with the same first coordinate but different second coordinates.

3.133 Yes, R *is* a function.

3.135 Yes, Z *is* a function (in effect the zero function).

3.137 Yes, the empty set $\varnothing \subseteq \varnothing \times B$ *is* a function from $\varnothing$ to B.

3.139 For each $X \in B$, $\mathbf{1}_A|_B (X) = 1 = \chi_B(X)$, so that $\mathbf{1}_A|_B$ coincides with $\chi_B(X)$ on B (but not on A).

3.141 If $V \neq W$, then $\chi_V \cup \chi_W$ is *not* a function, because of multiple values on $V \Delta W$, where one of χ_V or χ_W has the value 0 while the other has the value 1. Therefore, $\chi_V \cup \chi_W$ does *not* coincide with the *function* $\chi_{V \cup W}$.

3.143 Let

$$R := \{\varnothing\},$$
$$S := \{\{\varnothing\}\},$$
$$A := R \dot\cup S,$$
$$B := \{\varnothing\},$$
$$F : A \to B,$$
$$F := \{(\varnothing, \varnothing), (\{\varnothing\}, \varnothing)\}.$$

Then

$$R \cap S = \varnothing,$$
$$F"(R \cap S) = F"(\varnothing) = \varnothing,$$
$$F"(R) = F"(\{\varnothing\}) = \{F(\varnothing)\} = \{\varnothing\} = B,$$
$$F"(S) = F"(\{\{\varnothing\}\}) = \{F(\{\varnothing\})\} = \{\varnothing\} = B,$$
$$F"(R) \cap F"(S) = B \cap B = B \neq F"(R \cap S).$$

3.145 For all sets and for every function, $[F"(K)] \setminus [F"(H)] \subseteq F"(K \setminus H)$.

3.147 For each $X \in A$, from $\{X\} \subseteq A$ it follows that $F"(\{X\}) \subseteq B$, with $F"(\{X\}) = \{F(X)\}$ by definition of $F"$. Consequently, for each $X \in A$, the pair $(\{X\}, F"(\{X\})) \in \mathscr{P}(A) \times \mathscr{P}(B)$ corresponds to the pair $(X, F(X)) \in F \subseteq A \times B$.

3.149 $F"(\{\varnothing, \{\varnothing\}\}) = \{F(\varnothing), F(\{\varnothing\})\} = \{F(0), F(1)\} = \{2\}$.
$F \circ -1"(\{\varnothing, \{\varnothing\}\}) = F \circ -1"(\{0, 1\}) = \{2\}$.
$F(\{\varnothing, \{\varnothing\}\}) = F(2) = 0 = \varnothing$.
$F \circ -1"(\{\{\varnothing, \{\varnothing\}\}\}) = F \circ -1"(\{2\}) = \{0, 1\}$.

3.151 For each $X \in A$, $(I_B \circ F)(X) = I_B[F(X)] = F(X)$.

3.153 $\varnothing \subseteq F \circ \varnothing \subseteq \varnothing \times B \subseteq \varnothing$.

3.155 Define a function $G : F"(A) \to A$ as follows. For each $Y \in F"(A)$ there exists exactly one $X \in \mathscr{D}(F) \subseteq A$ with $Y = F(X)$; define $G(Y) := X$. Then $G[F(X)] = X$, so that $G \circ F = I_{\mathscr{D}(F)}$, with $\mathscr{D}(G) = F"(A) \subseteq B$.

3.157 Let

$$A := \{\varnothing, \{\varnothing\}\},$$
$$B := \{\varnothing\},$$
$$F := \{(\varnothing, \varnothing), (\{\varnothing\}, \varnothing)\},$$
$$G := \{(\varnothing, \varnothing)\}.$$

Then $(F \circ G)(\varnothing) = \varnothing$, so that $F \circ G = I_B$, and hence G is a right inverse for F. Yet F has no left inverse, because F is not injective.

3.159 Let

$$A := \{0, 1\},$$
$$B := \{0, 1, 2\},$$
$$F := \{(0, 0), (1, 1)\},$$

$$G := \{(0,0), (1,1), (2,0)\},$$
$$H := \{(0,0), (1,1), (2,1)\}.$$

Then $G \circ F = I_A$, and $H \circ F = I_A$.

3.161 The empty relation is vacuously reflexive, symmetric, and transitive because $(X, Y) \in \varnothing$ is False>
Reflexivity:

$$\forall X\{(X \in \varnothing) \Rightarrow [(X, X) \in \varnothing]\}$$

is True (a theorem) by the tautology (theorem) (False) $\Rightarrow$ (P).
Symmetry:

$$\forall X \forall Y\{[(X, Y) \in \varnothing] \Rightarrow [(Y, X) \in \varnothing]\}$$

is True (a theorem) by the tautology (theorem) (False) $\Rightarrow$ (P).
Transitivity:

$$\forall X \forall Y \forall Z\big(\{[(X, Y) \in \varnothing] \wedge [(Y, Z) \in \varnothing]\big) \Rightarrow [(X, Z) \in \varnothing]\}$$

is True (a theorem) by the tautology (theorem) (False) $\Rightarrow$ (P).

3.163 The relation of strict inclusion is vacuously anti-symmetric, because the hypothesis $(V \subsetneqq W) \wedge (W \subsetneqq V)$ is False for all sets V and W:

$(V \subsetneqq W) \wedge (W \subsetneqq V)$

$\updownarrow$ definition of $\subsetneqq$,

$[(V \subseteq W) \wedge (V \neq W)] \wedge [(W \subseteq V) \wedge (W \neq V)]$

$\updownarrow$ definitions of $\subseteq$ and $=$,

$\{\forall X[(X \in V) \Rightarrow (X \in W)]\} \wedge \{\exists Y[(Y \in W) \wedge (Y \notin V)]\}$
$\wedge\{\forall X[(X \in W) \Rightarrow (X \in V)]\} \wedge \{\exists Z[(Z \in V) \wedge (Z \notin W)]\}$

$\updownarrow: [\exists Y(P)] \Leftrightarrow \{\neg \forall Y[\neg(P)]\}$

$\{\forall X[(X \in V) \Rightarrow (X \in W)]\} \wedge \{\neg \forall Y \neg[(Y \in W) \wedge (Y \notin V)]\}$
$\wedge\{\forall X[(X \in W) \Rightarrow (X \in V)]\} \wedge \{\neg \forall Z \neg[(Z \in V) \wedge (Z \notin W)]\}$

$\updownarrow$ de Morgan's Law,

$$\{\forall X[(X \in V) \Rightarrow (X \in W)]\} \wedge \{\neg \forall Y([\neg(Y \in W)] \vee [\neg(Y \notin V)])\}$$
$$\wedge \{\forall X[(X \in W) \Rightarrow (X \in V)]\} \wedge \{\neg \forall Z([\neg(Z \in V)] \vee [\neg(Z \notin W)])\}$$

$\updownarrow$: $[(P) \Rightarrow (Q)] \Leftrightarrow \{[\neg(P)] \vee (Q)\}$,

$$\{\forall X[(X \in V) \Rightarrow (X \in W)]\} \wedge \{\neg \forall Y[(Y \in W) \Rightarrow (Y \in V)]\}$$
$$\wedge \{\forall X[(X \in W) \Rightarrow (X \in V)]\} \wedge \{\neg \forall Z[(Z \in V) \Rightarrow (Z \in W)]\}$$

$\updownarrow$ substitutions,

$$\{\forall X[(X \in V) \Rightarrow (X \in W)]\} \wedge \{\neg \forall Y[(Y \in W) \Rightarrow (Y \in V)]\}$$
$$\wedge \{\forall Y[(Y \in W) \Rightarrow (Y \in V)]\} \wedge \{\neg \forall X[(X \in V) \Rightarrow (X \in W)]\}$$

$\updownarrow$: $(P) \wedge [\neg(P)]$ is False,

False

Consequently, with $(V \subsetneqq W) \wedge (W \subsetneqq V)$ False, the implication

$$[(V \subsetneqq W) \wedge (W \subsetneqq V)] \Rightarrow (V = W)$$

is True.

3.165 Here are the equivalence classes:

$$A/\mathscr{R} = \{[0], [1]\},$$
$$[0] = \{0, 2, 4\},$$
$$[1] = \{1, 3, 5\}.$$

3.167 Verify that $\bigcup \mathscr{F} = B$ and that the elements of $\mathscr{F}$ are pairwise disjoint:

$$\bigcup \mathscr{F} = \{0, 2, 4, 6\} \cup \{1, 3, 5, 7\} = \{0, 1, 2, 3, 4, 5, 6, 7\} = B,$$

$$\{0, 2, 4, 6\} \cap \{1, 3, 5, 7\} = \varnothing.$$

Here is the corresponding equivalence relation:

$$
\mathscr{R}_{\mathscr{F}} = \left\{ \begin{array}{cccc}
& (1,7) & (3,7) & (5,7) & (7,7) \\
(0,6) & (2,6) & (4,6) & (6,6) & \\
& (1,5) & (3,5) & (5,5) & (7,5) \\
(0,4) & (2,4) & (4,4) & (6,4) & \\
& (1,3) & (3,3) & (5,3) & (7,3) \\
(0,2) & (2,2) & (4,2) & (6,2) & \\
& (1,1) & (3,1) & (5,1) & (7,1) \\
(0,0) & (2,0) & (4,0) & (6,0) &
\end{array} \right\}.
$$

3.169 Outline: For each equivalence relation $\mathscr{R}$, the partition $\mathscr{F}_{\mathscr{R}}$ consists of all the equivalence classes corresponding to $\mathscr{R}$. The relation $\mathscr{R}_{\mathscr{F}_{\mathscr{R}}}$ then consists of all the pairs from all the equivalence classes of $\mathscr{R}$, which is then $\mathscr{R}$ again.

3.171 The relation $\mathscr{Q}$ is

antisymmetric, vacuously, because it does not contain any two pairs (X, Y) and (Y, X),

asymmetric, because it does not contain any two pairs (X, Y) and (Y, X),

not connected, because $(2, 9) \notin \mathscr{Q}$ and $(9, 2) \notin \mathscr{Q}$,

irreflexive, because it does not contain any element on the diagonal,

not reflexive, because $(0, 0) \notin \mathscr{Q}$,

not strongly connected, because $(2, 9) \notin \mathscr{Q}$ and $(9, 2) \notin \mathscr{Q}$,

not symmetric, because $(0, 1) \in \mathscr{Q}$ but $(1, 0) \notin \mathscr{Q}$,

not transitive, because $(2, 6) \in \mathscr{Q}$ and $(6, 9) \in \mathscr{Q}$ but $(2, 9) \notin \mathscr{Q}$.

3.173 The relation $\mathscr{S}$ is

not antisymmetric, because $(0, 1) \in \mathscr{S}$ and $(1, 0) \in \mathscr{S}$ but $0 \neq 1$,

not asymmetric, because it does not contain any two pairs $(0, 1) \in \mathscr{S}$ and $(1, 0) \in \mathscr{S}$,

not connected, because $(8, 9) \notin \mathscr{S}$ and $(9, 8) \notin \mathscr{S}$,

irreflexive, because it does not contain any element on the diagonal,

not reflexive, because $(0, 0) \notin \mathscr{S}$,

not strongly connected, because $(8, 9) \notin \mathscr{S}$ and $(9, 8) \notin \mathscr{S}$,

symmetric, because if $(X, Y) \in \mathscr{S}$ then $(Y, X) \notin \mathscr{S}$,

not transitive, because $(9, 6) \in \mathscr{S}$ and $(6, 8) \in \mathscr{S}$ but $(9, 8) \notin \mathscr{S}$.

3.175 The relation $\mathscr{V}$ is

not antisymmetric, $(0, 1) \in \mathscr{V}$ and $(1, 0) \in \mathscr{V}$ but $0 \neq 1$,

not asymmetric, because it does not contain any two pairs $(0, 1) \notin \mathscr{V}$ and $(1, 0) \notin \mathscr{V}$,

not connected, because $(0, 2) \notin \mathscr{V}$ and $(2, 0) \notin \mathscr{V}$,

not irreflexive, because it contains at least one element on the diagonal,

not reflexive, because $(0, 0) \notin \mathscr{V}$,

not strongly connected, because $(0, 2) \notin \mathscr{V}$ and $(2, 0) \notin \mathscr{V}$,

not symmetric, because $(1, 2) \in \mathscr{V}$ but $(2, 1) \notin \mathscr{V}$,

not transitive, because $(0, 1) \in \mathscr{V}$ and $(1, 2) \in \mathscr{V}$ but $(0, 2) \notin \mathscr{V}$.

3.177 The empty relation is vacuously a partial order.

3.179 See the definitions of the "diagonal" (example 3.68) and "irreflexive" (definition 3.165).

3.181 By definition, every strict partial order is irreflexive and transitive. If it contained (X, Y) and (Y, X), then it would also contain (X, X) by transitivity, but it cannot contain (X, X) by irreflexivity; consequently, it cannot contain both (X, Y) and (Y, X), which makes it asymmetric.

3.183 By definition 3.176, an asymmetric relation cannot contain both (X, Y) and (Y, X), so that the hypothesis in the definition (3.173) of "anti-symmetric" is vacuously False.

3.185 By the preceding exercise (3.184), every asymmetric relation is also irreflexive; consequently, an asymmetric and transitive relation is also irreflexive and transitive, which is the definition of a strict partial order. Conversely, the solution to exercise 3.181 shows that every strict partial order is asymmetric, whence also asymmetric and transitive.

3.187 The empty relation is vacuously asymmetric and strongly connected.

3.189 The empty subset of the empty set is vacuously a chain with respect to the relation of inclusion.

Exercises from Chapter 4

4.1

$$5 = \left\{ \varnothing, \{\varnothing\}, \{\varnothing,\{\varnothing\}\}, \left\{\varnothing,\{\varnothing\},\{\varnothing,\{\varnothing\}\}\right\}, \left\{ \varnothing, \{\varnothing\}, \{\varnothing,\{\varnothing\}\}, \left\{\varnothing,\{\varnothing\},\{\varnothing,\{\varnothing\}\}\right\} \right\} \right\}$$

4.3 For each set, $X \in \{X\}$ by pairing, whence $X \in (X \cup \{X\})$ by union.
Moreover, for each set, $X \subseteq X$, whence $X \subseteq (X \cup \{X\})$, also by union.

4.5 The equation may fail. Let $A := \varnothing$ and $B := \left\{ \{\varnothing\} \right\}$. Then $A \subset B$ but

$$A \cup \{A\} = \varnothing \cup \{\varnothing\} = \{\varnothing\},$$
$$B \cup \{B\} = \{\{\varnothing\}\} \cup \left\{ \left\{ \{\varnothing\} \right\} \right\} = \left\{ \{\varnothing\}, \left\{ \{\varnothing\} \right\} \right\},$$

but $\varnothing \notin \left\{ \{\varnothing\}, \{\{\varnothing\}\} \right\}$ whence

$$\{\varnothing\} \not\subseteq \left\{ \{\varnothing\}, \{\{\varnothing\}\} \right\},$$

so that $A \cup \{A\} \not\subseteq B \cup \{B\}$.

4.7 The equation may fail. Let $A := \varnothing$ and $B := \{\{\varnothing\}\}$. Then

$$(A \cup \{A\}) \cap (B \cup \{B\}) = (\varnothing \cup \{\varnothing\}) \cap \left(\{\{\varnothing\}\} \cup \left\{ \{\{\varnothing\}\} \right\} \right) = \varnothing,$$

$$(A \cap B) \cup \{A \cap B\} = \left(\varnothing \cap \{\{\varnothing\}\} \right) \cup \left\{ \varnothing \cap \{\{\varnothing\}\} \right\} = \{\varnothing\},$$

so that $(A \cup \{A\}) \cap (B \cup \{B\}) \neq (A \cap B) \cup \{A \cap B\}$.

4.9 This proof proceeds by induction with M.

Initial Step

If $M = 0$, then $M = \varnothing$, whence the hypothesis "$K \in M$ and $L \in M$" is False for all $K, L \in \mathbb{N}$, and hence the implication is True.

Induction hypothesis

Assume that there exists $N \in \mathbb{N}$ such that the theorem holds for $M := N$, so that for all $K, L, \in \mathbb{N}$, if $K \in N$ and $L \in N$, then $K \cup L \in N$.

Induction step

For all $K, L, \in \mathbb{N}$, if $K \in (N \cup \{N\})$ and $L \in (N \cup \{N\})$, then only four cases can arise.

If $K \in N$ and $L \in N$, then $(K \cup L) \in N$ by induction hypothesis; from $N \subseteq N \cup \{N\}$ it then follows that $(K \cup L) \in (N \cup \{N\})$.

If $K \in \{N\}$ and $L \in N$, then $K = N$ and $L \subseteq N$, whence $K \cup L = N \cup L = N$, and hence $(K \cup L) = N \in \{N\} \subseteq (N \cup \{N\})$; swapping K and L yields a proof for $K \in N$ and $L \in \{N\}$.

If $K \in \{N\}$ and $L \in \{N\}$, then $K = N$ and $L = N$, whence $K \cup L = N$, and hence $(K \cup L) = N \in \{N\} \subseteq (N \cup \{N\})$.

4.11 Let $H := \mathbb{I}$ be as defined by the Axiom of Infinity.

$$H \in \mathscr{F} \qquad \textbf{yet unproved,}$$

$$\Updownarrow \quad \text{definition of } \mathscr{F},$$

$$[H \in \mathscr{P}(H)]$$
$$\wedge \big[(\varnothing \in H) \wedge \big(\forall X \{ (X \in H) \Rightarrow [(X \cup \{X\}) \in H] \} \big) \big]$$

which is True by the definition of H and the Axiom of Infinity.

4.13

$$\forall X \{ (X \in \mathbb{N}) \Rightarrow [(X \cup \{X\}) \in \mathbb{N}] \} \qquad \textbf{yet unproved,}$$

$$\Updownarrow \quad \text{definition of } \mathbb{N},$$

$$\forall X \{ (X \in \bigcap \mathscr{F}) \Rightarrow [(X \cup \{X\}) \in \bigcap \mathscr{F}] \}$$

$$\Updownarrow \quad \text{definition of } \bigcap,$$

$$\forall X \big[\{ \forall B [(B \in \mathscr{F}) \Rightarrow (X \in B)] \}$$
$$\Rightarrow \big(\forall B \{ (B \in \mathscr{F}) \Rightarrow [(X \cup \{X\}) \in B] \} \big) \big]$$

$$\Uparrow$$

$$\forall X \big[\forall B \big(\{ [(B \in \mathscr{F}) \Rightarrow (X \in B)] \}$$
$$\Rightarrow \{ (B \in \mathscr{F}) \Rightarrow [(X \cup \{X\}) \in B] \} \big) \big]$$

$$\Uparrow \quad \text{axiom P2,}$$
$$\forall X \{ \forall B \big[(B \in \mathscr{F})$$
$$\Rightarrow \{ (X \in B) \Rightarrow [(X \cup \{X\}) \in B] \} \big] \}$$

which is True by the definition of $\mathscr{F}$. Moreover, $\varnothing \in \mathbb{N}$ by theorem 4.8.

4.15

$$\vdash \mathbb{I} \in \mathscr{F} \qquad \text{theorem 4.4,}$$
$$\vdash (\mathbb{I} \in \mathscr{F}) \Rightarrow (\bigcap \mathscr{F} \subseteq \mathbb{I}) \qquad \text{theorem 3.48,}$$
$$\bigcap \mathscr{F} \subseteq \mathbb{I} \qquad \textit{Modus Ponens;}$$

$$\vdash (S \subseteq \mathbb{N}) \Leftrightarrow (S \subseteq \bigcap \mathscr{F}) \qquad \text{definition of } \mathscr{F},$$
$$S \subseteq \mathbb{I} \qquad \text{transitivity of } \subseteq;$$

$$\vdash (S \subseteq \mathbb{N}) \wedge \big[(\varnothing \in S) \wedge \big(\forall X \{ (X \in S) \Rightarrow [(X \cup \{X\}) \in S] \} \big) \big] \qquad \text{hypothesis on } S,$$
$$\vdash (\varnothing \in S) \wedge \big(\forall X \{ (X \in S) \Rightarrow [(X \cup \{X\}) \in S] \} \big) \qquad [(P) \wedge (Q)] \Rightarrow (Q) \text{ and } M.P.,$$
$$\vdash \big[(\varnothing \in S) \wedge \big(\forall X \{ (X \in S) \Rightarrow [(X \cup \{X\}) \in S] \} \big) \big] \Rightarrow (S \in \mathscr{F}) \qquad \text{definition of } \mathscr{F},$$
$$\vdash S \in \mathscr{F} \qquad M.P.$$
$$\vdash (S \in \mathscr{F}) \Rightarrow (\bigcap \mathscr{F} \subseteq S) \qquad \text{theorem 3.48,}$$
$$\bigcap \mathscr{F} \subseteq S \qquad \textit{Modus Ponens;}$$

$$\bigcap \mathscr{F} = S \qquad \bigcap \mathscr{F} \subseteq S \text{ and } S \subseteq \bigcap \mathscr{F}.$$
$$\vdash S = \mathbb{N} \qquad \text{definition of } \mathbb{N}.$$

4.17 Because $\varnothing \in \mathbb{N}$ it follows that $\bigcap \mathbb{N} \subseteq \varnothing$ whence $\bigcap \mathbb{N} = \varnothing$.

4.19 Let $S := V \dot{\cup} \{\varnothing, \{\varnothing\}\}$. This proof shows that $S = \mathbb{N}$, whence $V = S \setminus \{\varnothing, \{\varnothing\}\} = \mathbb{N} \setminus \{\varnothing, \{\varnothing\}\}$.

For each $X \in S$, either $X \in \{\varnothing, \{\varnothing\}\}$ or $X \notin \{\varnothing, \{\varnothing\}\}$.

If $X \in \{\varnothing, \{\varnothing\}\}$, then either $X = \varnothing$ or $X = \{\varnothing\}$.

In the first case (if $X = \varnothing$), then $\varnothing \in S$, by pairing and union. Moreover, $\varnothing \cup \{\varnothing\} = \{\varnothing\} \in V$ by hypothesis, whence $\varnothing \cup \{\varnothing\} \in S$.

In the second case (if $X = \{\varnothing\}$), then $\{\varnothing\} \cup \{\{\varnothing\}\} \in V$ whence $\{\varnothing\} \cup \{\{\varnothing\}\} \in S$.

Otherwise (if $X \in S \setminus \{\varnothing, \{\varnothing\}\}$), then $X \in V$, whence $X \cup \{X\} \in V$ by hypothesis, and hence $X \cup \{X\} \in S$ by union.

Thus, $X \cup \{X\} \in S$ for each $X \in S$, and $\varnothing \in S$. Consequently, $S = \mathbb{N}$ by the Principle of Mathematical Induction.

Therefore, $V = S \setminus \{\varnothing\} = \mathbb{N} \setminus \{\varnothing\}$.

4.21 $1 + 1 = 1 \cup \{1\}$ by definition of $M + 1 = M \cup \{M\}$, and $1 \cup \{1\} = 2$ by definition of 2.

4.23 $3 + 1 = 3 \cup \{3\}$ by definition of $M + 1 = M \cup \{M\}$, and $3 \cup \{3\} = 4$ by definition of 4.

4.25 $5 + 1 = 5 \cup \{5\}$ by definition of $M + 1 = M \cup \{M\}$, and $5 \cup \{5\} = 6$ by definition of 6.

4.27 $7 + 1 = 7 \cup \{7\}$ by definition of $M + 1 = M \cup \{M\}$, and $7 \cup \{7\} = 8$ by definition of 8.

4.29 $2 + 2 = 2 + (1 + 1) = (2 + 1) + 1$ by definition of $M + (N + 1) = (M + N) + 1$, and $(2 + 1) + 1 = 3 + 1 = 4$ by the preceding exercises.

4.31 $4+2 = 4+(1+1) = (4+1)+1$ by definition of $M+(N+1) = (M+N)+1$, and $(4+1)+1 = 5+1 = 6$ by the preceding exercises.

4.33 $6+2 = 6+(1+1) = (6+1)+1$ by definition of $M+(N+1) = (M+N)+1$, and $(6+1)+1 = 7+1 = 8$ by the preceding exercises.

4.35 $3+3 = 3+(2+1) = (3+2)+1$ by definition of $M+(N+1) = (M+N)+1$, and $(3+2)+1 = 5+1 = 6$ by the preceding exercises.

4.37 $5+3 = 5+(2+1) = (5+2)+1$ by definition of $M+(N+1) = (M+N)+1$, and $(5+2)+1 = 7+1 = 8$ by the preceding exercises.

4.39 $4+4 = 4+(3+1) = (4+3)+1$ by definition of $M+(N+1) = (M+N)+1$, and $(4+3)+1 = 7+1 = 8$ by the preceding exercises.

4.41 $2*2 = 2*(1+1) = (2*1)+2$ by definition of $M*(N+1) = (M*N)+M$, and $2*1 = 2$ by definition of $M*1 = M$, whence $(2*1)+2 = 2+2 = 4$ by the preceding exercises.

4.43 $4*2 = 4*(1+1) = (4*1)+4$ by definition of $M*(N+1) = (M*N)+M$, and $4*1 = 4$ by definition of $M*1 = M$, whence $(4*1)+4 = 4+4 = 8$ by the preceding exercises.

4.45 No, because $1+(1*1) = 1+1 = 2$ but $(1+1)*(1+1) = 2*2 = 4$.

4.47 Let $C := \mathbb{N} \times \mathbb{N}$ and specify $G : C \to C$ by $G(K,L) := (K*L, L+1)$. Then define $F : \mathbb{N} \to C$ by $F(0) := (1,1)$ and $F(I+1) := G[F(I)]$. Thus,

$$F(0) = (0,1),$$
$$F(1) = G[F(0)] = G(1,1) = (1*1, 1+1) = (1,2),$$
$$F(2) = G[F(1)] = G(1,2) = (1*2, 2+1) = (2,3),$$
$$F(3) = G[F(2)] = G(2,3) = (2*3, 3+1) = (6,4),$$
$$\vdots$$
$$F(I+1) = G[F(I)] = G(I!, I+1) = ((I!)*(I+1), (I+1)+1)$$
$$= ((I+1)!, I+2),$$
$$\vdots$$

Thus $N!$ equals the first projection of $F(N)$.

4.49 No, because $(1+2)! = 3! = 6$ but $(1!)+(2!) = 1+2 = 3$.

4.51 $2 < 5$ because $2 \in 5$:

$$\{\varnothing, \{\varnothing\}\} \in$$

$$\left\{ \varnothing, \{\varnothing\}, \{\varnothing,\{\varnothing\}\}, \left\{\varnothing,\{\varnothing\},\{\varnothing,\{\varnothing\}\}\right\}, \left\{ \varnothing, \{\varnothing\}, \{\varnothing,\{\varnothing\}\}, \left\{\varnothing,\{\varnothing\},\{\varnothing,\{\varnothing\}\}\right\} \right\} \right\}$$

4.53 $2 < 5$ and $5 < 7$ by previous exercises whence $2 < 7$ by transitivity.

4.55 $3 < 4$, and $4 < 5$, by definition of $<$, $5 < 7$ by previous exercises, whence $3 < 7$ by transitivity.

4.57 $2 < 7$ by previous exercises, and $7 < 8$ by definition of $<$ whence $2 < 8$ by transitivity.

4.59 If $1 < K$ and $1 < L$, then by theorem 4.58 there exist $I \in \mathbb{N}^*$ and $J \in \mathbb{N}^*$ with $K = I + 1$ and $L = J + 1$. Hence $K * L = (I + 1) * (J + 1) = I * J + I + J + 1 > 0 + 1 + 1 + 1$.

4.61 If $3 < K$ and $3 < L$, then $3 * 3 < K * L$, whence $7 < 9 < K * L$; thus $K < 4$ and $L < 4$. The only possibilities with $1 < K < 4$ and $1 < L < 4$ are $K \in \{2, 3\}$ and $L \in \{2, 3\}$. Yet $2 * 2 = 4 \neq 7$, $2 * 3 = 3 * 2 = 6 \neq 7$, and $3 * 3 = 9 \neq 7$.

4.63 Similarly, $K * L = 9$ if and only if $(K, L) \in \{(1, 9), (3, 3), (9, 1)\}$.

4.65 If $S \neq \varnothing$ and $S \subseteq \mathbb{N}$ then S has a smallest element $I \in S$, by the Well-Ordering Principle. Consequently, for every $J \in S$ it follows that $I \leq J$, which means that $(I = J) \vee (I < J)$, whence $(I < J)$ for $I \neq J$. Thus $\neg(J < I)$ because only one of the three relations $(<, =, >)$ holds. From $\neg(J \in I)$ for every $J \in S$ it follows that $I \cap S = \varnothing$.

4.67 This proof proceeds by contradiction. Negating the conclusion gives

$$\neg[(I \notin K) \vee (K \notin L) \vee (L \notin I)] = [\neg(I \notin K)] \wedge [\neg(K \notin L)] \wedge [\neg(L \notin I)]$$
$$= (I \in K) \wedge (K \in L) \wedge (L \in I)$$

If $(I \in K) \wedge (K \in L) \wedge (L \in I)$ for some $I, K, L \in \mathbb{N}$, then $I \in I$ by transitivity of $\in$ on $\mathbb{N}$, but $I \notin I$ for every $I \in \mathbb{N}$. Therefore the negation of the conclusion is False.

4.69 This proof proceeds by contradiction.

First, $I \notin I$ for every $I \in \mathbb{N}$: $\vdash \forall I \{(I \in \mathbb{N}) \Rightarrow [\neg(I \in I)]\}$.

Second, $\mathrm{Subf}_{\mathbb{N}}^{I}$ gives $(\mathbb{N} \in \mathbb{N}) \Rightarrow [\neg(\mathbb{N} \in \mathbb{N})]$, which has the pattern $(P) \Rightarrow [\neg(P)]$.

From the tautology $\{(P) \Rightarrow [\neg(P)]\} \Rightarrow [\neg(P)]$ it then follows that $\neg(\mathbb{N} \in \mathbb{N})$.

4.71 Calculate $[(2, 4)]_{\backsim} + [(6, 3)]_{\backsim} = [(0, 2)]_{\backsim} + [(3, 0)]_{\backsim} = [(0 + 3, 2 + 0)]_{\backsim} = [(3, 2)]_{\backsim} = [(1, 0)]_{\backsim}$.

4.73 Calculate $[(3, 1)]_{\backsim} - [(5, 2)]_{\backsim} = [(3, 1)]_{\backsim} + [(2, 5)]_{\backsim} = [(2, 0)]_{\backsim} + [(0, 3)]_{\backsim} = [(2 + 0, 0 + 3)]_{\backsim} = [(2, 3)]_{\backsim} = [(0, 1)]_{\backsim}$.

4.75 Kunen's addition of pairs of natural numbers commutes.

$$([(M, N)]_\simeq) + ([(P, Q)]_\simeq) = [(M + N, P + Q)]_\simeq$$
$$= [(N + M, Q + P)]_\simeq$$
$$= ([(P, Q)]_\simeq) + ([(M, N)]_\simeq).$$

4.77 Kunen's addition of pairs of natural numbers is associative.

$$([(K, L)]_\simeq + [(M, N)]_\simeq) + [(P, Q)]_\simeq = [(K + M, L + N)]_\simeq + [(P, Q)]_\simeq$$
$$= [((K + M) + P, (L + N) + Q)]_\simeq$$
$$= [(K + (M + P), L + (N + Q))]_\simeq$$
$$:= [(K, L)]_\simeq + [(N + M, Q + P)]_\simeq$$
$$:= [(K, L)]_\simeq + ([(P, Q)]_\simeq + [(M, N)]_\simeq).$$

4.79 Kunen's multiplication of pairs of natural numbers commutes.

$$([(M, N)]_\simeq) * ([(P, Q)]_\simeq) = [(M * P + N * Q, \ M * Q + N * P)]_\simeq$$
$$= [(P * M + Q * N, \ Q * M + P * N)]_\simeq$$
$$= [(P * M + Q * N, \ P * N + Q * M)]_\simeq$$
$$= ([(P, Q)]_\simeq) * ([(M, N)]_\simeq)$$

4.81 Kunen's multiplication of pairs of natural numbers is associative.

$$[(K, L)]_\simeq * ([(M, N)]_\simeq) * ([(P, Q)]_\simeq)$$

$$= [(K, L)]_\simeq * [(M * P + N * Q, \ M * Q + N * P)]_\simeq$$

$$= [(K * (M * P + N * Q) + L * (M * Q + N * P),$$
$$K * (M * Q + N * P) + L * (M * P + N * Q))]_\simeq$$

$$= [((K * P + L * Q) * M + (K * Q + L * P) * N,$$
$$(K * P + L * Q) * N + (K * Q + L * P) * M)]_\simeq$$

$$= [(K * P + L * Q, K * Q + L * P)]_\simeq * [(M, N)]_\simeq$$

$$= ([(K, L)]_\simeq * [(P, Q)]_\simeq) * [(M, N)]_\simeq.$$

4.83 Kunen's multiplication of pairs of natural numbers distributes over addition.

$$[(K, L)]_\simeq * ([(M, N)]_\simeq + [(P, Q)]_\simeq)$$

$$= [(K, L)]_\simeq * [(M + P, N + Q)]_\simeq$$

$$= [(K * (M + P) + L * (N + Q), K * (N + Q) + L * (M + P)]_\simeq$$

$$= [(K * P + L * Q + K * M + L * N, K * Q + L * P + K * N + L * M]_\simeq$$

$$= [(K * P + L * Q, K * Q + L * P)]_\simeq + [(K * M + L * N, K * N + L * M)]_\simeq$$

$$= ([(K, L)]_\simeq * [(P, Q)]_\simeq) + ([(K, L)]_\simeq * [(M, N)]_\simeq).$$

4.85 Subtraction does *not* commute:

$$[(0, 0)]_\simeq - [(1, 0)]_\simeq = [(0, 0)]_\simeq + [(0, 1)]_\simeq$$
$$= [(0 + 0, 0 + 1)]_\simeq$$
$$= [(0, 1)]_\simeq;$$

$$[(1, 0)]_\simeq - [(0, 0)]_\simeq = [(1, 0)]_\simeq + [(0, 1)]_\simeq$$
$$= [(1 + 0, 0 + 0)]_\simeq$$
$$= [(1, 0)]_\simeq$$

where $[(0, 1)]_\simeq \neq [(1, 0)]_\simeq$ because $M + Q = 0 + 0 \neq 1 + 1 = N + P$.

4.87 Kunen's multiplication of pairs of natural numbers distributes over subtraction.

$$X * (Y - Z) = X * [Y + (-Z)] = (X * Y) + [X * (-Z)] = (X * Y) + [-(X * Z)]$$
$$= (X * Y) - (X * Z).$$

4.89 Subtraction does *not* distribute over addition.

$$1 - (1 + 1) = 1 - 2 = -1,$$
$$(1 - 1) + (1 - 1) = 0 + 0 = 0.$$

4.91 By definition 4.102, $(2/3) + (7/5) = [(2 * 5) + (3 * 7)]/(3 * 5) = 31/15$.

4.93 By theorem 4.110, $(7/3) - (2/5) = (7/3) + [(-2)/5] = \{(7 * 5) + [3 * (-2)]\}/(3 * 5) = 29/15$.

4.95 By definition 4.102, $(2/3) * (7/5) = (2 * 7)/(3 * 5) = 14/15$.

4.97 By definition 4.117, $(2/3) \div (7/5) = (2/3) * (5/7) = (2 * 5)/(3 * 7) = 10/21$.

4.99 By exercise 4.97, $(2/3) \div (7/5) = 10/21$. Similarly, $(7/5) \div (2/3) = (7/5) * (3/2) = (7 * 3)/(5 * 2) = 21/10$. However, $10/21 \neq 21/10$, because, according to the relation in definition 557, $(10, 21) \not\equiv (21, 10)$ since $10 * 10 \neq 21 * 21$ in $\mathbb{N}$.

4.101 $(2/3) \div [(7/5) + (1/1)] = (2/3) \div \{[(7 * 1) + (1 * 5)]/(5 * 1)\} = (2/3) \div (12/5) = (2/3) * (5/12) = (2 * 5)/(3 * 12) = 5/18$. However, $[(2/3) \div (7/5)] + [(2/3) \div (1/1)] = (21/10) + (2/3) = [(21 * 3) + (10 * 2)]/(10 * 3) = 83/30$, and $83/30 \neq 5/18$ because $83 * 18 \neq 30 * 5$.

4.103 $(1/1) + [(2/3) \div (7/5)] = (1/1) + (10/21) = [(1 * 21) + (1 * 10)]/(1 * 21) = 31/21$. However, $[(1/1) + (2/3)] \div [(1/1) + (7/5)] = (5/3) \div (12/5) = (5/3) * (5/12) = 25/36$, and $25/36 \neq 31/21$ because $25 * 21 \neq 36 * 31$.

4.105 If $0 < (I/J)$ and $0 < (P/Q)$, then $0 < I * J$ and $0 < P * Q$. Also, $(I/J) + (P/Q) = (I * Q + J * P)/(J * Q)$, with

$$(I * Q + J * P) * (J * Q) = (I * J) * (Q * Q) + (J * J) * (P * Q) > 0$$

because $Q * Q > 0$ and $J * J > 0$. Hence $0 < [(I/J) + (P/Q)]$.

4.107 $(P/1)/(Q/R) = (P/1) * (R/Q) = (P * R)/(1 * Q) = (P * R)/Q$.

4.109 If there exists $M \in \mathbb{N}$ such that $K/L = M/1$, then $N = M$. In the alternative, by theorem 4.133, for each $K/L \in \mathbb{Q}$, there exists $N \in \mathbb{N}$ with $K/L < N/1$. Consequently, the set $S := \{N \in \mathbb{N} : K/L < N/1\}$ is not empty. By the Well-Ordering Principle, S contains a smallest element, denoted here by N.

4.111 By example 4.135 with $N := \emptyset$ it follows that $\#(\emptyset) = \emptyset = 0$.
    Alternatively, $\#(\emptyset) = 0$ because of the bijection $\emptyset : \emptyset \to \emptyset$ and $\emptyset = 0$.

4.113 By example 3.24, $\mathscr{P}(\{\emptyset\}) = \{\emptyset, \{\emptyset\}\} = 2$. By definition 4.140 with $N := 2$, it follows that $\#(2) = 2$.

4.115 From $\#(\mathscr{P}[\mathscr{P}(\emptyset)]) = 2$ follows $\#[\mathscr{P}(\mathscr{P}[\mathscr{P}(\emptyset)])] = 4$ whence $\#\{\mathscr{P}[\mathscr{P}(\mathscr{P}[\mathscr{P}(\emptyset)])]\} = 16$.

Alternatively, $\#\big[\mathscr{P}\big(\mathscr{P}\{\mathscr{P}[\mathscr{P}(\varnothing)]\}\big)\big] = 16$ because

$$\mathscr{P}\big(\mathscr{P}\{\mathscr{P}[\mathscr{P}(\varnothing)]\}\big)$$
$$= \mathscr{P}\big(\mathscr{P}\{\mathscr{P}[\{\varnothing\}]\}\big)$$
$$= \mathscr{P}\Big[\big\{\varnothing, \{\varnothing\}, \{\{\varnothing\}\}, \{\varnothing, \{\varnothing\}\}\big\}\Big]$$
$$= \Big\{\varnothing,$$
$$\{\varnothing\}, \{\{\varnothing\}\}, \big\{\{\{\varnothing\}\}\big\}, \big\{\{\varnothing, \{\varnothing\}\}\big\},$$
$$\{\varnothing, \{\varnothing\}\}, \big\{\varnothing, \{\{\varnothing\}\}\big\}, \big\{\varnothing, \{\varnothing, \{\varnothing\}\}\big\},$$
$$\big\{\{\varnothing\}, \{\{\varnothing\}\}\big\}, \big\{\{\varnothing\}, \{\varnothing, \{\varnothing\}\}\big\}, \big\{\{\{\varnothing\}\}, \{\varnothing, \{\varnothing\}\}\big\},$$
$$\big\{\varnothing, \{\varnothing\}, \{\{\varnothing\}\}\big\}, \big\{\varnothing, \{\varnothing\}, \{\varnothing, \{\varnothing\}\}\big\},$$
$$\big\{\varnothing, \{\{\varnothing\}\}, \{\varnothing, \{\varnothing\}\}\big\}, \big\{\{\varnothing\}, \{\{\varnothing\}\}, \{\varnothing, \{\varnothing\}\}\big\},$$
$$\equiv \{0, 1, 2, 3, 4, 5, 6, 7, 8, 9, 10, 11, 12, 13, 14, 15\}.$$

4.117 No, not all ordered pairs need to have the same cardinality. If $X \neq Y$, then $(X, Y) = \{\{X\}, \{X, Y\}\}$ with $\{X\} \neq \{X, Y\}$. In contrast, if $X = Y$, then $(X, X) = \{\{X\}, \{X, X\}\} = \{\{X\}, \{X\}\} = \{\{X\}\}$. There exist only two functions from (X, X) to (X, Y), neither of which is a bijection: $F : \{X\} \mapsto \{X\}$, and $G : \{X\} \mapsto \{X, Y\}$. Consequently, there does not exist any bijection from (X, X) to (X, Y), whence (X, X) and (X, Y) do not have the same cardinality.

4.119 By theorem 4.150 and by associativity of $*$, $\#[(A \times B) \times C] = [\#(A \times B)] * [\#(C)] = \{[\#(A)]*[\#(B)]\}*[\#(C)] = [\#(A)]*\{[\#(B)]*[\#(C)]\} = [\#(A)]*[\#(B \times C)] = \#[A \times (B \times C)]$.

4.121 Apply contraposition to theorem 4.146.

4.123 If A is denumerable, then there is a bijection $F : \mathbb{N} \to A$. Consequently, the function $G\ \mathbb{N}^* \to A$ defined by $G(N) := F(N + 1)$ is also a bijection.

If moreover $X \notin A$, then $H : \{0\} \to \{X\}$ is also a bijection.

Thus, $G \cup H$ is a bijection $(G \cup H) : (\{0\} \cup \mathbb{N}) \to (\{X\} \cup A)$. Hence $A \cup \{X\}$ is denumerable.

Now proceed by induction with the number of elements in B.

4.125 Use theorem 4.160 and $(A \cup B) = (A \setminus B) \cup (B \setminus A)$.

4.127 Apply theorem 4.156 and exercise 4.126.

4.129 Apply theorems 4.156 and 4.167.

4.131 Any non-surjective injection $F : X \hookrightarrow Y$ induces a non-surjective injection $G : 2^X \hookrightarrow 2^Y$ by restriction from Y to X.

4.133 $\#(2^X) = 2^{\#(X)} > \#(X)$ by induction with $\#(X) \in \mathbb{N}$.

4.135 If A is denumerable, then there exists a bijection $F : A \hookrightarrow \mathbb{N}$, which restricts to an injection on each subset $S \subseteq A$.

4.137 Apply theorem 4.166 twice.

4.139 The existence of every function F_I relies on the Axiom of Choice.

Exercises from Chapter 5

5.1 For each $X \in \mathscr{F}$ let $F(X)$ be the smallest element of X.

5.3 The set $\mathbb{Z}$ is nonempty but has *no* smallest element relative to $<$ or $\leq$.

5.5 With $M := 2$, define $[0]_2 \prec [1]_2$. Let $[I]_M := [1]_2$, $[K]_M := [0]_2$, $[L]_M := [1]_2$. Then $[0]_2 \prec [1]_2$ but $[1]_2 + [0]_2 \not\prec [1]_2 + [1]_2$.

5.7 For each countable set C there exists an injection $F : C \to \mathbb{N}$. Define a relation R on C by XRY if and only if $F(X) \leq F(Y)$.

5.9 If W does *not* contain a last element,

$$\neg\{\exists Z \forall Y[(Y \prec Z) \vee (Y = Z)]\}$$

whence

$$\forall Z \exists Y\{[\neg(Y \prec Z)] \wedge (Y \neq Z).\}$$

In other words, for each $Z \in W$ there exists $Y \in W$ such that $\neg(Y \prec Z)$ and $Y \neq Z$. In particular, $Y \notin W_Z$. In yet other words, there does *not* exist any $Z \in W$ such that $W = W_Z$. Because $\prec$ totally orders W, however, it follows that $Z \prec Y$ must hold. Let $\mathscr{G}$ consists of all the initial intervals in W. Then $\bigcup \mathscr{G} = W$, but there does not exist any $Z \in W$ such that $W = W_Z$. Consequently, $\bigcup \mathscr{G}$ is *not* an initial interval of W.

5.15 There is a transitive set on which $\in$ is *not* a transitive relation. For example, the set

$$A := \Big\{\varnothing, \{\varnothing\}, \{\{\varnothing\}\}\Big\}$$

is transitive, because every element of A is also a subset of A. Nevertheless, the relation $\in$ is *not* transitive on A, because $\varnothing \notin \{\{\varnothing\}\}$, even though

$$\varnothing \in \{\varnothing\}, \quad \{\varnothing\} \in \{\{\varnothing\}\}.$$

Exercises from Chapter 6

6.1 Each finite set *of pairwise disjoint sets* $\mathscr{F} = \{A_0, \ldots, A_{N-1}\}$ where each element A_j has exactly N_j elements has exactly $\prod_{k=0}^{N-1}(N_k)$ choice functions.

6.3 Theorem 6.6 shows that each finite set of nonempty sets has a choice function. The proof of theorem 6.20 shows that each choice function corresponds to a choice set.

6.5 The Choice-Set Principle 6.17 translates into formula (1):

$$\forall \mathscr{F}\Bigg(\{\forall A\,[(A \in \mathscr{F}) \Rightarrow (A \neq \varnothing)]\} \tag{1}$$

$$\Rightarrow \Big\{\exists S \Big[\Big(S \subseteq \bigcup \mathscr{F}\Big) \wedge \big(\forall A\,\{(A \in \mathscr{F}) \Rightarrow [\exists X (S \cap A = \{X\})]\}\big)\Big]\Big\}\Bigg).$$

6.7 For each set $\mathscr{F}$ of nonempty sets, define a relation $R \subseteq \mathscr{F} \times \bigcup \mathscr{F}$ by $R := \{(A, X) : (A \in \mathscr{F}) \wedge (X \in A)\}$. If the Choice-Relation Principle 6.15 holds, then there exists a function $F \subseteq R$ with the same domain as that of R. Hence F is a choice function. Thus the Choice-Function Principle 6.8 holds.

 Conversely, for each relation $R \subseteq A \times B$, and for each set X in the domain of R, define the vertical section of R at X by $R_X := \{Y \in B : (X, Y) \in R\}$. In particular, $R_X \neq \varnothing$ because X is in the domain of R. If the Choice-Function Principle 6.8 holds, then there exists a choice function $C : \mathscr{F} \to \bigcup \mathscr{F}$ for $\mathscr{F} := \{R_X : \exists Y[(X, Y) \in R]\}$. Thus $C(R_X) \in R_X$ for every X. Hence parametrize the vertical sections of R by $S : A \to \mathscr{P}(B)$ with $S(X) := R_X$, and set $F := C \circ S$. Thus F is a function, with $F(X) = C[S(X)] = C(R_X) \in R_X$, whence $F \subseteq R$. Thus the Choice-Relation Principle 6.15 holds.

6.9 In every partially ordered set A the empty subset $\varnothing \subseteq A$ is a chain. If the hypothesis of Zorn's Maximal-Element Principle 6.32 holds, then A contains an upper bound for $\varnothing$. Hence $A \neq \varnothing$ [88, p. 118].

6.11 Zorn's Maximal-Set Principle 6.34 translates into formula (3):

$$\forall \mathscr{F}\Bigg(\{(\mathscr{F} \neq \varnothing) \wedge \forall \mathscr{G}\,\big[\{[\mathscr{G} \in \mathscr{P}(\mathscr{F})] \wedge \big(\forall A \forall B\,\{[(A \in \mathscr{G}) \wedge (B \in \mathscr{G})]$$

$$\Rightarrow [(A \subseteq B) \vee (B \subseteq A)]\}\big)\} \Rightarrow \Big(\bigcup \mathscr{G} \in \mathscr{F}\Big)\big]\} \tag{2}$$

$$\Rightarrow \{\exists S\big[(S \in \mathscr{F}) \wedge \{\forall A\,[(A \in \mathscr{F}) \Rightarrow (S \not\subseteq A)]\}\big]\}\Bigg).$$

6.15 The Multiplicative Principle 6.36 is formula (3):

$$\forall \mathscr{F} \forall \mathscr{I} \forall I \Big\{ \big[(\mathscr{I} \neq \varnothing) \wedge (I : \mathscr{I} \to \mathscr{F}) \wedge \big(\forall i \{ (i \in \mathscr{I}) \Rightarrow [I(i) \neq \varnothing] \}\big) \big]$$

$$\tag{3}$$

$$\Rightarrow \big[\big(\textstyle\prod_{i \in \mathscr{I}} I(i) \big) \neq \varnothing \big] \Big\}.$$

6.17

(T.A) $\varnothing \in \mathscr{T}$ by example 6.50;

(T.B) if $\mathscr{A} \subseteq \mathscr{T}$ is linearly ordered by inclusion, and if $U, W \in \bigcup \mathscr{A}$, then there exist intervals $H, K \in \mathscr{A}$ such that $U \in H$ and $W \in K$, with $H \subseteq K$ or $K \subseteq H$. In the first case, if $H \subseteq K$, then $U, W \in K$. If also $V \in \mathbb{N}$ and $U \leq V \leq W$, then $V \in K \subseteq \bigcup \mathscr{A}$; Thus $\bigcup \mathscr{A} \in \mathscr{T}$; the second case is similar;

(T.C) if $A \in \mathscr{T}$, then either $A = \varnothing$ and $A \cup \{F(A)\} = \{0\} \in \mathscr{T}$, or $A \neq \varnothing$ and $A \cup \{F(A)\} = A \in \mathscr{T}$, because $F(A) = \min(A) \in A$.

6.19

(T.A) $\varnothing \in \mathscr{P}(E)$;

(T.B) if $\mathscr{A} \subseteq \mathscr{P}(E)$ then $\bigcup \mathscr{A} \subseteq E$ whence $\bigcup \mathscr{A} \in \mathscr{P}(E)$;

(T.C) if $A \in \mathscr{P}(E)$, then $A \cup \{F(A)\} \in \mathscr{P}(E)$, because $F(A) \in E$.

6.21 For each denumerable element D_ℓ of a denumerable family $\mathscr{F} = \{D_k \, k \in \mathbb{N}\}$ there exists a bijection $F : \mathbb{N} \to D_\ell$. Thus the set B_ℓ of all such bijections is not empty. Let $\mathscr{B} := \{B_\ell \, \ell \in \mathbb{N}\}$. In the Zermelo-Frænkel-Choice set theory, there exists a family choice-function $F : \mathbb{N} \to \bigcup \mathscr{B}$ such that $F_\ell := F(\ell) \in B_\ell$ for each $\ell \in \mathbb{N}$; thus each $F_\ell : \mathbb{N} \to D_\ell$ is a bijection. Hence the function $\mathbb{N} \times \mathbb{N} \to \bigcup \mathscr{B}$ with $(k, \ell) \mapsto F_\ell(k)$ is a bijection.

Exercises from Chapter 7

7.1 There are no dominant pure strategies in The Battle of the Sexes.

7.3 The two positions where they both go to the same show are Nash equilibria.

7.5 There are two-person games with a Nash equilibrium but without any dominant strategy and hence without dominant strategy equilibrium, for instance, The Battle of the Sexes (exercises 7.1 and 7.3).

7.7 For all $x \in A$ and $y \in B$,

$$\min\{f(x, y) : y \in B\} \leq f(x, y) \leq \max\{f(x, y) : x \in A\}.$$

Consequently,

$$\min\{f(x, y) : y \in B\} \leq \max\{f(x, y) : x \in A\},$$

where the left-hand side depends only on x while the right-hand side depends only on y. Therefore,

$$\max\{\min\{f(x,y) : y \in B\} : x \in A\} \leq \min\{\max\{f(x,y) : x \in A\} : y \in B\}.$$

7.9 For each strategy S_{Al} available to Al, Al identifies the worst payoff $p^{Al}_{S_{Al}, S_{Bo}}$ under every strategy S_{Bo} available to Bo:

$$\min\{p^{Al}_{S_{Al}, S_{Bo}} : S_{Bo} \in \text{Bo's strategies}\}.$$

To avoid the worst of the worst, Al plays the strategy S_{Al} that returns the best among the worst payoffs:

$$\text{payoff to Al} = \max\{\min\{p^{Al}_{S_{Al}, S_{Bo}} : S_{Bo} \in \text{Bo's strategies}\} : S_{Al} \in \text{Al's strategies}\}.$$

7.11 If a third player Ci has a single weakly dominant strategy, then Ci will play that strategy, leaving Al and Bo with a game for two players and any number of pure strategies, which need not have any Nash equilibrium.

7.13 To avoid failure, the Blue commander must play Middle. Thus the Red commander knows that Blue plays Middle; to avoid failure, Red must then play either Left or Right. In either case, (M,L) or (M,R), both commanders get a Draw.

7.15 Each of the $N-1$ girls who rejected him did so because she received and holds a better proposal. In particular, the last girl who rejected him received at least two proposals. Consequently, there is still at least one girl c who has not yet received any proposal. In the next round, b must then propose to c, which terminates the algorithm, because all the other girls have received at least one proposal (from b) and therefore hold one.

7.17 Depending on the beggars' and choosers' rankings, infinite sets need not admit of any total stable relations. Let the beggars and choosers consist of all negative and positive integers, indexed by their values:

$$B := \mathbb{Z}^*_- = \{\dots, -3, -2, -1\}; \qquad C := \mathbb{Z}^*_+ = \{1, 2, 3, \dots\};$$
$$b_{-m} := -m; \qquad\qquad\qquad c_n := n.$$

Because beggar b_{-m} begs for m, every chooser c ranks all beggars with the same well-order $b_{-k} \overset{\succ}{_c} b_{-\ell}$ if and only if $k < \ell$. In contrast, beggar b_{-m} ranks choosers by swapping $1, \dots, 2m$ with $2m+1, \dots, 4m$, and otherwise leaving the natural order on $\mathbb{Z}^*_+$ unchanged:

$$2m + 1, \dots, 4m, 1, \dots, 2m, 4m + 1, 4m + 2, \dots$$

With such rankings, there are no total stable relations. The proof proceeds by contraposition, showing that every total relation is unstable. Indeed, if a relation $M \subseteq B \times C$ is total, then there exists $k \in \mathbb{Z}_+^*$ with $(b_{-k}, c_1) \in M$, preceded by $k - 1 \geq 0$ couples

$$(b_{-1}, c_{\ell_1}), \ldots, (b_{1-k}, c_{\ell_{1-k}}), (b_{-k}, c_1), \ldots (b_{-n}, c_{\ell_n}), \ldots$$

Yet b_{-k} prefers $2k > k - 1$ choosers $c_{2k+1}, \ldots, c_{4k}$ to c_1. Hence there exists $n > k$ such that $(b_{-n}, c_{\ell_n}) \in M$ with $2k + 1 \leq \ell_n \leq 4k$, whence $(b_{-n}, c_{\ell_n}) \bowtie (b_{-k}, c_1)$, so that M is unstable.

7.19 Extend algorithm 7.31 (with a match maker) to stable relations that may relate each chooser to more than one beggar, with a quota q_c of beggars for chooser c (polygamy, polyandry, schools admitting more than one student, etc.). See Gale & Shapley's references [40, 41].

7.21 For all $(b_1, c_1), (b_2, c_2) \in U_{\mathcal{T}}$ there exist stable relations $M_1, M_2 \in \mathcal{T}$ with $(b_1, c_1) \in M_1$ and $(b_2, c_2) \in M_2$. Yet $M_1 \subseteq M_2$ or $M_2 \subseteq M_1$ because $\mathcal{T}$ is a chain. Thus $(b_1, c_1), (b_2, c_2) \in M := \max\{M_1, M_2\} \in \mathcal{T}$, but M is stable, so that none of the three conditions in definition 7.27 can hold. Therefore $U_{\mathcal{T}}$ is stable.

7.25 From $(X, X) \in \mathcal{R}^{\circ -1}$ if and only if $(X, X) \in \mathcal{R}$ follows $(X, X) \notin \mathcal{R} \setminus \mathcal{R}^{\circ -1}$.

7.27 If $(X, Y), (Y, Z) \in P_{\mathcal{R}} = \mathcal{R} \setminus \mathcal{R}^{\circ -1} \subseteq \mathcal{R}$, then $(X, Z) \in \mathcal{R}$, by transitivity of $\mathcal{R}$.

If also $(X, Z) \in \mathcal{R}^{\circ -1}$, then $(Z, X) \in \mathcal{R}$, but $(X, Y) \in \mathcal{R}$ by hypothesis, whence $(Z, Y) \in \mathcal{R}$ by transitivity of $\mathcal{R}$, so that $(Y, Z) \in \mathcal{R}^{\circ -1}$, contradicting the hypothesis that $(Y, Z) \notin \mathcal{R}^{\circ -1}$.

7.29 Combine the preceding solutions.

7.31 With at least two voters Val and Vic and at least two alternatives Al and Bo, there is a voters' profile r such that Val prefers Al to Bo while Vic prefers Bo to Al:

$$r(\text{Val}) = \{(\text{Al}, \text{Bo})\}; \qquad r(\text{Vic}) = \{(\text{Bo}, \text{Al})\}.$$

If σ is a permutation that swaps Val and Vic, then

$$(r \circ \sigma)(\text{Val}) = r[\sigma(\text{Val})] = r(\text{Vic}) = \{(\text{Bo}, \text{Al})\};$$

$$(r \circ \sigma)(\text{Vic}) = r[\sigma(\text{Vic})] = r(\text{Val}) = \{(\text{Al}, \text{Bo})\}.$$

If Val is a dictator, so that $\mathscr{F}(s) = s(\text{Val})$ for every voters' profile s, then $\mathscr{F}(r) = r(\text{Val}) = \{(\text{Al}, \text{Bo})\}$ while $\mathscr{F}(r \circ \sigma) = (r \circ \sigma)(\text{Val}) = \{(\text{Bo}, \text{Al})\}$. Thus $\mathscr{F}(r) \neq \mathscr{F}(r \circ \sigma)$, so that $\mathscr{F}$ is not invariant under permutations.

7.33 Weak unanimity implies unanimity.

References

1. L.V. Ahlfors, *Complex Analysis*, 3rd edn. (McGraw-Hill, New York, 1979). ISBN 0-07-000657-1. QA331.A45 1979. LCCC No. 78-17078. 515'.93; MR0510197 (80c:30001)
2. Anonymous (ed.), 'Three-parent' therapy approved. Science **347**(6222), 592 (2015). http://scim.ag/_mtvote
3. V.I. Arnold, From Hilbert's superposition problem to dynamical systems. Am. Math. Mon. **111**(7), 608–624 (2004). ISSN 0002-9890
4. K.J. Arrow, A difficulty in the concept of social welfare. J. Polit. Econ. **58**(4), 328–346 (1950)
5. I. Asimov, *The Gods Themselves* (Spectra, New York, 1990). ISBN-10: 0553288105. ISBN-13: 978-0553288100
6. S. Barberá, Pivotal voters: a new proof of Arrow's theorem. Econ. Lett. **6**(1), 13–16 (1980)
7. S. Becattini1, D. Latorre1, F. Mele1, M. Foglierini1, C. De Gregorio1, A. Cassotta1, B. Fernandez1, S. Kelderman, T.N. Schumacher, D. Corti1, A. Lanzavecchia1, F. Sallusto1, Functional heterogeneity of human memory cd4+ t cell clones primed by pathogens or vaccines. Science **347**(6220), 400–406 (2015). doi:10.1126/science.1260668. http://www.sciencemag.org/content/347/6220/400.long
8. P. Bernays, *Axiomatic Set Theory with a Historical Introduction by Abraham A. Fraenkel* (Dover, Mineola, 1991). ISBN 0-486-66637-9. QA248.B47 1991. LCCC No. 90-25812. 511.3'22–dc20
9. G.D.W. Berry, Peirce's contributions to the logic of statements and quantifiers, in *Studies in the Philosophy of Charles Sanders Peirce: First Series*, ed. by P.P. Wiener, F.H. Young (Harvard University Press, Cambridge, 1952), pp. 153–165
10. B. Bolzano, Paradoxien des Unendlichen, in *Essays on the Theory of Numbers*, ed. by R. Dedekind (Dover, Leipzig, 1851)
11. R.T. Boute, How to calculate proofs: bridging the cultural divide. Not. Am. Math. Soc. **60**(2), 173–191 (2013)
12. S.J. Brams, P.D. Straffin Jr., Prisoner's dilemma and professional sports drafts. Am. Math. Mon. **86**(2), 80–88 (1979)
13. S.J. Brams, M.D. Davis, P.D. Straffin Jr., *Module 311: Geometry of the Arm Race* (COMAP, Lexington, 1978)
14. W.L. Briggs, V.E. Henson, *The DFT: An Owner's Manual for the Discrete Fourier Transform* (Society for Industrial and Applied Mathematics, Philadelphia, 1995). ISBN 0-89871-342-0; QA403.5.B75 1995; 95-3232; 515'.723–dc20
15. D.R. Chalice, How to teach a class by the modified Moore method. Am. Math. Mon. **102**(4), 317–321 (1995)

© Springer Science+Business Media New York 2015
Y. Nievergelt, *Logic, Mathematics, and Computer Science*,
DOI 10.1007/978-1-4939-3223-8

16. A. Church, Correction to a note on the Entscheidungsproblem. J. Symb. Log. **1**(3), 101–102 (1936)
17. A. Church, A note on the Entscheidungsproblem. J. Symb. Log. **1**(1), 40–41 (1936)
18. A. Church, *Introduction to Mathematical Logic*. Princeton Landmarks in Mathematics and Physics (Princeton University Press, Princeton, 1996). Tenth printing. ISBN 0-691-02906-7
19. P.J. Cohen, The independence of the continuum hypothesis. Proc. Natl. Acad. Sci. U. S. A. **50**(6), 1143–1148 (1963)
20. P.J. Cohen, The independence of the continuum hypothesis, II. Proc. Natl. Acad. Sci. U. S. A. **51**(1), 105–110 (1964)
21. P.J. Cohen, *Set Theory and the Continuum Hypothesis*. (W. A. Benjamin, New York, 1966) QA9.C69
22. P.J. Cohen, *Set Theory and the Continuum Hypothesis* (Dover, Mineola, 2008). ISBN-10 0-486-46921-2; QA248.C614 2008; LCCC No. 2008042847 511.3'22—dc22
23. L. Dadda, Some schemes for parallel multipliers. Alta Frequenza **34**(2), 349–356 (1965)
24. C. Davis, The role of the untrue in mathematics. Math. Intell. **31**(3), 4–8 (2009). Text of a talk presented at the Joint Mathematics Meeting in Toronto on Monday 5 January 2009
25. R. Dedekind, *Essays on the Theory of Numbers* (Dover, New York, 1963). ISBN 0-486-21010-3; 63-3681. (Also Open Court Publishing, Chicago, 1901; QA248.D3)
26. M. Dresher, *Games of Strategy: Theory and Applications*. Prentice-Hall Applied Mathematics Series (Prentice-Hall Inc., Englewood Cliffs, 1961)
27. J.L.E. Dreyer, *A History of Astronomy from Thales to Kepler*, 2nd edn. (Dover, New York, 1953). SBN 486-60079-3; QB15.D77 1953; 53-12387
28. U. Dudley, *Mathematical Cranks*. MAA Spectrum (Mathematical Association of America, Washington, DC, 1992)
29. U. Dudley, *The Trisectors*. MAA Spectrum, revised edition (Mathematical Association of America, Washington, DC, 1994)
30. J. Dugundji, *Topology* (Allyn and Bacon, Boston, 1966). ISBN 0-205-00271-4. LCCC No. 66-10940
31. H.B. Enderton, *A Mathematical Introduction to Logic* (Academic Press, New York, 1972). LCCC No. 78-182659
32. J. Evans, *The History and Practice of Ancient Astronomy* (Oxford University Press, New York, 1998). ISBN 0-19-509539-1; QB16.E93 1998; LCCC No. 97-16539; 520'. 938–dc21
33. S. Feferman, Does mathematics need new axioms? Am. Math. Mon. **106**(2), 99–111 (1999)
34. G.B. Folland, *Real Analysis: Modern Techniques and Their Applications* (Wiley, New York, 1984)
35. A. Fraenkel, Axiomatische theorie der geordneten mengen. (untersuchungen über die grundlagen der mengenlehre. ii.). J. Reine Angew. Math. (Crelle's Journal) **1926**(155), 129–158 (1926)
36. A. Fraenkel, *Einleitung in die Mengenlehre*, volume IX of *Die Grundlehren der mathematischen Wissenschaften in Einzerldarstellungen mit besonderer Berücksichtigung der Anwendungsgebiete* (Dover, New York, 1946). First edition published by Julius (Springer, Berlin, 1919)
37. A. Fraenkel, *Abstract Set Theory* (North-Holland, Amsterdam, 1953)
38. G. Frege, *Begriffsschrift, eine der arithmetischen nachgebildete Formelsprache des reinen Denkens* (Nebert, Halle, 1879)
39. G. Frege, *Conceptual Notation, and Related Articles; Translated [from the German] and Edited with a Biography and Introduction by Terrell Ward Bynum* (Clarendon Press, Oxford, 1972). ISBN 0198243596; B3245.F22 B94
40. D. Gale, L.S. Shapley, College admissions and the stability of marriage. Am. Math. Mon. **69**(1), 9–15 (1962)
41. D. Gale, L.S. Shapley, College admissions and the stability of marriage. Am. Math. Mon. **120**(5), 386–391 (2013). Reprint of MR1531503

42. J. Geanakoplos, Three brief proofs of Arrow's impossibility theorem. Econ. Theory **26**(1), 211–215 (2005)
43. K. Gödel, Über formal unentscheidbare Sätze der Principia Mathematica und verwandter Systeme I. Mon. Math. Phys. **38**, 173–198 (1931) Leipzig
44. K. Gödel, The consistency of the axiom of choice and of the generalized continuum hypothesis. Proc. Natl. Acad. Sci. U. S. A. **24**, 556–557 (1938)
45. K. Gödel, *The Consistency of the Axiom of Choice and of the Generalized Continuum Hypothesis with the Axioms of Set Theory*. Annals of Mathematics Studies, vol. 3 (Princeton University Press, Princeton, 1940). QA9.G54
46. K. Gödel, *On formally Undecidable Propositions of Principia Mathematica and Related Systems* (Dover, Mineola, 1992). ISBN 0-486-66980-7. QA248.G573 1992. LCCC No. 91-45947. 511.3–dc20
47. G. Gonthier, Formal proof—the four-color theorem. Not. Am. Math. Soc. **55**(11), 1382–1393 (2008)
48. C. Gram, O. Hestvik, H. Isaksson, P.T. Jacobsen, J. Jensen, P. Naur, B.S. Petersen, B. Svejgaard, GIER—a Danish computer of medium size. IEEE Trans. Electron. Comput. **EC-12**(5), 629–650 (1963)
49. W. Grassmann, J.-P. Tremblay, *Logic and Discrete Mathematics: A Computer Science Perspective*. (Prentice Hall, Upper Saddle River, 1996). ISBN 0-13-501296-6; QA76.9.M35G725 1996; LCCC No. 95-38351; 005.1'01'5113–dc20
50. T.C. Hales, Formal proof. Not. Am. Math. Soc. **55**(11), 1370–1380 (2008)
51. E.C. Hall, MIT's role in project Apollo: final report on contracts NAS 9-163 and NAS 94065. Technical report R-700 (Charles Stark Draper Laboratory, MIT, Cambridge) (1972)
52. P. Hamburger, R.E. Pippert, Venn said it couldn't be done. Math. Mag. **73**(2), 105–110 (2000)
53. P. Hamburger, R.E. Pippert, A symmetrical beauty: a non-simple 7-Venn diagram with a minimum vertex set. Ars Combinatoria **66**, 129–137 (2003)
54. J. Harrison, Formal proof—theory and practice. Not. Am. Math. Soc. **55**(11), 1395–1406 (2008)
55. J.R. Harrison, *Handbook of Practical Logic and Automated Reasoning* (Cambridge University Press, London, 2009). ISBN-13: 9780521899574
56. F. Hausdorff, Bemerkung §ber den Inhalt von Punktmengen. Math. Ann. **75**(3), 428–433 (1914)
57. F. Hausdorff, *Set Theory*, 2nd edn. (Chelsea, New York, 1962). Translated from the German by John R. Aumann, et al. QA248.H353 1962; LCCC No. 62-19176
58. M. Henle, *Which Numbers Are Real?* Classroom Resource Materials Series (Mathematical Association of America, Washington, DC, 2012)
59. P. Henrici, *Essentials of Numerical Analysis with Pocket Calculator Demonstrations* (Wiley, New York, 1982). ISBN 0-471-05904-8; QA297.H42; 81-10468; 519.4
60. E. Hewitt, K. Stromberg, *Real and Abstract Analysis* (Springer, New York, 1965). LCCC No. 65-26609
61. D. Hilbert, *Foundations of Geometry*, 10th edn. (Open Court, La Salle, 1971). Eighth printing, 1996. Translated by Leo Unger and revised by Paul Bernays. ISBN 0-87548-164-7; LCCC No. 73-110344
62. D. Hilbert, W. Ackermann, *Principles of Mathematical Logic* (Chelsea Publishing Company, New York, 1950). BC135.H514. (Translation of David Hilbert & Wilhelm Ackermann, *Grundzüge der theoretischen Logik*, Julius Springer, Berlin, Germany, 1928 & 1938)
63. T.W. Hungerford, *Algebra* (Holt, Rinehart and Winston, New York, 1974). ISBN 0-03-086078-4; QA155.H83; LCCC No. 73-15693; 512
64. International Business Machines Corporation, 590 Madison Avenue, New York 22, NY. *IBM 604 Electronic Calculating Punch Manual of Operation*, 1954. http://www.bitsavers.org/pdf/ibm/604
65. International Business Machines Corporation, 590 Madison Avenue, New York 22, NY. *IBM 650 Magnetic-Drum Data Processing Machine Manual of Operation*, revised edition, June 1955. http://www.bitsavers.org/pdf/ibm/650

66. T.J. Jech, *The Axiom of Choice, II*. Studies in Logic and the Foundations of Mathematics, vol. 75 (North Holland, Amsterdam, 1973). ISBN 0-444-10484-4; QA248.J4 1973; LCCC No. 73-75535

67. M. Kaku, Venus envy. Wall Street J. **CCXLVIII**(40), A8 (2006). Western Edition

68. E. Kalai, Foreword: the high priest of game theory. Am. Math. Mon. **120**(5), 384–385 (2013)

69. H. Karttunen, P. Kröger, H. Oja, M. Poutanen, K.J. Donner (eds.), *Fundamental Astronomy* (Springer, New York, 1994). ISBN 0-387-57203; QB43.2.T2613 1993; 93-31098; 520–dc20 (second enlarged edition)

70. L. Keen, Julia sets, in *Chaos and Fractals: The Mathematics Behind the Computer Graphics* (American Mathematical Society, Providence, 1989), pp. 57–74

71. C.T. Kelley, in *Iterative Methods for Linear and Nonlinear Equations*. Frontiers in Applied Mathematics, vol. 16 (Society for Industrial and Applied Mathematics, Philadelphia, 1995). 0-89871-352-8; QA297.8.K45 1995; 95-32249; 519.4–dc20

72. S.C. Kleene, *Mathematical Logic* (Dover, Mineola, 2002). ISBN 0-486-42533-9; QA9.A1 K54 2002; LCCC No. 2002034823 (originally published by Wiley in 1967)

73. S. Kortenkamp, *Why Isn't Pluto a Planet?: A Book about Planets* (Capstone Press, Mankato, 2007). ISBN 0-7368-6753-8; QB701.K57 2007; LCCC No. 2006025648; 523.4–dc22

74. K. Kunen, *Set Theory: An Introduction to Independence Proofs* (North Holland, Amsterdam, 1980). ISBN 0-444-85401-0. QA248.K75 1980. LCCC No. 80-20375. 510.3'22

75. E. Lakdawalla, Pluto and the Kuiper belt. Sky Telesc. **127**(2), 18–25 (2014)

76. E. Landau, *Foundations of Analysis* (The Tan Chiang Book Co., 1951)

77. S. Lang, *Algebra* (Addison-Wesley, Reading, 1965). LCCC No. 65-23677

78. L.S. LaPiana, F.H. Bauer, Mars Climate Orbiter mishap investigation board phase I report. Technical report, National Aeronautics and Space Administration, 10 Nov 1999. http://sse. jpl.nasa.gov.news

79. J. Łukasiewicz, Zur vollen dreiwertigen Aussagenlogik. Erkenntniss **5**, 176 (1935)

80. J. Łukasiewicz, Zur Geschichte der Aussagenlogik. Erkenntniss **5**, 111–131 (1935)

81. Y.I. Manin, *A Course in Mathematical Logic* (Springer, New York, 1977). ISBN 0-387-90243-0; QA9.M296; LCCC No. 77-1838; 511'.3

82. M.F. Mann, R.R. Rathbone, J.B. Bennett, *Whirlwind I Operation Logic*, vol. 29 (Massachussetts Institute of Technology, Digital Computer Laboratory, Cambridge, 1954). http://www. bitsavers.org/pdf/mit/whirlwind

83. V.W. Marek, J. Mycielski, Foundations of mathematics in the twentieth century. Am. Math. Mon. **108**(5), 449–468 (2001). ISSN 0002-9890

84. A. Margaris, *First Order Mathematical Logic* (Dover Publications Inc., New York, 1990). Corrected reprint of the 1967 edition

85. B. Mates, *Elementary Logic*, 2nd edn. (Oxford University Press, New York, 1972). BC135.M37 1972. LCCC No. 74-166004

86. R. Miller, Computable fields and Galois theory. Not. Math. Am. Soc. **55**(7), 798–807 (2008). http://www.ams.org/notices

87. K. Mitchell, J. Ryan, Game theory and models of animal behavior, in *UMAP/ILAP Modules 2002–03: Tools for Teaching*, ed. by P.J. Campbell (COMAP, Bedford, 2003), pp. 1–48. Reprinted as Module 783, COMAP, Bedford, MA, 2003

88. J.D. Monk, *Introduction to Set Theory* (McGraw-Hill, New York, 1969). LCCC No. 68-20056

89. J.D. Monk, *Mathematical Logic* (Springer, New York, 1976). ISBN 0-387-90170-1. QA9.M68

90. M.R. Mugnier, G.A.M. Cross, F.N. Papavasiliou, The in vivo dynamics of antigenic variation in *Trypanosoma brucei*. Science **347**(6229), 1470–1473 (2015). doi:10.1126/science.aaa4502. http://www.sciencemag.org/content/347/6229/1470.full

91. J.R. Munkres, *Topology*, 2nd edn. (Prentice-Hall, Englewood Cliffs, 2000). ISBN 0-13-181629-2; QA611.M82 2000; 514–dc21; LCCC No. 99-052942

92. R.B. Myerson, *An Introduction to Game Theory*. Studies in Mathematical Economics. MAA Studies in Mathematics, vol. 25 (Mathematical Association of America, Washington, DC, 1986), pp. 1–61

93. J. Nash, Non-cooperative games. Ann. Math. Second Ser. **54**(2), 286–295 (1951)
94. J.F. Nash Jr., The bargaining problem. Econometrica **18**(2), 155–162 (1950)
95. J.F. Nash Jr., Equilibrium points in *n*-person games. Proc. Natl. Acad. Sci. U. S. A. **36**(1), 48–49 (1950)
96. National Resident Matching Program, http://www.nrmp.org/match-process/match-algorithm/
97. O. Neugebauer, *The Exact Sciences in Antiquity*, 2nd edn. (Dover, New York, 1969). SBN 486-22332-0; LCCC No. 69-20421. Reprint of the 1957 second edition from Brown University Press
98. O. Neugebauer, *A History of Ancient Mathematical Astronomy. Part I.* Studies in the History of Mathematics and Physical Sciences, vol. 1 (Springer, New York, 1975)
99. O. Neugebauer, *A History of Ancient Mathematical Astronomy. Part II.* Studies in the History of Mathematics and Physical Sciences, vol. 1 (Springer, New York, 1975)
100. O. Neugebauer, *A History of Ancient Mathematical Astronomy. Part III.* Studies in the History of Mathematics and Physical Sciences, vol. 1 (Springer, New York, 1975)
101. Y. Nievergelt, The truth table of the logical implication. Math. Gaz. **94**(531), 509–513 (2010). ISSN 0025-5572
102. Y. Nievergelt, H. Sullivan, Undecidability in fuzzy logic. UMAP J. **31**(4), 323–359 (2010). Also reprinted as UMAP Unit 804
103. Packard Bell Electronics, Packard Bell Computer, 1905 Armacost Avenue, Los Angeles 25, California. pb 250 Technical Manual, Volume 1, 15 July 1961. `http://www.bitsavers.org/pdf/packardBell`
104. C.S. Peirce, On the algebra of logic: a contribution to the philosophy of notation. Am. J. Math. **7**(2), 180–196 (1885)
105. F. Quinn, A revolution in mathematics? What really happened a century ago and why it matters today. Not. Am. Math. Soc. **59**(1), 31–37 (2012)
106. A. Rapoport, *Two-Person Game Theory: The Essential Ideas* (The University of Michigan Press, Ann Arbor, 1966)
107. A. Rapoport, *Two-Person Game Theory* (Dover Publications Inc., Mineola, 1999). Reprint of the 1966 original [Univ. Michigan Press, Ann Arbor, MI, 1966; MR0210463 (35 #1356)]
108. J.W. Robbin, *Mathematical Logic: A First Course* (Dover, Mineola, 2006). SBN 486-61272-4; LCCC No. 65-12253 (originally published by W. A. Benjamin, New York, 1969)
109. R.M. Robinson, Primitive recursive functions. Bull. Am. Math. Soc. **53**, 925–942 (1947)
110. J.B. Rosser, *Logic For Mathematicians* (McGraw-Hill, New York, 1953). BC135.R58; LCCC No. 51-12640
111. H. Rubin, J.E. Rubin, in *Equivalents of the Axiom of Choice, II.* Studies in Logic and the Foundations of Mathematics, vol. 116 (North Holland, Amsterdam, 1980). ISBN 0-444-87708-8; QA248.R8 1985; LCCC No. 84-28692; 511.3'22
112. D.G. Saari, *Chaotic Elections: A Mathematician Looks at Voting* (American Mathematical Society, Providence, 2001). ISBN 0-8218-2847-9
113. D.G. Saari, Mathematics and voting. Not. Am. Math. Soc. **55**(4), 448–455 (2008). http://www.ams.org/notices
114. E. Schechter, *Classical and Nonclassical Logics* (Princeton University Press, Princeton 2005). ISBN 0-691-12279-2; QA9.3.S39 2005; 2004066030; 160–dc22
115. R. Serrano, The theory of implementation of social choice rules. SIAM Rev. **46**(3), 377–414 (2004) (electronic)
116. T.A. Skolem, Einige Bemerkungen zur axiomatischen Begründung der Mengenlehre, in *Matematikerkongressen i Helsingfors den 4–7 Juli 1922, Den femte skandinaviska matematikercongressen, Redogšrelse* (Helsinki, SF, 1923), pp. 217–237. Akademiska Bokhandeln. Cited in [36] and [128]
117. R.M. Smullyan, *First-Order Logic* (Dover, Mineola, 1995). ISBN 0-486-68370-2. QA9.S57 1994. LCCC No. 94-39736. 511.3–dc20
118. M. Spivak, *Calculus*, 2nd edn. (Publish or Perish, Wilmington, DE, 1980). ISBN 0-914098-77-2. LCCC No. 80-82517

119. S. Stahl, *A Gentle Introduction to Game Theory*. Mathematical World, vol. 13 (American Mathematical Society, Providence, 1999)
120. J. Stillwell, *The Real Numbers. Undergraduate Texts in Mathematics. An Introduction to Set Theory and Analysis* (Springer, Cham, 2013)
121. J. Stoer, R. Bulirsch, *Introduction to Numerical Analysis*, 2nd edn. (Springer, New York, 1993). ISBN 0-387-97878-X; QA297.S8213 1992; 92-20536; 519.4–dc20
122. R.R. Stoll, *Sets, Logic, and Axiomatic Theories* (W. H. Freeman, San Francisco, 1961). LCCC No. 61-6784
123. A.A. Stolyar, *Introduction to Elementary Mathematical Logic* (Dover, Mineola, 1983). ISBN 0-486-64561-4. BC135.S7613 1983. LCCC No. 83-5223. 511.3
124. P.D. Straffin Jr., The prisoner's dilemma. UMAP J. **1**(1), 102–103 (1980)
125. P.D. Straffin Jr., Changing the way we think about the social world. Two Year Coll. Math. J. **14**(3), 229–232 (1983)
126. P.D. Straffin Jr., Game theory and nuclear deterrence. UMAP J. **1**(1), 87–92 (1989)
127. D.J. Struik, *A Concise History of Mathematics*, 4th edn. (Dover, New York, 1987). ISBN 0-486-60255-9; QA21.S87 1987; LCCC No. 86-8855; 510'.09 (revised edition)
128. P. Suppes, *Axiomatic Set Theory* (Dover, Mineola, 1972). ISBN 0-486-61630-4. LCCC No. 72-86226
129. A. Tarski, *Introduction to Logic and to the Methodology of Deductive Sciences* (Oxford University Press, Oxford, 1941)
130. A.E. Taylor, *Introduction to Functional Analysis* (Wiley, New York, 1958) 58-12704
131. A.W. Tucker, The prisoner's dilemma: a two-person dilemma. UMAP J. **1**, 101 (1980). Dated "Stanford, May 1950"
132. A.W. Tucker, A two-person dilemma: the prisoner's dilemma. Two Year Coll. Math. J. **14**(3), 228 (1983). Dated "Stanford, May 1950"
133. G. van Brummelen, *The Mathematics of the Heavens and the Earth: The Early History of Trigonometry* (Princeton University Press, Princeton, 2009). ISBN 978-0-691-12973-0; QA24.V36 2009; 516.2409–dc22; LCCC No. 2008032521
134. B.L. van der Waerden, *Geometry and Algebra in Ancient Civilizations* (Springer, Berlin, 1983). ISBN 3-540-12159-5. QA151.W34 1983. LCCC No. 83-501. 512'.009
135. J. von Neumann, Zur einführung der transfiniten Zahlen. Acta Szeged **1**, 199–208 (1923) Cited in [128, p. 129]
136. R.J. Walker, *Algebraic Curves* (Springer, New York, 1978). ISBN 0-387-90361-5. QA564.W35 1978. LCCC No. 78-11956. 516'.352. (First published by Princeton University Press, Princeton, 1950)
137. D.A. Weintraub, *Is Pluto a Planet?: A Historical Journey through the Solar System* (Princeton University Press, Princeton, 2007). ISBN 0-691-12348-9; QB602.9.W45 2007; LCCC No. 2006929630
138. Westinghouse Electric Corporation, Surface Division, P.O. Box 1897, Baltimore, MD 21203. *Technical Manual for DPS-2402 Computer, Volume I*, 1 February 1966. http://www.bitsavers.org/pdf/westinghouse
139. F. Wiedijk, Formal proof—getting started. Not. Am. Math. Soc. **55**(11), 1408–1417 (2008)
140. R.L. Wilder, *Introduction to the Foundations of Mathematics*, 2nd edn. (Dover, Mineola, 2012)
141. R.S. Wolf, *A Tour Through Mathematical Logic*. The Carus Mathematical Monographs, vol. 30. (Mathematical Association of America, Washington, DC, 2005). ISBN 0-88385-036-2; LCCC No. 2004113540
142. S. Wolfram, *The Mathematica Book*, 3rd edn. (Wolfram Media, Champaign, 1996). ISBN 0-9650532-1-0; QA76.95.W65 1996; LCCC No. 96-7218; 510'.285'53–dc20
143. N.N. Yu, A one-shot proof of Arrow's impossibility theorem. Econ. Theory **50**(2), 523–525 (2012)
144. E. Zermelo, Beweis, daß jede Menge wohlgeordnet werden kann. Math. Ann. **59**(4), 514–516, (1904)

145. E. Zermelo, Untersuchungen über die Grundlagen der Mengenlehre, I. Math. Ann. **65**(2), 261–281 (1908). Cited in [8, p. 21]
146. D.E. Zitarelli, The origin and early impact of the Moore method. Am. Math. Mon. **111**(6), 465–486 (2004)
147. M. Zorn, A remark on method in transfinite algebra. Bull. Am. Math. Soc. **41**(10), 667–670 (1935)
148. G. Zukav, *The Dancing Wu Li Masters: An Overview of the New Physics* (Morrow, New York, 1979). ISBN: 0-688-03402-0; QC173.98.Z84 1979; 530.1'2; LCCC No. 78-25827

Index

© Springer Science+Business Media New York 2015

Y. Nievergelt, *Logic, Mathematics, and Computer Science*,

DOI 10.1007/978-1-4939-3223-8

Printed in the United States
By Bookmasters